Optimization Theory and the Design of Feedback Control Systems

McGRAW-HILL ELECTRONIC SCIENCES SERIES

ABRAMSON Information Theory and Coding
BATTIN Astronautical Guidance
BREMER Superconductive Devices
BROXMEYER Inertial Navigation Systems
GILL Introduction to the Theory of Finite-state Machines
HUELSMAN Circuits, Matrices, and Linear Vector Spaces
KELSO Radio Ray Propagation in the Ionosphere
MERRIAM Optimization Theory and the Design of Feedback Control Systems
PAPOULIS The Fourier Integral and Its Applications
STEINBERG AND LEQUEUX (TRANSLATOR R. N. BRACEWELL)
Radio Astronomy

Optimization Theory and the Design of Feedback Control Systems

C. W. Merriam III

General Electric Research Laboratory
Schenectady, New York

McGraw-Hill Book Company

New York San Francisco Toronto London

Optimization Theory and the Design of Feedback Control Systems

Library of Congress Catalog Card Number 63-19315

41494

II

*to **Stephen** and **Gail***

Preface

During the past ten years, interest in the mathematics associated with the variational aspects or optimization of dynamic systems has been reawakened. Most likely, this interest can be traced to a growing concern for the economic operation of industrial and manufacturing processes and for the performance of military weapon systems. This area of mathematics is burgeoning at the present time and will, in the author's view, have a significant impact on the design and operation of many engineering and economic systems.

A key event in the resurgence of interest in variational mathematics was the development and proliferation of discrete dynamic programming by Dr. Richard Bellman and his associates. In addition to numerous mathematical contributions, Dr. Bellman's work illuminated two practical considerations not previously associated with variational mathematics. First, the numerical solutions of variational problems give rise to a rather unique and very difficult computational problem. Second, the structure of the optimum policy or control equation is an important part of any variational problem and must be examined in detail before the problem can be considered resolved.

Dynamic programming played an important role in the early

work on the application of optimization theory to feedback-control-system design. Specifically, the visualization of the control problem as a sequential decision process and dynamic programming as a method of generating an optimum sequence of decisions is completely compatible with the intuitive view of prediction and stability normally possessed by the control engineer. In addition, the direct formulation of dynamic programming in terms of the initial state or conditions of the process to be controlled is appealing and plausible to the control engineer who is familiar with ordinary differential equations. Based on teaching a course at Rensselaer Polytechnic Institute and two courses for practicing engineers at the General Electric Research Laboratory, the decision to emphasize the dynamic-programming format was made even though many results can be derived from the calculus of variations.

The usefulness of optimization theory in the design of feedback control systems is derived primarily from the flexibility of the approach as a direct method for the time-domain synthesis of high-performance systems. The synthesis problem associated with feedback controls has two facets. In this book, control-equation synthesis is emphasized. However, trajectory synthesis is an important adjunct to structuring the control equation and is treated primarily in later chapters. The usefulness of optimization theory is, of course, dependent upon the engineer's ability to obtain numerical solutions in addition to structuring the control equation in terms of available components. In regard to numerical solutions, optimization theory used for design purposes generally requires access to a large-scale digital computer. Today and in the future, this requirement imposes few if any restrictions. In fact, the balance of engineering and computer costs in many design problems favors the use of optimization theory, and this balance will become more favorable in the future.

The exposition of this book is directed to the engineer, and the engineering interpretation of the mathematical theory when used for design purposes is stressed. Hence, undue mathematical rigor is avoided, although exceptions to the derived results are noted carefully. Also, the amount of material presented for various classes of optimum control systems is in approximate proportion to the degree that these classes of systems can be reduced to practice. Reduction to practice mainly involves the availability and cost of components for constructing the optimum control equation plus the cost and time required to obtain a numerical solution. Also, the book is restricted to continuous systems for the sake of book length. Because of the emphasis on engineering in this book, the reader who is particularly interested in research is urged to read "Dynamic Programming," by R. Bellman, and "The Mathematical Theory of Optimal Processes," by L. S. Pontryagin et al.

In order to keep this book readily accessible to practicing engineers, mathematical short-hand notations have been used sparingly. The vector-matrix notation currently used in much control-theory literature could have been used to some advantage. However, the compactness of this notation also tends to obscure the inherent difficulties and complexities of applying the results to actual design problems.

This book is suitable for a two-semester graduate course in feedback control if the material is to be covered in depth. Such a course would presuppose an extensive background on the part of the students in the fundamental concepts of feedback systems and conventional design methods. If the introductory chapters are omitted and the remaining material is covered in less detail, this book would be suitable for a one-semester graduate course.

The contents of this book can be divided conveniently into four parts. The first part is introductory and is intended to review and shed perspective on the chronological developments in optimization theory applied to control-system design. Parameter Optimization and Impulse-response Optimization, Chapters 2 and 3, respectively, are intended to clarify design constraints in control-system optimization, some of which were introduced initially as mathematical necessities and not as engineering requirements. Except for the notation introduced in Chapter 1 and Appendix C, this first part may be omitted if the reader has had previous exposure to books such as "Analytical Design of Linear Feedback Controls," by G. C. Newton, Jr. et al. The second part, Chapters 4 through 6, deals with the introductory mathematical background required for control-system optimization. The emphasis of this material is on the mathematical origin and physical interpretation of boundary conditions and on the relationships between the calculus of variations and dynamic programming. Also, the consequences of selecting various mathematical forms for weighting errors are discussed via examples. The previously mentioned books by Bellman and Pontryagin would provide alternative or supplementary reading on this material. The third part, Chapters 7 and 8, deals with linear optimum controls. This material serves to introduce the reader to the control-equation synthesis problem. Considerable material also is directed to a number of practical considerations that are important to the engineer when the mathematical theory is used in specific design problems. These considerations form a basis for making a decision as to whether to use optimization theory as the method of design. The final three chapters are on nonlinear control systems. Two approaches to the control-equation synthesis problem are discussed in Chapters 9 and 11. The first approach results in noncomputing control equipment, whereas the latter approach requires the use of an on-line computer. Chapter 10 deals with computational methods associated with trajectory synthesis, which have reached a high degree of flexibility in the past two years.

The examples appearing in the text, which are used to illustrate the derivations and results, are artificial and somewhat irrelevant to actual engineering design problems. The selection of these problems was dictated by the decision to obtain closed-form analytical solutions. Therefore, the reader is urged to study carefully the detailed design problem in Appendix E. Also, the problems included in Appendix H should be helpful in illustrating the text.

The author is deeply indebted to many sources for text material and criticisms. The contributions of Messrs. F. J. Ellert and R. J. Ringlee of the General Electric Company warrant particular mention. The author has benefited greatly from long-term and fruitful associations with Messrs. Ellert and Ringlee. Appendix E was prepared jointly with Mr. Ellert, who also contributed a number of ideas in Chapter 8. Mr. Ringlee contributed and clarified a number of ideas included in Chapters 6 and 10. Dr. I. Lee of the General Electric Research Laboratory carefully read the final manuscript and made a number of helpful suggestions.

The author is grateful to Dr. R. L. Shuey and the management of the General Electric Research Laboratory for their support and encouragement, which were necessary to complete this book. The facilities of this laboratory were used extensively. Mrs. G. A. Croteau wrote many of the computer programs used for data generation. Mrs. E. S. Oros's accurate typing from handwritten notes was invaluable, and the final typing and proofreading of the manuscript were supervised by Mr. J. A. Pauze and Mrs. M. C. Carlo.

Without the patience and understanding of my wife Carolyn, this book would never have been completed.

C. W. Merriam III

Contents

1

Introduction

1-1 Evolutions in automatic-control theory

The field of automatic control underwent astounding change and growth during World War II. The needs for weapon systems and manufacturing facilities placed great pressure on the engineer and applied mathematician to develop practical theories for the design of control systems.

During this era, automatic-control theory, as well as practice, made great strides. In particular, the concepts of a feedback system, as opposed to a cascaded system, were developed to a high degree of sophistication. The problems of feedback-system sensitivity, stability, and performance were formulated mathematically; and the utilization of transform and complex-variable theories made practical a detailed analysis of the design problem. Also, the development of various analytical and graphical procedures made possible the frequency-domain analysis of improved system performance by the introduction of compensating networks.[1–4]§

At the end of the World War II era, engineers and researchers recognized the impact that had been made by automatic-control

§ Superscript numerals are keyed to the References in Appendix B.

theory in the development of high-performance systems. As a result, continuing efforts were devoted to broadening the mathematical base and the applications of automatic-control theory. Phase-plane and describing-function methods of analysis were developed for non-linear systems.[5,6] Analog-computer techniques were improved for the simulation of very complex systems.[7,8] The advent of digital computers as control-system components precipitated major developments in the theory of pulsed-data systems.[9,10] Also, statistical-signal analysis, originating in the communications field, was incorporated into control theory.[11] Even strictly analytical techniques for linear systems received considerable attention from theoreticians.[12] All these facets of control theory have had an important impact on the growth of the control field, and the engineer cannot neglect them in designing adequate systems for the wide range of control problems encountered in practice.

At the present time, the need for control systems in outer space, increased economic competition in manufacturing processes, and the availability of high-speed digital computers have set the stage for another evolutionary expansion in the control field. The desire to operate systems in outer space has created many new problems in both theory and componentry. Manufacturing process control requires analysis methods for systems of staggering complexity. On the other hand, high-speed digital computers are providing a new means of great flexibility and capacity for both the analysis and the construction of control systems in these modern applications.

Currently, the applied mathematician is succeeding in formulating feedback-control theory on a profound mathematical and conceptual basis. The stability theory of Lyapunov,[13] the topics of observability and controllability originated by Kalman,[14] and the mathematical optimization theories of Bellman[15] and Pontryagin[16] are notable contributions to the mathematical theory of automatic control. Perhaps developments in theory such as these will prove to be the most spectacular but also the most difficult for the engineer to assimilate. Furthermore, these theoretical contributions will become contributions to the control engineer only to the degree that they improve system performance and can be used efficiently as a basis for design. The efficient use of a theory for design, of course, is related intimately to whether design specifications are represented accurately and numerical solutions are obtained readily. No theoretical developments can escape scrutinizing examination from these points of view.

The notion of using mathematical optimization theory as a basis for design is an important departure from the other theoretical concepts mentioned previously for both philosophical and practical reasons. Philosophically, optimization theory is an attempt to provide a means for *direct system synthesis* as opposed to system synthesis via repeated analyses of controllers selected on a trial-and-error basis. From a practical standpoint, system design specifications are stated in the time domain for most engineering control problems. Mathematical optimization theory is formulated in the time domain and hence represents the engineering problem accurately. On the other hand, widely used frequency-domain design methods are difficult to interpret in terms of time-domain specifications; hence trial and error are required to obtain an accurate time-domain correspondence to the design-problem specifications. Furthermore, frequency-domain methods essentially are limited to linear time-invariant systems.

1-2 Time-domain formulation of a control problem

The time-domain aspects of control problems must be firmly in mind before the generalized mathematical formulation is presented. Therefore a simplified version of the homing-missile control problem is used as an illustration.

The terminal control of a homing missile is simplified by two assumptions. The maneuverings of both the missile and the target occur in a single geometrical plane. This eliminates the inertial cross-coupling in the missile caused by high roll rates which arise in "roll-to-turn" maneuvers. Also, the missile has only a small speed advantage over the target so that interception cannot be achieved with large line-of-sight errors.§ Small line-of-sight errors simplify the geometry because the time rate of change of the missile-target range (distance) is approximately constant.

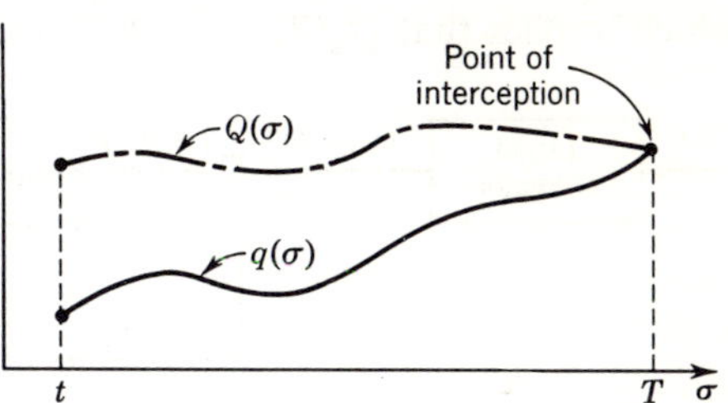

Fig. 1-1 Homing-missile geometry.

Therefore a fixed time for interception is established. For these assumptions, the homing-missile control problem is stated geometrically in Fig. 1-1. The definitions pertinent to the terminal-control problem are

$q(\sigma)$ = position of missile
$Q(\sigma)$ = position of target
t = real time corresponding to present position
T = time of interception
σ = dummy time variable

The goal of a terminal controller is to make the terminal miss distance

$$|Q(T) - q(T)|$$

as small as possible. Although the terminal miss distance shown in Fig. 1-1 is zero, generally sensor errors, target maneuvers, and limited acceleration capabilities of the missile cause error at the point of interception. However, the controller is designed to counteract the disturbances and dynamic constraints so that small terminal miss distances are obtained. Because the terminal or interception point is the major design consideration, the error signal

$$e_c(t) = Q_p(T, t) - q_e(T, t) \tag{1-1}$$

is proposed heuristically according to the following definitions:

$Q_p(T, t)$ = predicted value of $Q(T)$ given measurements of target position, velocity, etc., and information concerning likely target maneuvers

§ Actually, a collision course is established when the time rate of change of the line-of-sight angle is zero.

$q_e(T, t) =$ extrapolated value of $q(T)$ given measurements of missile position, velocity, etc., and information concerning missile dynamics; furthermore, the extrapolation is computed on the basis that no future control effort is applied to the missile on the time interval $t < \sigma \leqq T$

Therefore, the error signal $e_c(t)$ can be thought of as the computed terminal miss distance, assuming no subsequent control effort. Figure 1-2 is a block diagram of the proposed terminal-control system where the function F is the gain element, which may or may not be nonlinear, and $m(t)$ is the missile control signal. This control system has been proposed and studied by a number of researchers.[17–19]

The arguments leading to this particular system for terminal control are simple. As a hypothetical case, the assumption is made that perfect target prediction occurs, which implies that $Q_p(T, t) = Q(T)$ and hence $Q_p(T, t)$ is a known constant. If the

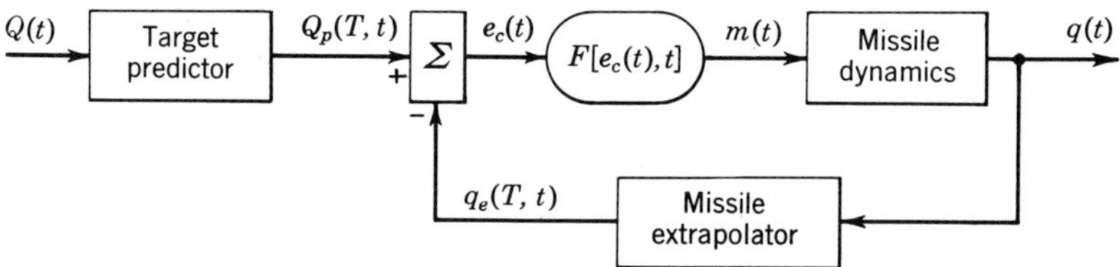

Fig. 1-2 Terminal-control-system block diagram.

extrapolated response is such that $q_e(T, t) = Q(T)$, then $m(t) = 0$ on the interval $t \leqq T$ according to the defined method for computing the extrapolated response. In other words, once a collision course is established, no more control effort is required. Then maximum control effort in either direction always is available to counter target maneuvers, etc., which is the chief concept of this system.

As the reader may suspect, desired-response prediction and response extrapolation are related intimately to system performance and system stability. These relationships are examined for the special case where the missile dynamics are linear and hence can be represented by the superposition integral[20]

$$q(t) = \int_{-\infty}^{t} w(t, \sigma) m(\sigma)\, d\sigma \tag{1-2}$$

The weighting function $w(t, \sigma)$, called the impulse response, is the response of the missile at time t due to a unit-impulse control signal at time σ. In addition, the impulse response is of the general form §

$$w(t, \sigma) = \sum_{n=1}^{N} F_n(t) G_n(\sigma) \tag{1-3}$$

for an Nth-order system of ordinary differential equations. Therefore the extrapolated response for the terminal-control-problem definition is

$$q_e(T, t) = \int_{-\infty}^{t} w(T, \sigma) m(\sigma)\, d\sigma \tag{1-4}$$

§ The $F_n(t)$ functions are linearly independent solutions of the homogeneous system of differential equations, and the $G_n(\sigma)$ functions are parameters[21] specialized for a unit impulse.

By using this relationship and differentiating (1-1), the differential equation defining terminal miss distance becomes

$$e_c'(t) + w(T, t)F[e_c(t), t] = \frac{\partial Q_p(T, t)}{\partial t} \tag{1-5}$$

Furthermore, perfect prediction implies that $Q_p(T, t)$ is a constant with respect to real time t, so that (1-5) reduces to

$$e_c'(t) + w(T, t)F[e_c(t), t] = 0 \tag{1-6}$$

The stability of this homogeneous first-order differential equation cannot be defined in terms of the frequency-domain concepts widely used in the control field because of the nonlinear and time-varying nature of the equation. However, suitable structural restrictions on F can be imposed so that $e_c(t)$ decreases monotonically to zero and remains zero thereafter. This monotonic behavior is a very strong form of stability and is highly desirable for the terminal-control problem. The stability of (1-6) must be expressed completely in terms of $w(T, t)$, which arises from response extrapolation only. Hence the conclusion is drawn that *response extrapolation is the basis for system stability.*

If target prediction is not perfect, the forcing term appearing in (1-5) is not zero. Target-prediction errors cause generally increased values of $e_c(t)$ and hence an increased terminal miss distance if, for instance, the target maneuvers are always evasive. Because terminal miss distance is the measure of system performance, target prediction is important. In fact, the conclusion is drawn that *desired-response prediction is the basis for system performance.*§

Although the concepts of response extrapolation and desired-response prediction have been introduced in terms of a special control problem, they are the crux of all control problems. These concepts, as alternatively stated in terms of stability and performance, are not new but are related only indirectly to conventional design methods. On the other hand, these concepts are related directly to optimization theory for control-system synthesis. This direct relationship gives credence to the claim that mathematical optimization theory can be utilized as a direct-synthesis procedure.

1-3 Mathematical representation of system performance

The terminal-control problem is an instructive example for reviewing certain time-domain aspects of automatic control. This problem is also useful for introducing the mathematical representation of control-system performance used in optimization theory.

The performance consideration pertinent to the homing-missile control problem is terminal miss distance. Therefore, the design engineer needs a measure of error for evaluation purposes. In this case, an error index might be taken to be

$$e(t) = F_q[Q(T) - q(T)] \tag{1-7}$$

§ Of course, stability is an important requisite for acceptable system performance in conventional design procedures. Here system performance implies the degree to which the system achieves a desired response in spite of noise and other disturbances.

where acceptable choices for the function F_q would be $F_q(x) = |x|$, $F_q(x) = x^2$, etc. The error index e is shown to be a function of real time t because the initial conditions in the missile determine in part the value of the error index that is achieved during an interception. An alternative but equivalent representation of the error index is found by introducing the familiar unit-impulse function u_0. This representation is expressed as the integral

$$e(t) = \int_t^T \phi(\sigma) F_q[Q(\sigma) - q(\sigma)]\, d\sigma \tag{1-8}$$

where

$$\phi(\sigma) = u_0(\sigma - T) \tag{1-9}$$

The equivalence is established from the property

$$\int u_0(\sigma - \sigma_0) f(\sigma)\, d\sigma = f(\sigma_0) \tag{1-10}$$

The upper limit of integration T is the time where the system operation terminates. The lower limit of integration t is real time in the system. Therefore, the integration, with respect to the dummy time variable σ appearing in (1-8), is over all future and present time, $t \leqq \sigma \leqq T$, where the control system operates.

The advantage of expressing the error index in terms of an integral is that a variety of control problems can be represented directly in this form. For instance, a curve-following servo is designed so that $q(t) \approx Q(t)$ for all values of real time. In this case, errors are weighted equally over all future and present time. This is accomplished by specializing (1-8) so that $\phi(\sigma) = 1$ and $T = \infty$.

In most control-system design problems, adequate system performance is stated in terms of multiple design specifications. In the homing-missile problem, for instance, tactical and sensor considerations may require that the interception be accomplished as a tail chase, $Q'(T) = q'(T)$. This added consideration can be introduced in terms of the error index

$$e(t) = \int_t^T \phi(\sigma)\{F_q[Q(\sigma) - q(\sigma)] + \alpha F_{q'}[Q'(\sigma) - q'(\sigma)]\}\, d\sigma \tag{1-11}$$

where $\phi(\sigma)$ is given by (1-9). The multiplier α determines the compromise between allowable miss distance and the degree of tail chase. The case $\alpha = 0$ weights miss distance only; the case $\alpha = \infty$ weights tail-chase conditions only. Furthermore, the missile has finite acceleration capabilities, which must be observed during the trajectory. If the missile is described by the differential equation

$$q''(t) = m(t) \tag{1-12}$$

then the error index is expressed as

$$e(t) = \int_t^T \{\phi(\sigma) F_q[Q(\sigma) - q(\sigma)] + \alpha\phi(\sigma) F_{q'}[Q'(\sigma) - q'(\sigma)] + F_m[m(\sigma)]\}\, d\sigma \tag{1-13}$$

If the maximum acceleration magnitude is defined to be M, then the weighting of control effort, which is equivalent to acceleration in this case, may be taken to be

$$F_m[m(\sigma)] = \begin{cases} \infty & M \leqq m(\sigma) \\ 0 & -M < m(\sigma) < M \\ \infty & m(\sigma) \leqq -M \end{cases} \tag{1-14}$$

or, alternatively,

$$F_m[m(\sigma)] = \left[\frac{m(\sigma)}{M}\right]^{2n} \quad n \to \infty \tag{1-15}$$

In any event, the criterion for system design is to construct the controller which minimizes (1-13) for all initial conditions in the missile. The infinite penalty for exceeding acceleration limits ensures that the missile does not disintegrate when (1-13) is a minimum.

1-4 Definitions and nomenclature for process description

In order to facilitate the presentation of the remaining material, a number of definitions are introduced here. Also, most of the notation needed in later chapters is specified here. An attempt has been made to select notation that corresponds to both the standards of conventional control literature, as recommended by professional societies, and the mathematical literature pertinent to optimization theory.

In control-system design problems, the engineer is given the process or fixed member of the system. The fixed member of the system is viewed as being not only the device to be controlled, such as an aircraft or a chemical plant, but also the actuators and sensors associated with that device. The mathematical description of the fixed-member dynamics is called the *dynamic process*.

The inputs, or independent variables, of the dynamic process are the *control signals* $m_1(t)$, $m_2(t)$, . . . , $m_M(t)$. The outputs, or dependent variables, of the dynamic process are the *response signals* $q_1(t)$, $q_2(t)$, . . . , $q_Q(t)$. These response signals may not always be physical variables but instead may be economic variables. For instance, the primary response variable for a gasoline engine may be efficiency, which is computed in terms of the physical variables manifold pressure, engine rpm, and rate of fuel flow.

Because response signals may not be physical variables, a second set of outputs is associated with the dynamic process. These outputs are called the *state signals* $x_1(t)$, $x_2(t)$, . . . , $x_N(t)$. State signals require a careful definition here because generally they have not been associated with frequency-domain design techniques. All dynamic processes discussed in this book are defined in terms of ordinary differential equations, as opposed to partial differential equations, although pure time delay is discussed as an exception. Also, the dynamic processes under consideration never involve hysteresis and similar phenomena. For these restrictions, the dynamic process is called *state-determined*, and a finite number of variables uniquely determine the distribution of energy or state of the system. If the state of these systems is known at $t = t_0$, then the contribution to future states of the dynamic process, $t > t_0$, due to past inputs $t < t_0$, is defined completely. The minimal number of signals required to define the state of these dynamic processes is equal to the order N of the system of differential equations.§ For readers who are familiar with

§ The measurement of state signals is required in order to construct the optimum feedback control systems with physical components. The selection of state signals is not unique; hence the engineer formulates the dynamic process so that the state signals are measurable wherever possible. The problem of the incomplete measurement of state is discussed in a later chapter.

analog-computer simulation, the output signals of the integrators used to solve a system of differential equations are state signals. This particular set of state signals is the set commonly used in the remaining chapters.

An important class of disturbances encountered in practice is referred to as load disturbances. Load disturbances occur as thermally induced torques on a gyro,

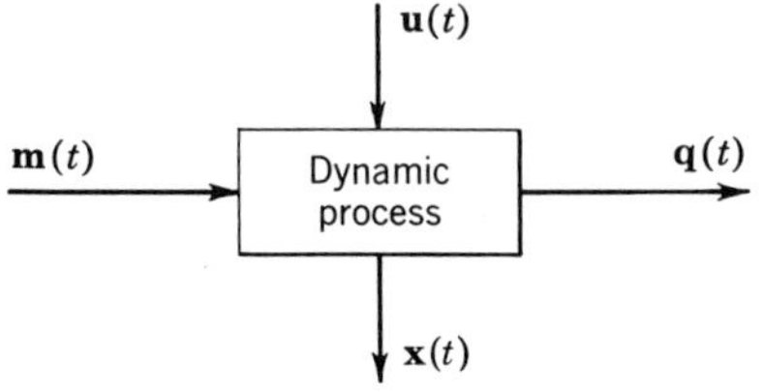

Fig. 1-3 Diagram of the dynamic process.

wind-gust accelerations of an aircraft, and the like. Therefore, a set of *load-disturbance signals* $u_1(t), u_2(t), \ldots, u_N(t)$ is included as a second set of inputs to the dynamic process.

The block diagram of the input and output signals of the dynamic process is shown in Fig. 1-3. The signals are represented vectorially as $\mathbf{m}(t)$, $\mathbf{q}(t)$, $\mathbf{u}(t)$, $\mathbf{x}(t)$ for notational convenience. These vectors have M, Q, and N components such that

$$\mathbf{m}(t) = \begin{bmatrix} m_1(t) \\ m_2(t) \\ \cdot \\ \cdot \\ \cdot \\ m_M(t) \end{bmatrix} \qquad \mathbf{q}(t) = \begin{bmatrix} q_1(t) \\ q_2(t) \\ \cdot \\ \cdot \\ \cdot \\ q_Q(t) \end{bmatrix}$$
$$\mathbf{x}(t) = \begin{bmatrix} x_1(t) \\ x_2(t) \\ \cdot \\ \cdot \\ \cdot \\ x_N(t) \end{bmatrix} \qquad \mathbf{u}(t) = \begin{bmatrix} u_1(t) \\ u_2(t) \\ \cdot \\ \cdot \\ \cdot \\ u_N(t) \end{bmatrix} \tag{1-16}$$

A convenient format for writing the differential equations defining the state signals is

$$\mathbf{x}'(t) = \mathbf{f}[\mathbf{x}(t), \mathbf{m}(t), t] \tag{1-17}$$

which is equivalent to the set of first-order differential equations

$$x_n'(t) = f_n[\mathbf{x}(t), \mathbf{m}(t), t] \qquad n = 1, 2, \ldots, N \tag{1-18}$$

The load disturbances are represented by the explicit dependence of $\mathbf{f}$ on time t in (1-17). This explicit time dependence also arises because of time-varying coefficients of the differential equations. Equation (1-17) generally is referred to as the *state equation* of the dynamic process. Also, the *response equation*

$$\mathbf{q}(t) = \mathbf{g}[\mathbf{x}(t), t] \tag{1-19}$$

is used for the dynamic processes treated in later chapters and is applicable to most control problems. Finally, many control problems involve signal saturation so that the *saturation equations*

$$\mathbf{m}(t) \in \mathscr{M}(t) \qquad \mathbf{x}(t) \in \mathscr{X}(t) \tag{1-20}$$

may also be included in the dynamic process. The notation $\mathbf{m}(t) \in \mathscr{M}(t)$ is a statement that the vector $\mathbf{m}(t)$ lies within or on the boundary of a closed region $\mathscr{M}(t)$ of the vector space. An example of a saturation equation is $-L_n \leqq m_n(t) \leqq L_n$ for $n = 1, 2, \ldots, M$.

For linear dynamic processes, the introduction of additional notation is convenient. In particular, the state equation is written as

$$\mathbf{x}'(t) = B\mathbf{x}(t) + C\mathbf{m}(t) + \mathbf{u}(t) \tag{1-21}$$

and the response equation is written as

$$\mathbf{q}(t) = A\mathbf{x}(t) \tag{1-22}$$

If the dynamic process is linear, saturation does not occur and hence the saturation equations simply are omitted. In (1-21) and (1-22), the matrices A, B, and C are of the form

$$A = \begin{bmatrix} a_{11}(t) & a_{12}(t) & \cdots & a_{1N}(t) \\ a_{21}(t) & a_{22}(t) & \cdots & a_{2N}(t) \\ \cdot & \cdot & \cdot & \cdot \\ a_{Q1}(t) & a_{Q2}(t) & \cdots & a_{QN}(t) \end{bmatrix} \qquad B = \begin{bmatrix} b_{11}(t) & b_{12}(t) & \cdots & b_{1N}(t) \\ b_{21}(t) & b_{22}(t) & \cdots & b_{2N}(t) \\ \cdot & \cdot & \cdot & \cdot \\ b_{N1}(t) & b_{N2}(t) & \cdots & b_{NN}(t) \end{bmatrix}$$

$$C = \begin{bmatrix} c_{11}(t) & c_{12}(t) & \cdots & c_{1M}(t) \\ c_{21}(t) & c_{22}(t) & \cdots & c_{2M}(t) \\ \cdot & \cdot & \cdot & \cdot \\ c_{N1}(t) & c_{N2}(t) & \cdots & c_{NM}(t) \end{bmatrix} \tag{1-23}$$

Vector-matrix notation for differential and algebraic equations is used commonly in mathematical literature[22] and is very convenient in later chapters.

In order to clarify the use of the notation previously introduced, suppose that the fixed member of the control system is described by the transfer function

$$\frac{q_1(s)}{m_1(s)} = \frac{\sum_{n=0}^{N-1} c_n s^n}{\sum_{n=0}^{N} d_n s^n} \tag{1-24}$$

where $q_1(s)$ and $m_1(s)$ are the Laplace transforms of $q_1(t)$ and $m_1(t)$, respectively, and s is the complex-frequency variable.§ No generality is sacrificed by choosing

§ Strictly speaking, the use of $q(s)$ to denote the Laplace transform of $q(t)$ is incorrect function notation. However, this notation is common in the control field and will be adopted here, with reservations, for the transforms of signals only.

$d_N = 1$. By cross-multiplying and using the properties of the transform, (1-24) is equivalent to the differential equation

$$\sum_{n=0}^{N} d_n q_1^{(n)}(t) = \sum_{n=0}^{N-1} c_n m_1^{(n)}(t) \tag{1-25}$$

in the time domain. Furthermore, a new variable $x_N(t)$ can be introduced so that (1-25) is equivalent to

$$\sum_{n=0}^{N} d_n x_N^{(n)}(t) = m_1(t) \tag{1-26}$$

and

$$\sum_{n=0}^{N-1} c_n x_N^{(n)}(t) = q_1(t) \tag{1-27}$$

Finally, the additional variables are introduced so that

$$x_n'(t) = x_{n-1}(t) \qquad n = 2, 3, \ldots, N \tag{1-28}$$

and (1-26) then reduces to

$$x_1'(t) = \sum_{n=1}^{N} [-d_{N-n} x_n(t)] + m_1(t) \tag{1-29}$$

Similarly, (1-27) reduces to

$$q_1(t) = \sum_{n=1}^{N} c_{N-n} x_n(t) \tag{1-30}$$

Clearly, (1-28) and (1-29) comprise the state equation and (1-30) the response equation of the fixed member originally defined in (1-24). For this example, the matrices defined in (1-23) are

$$A = [c_{N-1} \quad c_{N-2} \quad \cdots \quad c_0]$$

$$B = \begin{bmatrix} -d_{N-1} & -d_{N-2} & \cdots & \cdots & \cdots & \cdots & -d_0 \\ 1 & 0 & 0 & \cdots & 0 & 0 & 0 \\ 0 & 1 & 0 & \cdots & 0 & 0 & 0 \\ 0 & 0 & 1 & \cdots & 0 & 0 & 0 \\ \cdot & \cdot & \cdot & \cdot & \cdot & \cdot & \cdot \\ 0 & 0 & 0 & \cdots & 1 & 0 & 0 \\ 0 & 0 & 0 & \cdots & 0 & 1 & 0 \end{bmatrix} \tag{1-31}$$

$$C = \begin{bmatrix} 1 \\ 0 \\ \cdot \\ \cdot \\ \cdot \\ 0 \end{bmatrix}$$

An analog-computer simulation of this fixed member is shown in Fig. 1-4, where the state signals have been indicated.

1-5 Definitions and nomenclature for performance description

In most control-system design problems, the engineer has certain performance specifications which must be satisfied. Traditionally, control engineers have used performance specifications such as damping ratio and natural frequency which are associated closely on a conceptual basis with linear second-order systems. These widely used performance specifications primarily shape the transient response of the system without regard to the particular signals occurring in the system. Also, such specifications are only meaningful in a mathematical sense, and probably in a conceptual sense, for linear time-invariant systems of low order. On the other hand,

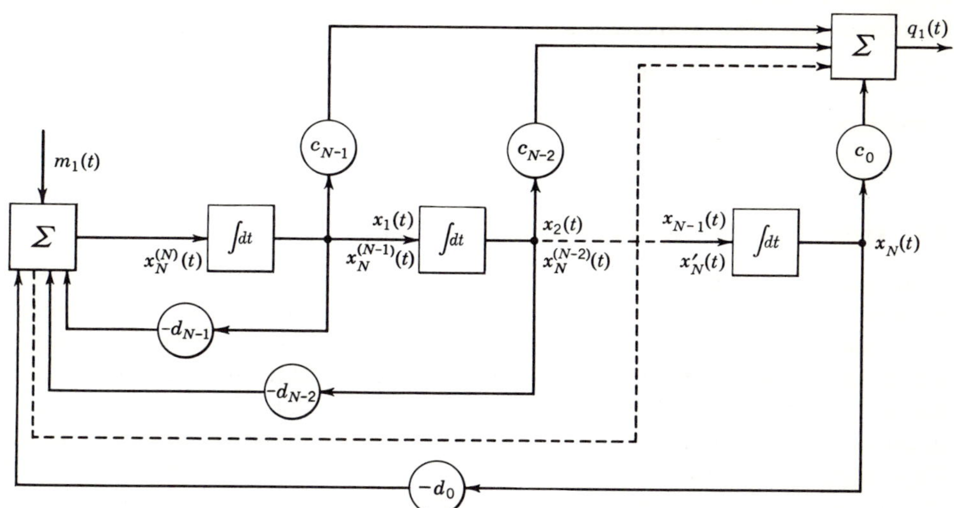

Fig. 1-4 Analog-computer simulation of example dynamic process.

the performance specifications used throughout this book pertain to the approximating properties of the system in achieving the best possible performance under given design conditions. Also, the assumption is made that the performance specifications are given, even though the engineer may have to formulate them.

The performance specifications are assumed to be written in terms of errors between actual and desired responses as a matter of choice. Therefore, the criterion used in this book for design is the selection of the control system that minimizes these errors as calculated for a fixed member and performance specifications.

The instantaneous errors incurred in the system are calculated in terms of an *error measure* $e_m(t)$. The error measure is written in terms of the response and control signals in the general form

$$e_m(t) = h[\mathbf{q}(t), \mathbf{m}(t), t] \tag{1-32}$$

The *error index* $e(t)$ is the total error incurred in the system over the present and future time, $t \leq \sigma \leq T$, where the control system operates. The error index is found by integrating the error measure according to

$$e(t) = \int_t^T e_m(\sigma)\, d\sigma \tag{1-33}$$

A commonly used form of error measure expresses the errors in each control and response signal separately.[22] This particular form is written as

$$e_m(t) = \sum_{n=1}^{Q} \phi_n(t)\Phi_n[Q_n(t) - q_n(t)] + \sum_{n=1}^{M} \psi_n(t)\Psi_n[M_n(t) - m_n(t)] \tag{1-34}$$

The signals $Q_1(\sigma)$, $Q_2(\sigma)$, . . . , $Q_Q(\sigma)$ are called *desired-response signals*, and $M_1(\sigma)$, $M_2(\sigma)$, . . . , $M_M(\sigma)$ are called *desired-control signals*. Also, the factors $\phi_1(\sigma)$, $\phi_2(\sigma)$, . . . , $\phi_Q(\sigma)$ and $\psi_1(\sigma)$, $\psi_2(\sigma)$, . . . , $\psi_M(\sigma)$ are termed *weighting factors*. These weighting factors may be treated as Lagrange multipliers,[23] but for purposes of discussion they generally are considered to be fixed by the engineer. Also, the functions Φ_n and Ψ_n are assumed to be specified by the engineer for each design problem. The error measure is said to be *quadratic* when

$$\Phi_n(y) = \Psi_n(y) = y^2 \qquad \text{for all } n \tag{1-35}$$

Considerable material in later chapters is devoted to the implications of choosing various forms of the error measure.

1-6 Definitions for optimum control systems

A number of connotations of the word *optimum* are associated with system design in technical literature. As a result, confusion often arises. For the purposes of this book, the *optimum control system* is defined as the control system that minimizes a given error index for a given dynamic process and subject to given design constraints. Design constraints could be restrictions stating that the control system must be linear or that the control system must have a specific configuration of components. This definition is worded carefully in order to avoid possible confusion. The reader should note that the optimum system is defined in terms of a mathematical model of the design problem. If the mathematical model is altered, then the minimization leads to a different control system. A typical change in the mathematical model arises from the adjustment of weighting factors in the error index, a procedure which may be required in order to satisfy all the design specifications simultaneously.

The historical development of control-system optimization theory has led to ever-increasing generality in the concept of optimum systems. As a result, care must be exercised in distinguishing between various classes of optimum systems.

The least general class of systems arises when the configuration of the controller is assumed fixed and a limited number of parameters are adjusted to minimize the error index. Such parameters may be gains and time constants. *Parameter optimization* is the selection of these parameters to minimize the error index.

A more general class of systems occurs as the result of optimizing the impulse response of a system. This class is limited to linear dynamic processes and quadratic error measures. *Impulse-response optimization* is the selection of the composite impulse response of both the dynamic process and the controller to minimize the error index. Here the mathematics of linear systems is used to determine the impulse response or transfer function of the controller from the dynamic process and composite impulse response.

Chronologically, parameter optimization and impulse-response optimization are the forerunners of the material in this book. Some review is included, however, in order to provide a transition and clarify the concepts. *System optimization* is the major topic and is the selection of both the configuration and the component values of the controller to minimize the error index. In other words, the complete characteristics of the controller are found without restrictions or constraints on the linearity, configuration, or parameters of the controller. System optimization provides the most general class of optimum controllers and gives the complete solution to the minimization problem.

An important point of distinction arises when considering linear systems in the context of optimization theory. When system optimization results in a linear controller, the controller is termed the *linear optimum controller.* On the other hand, when the optimum controller is in fact nonlinear but the engineer constrains himself to investigating only linear controllers, the resulting linear controller is termed the *optimum linear controller.* This distinction can cause confusion and hence is treated by a simple example in the next chapter for clarification purposes.

2

Parameter Optimization

2-1 Heuristic development of the method

As mentioned in the first chapter, the least general optimization procedure is to take a controller of fixed configuration and adjust the parameters of the controller so that an error index is minimized. The parameters would be gains and time constants in the case of a linear controller. As a vehicle for developing the mathematical methods pertinent to the fixed-configuration design problem, the simplest possible control system is considered first.

A simplified linear model of a position servo is shown in Fig. 2-1, where the control signal $m(t)$ is considered to be an exciting voltage and the response signal $q(t)$ is considered to be shaft position. The desired-response signal $Q(t)$ is defined as the desired shaft position, and the difference between $Q(t)$ and $q(t)$ is the error signal. As a start, the error measure is chosen to be the quadratic

$$e_m(t) = [Q(t) - q(t)]^2 \tag{2-1}$$

This error measure has convenient mathematical properties and is used in the pertinent literature.[1,12] The optimum controller is designed to minimize future errors as given by the error index

$$e(t) = \int_t^\infty e_m(\sigma)\, d\sigma \tag{2-2}$$

The transient errors are due to the existing shaft position of the servo, $q(t)$, and are assumed to be minimized by adjusting the gain k. In order to further simplify this design problem, the desired response is assumed to be a constant $Q(t) = Q$.

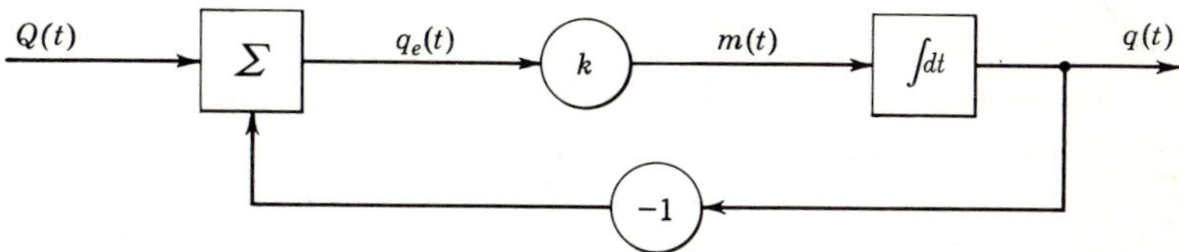

Fig. 2-1 Block diagram of a position servo.

An elementary transient analysis of the system shown in Fig. 2-1 gives the future system response

$$q(\sigma) = Q(1 - e^{-k(\sigma-t)}) + q(t)e^{-k(\sigma-t)} \qquad \sigma \geqq t \tag{2-3}$$

Then, from (2-1), the future value of the error measure is

$$e_m(\sigma) = [Q - q(t)]^2\, e^{-2k(\sigma-t)} \qquad \sigma \geqq t \tag{2-4}$$

and, from (2-2), the error index becomes

$$e(t) = \frac{[Q - q(t)]^2}{2k} \tag{2-5}$$

When the gain k is varied so as to minimize the value of $e(t)$, an increasing k decreases $e(t)$ monotonically. Therefore, the selection $k = \infty$ minimizes the error index. In this simple problem, this result could have been predicted without actually computing the error index because the system response approaches exponentially the desired response with no overshoot. Also, the system time constant decreases with increasing loop gain, as indicated in Fig. 2-2.

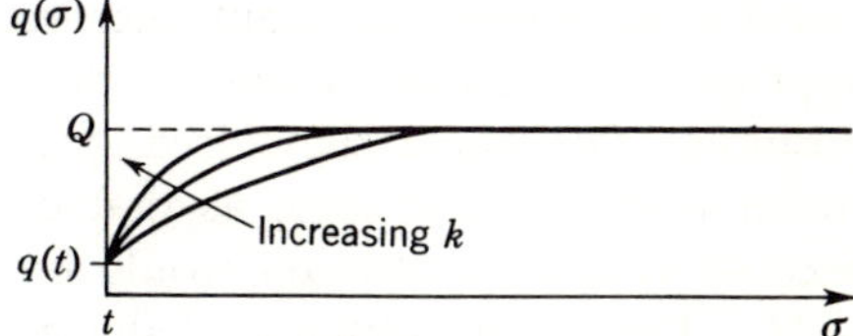

Fig. 2-2 Future response of position servo.

However, the maximum magnitude of the control signal increases with increasing k, and a possibility of velocity saturation in the system exists. If the amplitude of the excitation voltage corresponding to velocity saturation is defined as M, then a situation may exist such as the one depicted in Fig. 2-3. The future value of the control signal without saturation is calculated to be

$$m(\sigma) = [Q - q(t)]ke^{-k(\sigma-t)} \qquad \sigma \geqq t \tag{2-6}$$

In other words, the optimum value of k causes a control signal that results in velocity saturation, and a nonlinearity is experienced.

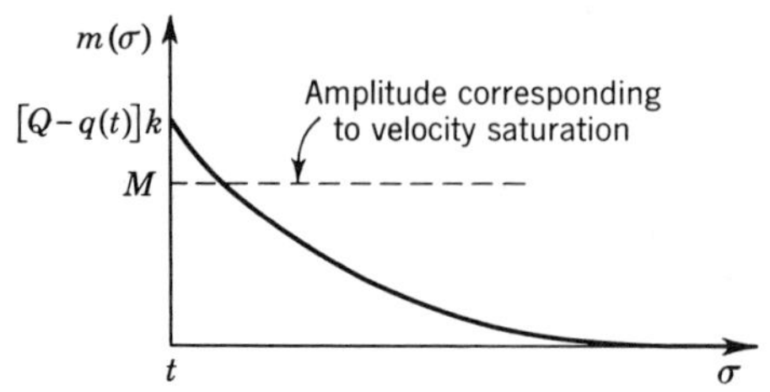

Fig. 2-3 Control signal of position servo without saturation.

In order to avoid the impending mathematical difficulties of nonlinear system behavior, a restrictive class of optimization problems can be investigated, namely, optimum linear systems. For this class, the gain k can be adjusted to minimize $e(t)$ subject to the restriction that the system remain linear. Because of the monotonic behavior of $e(t)$, the value of k is chosen to be as large as possible without velocity saturation. Therefore

$$k = \frac{M}{|Q - q(t)|} \tag{2-7}$$

gives the optimum linear system, as can be seen from Fig. 2-3. With this value of gain, the system error index becomes

$$e(t) = \frac{|Q - q(t)|^3}{2M} \tag{2-8}$$

for the optimum linear system.

So far, only the optimum selection of the gain k subject to the constraint that the control system remain linear has been discussed. The question arises at this time: Is there any nonlinear control system that performs better in the sense of minimizing (2-2) than the optimum linear system? Because the model of the servo is first-order, there are no stability problems introduced by considering a controller with the on-off gain element shown in Fig. 2-4.

The on-off gain element performs the function $m(t) = M$ when $[Q(t) - q(t)] > 0$, $m(t) = 0$ when $[Q(t) - q(t)] = 0$, and $m(t) = -M$ when $[Q(t) - q(t)] < 0$. This is called a *maximum-effort* controller because the maximum velocity is utilized in the appropriate direction whenever the shaft position differs from the desired shaft position. Of course, the closed-loop system response of the maximum-effort controller is exactly the same as that of the system shown in Fig. 2-1 when $k = \infty$, because velocity saturation would limit the velocity of the servo to exactly M and $-M$. The future response of the position servo with a maximum-effort controller

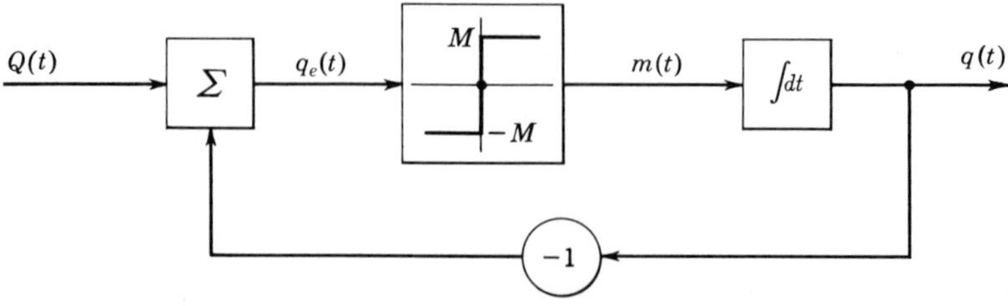

Fig. 2-4 Position servo with a maximum-effort controller.

is sketched in Fig. 2-5 for $Q(t) = Q$. The future value of the error measure defined in (2-1) is

$$e_m(\sigma) = \begin{cases} \{[Q - q(t)] - (\sigma - t)M\}^2 & t \leqq \sigma \leqq t + \dfrac{Q - q(t)}{M} \\ 0 & t + \dfrac{Q - q(t)}{M} \leqq \sigma \end{cases} \tag{2-9}$$

and the error index, given in (2-2), becomes

$$e(t) = \frac{|Q - q(t)|^3}{3M} \tag{2-10}$$

The value of the error index given in (2-10) for the maximum-effort controller represents a 50 per cent increase in performance compared with the value of the performance index given in (2-8) for the optimum linear controller. This demonstrates that a linear controller is not always the optimum controller. The maximum-effort controller happens to be the optimum controller in this case, a fact which has

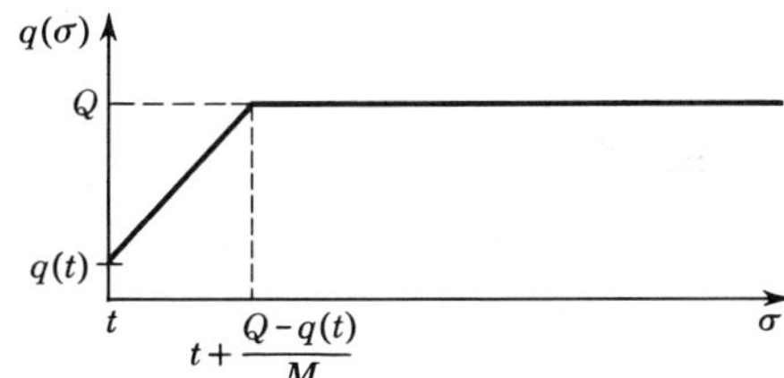

Fig. 2-5 Response of position servo with a maximum-effort controller.

not been proved as yet. More generally, when only response errors are weighted in the error index, the optimum controller is always of the maximum-effort variety.

However, now assume that, in addition to the need for reducing response errors in the position servo, there is also the need to keep the amount of power input to the motor as small as possible. The power input to the motor is assumed to be proportional to the square of the excitation voltage and therefore proportional to $m^2(t)$. If the "cost" of response error is assumed to be proportional to the response error squared, then the rate at which total cost is being incurred is given by

$$c(t) = c_1[Q(t) - q(t)]^2 + c_2 m^2(t) \tag{2-11}$$

The constants c_1 and c_2 are assumed known. If an error measure is defined as $e_m(t) = c(t)/c_1$, then

$$e_m(t) = [Q(t) - q(t)]^2 + \rho m^2(t) \tag{2-12}$$

where ρ is known and given by $\rho = c_2/c_1$. Clearly, the minimum of $e_m(t)$ is proportional to the minimum of $c(t)$; hence (2-12) can be used as the error measure. Now the integral given in (2-2) is evaluated under the conditions that the desired response is a constant Q and the position servo is linear. The error index now becomes

$$e(t) = \frac{[Q - q(t)]^2}{2}\left(\frac{1}{k} + \rho k\right) \tag{2-13}$$

If the value of the error index given in (2-13) is plotted vs. the gain k, then $e(t)$ is seen to be the convex function shown in Fig. 2-6. For this curve, a finite value of k, denoted as k^*, yields a minimum for $e(t)$. This minimum value is denoted as a_0.

The curve of $e(t)$ versus k actually could be plotted, and the minimum value of $e(t)$ plus the corresponding value of k could be located graphically. However, a more expeditious procedure is to utilize the geometrical properties of a convex curve, such as shown in Fig. 2-6.

In particular, the curve can be expanded in terms of a Taylor series about the optimum value of k, k^*. This Taylor series expansion is written as

$$e(t) = a_0 + a_1(k - k^*) + a_2(k - k^*)^2 + \cdots \tag{2-14}$$

where

$$a_0 = e(t)\,\big|_{k=k^*} \qquad a_1 = \left.\frac{\partial e(t)}{\partial k}\right|_{k=k^*} \qquad a_2 = \left.\frac{1}{2}\frac{\partial^2 e(t)}{\partial k^2}\right|_{k=k^*} \qquad \cdots \tag{2-15}$$

From the geometrical properties of a convex curve, such as shown in Fig. 2-6, the slope of the curve at its minimum point is zero and the second derivative of the curve

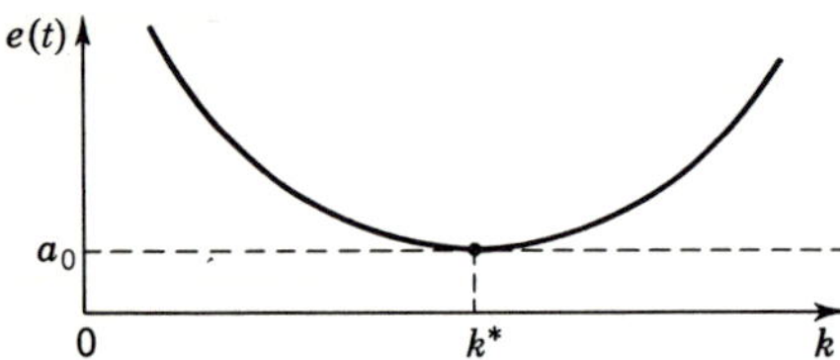

Fig. 2-6 Error index vs. gain for linear position servo when input power is weighted.

at this point must be positive. Therefore, the condition for a minimum is $a_1 = 0$ and $a_2 > 0$; hence

$$\left.\frac{\partial e(t)}{\partial k}\right|_{k=k^*} = 0 \qquad \left.\frac{\partial^2 e(t)}{\partial k^2}\right|_{k=k^*} > 0 \tag{2-16}$$

These conditions for finding the minimum of a convex curve can be applied to (2-13) by evaluating the partial derivatives. The conditions for minimum error become, respectively,

$$k^* = \pm\rho^{-\frac{1}{2}} \qquad k^* > 0 \tag{2-17}$$

Hence, the optimum value of the gain, $k^* = \rho^{-\frac{1}{2}}$, is determined uniquely from (2-16). Finally, the minimum value of the error index, a_0, is found by substituting the optimum value of the gain into (2-13); and this result is

$$e(t) = \frac{[Q - q(t)]^2}{\rho^{-\frac{1}{2}}} \tag{2-18}$$

When the control effort is weighted in the error index, the optimum value of the gain is no longer infinite. Therefore the optimum system is linear over a finite range of values for the existing response error. From (2-17) and (2-6), linearity exists when

$$\rho^{\frac{1}{2}} > \frac{|Q - q(t)|}{M} \tag{2-19}$$

For this condition, the region of the k axis where saturation would occur lies entirely to the right of the minimum point in Fig. 2-6. On the other hand, if

$$\rho^{\frac{1}{2}} \leq \frac{|Q - q(t)|}{M} \tag{2-20}$$

then the region of the k axis where saturation would occur includes the minimum point in Fig. 2-6. When (2-20) exists, the gain that minimizes $e(t)$ subject to the constraint that the system be linear is $k = M/|Q - q(t)|$, as opposed to $k = \rho^{-\frac{1}{2}}$ when (2-19) exists. For optimum linear control, (2-13) becomes

$$e(t) = \frac{|Q - q(t)|^3}{2M} \left\{1 + \rho \frac{M^2}{[Q - q(t)]^2}\right\} \qquad \rho^{\frac{1}{2}} \leqq \frac{|Q - q(t)|}{M} \tag{2-21}$$

Thus far, the weighting factor ρ has been treated as a known constant. However, if ρ is treated as an adjustable parameter in the design problem, ρ can be adjusted so that $\rho^{\frac{1}{2}} = |Q - q(t)|/M$. This particular value of ρ minimizes $e(t)$ when power input has no cost but the optimum system is constrained to be linear, as can be seen from (2-7) and (2-8). This mathematical method of adjusting the weighting factor ρ in order to avoid saturation and hence obtain the optimum linear system is the method of Lagrange multipliers. This method was first introduced into control-system design by Newton.[12]

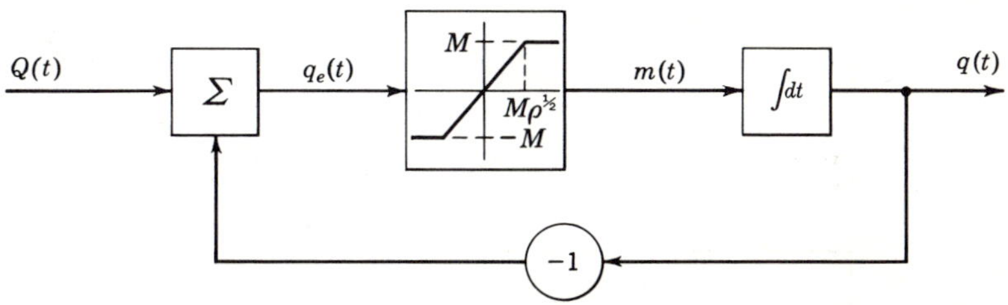

Fig. 2-7 Position servo with a linear zone in the saturating controller.

Now let us return and evaluate the performance of the maximum-effort controller depicted in Fig. 2-4 with the error measure given in (2-12) and ρ fixed. For this system and measure, the index given by (2-2) is evaluated to be

$$e(t) = \frac{|Q - q(t)|^3}{3M} \left\{1 + 3\rho \frac{M^2}{[Q - q(t)]^2}\right\} \tag{2-22}$$

If (2-21) and (2-22) are compared under the condition that $|Q - q(t)|/M$ is very large with respect to $\rho^{\frac{1}{2}}$, the maximum-effort system gives less error than the optimum linear system. On the other hand, if (2-18) and (2-22) are compared under the condition that $|Q - q(t)|/M$ is very small with respect to $\rho^{\frac{1}{2}}$, then the error of the maximum-effort system becomes arbitrarily large with respect to that of the linear system.

The question now arises as to whether there is another nonlinear system which performs as well as or better than both the optimum linear and the maximum-effort systems for all values of $|Q - q(t)|/M$. The previous comparisons immediately suggest a control system that has a linear zone for small errors and saturation for large errors, such as the one depicted in Fig. 2-7. For this system, the error index with power input weighted according to (2-12) is

$$e(t) = \begin{cases} \dfrac{[Q - q(t)]^2}{\rho^{-\frac{1}{2}}} & \rho^{\frac{1}{2}} > \dfrac{|Q - q(t)|}{M} \\[2ex] \dfrac{[Q - q(t)]^2}{\rho^{-\frac{1}{2}}} \left\{\dfrac{|Q - q(t)|}{3M\rho^{\frac{1}{2}}} + \dfrac{M\rho^{\frac{1}{2}}}{|Q - q(t)|} - \dfrac{M^2\rho}{3[Q - q(t)]^2}\right\} & \rho^{\frac{1}{2}} \leqq \dfrac{|Q - q(t)|}{M} \end{cases} \tag{2-23}$$

The performance of the optimum linear, maximum-effort, and saturating-with-linear-zone controllers can be compared easily from the sketch in Fig. 2-8.

Although not proved, the saturating controller with a linear zone, shown in Fig. 2-7, is the optimum controller for the error measure given in (2-12). The optimum linear controller is in fact optimum over a range of response errors, but this range vanishes as $\rho \to 0$. The optimum controller is a maximum-effort system when $\rho \to 0$.

The detailed treatment of this simple problem has been included in order to identify clearly the difference between an optimum controller and an optimum

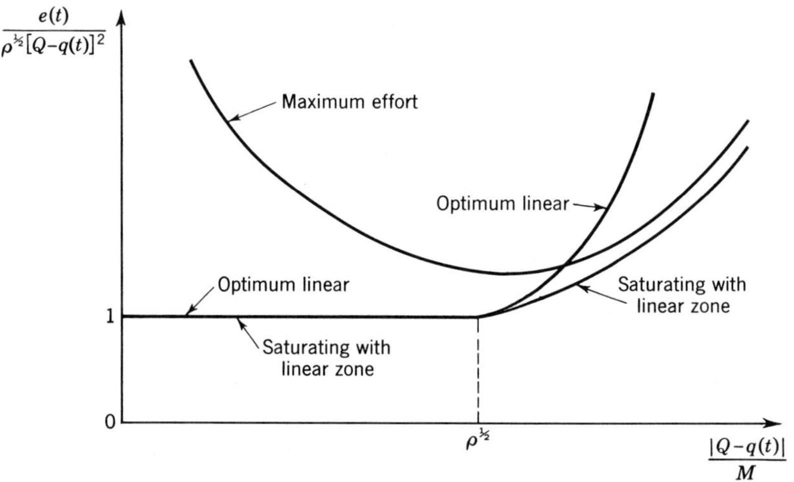

Fig. 2-8 Performance of optimum linear and nonlinear controllers for position servo when power input is weighted.

linear controller. Also, this example suggests the use of Lagrange multipliers as a mathematical method of optimizing linear controllers as a subclass of all possible controllers.

2-2 Mathematical development of the method for linear systems

In order to make the parameter-optimization method applicable to a wider range of problems than the simple one discussed previously, a number of useful mathematical methods are introduced here. First of all, when the fixed elements of the control system are represented by a differential equation of high order, the analytical method of writing down the exact transient response of the system in terms of unspecified parameters becomes difficult or impossible. This difficulty arises because the natural frequencies of the closed-loop system cannot be determined unless all the parameters are specified numerically. Therefore, a method of expressing the value of the error index directly in terms of the undetermined parameters of the control system via a frequency-domain analysis is needed.

If the Fourier transform of an arbitrary function $f(t)$ is defined as

$$f(s) = \int_{-\infty}^{\infty} f(t)e^{-st}\,dt \tag{2-24}$$

where $s = j\omega$, then the inverse transform is

$$f(t) = \frac{1}{2\pi j} \int_{-j\infty}^{j\infty} f(s) e^{st}\, ds \tag{2-25}$$

A well-known equivalence, called Parseval's theorem,[24] between time and frequency domains is

$$\int_{-\infty}^{\infty} f^2(t)\, dt = \frac{1}{2\pi j} \int_{-j\infty}^{j\infty} f(s) f(-s)\, ds \tag{2-26}$$

Therefore, when an error index

$$e = \int_{-\infty}^{\infty} e_q{}^2(\sigma)\, d\sigma \tag{2-27}$$

is used,§ the index can be stated directly in terms of the transformed variables as

$$e = \frac{1}{2\pi j} \int_{-j\infty}^{j\infty} e_q(s) e_q(-s)\, ds \tag{2-28}$$

By using (2-28), time-domain analysis of the system transients can be avoided completely, which is necessary in order to treat systems of high order.

However, the explicit evaluation of (2-28) generally involves contour integration, and such integrations would prove time-consuming if performed for each design problem. Phillips[1] recognized this drawback and developed a method of tabulating (2-28) in terms of the coefficients of $e_q(s)$ when $e_q(s)$ is rational in s. The polynomial form of $e_q(s)$ is written as

$$e_q(s) = \frac{c(s)}{d(s)} \tag{2-29}$$

where

$$c(s) = \sum_{k=0}^{N-1} c_k s^k \tag{2-30}$$

and

$$d(s) = \sum_{k=0}^{N} d_k s^k \tag{2-31}$$

The first three integrals in the tabulation are

$$e = \frac{c_0{}^2}{2d_0 d_1} \qquad N = 1 \tag{2-32}$$

$$e = \frac{c_1{}^2 d_0 + c_0{}^2 d_2}{2d_0 d_1 d_2} \qquad N = 2 \tag{2-33}$$

and

$$e = \frac{c_2{}^2 d_0 d_1 + (c_1{}^2 - 2c_0 c_2)\, d_0 d_3 + c_0{}^2 d_2 d_3}{2d_0 d_3(-d_0 d_3 + d_1 d_2)} \qquad N = 3 \tag{2-34}$$

This tabulation[12] has been carried up to $N = 10$, but only the first three integrals are necessary for the purposes of this book. The reader should note that the tabulation is pertinent only when $c(s)$ is of degree of at least one less than $d(s)$ and when $d(s)$ has zeros only in the left half plane. With the use of these tabulated integrals, the designer can express the error index directly in terms of system parameters by frequency-domain methods.

§ At $\sigma = -\infty$, the initial conditions in the system are assumed to be zero; hence the time-function notation for the error index is dropped.

Some confusion may arise from the use of the notation s (normally reserved for the complex-frequency variable in Laplace transforms) with Fourier transforms. Throughout this book, similar to the treatment by Newton,[12] the tacit assumption that $s = j\omega$ is made when the bilateral transform is taken as in (2-24). On the other hand, when the one-sided transform

$$f(s) = \int_0^\infty f(t)e^{-st}\,dt \tag{2-35}$$

is taken, it is admissible to assume that $s = \alpha + j\omega$ and (2-35) is identically the Laplace transform. In the cases where the bilateral transform is used, such as in the derivation of (2-28) from Parseval's theorem, the exponential convergence of the Laplace transform is not needed. For instance, the Laplace transform of (2-3) with respect to $\sigma - t$ yields

$$q(s) = \frac{-q(t)s + Qk}{s(s + k)} \tag{2-36}$$

which does not satisfy the Dirichlet conditions and hence is not Fourier-transformable. However, the Laplace transform of $e_q(\sigma)$ appearing in (2-4) with respect to $\sigma - t$ yields

$$e_q(s) = \frac{Q - q(t)}{s + k} \tag{2-37}$$

which is Fourier-transformable, and therefore s can be interpreted as $s = j\omega$. The constant component in the system response is not Fourier-transformable, but a constant cannot appear in the error measure because the error index would be infinite. Finally, the use of the Laplace transform, with respect to $\sigma - t$, in evaluating $e_q(s)$ implies that $e_q(\sigma) = 0$ for $\sigma - t < 0$; hence the error index given in (2-27) automatically has been specialized to

$$e = \int_t^\infty e_q^{\,2}(\sigma)\,d\sigma \tag{2-38}$$

Therefore, the values of the indices computed from the tabulated integrals when $e_q(s)$ is found from the Laplace transform include only present and future errors. This is precisely what is wanted.

Now that a method of expressing the error index for complex linear systems with multiple parameters is available, the minimization procedure for the error index must be generalized. In particular, when a fixed-configuration system contains K adjustable parameters, $p_1, p_2, \ldots, p_K$, the error index can be written as a function

$$e(\mathbf{p}) = \int_t^\infty e_m(\sigma)\,d\sigma \tag{2-39}$$

where $\mathbf{p}$ is the column vector with K components:

$$\mathbf{p} = \begin{bmatrix} p_1 \\ p_2 \\ \cdot \\ \cdot \\ \cdot \\ p_K \end{bmatrix} \tag{2-40}$$

If there is a unique set of parameters which minimize the error index, then a Taylor series can be expanded around the values of the parameters that minimize the integral. By denoting the optimum value of $\mathbf{p}$ as $\mathbf{p}^*$ and the corresponding optimum component values as p_k^*, the Taylor series is written as

$$e(\mathbf{p}) = a_0 + \sum_{k=1}^{K} a_{1k}(p_k - p_k^*) + \sum_{j=1}^{K} \sum_{k=1}^{K} a_{2jk}(p_j - p_j^*)(p_k - p_k^*) + \cdots \tag{2-41}$$

With the same geometrical considerations as those used for the one-parameter function, the conditions for a minimum are given by the condition that $a_{1k} = 0$ and a_{2jk} forms a positive-definite matrix. When these coefficients are evaluated in terms of the partial derivatives of the index, the conditions for a minimum value of the error index are

$$\left.\frac{\partial e(\mathbf{p})}{\partial p_k}\right|_{\mathbf{p}=\mathbf{p}^*} = 0 \qquad 1 \leqq k \leqq K \tag{2-42}$$

and $[a_{2jk}]$ is positive definite, where

$$a_{2jk} = \left.\frac{1}{2}\frac{\partial^2 e(\mathbf{p})}{\partial p_j\,\partial p_k}\right|_{\mathbf{p}=\mathbf{p}^*} \tag{2-43}$$

The notation of evaluating the partial derivatives at $\mathbf{p} = \mathbf{p}^*$ is implied by satisfying the equations in (2-42); hence this notation can be dropped.

The conditions given in (2-42) and (2-43) are necessary for a minimum value of the error index. However, these conditions are not sufficient in the sense that the minimum may not be unique. Discontinuous functions and other ill-behaved functions for $e(\mathbf{p})$ could arise, and the application of (2-42) and (2-43) would result in ambiguous answers. A unique minimum exists if the error index is strictly convex. However, tests for the presence of a strictly convex function are difficult to carry out. Therefore, the designer generally must rely on his common sense to interpret the values for the parameters obtained from (2-42) and (2-43).

In order to include the physical constraints on the fixed-configuration control system in the mathematics of parameter optimization, the method of Lagrange multipliers is introduced. The avoidance of saturation, and thus the retaining of linearity, already has been pointed out as a use for Lagrange multipliers.

The development of the method used here is more expository than rigorous but does have the advantage of clarifying the concepts behind the method. Assume first that a general function $e(\mathbf{p})$ is minimized with respect to the K components of $\mathbf{p}$. Also, assume that the component values must be chosen subject to satisfying K constraint equations written as

$$e_{cn}(\mathbf{p}) \leqq c_n \qquad n = 1, 2, \ldots, K \tag{2-44}$$

The constants in (2-44) are quantities such as the integral square of control effort. Now a new function is defined arbitrarily as

$$e_c(\mathbf{p}, \boldsymbol{\lambda}) = e(\mathbf{p}) + \sum_{n=1}^{K} \lambda_n[e_{cn}(\mathbf{p}) - c_n] \tag{2-45}$$

and the Lagrange-multiplier vector is defined as

$$\boldsymbol{\lambda} = \begin{bmatrix} \lambda_1 \\ \lambda_2 \\ \cdot \\ \cdot \\ \cdot \\ \lambda_K \end{bmatrix} \tag{2-46}$$

The proposed procedure for minimizing $e(\mathbf{p})$ subject to the K constraints given in (2-44) is to minimize $e_c(\mathbf{p}, \boldsymbol{\lambda})$ with respect to both $\mathbf{p}$ and $\boldsymbol{\lambda}$. If this minimization procedure results in $\lambda_n[e_{cn}(\mathbf{p}) - c_n] = 0$ for $n = 1, 2, \ldots, K$, then $e(\mathbf{p})$ is minimized subject to the constraints given in (2-44). Two possibilities exist: If $e_{cn}(\mathbf{p}) = c_n$, then $\lambda_n \geqq 0$; if $e_{cn}(\mathbf{p}) < c_n$, then $\lambda_n = 0$.

More formally, the minimizing values of $\mathbf{p}$ and $\boldsymbol{\lambda}$ for (2-45) can be found by the same process used to minimize (2-39). The necessary conditions for a minimum which corresponds to (2-42) are

$$\frac{de_c(\mathbf{p}, \boldsymbol{\lambda})}{dp_k} = 0 \qquad \text{and} \qquad \frac{de_c(\mathbf{p}, \boldsymbol{\lambda})}{d\lambda_n} = 0 \qquad 1 \leqq k, n \leqq K \tag{2-47}$$

assuming that $\lambda_n > 0$. Total derivatives with respect to p_k and λ_n have been used in (2-47) because the functional independence of $\mathbf{p}$ and $\boldsymbol{\lambda}$ has not been established. If $\mathbf{p}$ and $\boldsymbol{\lambda}$ can be assumed to be functionally independent, the total derivatives then can be replaced by partial derivatives, and the two expressions in (2-47) become

$$\frac{\partial e(\mathbf{p})}{\partial p_k} + \sum_{n=1}^{K} \lambda_n \frac{\partial e_{cn}(\mathbf{p})}{\partial p_k} = 0 \qquad k = 1, 2, \ldots, K \tag{2-48}$$

and

$$e_{cn}(\mathbf{p}) - c_n = 0 \qquad n = 1, 2, \ldots, K \tag{2-49}$$

This assertion of independence now must be verified.

The equations given in (2-49) show that the K components of $\mathbf{p}$ are functions only of the K constraint constants. These functions can be substituted into (2-48), and the values of the Lagrange multipliers can be found as functions of only the constraint constants from (2-48). On the other hand, if there are only N constraints, $N < K$, placed on the system design, then only N Lagrange multipliers are nonzero. Therefore, (2-48) contains K equations and N unknown Lagrange multipliers. In this case, there must be $K - N$ functional relationships between the partial derivatives of e and e_{cn} if a unique solution exists.§ These $K - N$ relationships involve the components of $\mathbf{p}$ only. In addition, (2-49) gives N functional relationships between the components of $\mathbf{p}$ and the constraint constants because (2-49) now contains only N constraint equations. Therefore, a total of K functional relationships are found for the K components of $\mathbf{p}$ in terms of only N constraint constants. In conclusion, the assertion that $\mathbf{p}$ and $\boldsymbol{\lambda}$ are independent in the derivation of (2-48) and (2-49) has proved correct.

§ This uniqueness property is fundamental in the theory of linear algebraic equations.[23] Equation (2-48) is linear with respect to the components of $\boldsymbol{\lambda}$.

2-3 Example of the mathematical method

In order to illustrate the mathematical methods discussed in Sec. 2-2, an example is chosen here to demonstrate the use of the tabulated integrals and the Lagrange-multiplier technique. This example also is being introduced in order to discuss the usability of this method as a design tool.

The fixed-configuration controller under design is shown in Fig. 2-9. This is a position-servo controller with an additional loop included in order to compensate for the ramp response. The existing values of the system state variables are assumed

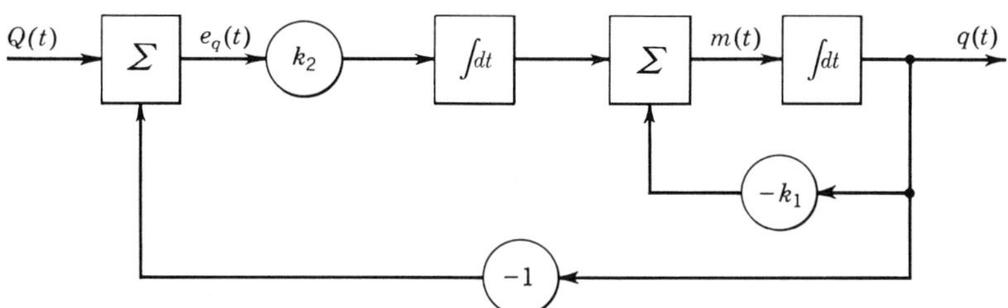

Fig. 2-9 Position servo with compensation for ramp response.

to be zero, and the present and future values of the desired shaft position are assumed to be a constant $Q(\sigma) = Q$. With the gain k_2 considered to be fixed, the future transient error, as measured by the index

$$e = \int_t^\infty [Q - q(\sigma)]^2 \, d\sigma \tag{2-50}$$

is to be minimized by adjusting the gain k_1.

The Laplace transform of $[Q - q(\sigma)]$ with respect to $\sigma - t$, when written in terms of the notation used in the tabulated integrals, is

$$e_q(s) = Q \frac{s + k_1}{s^2 + k_1 s + k_2} \tag{2-51}$$

Because $d(s)$ is of degree 2, (2-33) is used to evaluate (2-50) in terms of the control-system parameters. The result is

$$e(\mathbf{p}) = Q^2 \frac{k_1^2 + k_2}{2k_1 k_2} \tag{2-52}$$

By treating k_2 as fixed, the minimization procedure is carried out with respect to only the gain k_1. The application of (2-42) yields the minimizing condition

$$\frac{\partial e(\mathbf{p})}{\partial k_1} = \frac{4k_1^2 k_2 - (k_1^2 + k_2)2k_2}{4k_1^2 k_2^2} = 0 \tag{2-53}$$

or, finally,

$$k_1 = \sqrt{k_2} \tag{2-54}$$

This optimum value of the gain k_1 gives the minimum value of the error index

$$e = \frac{Q^2}{\sqrt{k_2}} \tag{2-55}$$

by substituting (2-54) into (2-52).

Now a constraint is added to the design problem. This constraint requires that the integral squared value of the control signal be bounded by the number M^2. In other words, a constraint given by

$$e_{c1}(\mathbf{p}) = \int_t^\infty m^2(\sigma)\,d\sigma \leqq M^2 \tag{2-56}$$

is imposed. The constraint index e_{c1} can be evaluated in the same fashion as that for the evaluation of e. The transform of the control signal, in terms of the notation associated with the tabulated integrals, is evaluated readily to be

$$m(s) = Q\frac{k_2}{s^2 + k_1 s + k_2} \tag{2-57}$$

Again, by use of (2-33), the constraint index is

$$e_{c1}(\mathbf{p}) = \frac{Q^2}{2}\frac{k_2}{k_1} \tag{2-58}$$

By use of the Lagrange-multiplier method, the function to be minimized [given by (2-45)] is

$$e_c(\mathbf{p}, \boldsymbol{\lambda}) = Q^2\frac{k_1{}^2 + k_2}{2k_1k_2} + \lambda_1\left(\frac{Q^2}{2}\frac{k_2}{k_1} - M^2\right) \tag{2-59}$$

The conditions for a minimum value of (2-59), if k_2 is treated as fixed, are found from (2-48) and (2-49) to be

$$k_1{}^2 = k_2 + \lambda_1 k_2{}^2 \tag{2-60}$$

and

$$k_1 = \frac{Q^2}{2M^2}k_2 \tag{2-61}$$

If (2-61) is substituted into (2-60), the Lagrange multipler is given by

$$\lambda_1 = \frac{Q^4}{4M^4} - \frac{1}{k_2} \qquad \lambda_1 \geqq 0 \tag{2-62}$$

If λ_1 is greater than zero, then the optimum gain is

$$k_1 = \frac{Q^2}{2M^2}k_2 \qquad \frac{Q^4}{4M^4} > \frac{1}{k_2} \tag{2-63}$$

and the minimum value of the response error index is

$$e = M^2\frac{(Q^4/4M^4)k_2 + 1}{k_2} \qquad \frac{Q^4}{4M^4} > \frac{1}{k_2} \tag{2-64}$$

On the other hand, if λ_1 is zero, the optimum gain is given by (2-54), and the minimum value of the response error index is given by (2-55). At the transition point between these two cases $Q^4/4M^4 = 1/k_2$, the optimum gains given by (2-54) and (2-63) are equal, and the minimum values of the response error indices given by (2-55) and (2-64) are equal. The case where $Q^4/4M^4 < 1/k_2$ need not be considered because (2-56) is automatically satisfied by the optimum gain found without the use of the Lagrange multiplier.

A procedure of setting all Lagrange multipliers equal to zero and determining the parameters from (2-48) may be helpful as a first step in solving more complicated constraint problems. Equating the multipliers to zero eliminates an equal number of variables and equations in the solution. If all the constraints given in (2-44) are satisfied with these parameters, the minimization problem is solved. If one or more of the constraints are not satisfied, a good indication of which multipliers are not zero is provided.

2-4 Parameter optimization with statistical signals

The input signals of many control systems are statistical as opposed to deterministic. Such signals occur with the presence of measurement noise, random load disturbances, and desired responses with random components. If the desired response of the control system shown in Fig. 2-1 is statistical, then the error signal $e_q(t)$ also is statistical. Furthermore, the value of the error index $e(t)$ depends on the particular member of the desired-response ensemble used in the integration. Therefore, the optimum gain depends on which member of the ensemble is chosen. However, the ensemble member which occurs cannot be identified at the time of design so that an impasse is reached unless an average set of conditions is used in the design.

The average used in parameter optimization is the ensemble mean (see Appendix C for a brief discussion of statistical signals). For statistical signals, the error index

$$e = \overline{[Q(t) - q(t)]^2} \tag{2-65}$$

is used, where the bar indicates the ensemble average of all possible members. For time-invariant systems and ergodic signals, the ensemble average is equal to the time average, so that the error index given in (2-65) is equivalent to

$$e = \lim_{T\to\infty} \frac{1}{2T} \int_{-T}^{T} [Q(\sigma) - q(\sigma)]^2 \, d\sigma \tag{2-66}$$

Therefore the error index used for statistical signals is related intimately to the error index used for deterministic signals. Furthermore, this error index leads to basically the same mathematical methods for finding the optimum system as those introduced for deterministic signals.

If the error signal $e_q(t)$ is defined as

$$e_q(t) = Q(t) - q(t) \tag{2-67}$$

then the error index is

$$e = \overline{e_q^{\,2}(t)} \tag{2-68}$$

This error index is equal to the autocorrelation function

$$\phi_{e_q e_q}(\tau) = \overline{e_q(t) e_q(t + \tau)} \tag{2-69}$$

under the condition that $\tau = 0$. Furthermore, the autocorrelation function can be written in terms of the spectral density:

$$\phi_{e_q e_q}(\tau) = \frac{1}{j} \int_{-j\infty}^{j\infty} \Phi_{e_q e_q}(s) e^{s\tau} \, ds \tag{2-70}$$

where

$$\Phi_{e_q e_q}(s) = \frac{1}{2\pi} \int_{-\infty}^{\infty} \phi_{e_q e_q}(\tau) e^{-s\tau} \, d\tau \tag{2-71}$$

Therefore, the error index can be written in terms of the spectral density as

$$e = \frac{1}{j} \int_{-j\infty}^{j\infty} \Phi_{e_q e_q}(s)\, ds \tag{2-72}$$

Because the autocorrelation function is always an even function of τ, the corresponding spectral density is also even and can be factored into the form

$$\Phi_{e_q e_q}(s) = E_q(s)E_q(-s) \tag{2-73}$$

The expression for the error index given in (2-72) is equivalent to tabulated integrals given in (2-28) except for the factor $1/2\pi$. This factor is supplied automatically from the spectral density defined in (2-71).

Finally, frequency-domain relationships for statistical signals and linear systems can be used to express the spectral density of $e_q(t)$ in terms of the spectral density of $Q(t)$. If the transfer function

$$W_e(s) = \frac{e_q(s)}{Q(s)} \tag{2-74}$$

is defined, then the spectral density of the error signal is

$$\Phi_{e_q e_q}(s) = W_e(s)W_e(-s)\Phi_{QQ}(s) \tag{2-75}$$

where $\Phi_{QQ}(s)$ is the spectral density of the desired-response signal. Equations (2-72) and (2-75) provide analytical means for treating statistical signals in the same manner as deterministic signals for parameter-optimization problems.

2-5 Example of the mathematical method with statistical signals

The design problem under consideration is the position servo shown in Fig. 2-9. The gain k_1 is adjusted to give a minimum mean-square error between the desired response and the actual response. Therefore, the error index is given in (2-65) for this problem. Furthermore, the desired response is a statistical signal with an autocorrelation function assumed to be

$$\phi_{QQ}(\tau) = Q^2 e^{-a|\tau|} \tag{2-76}$$

The statistical characteristics of this desired-response signal have two interesting limiting forms. In particular, under the condition that $a = 0$, the autocorrelation function is a constant and hence the desired-response signal must be a constant signal. The other limiting form occurs for $a = \infty$ when $Q^2 = aN^2/2$. Under this condition, the autocorrelation function of the desired response is an impulse function of area N^2 and hence is zero when $|\tau| > 0$. This second case corresponds to a signal which is called white noise. White noise has the property that its spectral density is constant over all frequencies.

The transfer function from the desired-response signal $Q(t)$ to the error signal $[Q(t) - q(t)]$ can be found readily from Fig. 2-9. This transfer function is found to be

$$W_e(s) = 1 - \frac{k_2}{s^2 + k_1 s + k_2} = \frac{s(s + k_1)}{s^2 + k_1 s + k_2} \tag{2-77}$$

Also, the Fourier transform of the autocorrelation function given in (2-76), as determined by (2-71), is

$$\Phi_{QQ}(s) = \frac{Q^2}{2\pi}\left(\frac{1}{s+a} + \frac{1}{-s+a}\right) = \frac{Q^2}{2\pi}\frac{2a}{(s+a)(-s+a)} \tag{2-78}$$

By using the relationship for the spectral density of the error signal given in (2-75), this spectral density is written as

$$\Phi_{e_q e_q}(s) = \frac{1}{2\pi}\frac{c(s)\,c(-s)}{d(s)\,d(-s)} \tag{2-79}$$

where
$$c(s) = \sqrt{2a}\;Q(s^2 + k_1 s) \tag{2-80}$$

and
$$d(s) = s^3 + (k_1 + a)s^2 + (k_2 + ak_1)s + ak_2 \tag{2-81}$$

From (2-80) and (2-81), the tabulated integral (2-34) yields the value of the error index

$$e = Q^2 a\,\frac{k_1^2 + ak_1 + k_2}{ak_1^2 + (k_2 + a^2)k_1} \tag{2-82}$$

Differential calculus now can be applied to minimize the error index given in (2-82). If the first derivative of e with respect to k_1 is set equal to zero, the minimizing value of the gain k_1 is

$$k_1 = \sqrt{a^2 + k_2} \tag{2-83}$$

In the special case where the desired-response signal is a constant, $a = 0$, the optimum value of the gain is precisely the optimum value of the gain found in the example problem with a deterministic signal. This result verifies that the treatments of statistical and deterministic signals in parameter optimization are equivalent. On the other hand, when the desired-response signal approaches white noise, $a \to \infty$, the optimum value of k_1 approaches infinity proportionately to a. The condition where the gain k_1 approaches infinity corresponds to opening the major feedback loop or, rather, simply turning off the position servo. This result is reasonable in the sense that an energy-storage device such as the position servo cannot follow white noise. Therefore, the minimum mean-square value of the error signal cannot be reduced by finite values of the gain k_1.

2-6 Summary

The parameter-optimization procedure discussed here applies to the design of linear time-invariant control systems when the error index is integral squared error for deterministic signals and mean-square error for statistical signals. The method results in the necessity of solving a set of nonlinear algebraic equations. Unfortunately, when the design problem becomes only slightly more complicated than the examples given, great difficulty arises in the solution of these algebraic equations. For these slightly more complicated problems, designers have found that an effective means of solving such equations is using graphical techniques. In other words, a reasonably effective method is to plot a number of points by inserting numerical values for the control-system gains and then interpolating. The interpolation gives a rough estimate of the optimum gains, and the procedure can be repeated until the desired accuracy is achieved. However, in considerably more complicated problems where more than one gain is adjusted for minimum conditions, even the graphical means become unwieldy. Unfortunately, in these more difficult problems, no effective means for solving these algebraic equations exists. Of course, the possibility of using computational aids and search techniques could be considered.

However, because the adjustment of parameters in a fixed configuration is the least general optimization procedure and because these practical difficulties exist, more general optimization procedures are investigated instead of a more detailed study of this procedure. The purpose of introducing the parameter-optimization procedure is to introduce the concept of optimization theory as a control-system design technique. Also, in elementary terms, the concept of investigating optimum and optimum linear systems as a subclass has been discussed in relationship to the fixed-configuration design problem.

3

Impulse-response Optimization

In this chapter the so-called free-configuration problem is treated. In particular, the dynamic process and the system controls are considered in composite form, as shown in Fig. 3-1.

The composite system is assumed to have an impulse response $w(t, \mu)$ with an input signal $v(t)$ and a response signal $q(t)$. The free-configuration problem is the problem of determining the optimum impulse response $w(t, \mu)$. The mathematical techniques used for finding the optimum impulse response are developed in this chapter for statistical signals. However, some special considerations are necessary for deterministic signals, and these considerations are presented in a subsequent section in this chapter. Because statistical signals are assumed, the so-called mean-square error index is assumed. Therefore, the error index used here is

$$e(t) = \overline{[Q(t) - q(t)]^2} \tag{3-1}$$

where the desired response of the system is $Q(t)$ and the actual response of the system is given by the superposition integral

$$q(t) = \int_{-\infty}^{t} w(t, \mu)v(\mu)\, d\mu \tag{3-2}$$

The statistical averaging used in (3-1) is with respect to the ensembles of both $Q(t)$ and $v(\mu)$.

Because the systems have been restricted to linear systems, as implied by the assumed impulse response, this chapter is restricted to the development of optimum linear control systems. However, when a special class of statistical signals is considered in a subsequent section, the optimum linear controls are shown to be the optimum controls. This special class of statistical signal is called gaussian.

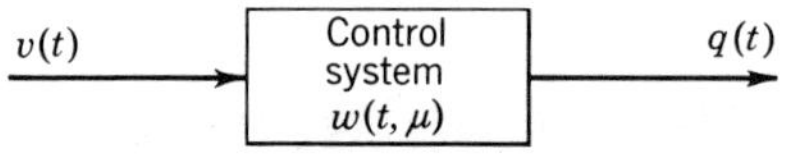

Fig. 3-1 Control-system block diagram in composite form.

3-1 Wiener-Hopf condition for statistical signals

The condition defining the optimum impulse response for the performance index given in (3-1) is derived here. In parameter optimization, the optimum parameters were determined by investigating the geometrical shape of the error index around the optimum value for the parameter. In an analogous fashion, the optimum impulse response is found by assuming an impulse response given as

$$w(t, \mu) = w^*(t, \mu) + \epsilon\eta(t, \mu) \tag{3-3}$$

The optimum impulse response $w^*(t, \mu)$ is assumed to be that impulse response that gives rise to minimum mean-square error. In (3-3), ϵ is considered to be an adjustable parameter and $\eta(t, \mu)$ is assumed to be an arbitrary perturbation in the weighting function $w(t, \mu)$. By using the assumed form given in (3-3), the error index given in (3-1) becomes

$$e(t) = \overline{\left\{Q(t) - \int_{-\infty}^{t} [w^*(t, \mu) + \epsilon\eta(t, \mu)]v(\mu)\, d\mu\right\}^2} \tag{3-4}$$

The conditions for a minimum value of the error index $e(t)$ are

$$\left.\frac{\partial e(t)}{\partial \epsilon}\right|_{\epsilon=0} = 0 \qquad \left.\frac{\partial^2 e(t)}{\partial \epsilon^2}\right|_{\epsilon=0} > 0 \tag{3-5}$$

if they are satisfied for all $\eta(t, \mu)$. The first condition in (3-5) states that the slope of the error index $e(t)$ is zero for all infinitesimal perturbations around the optimum impulse response. The second condition in (3-5) states that the error index must be a convex function of the perturbation. These conditions, of course, are similar to those for the optimum parameters found in Chap. 2. However, the arbitrary perturbation $\eta(t, \mu)$ ensures that all possible linear systems are considered in the selection of the optimum.

The conditions given in (3-5) can now be applied directly to the error index given in (3-4). The first condition in (3-5) yields

$$\left.\frac{\partial e(t)}{\partial \epsilon}\right|_{\epsilon=0} = -2\overline{\left[Q(t) - \int_{-\infty}^{t} w^*(t, \mu)v(\mu)\, d\mu\right]\int_{-\infty}^{t} \eta(t, \mu)v(\mu)\, d\mu} = 0 \tag{3-6}$$

The equality has been derived by commuting the linear mathematical operations of taking the partial derivative and the averaging process. The product of integrals in (3-6) can be written as a double integral, and then the averaging process can be commuted with the integrations, resulting in

$$\int_{-\infty}^{t} \eta(t, \alpha)\left[\overline{Q(t)v(\alpha)} - \int_{-\infty}^{t} w^*(t, \mu)\overline{v(\mu)v(\alpha)}\, d\mu\right] d\alpha = 0 \tag{3-7}$$

Because $\eta(t, \alpha)$ is an arbitrary function, (3-7) can be satisfied only if the integrand of this integral vanishes everywhere on the interval $-\infty < \alpha \leqq t$. Therefore, the necessary condition for minimum mean-square error is

$$\overline{Q(t)v(\alpha)} = \int_{-\infty}^{t} w^*(t, \mu)\overline{v(\mu)v(\alpha)}\, d\mu \qquad \alpha \leqq t \tag{3-8}$$

The integral equation given in (3-8) is called the Wiener-Hopf condition[25] specifying the optimum impulse response $w^*(t, \mu)$. However, this is the Wiener-Hopf condition as generalized by Booton[26] for impulse responses of time-varying systems and for nonstationary statistical signals. The Wiener-Hopf condition can be written in an alternative form, using the notation defined in Appendix C for correlation functions, as

$$\phi_{vQ}(\alpha, t) = \int_{-\infty}^{t} w^*(t, \mu)\phi_{vv}(\mu, \alpha)\, d\mu \qquad \alpha \leqq t \tag{3-9}$$

The derivation of the Wiener-Hopf condition thus far has been based only on the first condition for a minimum given in (3-5). The second condition in (3-5) ensures that a minimum has been achieved as opposed to just a stationary point on the error index $e(t)$. Application of the second condition in (3-5) yields

$$\left.\frac{\partial^2 e(t)}{\partial \epsilon^2}\right|_{\epsilon=0} = 2\overline{\left[\int_{-\infty}^{t} \eta(t, \mu)v(\mu)\, d\mu\right]^2} > 0 \tag{3-10}$$

which ensures that a minimum point has in fact been achieved because the square is always positive. In order to determine the minimum value of the error index $e(t)$, the optimum impulse response is substituted into (3-1), and then the square in (3-1) is expanded into the form

$$\begin{aligned} e(t) = \overline{Q^2(t)} - \int_{-\infty}^{t} w^*(t, \mu)\overline{Q(t)v(\mu)}\, d\mu \\ - \int_{-\infty}^{t} w^*(t, \alpha)\left[\overline{Q(t)v(\alpha)} - \int_{-\infty}^{t} w^*(t,\mu)\overline{v(\mu)v(\alpha)}\, d\mu\right] d\alpha \end{aligned} \tag{3-11}$$

However, the last term in (3-11) vanishes by virtue of the condition given in (3-8), and therefore the minimum value of the error index is

$$e(t) = \overline{Q^2(t)} - \int_{-\infty}^{t} w^*(t, \mu)\overline{Q(t)v(\mu)}\, d\mu \tag{3-12}$$

The minimum value of the error index can be rewritten in terms of correlation functions as

$$e(t) = \phi_{QQ}(t, t) - \int_{-\infty}^{t} w^*(t, \mu)\phi_{vQ}(\mu, t)\, d\mu \tag{3-13}$$

A slightly more general formulation of the optimum impulse response of the system occurs when the response of the control system is an approximation to either a future value or a past value of the desired-response signal. In other words, the response of the control system is a prediction or an interpolation of the desired-response signal. For this more general problem, the error index is

$$e(t) = \overline{[Q(t + \tau) - q(t)]^2} \tag{3-14}$$

where for prediction $\tau > 0$ and for interpolation $\tau < 0$ because real time is defined as t. However, the actual system response is given by the superposition integral

$$q(t) = \int_{-\infty}^{t} w(t, \mu)v(\mu)\, d\mu \tag{3-15}$$

where the input signal is weighted only over past and present time, as is required in any physical situation.

The condition for the minimum value of the error index $e(t)$ given in (3-14) is derived by mathematical steps analogous to those given in (3-3) through (3-7). Application of these procedures leads to the condition for minimum mean-square error

$$\overline{Q(t + \tau)v(\alpha)} = \int_{-\infty}^{t} w^*(t, \mu)\overline{v(\mu)v(\alpha)}\, d\mu \qquad \alpha \leqq t \tag{3-16}$$

The minimum value of the index given in (3-14) is found to be

$$e(t) = \overline{Q^2(t + \tau)} - \int_{-\infty}^{t} w^*(t, \mu)\overline{Q(t + \tau)v(\mu)}\, d\mu \tag{3-17}$$

by the same procedures used to derive (3-12). Under the condition $\tau = 0$, the Wiener-Hopf condition and the minimum mean-square error given in (3-16) and (3-17) reduce respectively to the conditions found in (3-8) and (3-12).

3-2 Approximating properties of minimum mean-square estimates

The derivations in Sec. 3-2 developed the condition for the impulse response that minimizes the mean-square error index. As stated previously, the optimum system may not be a linear system. However, the optimum system is linear for gaussian signals, a property that can be verified by considering the correlation between the system error $[Q(t) - q(t)]$ and the input signal $v(t)$. This correlation is written as

$$\overline{v(t)[Q(t) - q(t)]} = \overline{Q(t)v(t)} - \int_{-\infty}^{t} w(t, \mu)\overline{v(\mu)v(t)}\, d\mu \tag{3-18}$$

For the optimum linear system, (3-18) reduces to

$$\overline{v(t)[Q(t) - q^*(t)]} = 0 \tag{3-19}$$

by virtue of (3-8) so that the error and the input signals are uncorrelated. If gaussian signals are uncorrelated, they also are statistically independent. When the input signal and the desired-response signal are gaussian, the response signal $q(t)$ and hence the error signal are gaussian. Therefore the optimum error signal is statistically independent with respect to the input signal for gaussian signals. When statistical independence occurs, no further reduction of the error can be achieved because no additional properties of the desired-response signal can be extracted with averaging processes. Hence the conclusion is drawn that the optimum linear system is indeed the optimum system for gaussian input and desired-response signals.

Because the optimum linear system is an approximation to the optimum system, the question naturally arises as to what are the approximating properties of the linear systems. The approximating properties of the estimate $Q_e(t)$ of the desired-response signal $Q(t)$ are sought for the mean-square error index

$$e(t) = \overline{[Q(t) - Q_e(t)]^2}^{\,Q,v} \tag{3-20}$$

The notation Q, v denotes that the averaging is taken over both the input-signal ensemble and the desired-response-signal ensemble by using the joint-probability density. The minimization of (3-20) is accomplished by assuming

$$Q_e(t) = Q_e^*(t) + \epsilon\eta(t) \tag{3-21}$$

The minimizing estimate $Q_e^*(t)$ and the arbitrary perturbation $\eta(t)$ are functions of the input $v(\sigma)$ on the interval $-\infty < \sigma \leqq t$ but not functions of $Q(t)$. Also $Q_e^*(t)$ and $\eta(t)$ may be assumed to belong to any specific class of functionals§ involving $v(\sigma)$. Of course, only linear functionals generally are discussed in this chapter. The condition for minimizing (3-20) is

$$\left.\frac{\partial e(t)}{\partial \epsilon}\right|_{\epsilon=0} = 0 \tag{3-22}$$

which yields

$$\overline{\eta(t)[Q(t) - Q_e^*(t)]}^{Q,v} = 0 \tag{3-23}$$

The condition for minimum mean-square error can be rewritten by observing the relationship

$$\overline{f(x, y)}^{x,y} = \overline{\overline{f(x, y)}^{x|y}}^{y} \tag{3-24}$$

The notation $x \mid y$ denotes that the averaging is taken over x given y according to the conditional-probability density. The final averaging, denoted by y, is taken over y according to the amplitude-probability density of y. By using the property given in (3-24) and observing that $\eta(t)$ and $Q_e^*(t)$ are not dependent upon the desired-response-signal ensemble, the condition for minimum error becomes

$$\overline{\eta(t)[\overline{Q(t)}^{Q|v} - Q_e^*(t)]}^{v} = 0 \tag{3-25}$$

The form given in (3-25) for the minimum-mean-square error condition clarifies the approximating properties of the optimum estimate.

The exact optimum system occurs when

$$Q_e^*(t) = \overline{Q(t)}^{Q|v} \tag{3-26}$$

so that the "best" estimate is equal to the conditional mean of $Q(t)$ given $v(\sigma)$ on the interval $-\infty < \sigma \leqq t$. The optimum system occurs when $\eta(t)$ and $Q_e^*(t)$ are not restricted to a particular function class or when $\eta(t)$ and $Q_e^*(t)$ are chosen from the same class of functionals as $\overline{Q(t)}^{Q|v}$. For gaussian signals, the conditional mean of $Q(t)$ is linear with respect to $v(\sigma)$ on the interval $-\infty < \sigma \leqq t$, as demonstrated in Appendix C.

On the other hand, when $\eta(t)$ and $Q_e^*(t)$ are restricted to a class of functionals other than that of the conditional mean of $Q(t)$, then the condition given in (3-25) defines the optimum estimate within that particular restricted class of functionals. In other words, the error in the approximation of the conditional mean of $Q(t)$ must vanish when averaged over the entire ensemble with respect to all perturbations within the restricted class of functionals. Of course, the minimum value of the

§ The reader is referred to Appendix D for a brief discussion of functionals which is presented in the context of state-determined dynamic processes.

error index given in (3-20) is always increased when $Q_e^*(t)$ is chosen from a restricted class of functionals. The minimum mean-square error is found to be

$$e^*(t) = \overline{[Q(t) - \overline{Q(t)}^{Q|v}]^2}^{Q,v} + \overline{[\overline{Q(t)}^{Q|v} - Q_e^*(t)]^2}^{v} \qquad (3\text{-}27)$$

from evaluating (3-20) for $Q_e(t) = Q_e^*(t)$ and some manipulation. The second term is a positive-definite function of the error between the unrestricted and restricted mean-square error estimates which is averaged over the input-signal ensemble.

This section is intended to show the relationship between minimum mean-square error estimates and approximation theory for functions occurring in problems arising in many fields not related to statistical signals. Also, this section is intended to clarify the conceptual difference between the optimum system and the optimum linear system.

3-3 Example solutions for nonstationary signals

In order to develop insight into the interpretation of the Wiener-Hopf condition for the minimum-mean-square-error impulse response, a class of nonstationary signals is considered here. This particular class of nonstationary signals emphasizes the relationships between the time variables which appear in the Wiener-Hopf condition.

The optimization problem under consideration is defined partially by the control system shown in Fig. 3-2. The switch shown in this system closes at $t = t_0$, and therefore the input signal $v(t) = 0$ for $t < t_0$. Because the input signal is zero for $t < t_0$, the system impulse response weights signals only after time t_0. In other

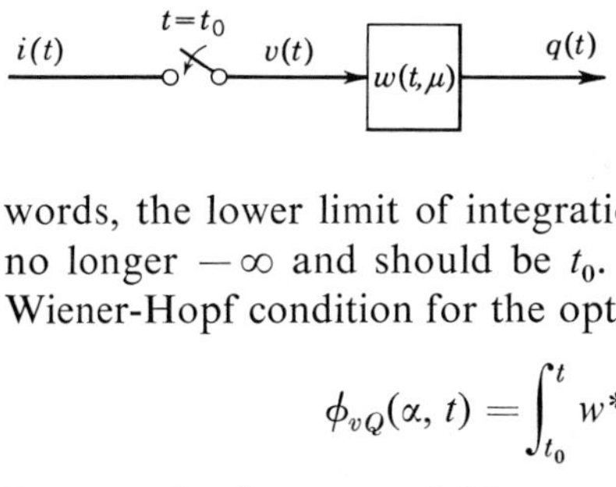

Fig. 3-2 Control system with zero input for $t < t_0$.

words, the lower limit of integration in the superposition integral given in (3-15) is no longer $-\infty$ and should be t_0. Therefore, the lower limit of integration in the Wiener-Hopf condition for the optimum impulse response is now given by

$$\phi_{vQ}(\alpha, t) = \int_{t_0}^{t} w^*(t, \mu)\phi_{vv}(\mu, \alpha)\, d\mu \qquad t_0 \leqq \alpha \leqq t \qquad (3\text{-}28)$$

Because the dummy variables μ and α range over the interval t_0 to t, the signal v in the autocorrelation and cross-correlation functions in (3-28) is replaced by the signal i. On the interval $t_0 \leqq \mu,\ \alpha \leqq t$, the signals v and i are identical, as can be seen in Fig. 3-2. Therefore, the Wiener-Hopf condition given in (3-28) is replaced by

$$\phi_{iQ}(\alpha, t) = \int_{t_0}^{t} w^*(t, \mu)\phi_{ii}(\mu, \alpha)\, d\mu \qquad t_0 \leqq \alpha \leqq t \qquad (3\text{-}29)$$

Now the Wiener-Hopf condition is stated in terms of the input signal $i(t)$, for which statistical characteristics are assumed.

The input signal $i(t)$ is assumed to be

$$i(t) = \sum_{n=1}^{M} I_n i_n(t) + n(t) \qquad (3\text{-}30)$$

Here the signals $i_n(t)$ are assumed to be deterministic functions of time. The coefficients of these time functions I_n are random parameters with known probability densities. The signal $n(t)$ in the input signal is assumed to be a random function of time and arises in practice as a noise signal. The desired-response signal is assumed to be

$$Q(t) = \sum_{n=1}^{M} I_n Q_n(t) \tag{3-31}$$

where the components of the desired-response signal $Q_n(t)$ are deterministic time functions which are linearly dependent, respectively, on the deterministic-input-signal components $i_n(t)$. In other words, $Q_n(t)$ is a signal generated by a linear operation on $i_n(t)$. The random parameters of the input and desired-response signals are defined by

$$\overline{I_n I_m} = \begin{cases} 1 & m = n \\ 0 & m \neq n \end{cases} \qquad \overline{I_n n(t)} = 0 \tag{3-32}$$

The autocorrelation function of the gaussian noise signal also is assumed to be

$$\overline{n(t)n(t+\tau)} = N(t)u_0(\tau) \tag{3-33}$$

Because $Q_n(t)$ is linearly dependent upon $i_n(t)$ and because $n(t)$ is gaussian, the minimum mean-square error system is linear and is defined by (3-29). For these assumed signals, the autocorrelation and cross-correlation functions are, respectively,

$$\phi_{ii}(\mu, \alpha) = \sum_{n=1}^{M} i_n(\mu)i_n(\alpha) + N(\mu)u_0(\alpha - \mu) \tag{3-34}$$

and

$$\phi_{iQ}(\alpha, t) = \sum_{n=1}^{M} i_n(\alpha)Q_n(t) \tag{3-35}$$

The correlation functions given in (3-34) and (3-35) now are substituted into the Wiener-Hopf condition given in (3-29). The result is

$$\sum_{n=1}^{M} Q_n(t)i_n(\alpha) = \int_{t_0}^{t} w^*(t, \mu)\left[\sum_{n=1}^{M} i_n(\mu)i_n(\alpha)\right] d\mu + N(\alpha)w^*(t, \alpha) \qquad t_0 \leqq \alpha \leqq t \tag{3-36}$$

where the impulse in the autocorrelation function of the input signal due to the noise signal has been integrated. The Wiener-Hopf condition given in (3-36) is rewritten as

$$w^*(t, \alpha) = \sum_{n=1}^{M} \frac{i_n(\alpha)}{N(\alpha)}\left[Q_n(t) - \int_{t_0}^{t} w^*(t, \mu)i_n(\mu)\, d\mu\right] \qquad t_0 \leqq \alpha \leqq t \tag{3-37}$$

The bracketed terms appearing in (3-37) are seen to be functions only of time t. Therefore, the following functions of time are defined:

$$k_n(t) = Q_n(t) - \int_{t_0}^{t} w^*(t, \mu)i_n(\mu)\, d\mu \qquad n = 1, 2, \ldots, M \tag{3-38}$$

These functions of time are substituted back into (3-37), and then the optimum impulse response is given by

$$w^*(t, \alpha) = \sum_{n=1}^{M} \frac{i_n(\alpha)}{N(\alpha)} k_n(t) \qquad t_0 \leqq \alpha \leqq t \tag{3-39}$$

This optimum impulse response is substituted into the definition of the time functions given in (3-38), and therefore these time functions are defined by the following set of linear algebraic equations:

$$k_n(t) + \sum_{m=1}^{M} k_m(t)\left[\int_{t_0}^{t} \frac{i_m(\mu)i_n(\mu)}{N(\mu)}\,d\mu\right] = Q_n(t) \qquad n = 1, 2, \ldots, M \tag{3-40}$$

The solution of the equations given in (3-40) represents a complete solution to the optimum impulse response as given in (3-39).

In order to demonstrate the realization of an optimum-impulse-response function, the position-servo control problem is considered again. The fixed element of the position servo is considered to be simply an integrator and is described by the differential equation

$$\theta'(t) = m(t) \tag{3-41}$$

where $\theta(t)$ is the shaft angle and $m(t)$ is the voltage excitation. Further, the input and the desired-response signals are specialized to

$$N(t) = N^2 \qquad i_1(t) = Q_1(t) = Q \qquad i_n(t) = Q_n(t) = 0 \qquad n = 2, 3, \ldots, M \tag{3-42}$$

Also, the assumption is made that the switch closes at $t_0 = 0$.

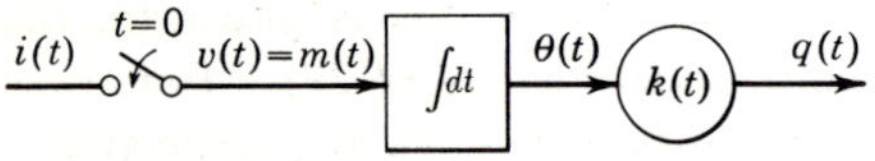

Fig. 3-3 Open-loop configuration for optimum position-servo control.

Because the input and desired-response signals each have one deterministic component as given by (3-42), the optimum impulse response given in (3-39) specializes to

$$w^*(t, \mu) = \frac{Q}{N^2} k_1(t) \qquad 0 \leqq \alpha \leqq t \tag{3-43}$$

The time function $k_1(t)$ is found from (3-40), when specialized to

$$k_1(t)\left(1 + \int_0^t \frac{Q^2}{N^2}\,d\mu\right) = Q \tag{3-44}$$

and hence is

$$k_1(t) = \frac{Q}{1 + (Q^2/N^2)t} \tag{3-45}$$

Therefore, the optimum impulse response is

$$w^*(t, \alpha) = \frac{Q^2/N^2}{1 + (Q^2/N^2)t} = k(t) \qquad 0 \leqq \alpha \leqq t \tag{3-46}$$

If this optimum impulse response is substituted into the superposition integral given in (3-2), the response of the system is

$$q(t) = k(t) \int_0^t v(\mu)\,d\mu \tag{3-47}$$

This superposition integral shows that the optimum system is simply an integrator with the output multiplied by the time-varying gain $k(t)$. Therefore, the position-servo control system could be realized in open-loop form, as shown in Fig. 3-3.

However, the open-loop configuration has many obvious disadvantages from a control standpoint. Load disturbances and initial-condition errors of the position servo, which is represented by the integrator, cause additional errors in the output signal $q(t)$. The reader should note that these disturbances and initial-condition errors have not been accounted for in the optimization procedure. This is an inherent disadvantage of impulse-response optimization. It is possible, however, to derive a feedback configuration for the position-servo control problem. If the response given in (3-47) is differentiated, the differential equation for the response signal is

$$\frac{1}{k(t)} q'(t) + q(t) = v(t) \tag{3-48}$$

A block diagram representing this differential equation is shown in Fig. 3-4, which is a feedback configuration for the optimum position-servo control. This feedback configuration does provide some control of initial-position errors and load disturbances.

However, from (3-46) it is seen that the gain $k(t)$ approaches zero as time t approaches infinity. Therefore after a period of time, the feedback loop gain of the

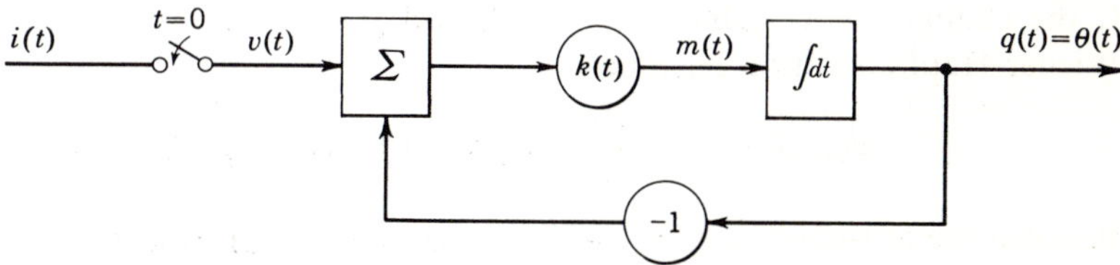

Fig. 3-4 Feedback configuration for optimum position-servo control.

system shown in Fig. 3-4 decreases to zero so that, after a period of time, the position-servo control system again is open-loop. Hence, the same problem arises with load disturbances in the servo motor.

The only way this undesirable open-loop effect can be avoided is to place a loop around the servo motor which would have an infinite gain and therefore cause the transfer function of the motor portion of the system to be unity. For this undesirable scheme, the input signal to the feedback portion of the system would be operated on by an integrator and time-varying gain, as shown in Fig. 3-3. The difficulty in impulse-response optimization is that no provision is made in the optimization procedure for specifying the effect of initial-condition errors and load disturbances. In other words, impulse-response optimization does not provide a specification of the configuration of the optimum system. The specification of the configuration of the optimum system is an extremely important requisite when optimization theory is to be used as a control-system-synthesis procedure. The method of Wiener and Lee does not provide this essential feature. Therefore, in subsequent chapters, a theory is developed to specify the optimum configuration of the control system.

The method for solving the class of time-varying impulse-response optimization problems defined by the signals given in (3-30) and (3-31) was developed by Booton.[28] This class of signals results in an impulse response that is described by an Mth-order ordinary linear differential equation because the impulse response given in (3-39) contains M additive component weighting functions. However, a determination of the system components may be very difficult because of the time variations in the system. More general classes of time-varying impulse-response optimization problems can be solved.[29] On the other hand, these optimum systems generally cannot be realized exactly with a finite number of components, and approximations to the exact optimum are not wholly satisfactory from a conceptual viewpoint, although important practically. Therefore, time-invariant impulse-response optimization problems are investigated now.

Under the condition that the infinite past is not weighted in the superposition integral, the optimum linear control system is time-varying. Therefore the question is raised at this time: Under what conditions is the optimum impulse response time-invariant? Certainly, the infinite past must be weighted in the superposition integral. Therefore, the control system must be in steady state with respect to initial conditions if the assumption is made that the infinite past is weighted. If the input and desired-response signals also are considered to be stationary, then the Wiener-Hopf equation given in (3-16) is written as

$$\phi_{vQ}(t + \tau - \alpha) = \int_{-\infty}^{t} w^*(t, \mu)\phi_{vv}(\alpha - \mu)\, d\mu \qquad \alpha \leqq t \tag{3-49}$$

If the changes of variables $\beta = t - \alpha$ and $\gamma = t - \mu$ are inserted into (3-49), the Wiener-Hopf condition becomes

$$\phi_{vQ}(\beta + \tau) = \int_{0}^{\infty} w^*(t, t - \gamma)\phi_{vv}(\beta - \gamma)\, d\gamma \qquad \beta \geqq 0 \tag{3-50}$$

Because the left-hand side of (3-50) is independent of time t, the right-hand side of (3-50) must also be independent of t. This independence can occur only when the optimum impulse response represents a linear time-invariant system. With the assumption that the optimum impulse response is time-invariant, $w^*(t, t - \gamma) = w^*(\gamma)$, (3-50) reduces to

$$\phi_{vQ}(\beta + \tau) = \int_{0}^{\infty} w^*(\gamma)\phi_{vv}(\beta - \gamma)\, d\gamma \qquad \beta \geqq 0 \tag{3-51}$$

The equation given in (3-51) is the traditional form of the Wiener-Hopf condition for optimum time-invariant impulse responses. Methods for solving integral equations of this type now are discussed.

3-4 Direct-substitution method of solution for stationary signals

In order to develop the concepts for solving (3-51), a simple case is investigated first. The assumption is made that the input signal is given as

$$v(t) = Q(t) + n(t) \tag{3-52}$$

where the correlation functions of the desired-response and noise signals are given as

$$\phi_{QQ}(\beta) = Q^2 e^{-a|\beta|} \qquad \phi_{nn}(\beta) = N^2 u_0(\beta) \qquad \phi_{Qn}(\beta) = 0 \tag{3-53}$$

Because the impulse response of a lumped-parameter linear system generally is composed of exponential components, the assumption is made that the optimum impulse response is

$$w^*(\gamma) = \sum_{n=1}^{M} W_n e^{-w_n \gamma} \tag{3-54}$$

We now attempt to find the coefficients and the exponents of the assumed impulse response by directly substituting (3-54) into (3-51) when the correlation functions given in (3-53) are also substituted.[11] Upon making these substitutions and performing the indicated integrations, the result is

$$Q^2e^{-a|\beta+\tau|} = \sum_{n=1}^{M} W_n\left(\frac{2aQ^2}{a^2 - w_n^2}e^{-w_n\beta} + \frac{Q^2}{w_n - a}e^{-a\beta} + N^2e^{-w_n\beta}\right) \qquad \beta \geqq 0 \quad (3\text{-}55)$$

If the assumption is made first that $\tau \geqq 0$, then $\beta + \tau \geqq 0$ and the absolute value appearing in (3-55) can be dropped. Because (3-55) must be satisfied for all $\beta \geqq 0$, the coefficients of the exponentials with equal exponents must vanish. If the terms involving $e^{-a\beta}$ and $e^{-w_n\beta}$ are collected, their respective coefficients must satisfy

$$Q^2\left(e^{-a\tau} - \sum_{n=1}^{M} \frac{W_n}{w_n - a}\right) = 0 \qquad \tau \geqq 0 \quad (3\text{-}56)$$

and

$$W_n\left(\frac{2aQ^2}{a^2 - w_n^2} + N^2\right) = 0 \qquad \tau \geqq 0 \quad (3\text{-}57)$$

if $w_n \neq a$. If (3-57) is solved for w_n and it is assumed that $W_n \neq 0$, there are only two values of w_n that satisfy (3-57). However, only a positive value of w_n is acceptable because the integration leading to (3-55) would not be bounded for a negative value of the exponent. In other words, only stable systems with decaying exponentials in the system impulse response are considered. Therefore, there is only one exponential in the impulse response and $M = 1$ with the exponent given as

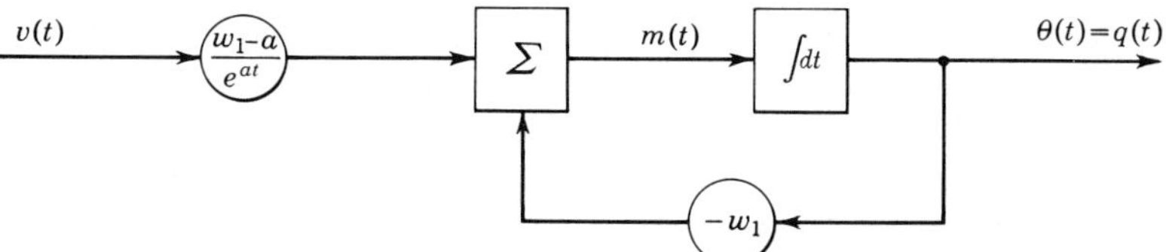

Fig. 3-5 Feedback configuration for optimum position-servo control, $\tau \geqq 0$.

$$w_1 = \sqrt{a^2 + 2a\frac{Q^2}{N^2}} \quad (3\text{-}58)$$

From (3-56), the coefficient of the exponential in the optimum impulse response is

$$W_1 = (w_1 - a)e^{-a\tau} \quad (3\text{-}59)$$

Finally, the optimum impulse response for this example is

$$w^*(\gamma) = (w_1 - a)e^{-a\tau}e^{-w_1\gamma} \qquad \tau \geqq 0 \quad (3\text{-}60)$$

In other words, the optimum impulse response represents a control system that has a simple-lag transfer function for $\tau \geqq 0$.

If the position-servo control problem is examined again, the feedback configuration shown in Fig. 3-5 can be arrived at by performing the same steps leading to the system shown in Fig. 3-4. The control system shown in Fig. 3-5 has an impulse response that represents the optimum mean-square predictor for the input and desired-response signals specified by the correlation functions given in (3-53) for $\tau \geqq 0$.

The solution for $\tau < 0$ is less obvious when this method of direct substitution is used. In this case, the left-hand side of (3-55) has an exponential $e^{-a\beta}$ for $\beta \geqq -\tau$ and an exponential $e^{a\beta}$ for $0 \leqq \beta < -\tau$, and the assumed form for the optimum impulse response does not satisfy (3-55) for all $\beta \geqq 0$. Because the selection of the form of the optimum impulse response is less obvious and somewhat subject to trial and error when $\tau < 0$, this case is treated by another method.

Finally, the minimum value of the error index is computed by inserting (3-60) into (3-17), and the result is

$$e^* = Q^2\left(1 - \frac{w_1 - a}{w_1 + a}\, e^{-2a\tau}\right) \qquad \tau \geqq 0 \tag{3-61}$$

From (3-61), it is seen that the mean-square error increases as the prediction interval τ increases. As τ increases, the error index e^* approaches Q^2, which is simply the mean-square value of the desired-response signal itself. This increase in mean-square error is, of course, due to the fact that a statistical signal $Q(t + \tau)$ becomes less and less correlated with its past. The control system measures only the past and present segment of the input signal containing $Q(t)$.

This direct-substitution method, as outlined in this section, can be extended to the general situation where the correlation functions of the pertinent signals are described by impulses and sums of exponential component terms. However, the notation pertinent to the more general situation within this class of problems is unnecessarily difficult with the direct-substitution method. Therefore, an alternative method of specifying the optimum impulse response that is stated in the frequency domain is developed in the next section. This second method of solving the integral equation of the type given in (3-51) is somewhat more difficult to understand because of the mathematical manipulations that are necessary to develop it.

3-5 Spectral-factorization method of solution for stationary signals

The frequency-domain method for solving the Wiener-Hopf condition for stationary statistical signals is developed by rewriting (3-51) as

$$y(\beta) = \phi_{vQ}(\beta + \tau) - \int_0^\infty w^*(\gamma)\phi_{vv}(\beta - \gamma)\, d\gamma \tag{3-62}$$

The function $y(\beta)$ satisfies

$$y(\beta) = 0 \qquad \beta \geqq 0 \tag{3-63}$$

because of the equality given in (3-51) on the interval $0 \leqq \beta$. The Fourier transform of this function $y(\beta)$ is defined as

$$Y(s) = \frac{1}{2\pi}\int_{-\infty}^{\infty} y(\beta)e^{-s\beta}\, d\beta \qquad s = j\omega \tag{3-64}$$

By transforming both sides of (3-62), $Y(s)$ is

$$Y(s) = \frac{1}{2\pi}\int_{-\infty}^{\infty} \phi_{vQ}(\beta + \tau)e^{-s\beta}\, d\beta - \int_0^\infty w^*(\gamma)\left[\frac{1}{2\pi}\int_{-\infty}^{\infty} \phi_{vv}(\beta - \gamma)e^{-s\beta}\, d\beta\right] d\gamma \tag{3-65}$$

The first term on the right-hand side of (3-65) is rewritten as

$$\frac{1}{2\pi}\int_{-\infty}^{\infty} \phi_{vQ}(\beta + \tau)e^{-s\beta}\, d\beta = e^{s\tau}\frac{1}{2\pi}\int_{-\infty}^{\infty} \phi_{vQ}(\alpha)e^{-s\alpha}\, d\alpha \tag{3-66}$$

by a simple transformation of variables. The integral on the right-hand side of (3-66) is simply the spectral density of $\phi_{vQ}(\alpha)$. By a similar transformation of variables, the integral with respect to β on the right-hand side of (3-65) is written in terms of the spectral density of $\phi_{vv}(\alpha)$. Therefore, (3-65) reduces to

$$Y(s) = e^{s\tau}\Phi_{vQ}(s) - \Phi_{vv}(s)\int_0^\infty w^*(\gamma)e^{-s\gamma}\,d\gamma \tag{3-67}$$

The integral remaining in (3-67) is simply the Fourier transform of the optimum impulse response; hence the transfer function of the optimum system is defined as

$$W^*(s) = \int_0^\infty w^*(\gamma)e^{-s\gamma}\,d\gamma \tag{3-68}$$

Finally, then, the Fourier transform of the time function $y(\beta)$ is

$$Y(s) = e^{s\tau}\Phi_{vQ}(s) - \Phi_{vv}(s)W^*(s) \tag{3-69}$$

The properties of the Fourier transform now are used to find an explicit expression for the optimum transfer function $W^*(s)$. The transform $Y(s)$ must have poles only in the right half plane because $y(\beta) = 0$ for $\beta \geq 0$ and $y(\beta)$ must be Fourier-transformable by assumption. In other words, $y(\beta)$ is a decaying negative-time function. The explicit solution of (3-69) is found by separating the right-hand side of (3-69) into positive- and negative-time functions because the positive-time function must vanish in order to satisfy (3-63). This separation is accomplished by factoring the spectral density $\Phi_{vv}(s)$ into two factors, one of which contains poles and zeros only in the left half plane and the other of which contains poles and zeros only in the right half plane. More generally, this factorization is made so that the factors contain all the singularities occurring in the left half and right half planes, respectively. Therefore, the spectral density $\Phi_{vv}(s)$ is written as

$$\Phi_{vv}(s) = \Phi_{vv}^{+}(s)\Phi_{vv}^{-}(s) \tag{3-70}$$

where the factors are defined so that

$$\begin{aligned} \Phi_{vv}^{+}(s) &= \text{factor containing poles and zeros only in left half plane} \\ \Phi_{vv}^{-}(s) &= \text{factor containing poles and zeros only in right half plane} \end{aligned} \tag{3-71}$$

Also, the ratio $e^{s\tau}\Phi_{vQ}(s)/\Phi_{vv}^{-}(s)$ is separated into two additive terms, one of which is the transform of the positive-time function of this ratio and the other of which is the transform of the negative-time portion of this ratio. In other words, this ratio is written as

$$\frac{e^{s\tau}\Phi_{vQ}(s)}{\Phi_{vv}^{-}(s)} = \left[\frac{e^{s\tau}\Phi_{vQ}(s)}{\Phi_{vv}^{-}(s)}\right]_+ + \left[\frac{e^{s\tau}\Phi_{vQ}(s)}{\Phi_{vv}^{-}(s)}\right]_- \tag{3-72}$$

where the two terms appearing in (3-72) are defined as

$$\begin{aligned} \left[\frac{e^{s\tau}\Phi_{vQ}(s)}{\Phi_{vv}^{-}(s)}\right]_+ &= \text{transform of time function which is zero for all negative time} \\ \left[\frac{e^{s\tau}\Phi_{vQ}(s)}{\Phi_{vv}^{-}(s)}\right]_- &= \text{transform of time function which is zero for all positive time} \end{aligned} \tag{3-73}$$

By using (3-70) and (3-72), the expression in (3-69) is rewritten as

$$Y(s) = \left\{\left[\frac{e^{s\tau}\Phi_{vQ}(s)}{\Phi_{vv}^{-}(s)}\right]_{+} - \Phi_{vv}^{+}(s)W^{*}(s)\right\}\Phi_{vv}^{-}(s) + \left[\frac{e^{s\tau}\Phi_{vQ}(s)}{\Phi_{vv}^{-}(s)}\right]_{-}\Phi_{vv}^{-}(s) \quad (3\text{-}74)$$

As stated previously, the transform $Y(s)$ contains poles only in the right half plane. Also, according to the definitions in (3-71) and (3-73), the second term in (3-74) has poles only in the right half plane. However, the term within the braces of (3-74) has poles only in the left half plane because of the definitions given in (3-71) and (3-73) and also because $W^*(s)$ can have poles only in the left half plane, as is seen from (3-68). Therefore, in order that (3-74) be satisfied on both sides of the equation with no poles in the left half plane, the terms within the braces of this equation must vanish. From (3-74), this condition yields the expression for the transfer function of the optimum impulse response:

$$W^*(s) = \frac{[e^{s\tau}\Phi_{vQ}(s)/\Phi_{vv}^{-}(s)]_{+}}{\Phi_{vv}^{+}(s)} \quad (3\text{-}75)$$

This expression gives an explicit relationship for the transfer function of the optimum impulse response in the frequency domain.[24]

In order to demonstrate the use of (3-75), the following simple example is considered. The input signal is assumed to be that given by (3-52), and the correlation functions of the input and desired-response signals are assumed to be those given in (3-53). For this example, the spectral densities of the correlation functions appearing in (3-75) are

$$\Phi_{vQ}(s) = \frac{Q^2}{2\pi}\frac{2a}{(s+a)(-s+a)} \qquad \Phi_{vv}(s) = \frac{N^2}{2\pi}\frac{(s+w_1)(-s+w_1)}{(s+a)(-s+a)}$$
$$w_1 = \sqrt{a^2 + 2a\frac{Q^2}{N^2}} \quad (3\text{-}76)$$

and are the Fourier transforms of the expressions in (3-53). By using (3-76), the spectral density $\Phi_{vv}(s)$ is factored so that

$$\Phi_{vv}^{+}(s) = \frac{s+w_1}{s+a} \qquad \Phi_{vv}^{-}(s) = \frac{N^2}{2\pi}\frac{-s+w_1}{-s+a} \quad (3\text{-}77)$$

Then the transform of the positive-time function defined in (3-73) is

$$\left[\frac{e^{s\tau}\Phi_{vQ}(s)}{\Phi_{vv}^{-}(s)}\right]_{+} = \frac{2aQ^2}{N^2}\left[\frac{e^{s\tau}}{(s+a)(-s+w_1)}\right]_{+} \quad (3\text{-}78)$$

For the assumption $\tau \geqq 0$, the function $e^{s\tau}$ has an essential singularity in the right half plane. For this case, the transform in (3-78) has a simple pole at $s = -a$ in the left half plane, and the positive-time function of this transform must be an exponential with an exponent $-a$. If the calculus of residues is applied to (3-78) under the condition $\tau \geqq 0$, the transform of the positive-time component is

$$\left[\frac{e^{s\tau}\Phi_{vQ}(s)}{\Phi_{vv}^{-}(s)}\right]_{+} = \frac{2aQ^2}{N^2}\frac{e^{-a\tau}}{w_1+a}\frac{1}{s+a} \qquad \tau \geqq 0 \quad (3\text{-}79)$$

Therefore, the transform of the optimum impulse response is

$$W^*(s) = (w_1 - a)e^{-a\tau}\frac{1}{s + w_1} \qquad \tau \geqq 0 \tag{3-80}$$

and the optimum impulse response for this case is found to be

$$w^*(\gamma) = (w_1 - a)e^{-a\tau}e^{-w_1\gamma} \qquad \tau \geqq 0 \tag{3-81}$$

by applying the inverse transform process. Of course, this is the result found by the direct-substitution method.

In order to illustrate further the solution of (3-75), this same example is solved under the condition $\tau \leqq 0$ so that the response of the optimum control system lags the desired response. For $\tau \leqq 0$, the function $e^{s\tau}$ now has an essential singularity in the left half plane. The procedure which is used to arrive at (3-79) is now not so obvious. In order to demonstrate the effects of the essential singularity in the left half plane when performing the operation indicated in (3-78) for finding the positive-time function, the following transform is introduced:

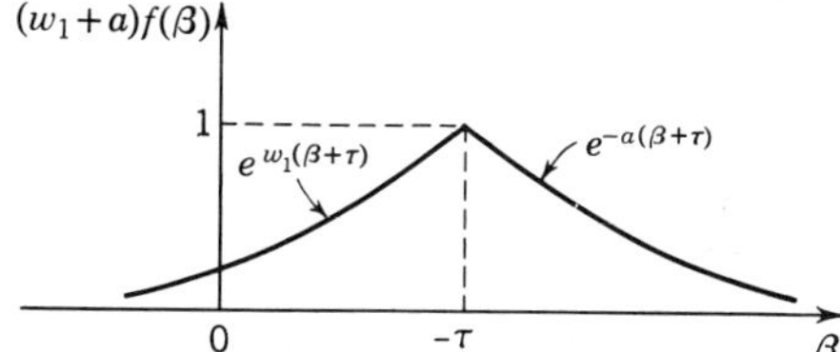

Fig. 3-6 Sketch of $f(\beta)$ for $\tau < 0$.

$$F(s) = \frac{e^{s\tau}}{(s + a)(-s + w_1)} = \frac{1}{w_1 + a}\left(\frac{e^{s\tau}}{s + a} + \frac{e^{s\tau}}{-s + w_1}\right) \tag{3-82}$$

The function $e^{s\tau}$ which appears in the transform $F(s)$ is a shifting operation in time. By using the separation into two time components of $F(s)$ as shown in (3-82), the inverse transform of $F(s)$ is

$$f(\beta) = \frac{1}{w_1 + a}\begin{cases} e^{-a(\beta+\tau)} & \beta + \tau \geqq 0 \\ e^{w_1(\beta+\tau)} & \beta + \tau \leqq 0 \end{cases} \tag{3-83}$$

This time function is sketched in Fig. 3-6 under the condition that $\tau < 0$. The positive-time portion of this time function is composed of two components, one on the interval $0 \leqq \beta \leqq -\tau$ and another on the interval $-\tau \leqq \beta$. The Fourier transform of the positive-time component of the function $f(\beta)$ is

$$[F(s)]_+ = \int_0^\infty f(\beta)e^{-s\beta}\,d\beta = \frac{1}{w_1 + a}\,\frac{(s + a)e^{w_1\tau} - (w_1 + a)e^{s\tau}}{(s - w_1)(s + a)} \qquad \tau \leqq 0 \tag{3-84}$$

and this transform is precisely the transform being sought in (3-78). By combining (3-84) and (3-75), the transfer function of the optimum impulse response is

$$W^*(s) = (w_1 - a)\frac{(s + a)e^{w_1\tau} - (w_1 + a)e^{s\tau}}{(s - w_1)(s + w_1)} \qquad \tau \leqq 0 \tag{3-85}$$

The inverse transform of (3-85) yields the optimum impulse response

$$w^*(\gamma) = \begin{cases} (w_1 - a)e^{w_1\tau}\left(\cosh w_1\gamma + \dfrac{a}{w_1}\sinh w_1\gamma\right) & 0 < \gamma \leqq -\tau \\ (w_1 - a)\left(\cosh w_1\tau - \dfrac{a}{w_1}\sinh w_1\tau\right)e^{-w_1\gamma} & -\tau \leqq \gamma \end{cases} \qquad \tau \leqq 0 \quad (3\text{-}86)$$

Finally, the minimum value of the error index given in (3-17) is found to be

$$e^* = Q^2\left(1 - \frac{w_1 - a}{w_1} + \frac{a}{w_1}\frac{w_1 - a}{w_1 + a}e^{2w_1\tau}\right) \qquad \tau \leqq 0 \qquad (3\text{-}87)$$

by substituting (3-86) and integrating. A comparison of (3-81) and (3-86) under the condition $\tau = 0$ shows that the two impulse responses found for $\tau \geqq 0$ and $\tau \leqq 0$ are identical.

The sketch of the minimum value of e versus τ in Fig. 3-7 shows that mean-square error decreases the more the system response lags the desired response.

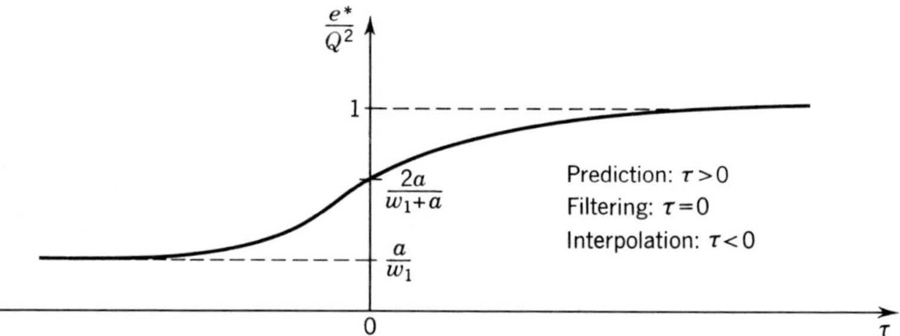

Fig. 3-7 Minimum mean-square error vs. time shift τ.

This sketch is plotted from (3-61) and (3-87) for the respective cases $\tau \geqq 0$ and $\tau \leqq 0$. The reader should note that there is an irretrievable error which cannot be recovered no matter how much the system response lags except in the special case where $a/w_1 = 0$. This special case occurs either when $a = 0$ or when $w_1 = \infty$. The desired-response signal is a constant when $a = 0$. On the other hand, the mean-square value of the noise signal is zero, $N^2 = 0$, when $w_1 = \infty$, as can be seen from (3-76).

A difficulty arises in the realization of the optimum transfer function given in (3-85) for $\tau < 0$ because of the term involving $e^{s\tau}$. The factor $e^{s\tau}$ is a shifting operation; hence the realization of the optimum system for $\tau < 0$ would involve the use of a pure time delay. Of course, a pure time delay can only be approximated with lumped-parameter components.

3-6 Wiener-Hopf condition for multiple statistical-signal inputs

The situation arises in many control-system design problems where more than one statistical input signal occurs in the system. In this section, the condition for the optimum impulse response of a system with M statistical input signals is developed. A general linear system with M input signals is shown in Fig. 3-8.

This linear system is to be optimized so that the mean-square error given in (3-14) is minimized, and the system response $q(t)$ now is

$$q(t) = \sum_{n=1}^{M} \int_{-\infty}^{t} w_n(t, \mu) v_n(\mu)\, d\mu \tag{3-88}$$

The optimum linear system is in fact the optimum system if all the signals are gaussian. The error index is minimized as in the single-statistical-input-signal case by defining the optimum impulse responses as

$$w_n(t, \mu) = w_n^*(t, \mu) + \epsilon_n \eta_n(t, \mu) \tag{3-89}$$

The conditions for the minimum value of this index are

$$\left.\frac{\partial e(t)}{\partial \epsilon_n}\right|_{\epsilon_m = 0} = 0 \qquad m, n = 1, 2, \ldots, M \tag{3-90}$$

which is equivalent to the condition given in (3-5). If these conditions are applied, the M conditions for minimum mean-square error are

$$\overline{Q(t + \tau) v_m(\alpha)} = \sum_{n=1}^{M} \int_{-\infty}^{t} w_n^*(t, \mu) \overline{v_n(\mu) v_m(\alpha)}\, d\mu \qquad \alpha \leqq t;\ m = 1, 2, \ldots, M \tag{3-91}$$

Therefore, the optimum linear system for M statistical input signals is defined by the M Wiener-Hopf conditions given in (3-91). However, the methods developed

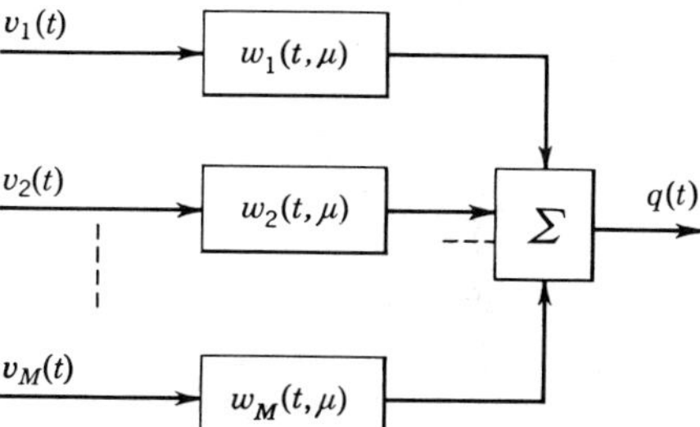

Fig. 3-8 General linear system with M input signals.

for solving a single Wiener-Hopf condition discussed in Secs. 3-4 and 3-5 cannot be applied directly to solving multiple Wiener-Hopf equations.

In order to develop a method for solving (3-91), a simpler case with two input signals is considered here such that

$$v_1(t) = Q(t) + n_1(t) \qquad v_2(t) = Q(t) + n_2(t) \qquad v_n(t) = 0 \qquad n = 3, 4, \ldots, M \tag{3-92}$$

Also, the problem is restricted further by assuming that the statistical characteristics of the input and desired-response signals are

$$\begin{aligned} \overline{Q(t)Q(t + \beta)} &= Q^2 e^{-q|\beta|} \\ \overline{n_1(t)n_1(t + \beta)} &= N_1^2 e^{-n_1|\beta|} \\ \overline{n_2(t)n_2(t + \beta)} &= N_2^2 e^{-n_2|\beta|} \end{aligned} \tag{3-93}$$

and

$$\overline{Q(t)n_1(t + \beta)} = \overline{Q(t)n_2(t + \beta)} = \overline{n_1(t)n_2(t + \beta)} = 0$$

For the case $\tau = 0$, the conditions given in (3-91) are

$$Q^2e^{-q\beta} = \int_0^\infty w_1^*(\gamma)(Q^2e^{-q|\beta-\gamma|} + N_1^2e^{-n_1|\beta-\gamma|})\,d\gamma + \int_0^\infty w_2^*(\gamma)Q^2e^{-q|\beta-\gamma|}\,d\gamma \qquad \beta \geqq 0 \quad (3\text{-}94)$$

and

$$Q^2e^{-q\beta} = \int_0^\infty w_1^*(\gamma)Q^2e^{-q|\beta-\gamma|}\,d\gamma + \int_0^\infty w_2^*(\gamma)(Q^2e^{-q|\beta-\gamma|} + N_2^2e^{-n_2|\beta-\gamma|})\,d\gamma \qquad \beta \geqq 0 \quad (3\text{-}95)$$

where the variable transformations $\beta = t - \alpha$ and $\gamma = t - \mu$ have been introduced for the stationary signals and time-invariant systems. The direct-substitution method now is attempted, where the impulse responses $w_1(\gamma)$ and $w_2(\gamma)$ are assumed to be

$$w_1^*(\gamma) = Au_0(\gamma) + \sum_{i=1}^{N} A_ie^{-a_i\gamma} \qquad (3\text{-}96)$$

and

$$w_2^*(\gamma) = Bu_0(\gamma) + \sum_{j=1}^{M} B_je^{-b_j\gamma} \qquad (3\text{-}97)$$

When these two impulse responses are substituted into (3-94) and (3-95) and the indicated integrations are performed, the conditions for minimum mean-square error become

$$Q^2e^{-q\beta} = \sum_{i=1}^{N} A_i\left(\frac{Q^2}{a_i - q}\,e^{-q\beta} - \frac{2qQ^2}{a_i^2 - q^2}\,e^{-a_i\beta} + \frac{N_1^2}{a_i - n_1}\,e^{-n_1\beta} - \frac{2n_1N_1^2}{a_i^2 - n_1^2}\,e^{-a_i\beta}\right) + \sum_{j=1}^{M} B_j\left(\frac{Q^2}{b_j - q}\,e^{-q\beta} - \frac{2qQ^2}{b_j^2 - q^2}\,e^{-b_j\beta}\right) + A(N_1^2e^{-n_1\beta} + Q^2e^{-q\beta}) + BQ^2e^{-q\beta} \qquad \beta \geqq 0 \quad (3\text{-}98)$$

and

$$Q^2e^{-q\beta} = \sum_{i=1}^{N} A_i\left(\frac{Q^2}{a_i - q}\,e^{-q\beta} - \frac{2qQ^2}{a_i^2 - q^2}\,e^{-a_i\beta}\right) + AQ^2e^{-q\beta} + B(Q^2e^{-q\beta} + N_2^2e^{-n_2\beta}) + \sum_{i=1}^{M} B_j\left(\frac{Q^2}{b_j - q} - \frac{2qQ^2}{b_j^2 - q^2}\,e^{-b_j\beta} + \frac{N_2^2}{b_j - n_2}\,e^{-n_2\beta} - \frac{2n_2N_2^2}{b_j^2 - n_2^2}\,e^{-b_j\beta}\right) \qquad \beta \geqq 0 \quad (3\text{-}99)$$

For the assumption $a_i, b_j \neq q$, the coefficients of $e^{-q\beta}$ in (3-98) and (3-99) must vanish by the same reasoning used in arriving at (3-57) for the single-input-signal case. When the coefficients of $e^{-q\beta}$ vanish, (3-98) and (3-99) respectively reduce to

$$\sum_{i=1}^{N} A_i\left(\frac{qQ^2}{a_i^2 - q^2} + \frac{n_1N_1^2}{a_i^2 - n_1^2}\right)e^{-a_i\beta} + \sum_{j=1}^{M} B_j\,\frac{qQ^2}{b_j^2 - q^2}\,e^{-b_j\beta} = 0 \qquad \beta \geqq 0 \quad (3\text{-}100)$$

and

$$\sum_{i=1}^{N} A_i\,\frac{qQ^2}{a_i^2 - q^2}\,e^{-a_i\beta} + \sum_{j=1}^{M} B_j\left(\frac{qQ^2}{b_j^2 - q^2} + \frac{n_2N_2^2}{b_j^2 - n_2^2}\right)e^{-b_j\beta} = 0 \qquad \beta \geqq 0 \quad (3\text{-}101)$$

However, in order to satisfy (3-100) and (3-101) for all $\beta \geqq 0$, the exponents of the exponentials appearing in these equations must be equal. In other words, (3-100) and (3-101) are satisfied for $\beta \geqq 0$ only when

$$a_k = b_k \qquad k = 1, 2, \ldots, N = M \tag{3-102}$$

Under the condition given in (3-102), the equalities in (3-100) and (3-101) are satisfied when the coefficients of each of these exponentials vanish; hence

$$A_k\left(\frac{qQ^2}{a_k^2 - q^2} + \frac{n_1N_1^2}{a_k^2 - n_1^2}\right) + B_k\frac{qQ^2}{a_k^2 - q^2} = 0 \qquad k = 1, 2, \ldots, N \tag{3-103}$$

and

$$A_k\frac{qQ^2}{a_k^2 - q^2} + B_k\left(\frac{qQ^2}{a_k^2 - q^2} + \frac{n_2N_2^2}{a_k^2 - n_2^2}\right) = 0 \qquad k = 1, 2, \ldots, N \tag{3-104}$$

The most general solutions for A_k and B_k occur only when the determinant of the coefficients of A_k and B_k in (3-103) and (3-104) vanishes. Hence, (3-103) and (3-104) reduce to

$$\left(\frac{qQ^2}{a_k^2 - q^2} + \frac{n_1N_1^2}{a_k^2 - n_1^2}\right)\left(\frac{qQ^2}{a_k^2 - q^2} + \frac{n_2N_2^2}{a_k^2 - n_2^2}\right) - \left(\frac{qQ^2}{a_k^2 - q^2}\right)^2 = 0$$

$$k = 1, 2, \ldots, N \tag{3-105}$$

This equation is solved for a_k, and algebraic manipulations yield

$$a_k^2 = \frac{qn_1n_2^2Q^2N_1^2 + qn_1^2n_2Q^2N_2^2 + q^2n_1n_2N_1^2N_2^2}{qn_1Q^2N_1^2 + qn_2Q^2N_2^2 + n_1n_2N_1^2N_2^2} \tag{3-106}$$

However, there is only one positive root of (3-106) and hence $N = 1$. Therefore, the exponents appearing in the exponential of the two impulse responses are

$$a_1, b_1 = \sqrt{\frac{qn_1n_2(n_2Q^2N_1^2 + n_1Q^2N_2^2 + qN_1^2N_2^2)}{qn_1Q^2N_1^2 + qn_2Q^2N_2^2 + n_1n_2N_1^2N_2^2}} \tag{3-107}$$

The coefficients of the exponentials and impulses appearing in (3-96) and (3-97) now are found. The relationship

$$A_1 + \frac{B_1}{1 + \dfrac{n_1N_1^2}{qQ^2}\dfrac{a_1^2 - q^2}{a_1^2 - n_1^2}} = 0 \tag{3-108}$$

is found from either (3-103) or (3-104). Furthermore, the relationships

$$\frac{A_1}{a_1 - n_1} + A = 0 \tag{3-109}$$

$$\frac{B_1}{a_1 - n_2} + B = 0 \tag{3-110}$$

and

$$\frac{A_1}{a_1 - q} + \frac{B_1}{a_1 - q} + A + B = 1 \tag{3-111}$$

are found by setting the coefficients of $e^{-n_1\beta}$, $e^{-n_2\beta}$, and $e^{-q\beta}$ equal to zero in (3-98) and (3-99). Finally, the minimum value of the error index is found by first substituting (3-96) and (3-97) into (3-88). Then, this expression for the system response

$q(t)$ is substituted into (3-1), and the averaging and integrating processes are performed. The result is

$$e^* = Q^2\left(1 - A - B - \frac{A_1 + B_1}{a_1 + q}\right) \tag{3-112}$$

In order to examine the results of this optimization procedure for a typical numerical example, the following parameters are assumed:

$$Q^2 = N_1^2 = 1 \qquad N_2^2 = 0.5 \qquad q = n_2 = 1 \qquad n_1 = 0.1 \tag{3-113}$$

When (3-107) through (3-112) are solved, these parameter values result in

$$\begin{aligned} w_1^*(\gamma) &= 0.57u_0(\gamma) - 0.272e^{-0.489\gamma} \\ w_2^*(\gamma) &= 0.353u_0(\gamma) + 0.181e^{-0.489\gamma} \\ e^* &= 0.038 \end{aligned} \tag{3-114}$$

The transforms of these optimum impulse responses yield the optimum transfer functions

$$W_1^*(s) = 0.57\frac{s + 0.016}{s + 0.489} \qquad W_2^*(s) = 0.353\frac{s + 1}{s + 0.489} \tag{3-115}$$

It is seen from (3-115) that one part of the optimum linear system is a lag-lead network and the other part is a lead-lag network.

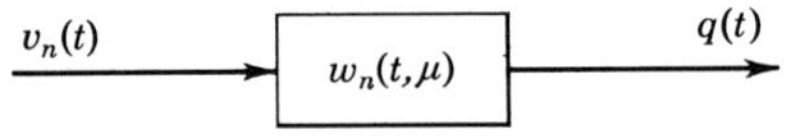

Fig. 3-9 Linear system for one input signal: $n = 1, 2$.

An interesting point of note occurs when each input signal to the linear system contains the desired response $Q(t)$. In this case, the poles of the various filters in the optimum linear system are identical. This result, obtained for two input signals, can be generalized for M input signals and for general forms of the statistical characteristics of the input and desired-response signals.

The method of spectral factorization can be applied to multiple Wiener-Hopf equations.[31] However, the method of spectral factorization in this case is not obvious and cannot be developed from direct algebraic manipulations such as used in the direct-substitution method. Therefore, the spectral-factorization method is not presented here for the case where the control system has two or more statistical input signals.

From an engineering standpoint, it is interesting to consider the optimum linear mixing of statistical signals such as indicated in Fig. 3-8 vs. simpler linear systems. Specifically, the statistical approximation properties of a linear system that operates on only one of the possible two input signals to generate the response $q(t)$ could be considered. Two such simple linear systems are indicated in Fig. 3-9. For the signals defined by (3-93) and (3-113), the two optimum impulse responses and the respective minimum values of e for these two simple linear systems are

$$\begin{aligned} w_1^*(\gamma) &= 0.76u_0(\gamma) - 0.164e^{-0.316\gamma} \qquad & e^* &= 0.365 \\ w_2^*(\gamma) &= 0.667u_0(\gamma) & e^* &= 0.333 \end{aligned} \tag{3-116}$$

From the values of the minimum error index for the two filters given in (3-116), it is seen that the optimum linear mixing of the two signals specified in (3-114) reduces the mean-square error in this case by an order of magnitude. The conclusion can be drawn from these numerical results that the optimum linear mixing, as opposed to the optimum filtering of each signal separately, results generally in a large decrease in mean-square error. This conclusion indicates that, in a number of design problems, a performance increase is obtained by designing more complicated linear control systems. Such mixing problems arise in typical problems such as the mixing of doppler radar and accelerometer signals used in the erection of a vertical gyro.

Often suggested in the mixing of two input signals, a distortionless linear system for two input signals could be assumed. Such a distortionless linear system is shown in Fig. 3-10. The response of the system shown in Fig. 3-10 is

$$q(t) = Q(t) + n_2(t) + \int_0^\infty w_1(\gamma)[n_1(t-\gamma) - n_2(t-\gamma)]\,d\gamma \qquad (3\text{-}117)$$

From (3-117), it is seen that the system response contains the desired response unchanged; hence this system is called a *distortionless* system. The optimum impulse

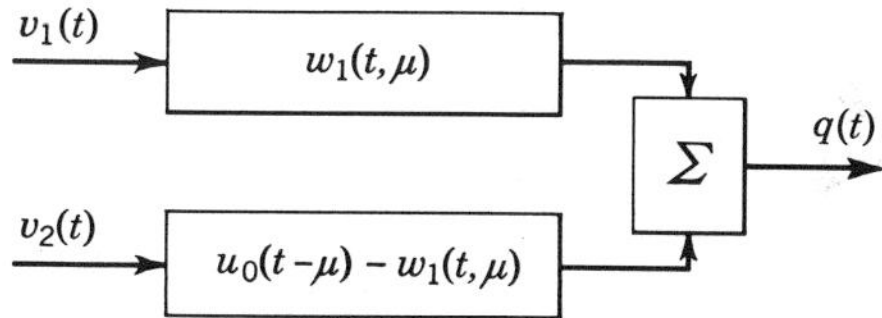

Fig. 3-10 Distortionless linear system for two input signals.

response $w_1^*(t, \mu)$ is found for the system shown in Fig. 3-10 by methods similar to those already used. The condition for the minimum mean-square error of the system shown in Fig. 3-10 is

$$\overline{n_2(t-\beta)n_2(t)} = \int_0^\infty w_1^*(\gamma)[\overline{n_1(t-\gamma)n_1(t-\beta)} + \overline{n_2(t-\gamma)n_2(t-\beta)}]\,d\gamma \qquad \beta \geq 0 \qquad (3\text{-}118)$$

Because the optimum distortionless system is described by a single Wiener-Hopf equation, (3-118) is solved by the techniques described in Secs. 3-4 and 3-5. The optimum impulse response of this filter and the minimum value of e are found to be

$$w_1^*(\gamma) = 0.646u_0(\gamma) - 0.205e^{-0.418\gamma} \qquad e^* = 0.25 \qquad (3\text{-}119)$$

From the values of the minimum mean-square errors given for these various cases, it is seen that the optimum linear mixing gives a performance which is considerably better than that for the optimum distortionless system. Therefore, in some engineering design problems, the optimum linear mixing gives considerably better performance than optimum linear systems of special configurations.

3-7 Saturation constraints for stationary statistical signals

As in the case of parameter optimization, saturation constraints are important in impulse-response optimization because of the limited magnitudes of the variables in the fixed elements of the control system.[12] For instance, the mean-square value of the control signal $m(t)$, shown in Fig. 3-5 for the design problem discussed in Sec. 3-4, is infinite. In other words, instantaneous values of this control signal can be infinite on an instantaneous basis for gaussian input signals. The block diagram of the control system for saturation-constraint problems§ is shown in Fig. 3-11. The composite impulse response of the control system is $w(t - \mu)$, and $w_s(t - \mu)$ is the impulse response from the system output to the saturation variable $q_s(t)$. The optimization problem in the case of saturation constraints is the minimization of the mean-square error response signal subject to the constraint

$$\overline{q_s^2(t)} \leq c_1 \tag{3-120}$$

Therefore, as in Chap. 2, the error index to be minimized is

$$e_c = \overline{[Q(t) - q(t)]^2} + \lambda_1[\overline{q_s^2(t)} - c_1] \tag{3-121}$$

In this section, the statistical signals are assumed to be stationary, and therefore the optimum linear system is time-invariant. For this case, the Lagrange multiplier λ_1 is a constant instead of a function of time. Also, the analysis is for linear systems, because λ_1 is adjusted so that actual nonlinearities almost never occur in the actual system and can be neglected.

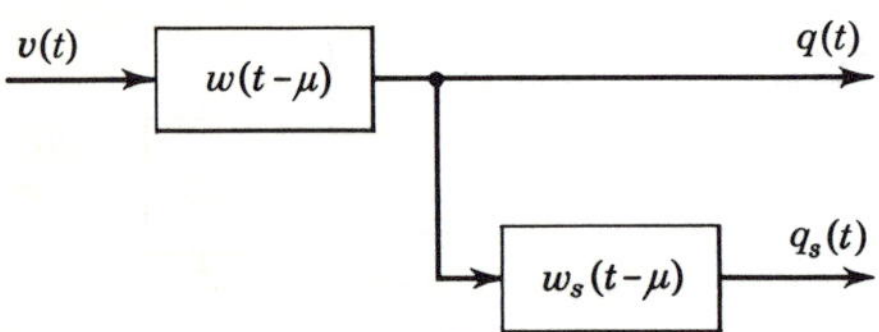

Fig. 3-11 Control-system block diagram for saturation constraint.

For a time-invariant system, the system response signal is written as

$$q(t) = \int_0^\infty w(\gamma)v(t - \gamma)\,d\gamma \tag{3-122}$$

and the saturation signal is written as

$$q_s(t) = \int_0^\infty w_s(\beta)q(t - \beta)\,d\beta = \int_0^\infty w(\gamma)\left[\int_0^\infty w_s(\beta)v(t - \beta - \gamma)\,d\beta\right] d\gamma \tag{3-123}$$

As in the design problem with no saturation constraint, the optimum impulse response is assumed to be defined by

$$w(\gamma) = w^*(\gamma) + \epsilon\eta(\gamma) \tag{3-124}$$

and the condition for minimum error is

$$\left.\frac{\partial e_c}{\partial \epsilon}\right|_{\epsilon=0} = 0 \qquad \left.\frac{\partial e_c}{\partial \lambda_1}\right|_{\epsilon=0} = 0 \tag{3-125}$$

When the first condition in (3-125) is applied to (3-121) after substituting (3-122), (3-123), and (3-124), the condition for minimum error is

$$\int_0^\infty \eta(\beta)\left\{\overline{Q(t)v(t - \beta)} - \int_0^\infty w^*(\gamma)\overline{v(t - \gamma)v(t - \beta)}\,d\gamma \right.$$
$$\left. -\lambda_1\int_0^\infty w^*(\gamma)\left[\int_0^\infty\int_0^\infty w_s(\eta)w_s(\mu)\overline{v(t - \eta - \gamma)v(t - \mu - \beta)}\,d\eta\,d\mu\right] d\gamma\right\} d\beta = 0 \tag{3-126}$$

§ The term saturation constraint is used here in conformance with Newton[12] but is a mathematical misnomer. The optimization problem treated is strictly linear, and only mean-square constraints are imposed. The usage of this term denotes the engineering motivation associated with the constraints.

With the same reasoning that led to (3-8), the condition for the optimum impulse response is

$$\phi_{vQ}(\beta) = \int_0^\infty w^*(\gamma)\left[\phi_{vv}(\beta-\gamma) + \lambda_1 \int_0^\infty \int_0^\infty w_s(\eta) w_s(\mu) \phi_{vv}(\beta - \gamma + \eta - \mu)\, d\eta\, d\mu\right] d\gamma$$

$$\beta \geqq 0 \qquad (3\text{-}127)$$

where the notation for correlation functions has been introduced.

However, a modified correlation function associated with the saturation-constraint problem is defined as

$$\phi_s(\beta-\gamma) = \phi_{vv}(\beta-\gamma) + \lambda_1 \int_0^\infty \int_0^\infty w_s(\eta) w_s(\mu) \phi_{vv}(\beta-\gamma+\eta-\mu)\, d\eta\, d\mu \qquad (3\text{-}128)$$

and hence (3-127) becomes

$$\phi_{vQ}(\beta) = \int_0^\infty w^*(\gamma)\phi_s(\beta-\gamma)\, d\gamma \qquad \beta \geqq 0 \qquad (3\text{-}129)$$

This condition defining the optimum impulse response subject to a constraint is precisely the Wiener-Hopf condition; hence the optimum transfer function is

$$W^*(s) = \frac{[\Phi_{vQ}(s)/\Phi_s^-(s)]_+}{\Phi_s^+(s)} \qquad (3\text{-}130)$$

as expressed in terms of spectral factorization. Frequency-domain relationships for statistical signals in linear systems are applied to the modified correlation function defined in (3-128), yielding

$$\Phi_s(s) = \Phi_{vv}(s) + \lambda_1 W_s(s) W_s(-s) \Phi_{vv}(s) \qquad (3\text{-}131)$$

Therefore, the solution of the saturation-constraint problem is identical to the case of no saturation constraint except for the adjustment of the Lagrange multiplier λ_1 in order to satisfy (3-120).

The adjustment of the Lagrange multiplier so that a saturation level defined as Q_s is not exceeded by the saturation signal $q_s(t)$ can only be done on a probabilistic basis. If the input signal to the control system is a gaussian statistical signal, then the saturation signal also is a gaussian signal. The probability density of the gaussian saturation signal is written as

$$p[q_s(t)] = \frac{1}{\sqrt{2\pi}\sigma_{q_s}} \exp\left\{-\frac{[q_s(t) - m_{q_s}]^2}{2\sigma_{q_s}^{\;2}}\right\} \qquad (3\text{-}132)$$

where

$$m_{q_s} = \overline{q_s(t)} \qquad \sigma_{q_s}^{\;2} = \overline{[q_s(t) - \overline{q_s(t)}]^2} \qquad (3\text{-}133)$$

The probability that the signal $q_s(t)$ will be within the specified range

$$-3\sigma_{q_s} + m_{q_s} \leqq q_s(t) \leqq 3\sigma_{q_s} + m_{q_s}$$

is

$$P[-3\sigma_{q_s} \leqq q_s(t) - m_{q_s} \leqq 3\sigma_{q_s}] = \int_{-3\sigma_{q_s}}^{3\sigma_{q_s}} p[q_s(t)]\, d[q_s(t) - m_{q_s}] \approx 0.99 \qquad (3\text{-}134)$$

for the probability density given in (3-132). In other words, the saturation signal is between this specified range with a probability of 0.99. Therefore, the amplitude of the saturation signal satisfies

$$|q_s(t)| \leqq 3\sigma_{q_s} + m_{q_s} \qquad (3\text{-}135)$$

with a probability of approximately 0.99. If the amplitude of $q_s(t)$ is adjusted to be less than the saturation level Q_s with a probability of approximately 0.99, then

$$3\sigma_{q_s} + m_{q_s} \leqq Q_s \tag{3-136}$$

The expressions given in (3-135) when squared are substituted into (3-136), and therefore the mean-square value of $q_s(t)$ satisfies

$$\overline{q_s^2(t)} \leqq c_1 = \tfrac{1}{9}[Q_s - \overline{q_s(t)}]^2 - [\overline{q_s(t)}]^2 \tag{3-137}$$

If the Lagrange multiplier λ_1 is adjusted so that the mean-square value of the saturation signal satisfies (3-137), then the magnitude of the gaussian saturation signal is less than the saturation level Q_s 99 per cent of the time. Therefore, the control system closely approximates linear behavior.§

In order to demonstrate the use of saturation constraints for design problems with statistical signals, a simple example is defined by

$$v(t) = Q(t) + n(t) \tag{3-138}$$

$$\phi_{QQ}(\beta) = Q^2 e^{-a|\beta|} \qquad \phi_{nn}(\beta) = N^2 u_0(\beta) \qquad \phi_{nQ}(\beta) = 0 \tag{3-139}$$

and

$$W_s(s) = s \tag{3-140}$$

This is a design problem with the gaussian input signal being the desired response plus additive noise and with the saturation constraint being a velocity constraint. By finding the spectral density of the input signal and then using (3-131), the spectral density of the modified correlation function is

$$\Phi_s(s) = \frac{\lambda_1 N^2}{2\pi} \frac{(s + w_1)(s + \lambda_1^{-\frac{1}{2}})(-s + w_1)(-s + \lambda_1^{-\frac{1}{2}})}{(s + a)(-s + a)} \tag{3-141}$$

where

$$w_1 = \sqrt{a^2 + 2a\frac{Q^2}{N^2}} \tag{3-142}$$

The spectral factorization is chosen so that

$$\Phi_s^+(s) = \frac{(s + w_1)(s + \lambda_1^{-\frac{1}{2}})}{s + a} \qquad \Phi_s^-(s) = \frac{\lambda_1 N^2}{2\pi} \frac{(-s + w_1)(-s + \lambda_1^{-\frac{1}{2}})}{-s + a} \tag{3-143}$$

and hence

$$\left[\frac{\Phi_{vQ}(s)}{\Phi_s^-(s)}\right]_+ = \frac{2aQ^2}{\lambda_1 N^2}\left[\frac{1}{(s + a)(-s + w_1)(-s + \lambda_1^{-\frac{1}{2}})}\right]_+ = \frac{w_1 - a}{\lambda_1(a + \lambda_1^{-\frac{1}{2}})} \frac{1}{s + a} \tag{3-144}$$

Therefore, the transfer function of the optimum system is

$$W^*(s) = \frac{w_1 - a}{(a\lambda_1^{\frac{1}{2}} + 1)\lambda_1^{\frac{1}{2}}} \frac{1}{(s + w_1)(s + \lambda_1^{-\frac{1}{2}})} \tag{3-145}$$

as written in terms of the Lagrange multiplier λ_1.

In order to compute λ_1, first the spectral density of the saturation signal is derived according to

$$\Phi_{q_s q_s}(s) = W_s(s) W_s(-s) W^*(s) W^*(-s) \Phi_{vv}(s) \tag{3-146}$$

§ This conclusion generally is valid for control-signal saturation. However, this conclusion is not valid for state-signal saturation in many situations.

For the example under consideration here, this spectral density becomes

$$\Phi_{q_sq_s}(s) = \frac{1}{2\pi} N^2 \frac{(w_1 - a)^2}{(a\lambda_1^{\frac{1}{2}} + 1)^2\lambda_1} \frac{(s)(-s)}{(s + a)(s + \lambda_1^{-\frac{1}{2}})(-s + a)(-s + \lambda_1^{-\frac{1}{2}})} \tag{3-147}$$

From this spectral-density function, the mean value of the saturation signal is seen to be zero because $\phi_{q_sq_s}(\infty) = 0$. Also, the mean-square value of the saturation signal is found from the tabulated integrals to be

$$\overline{q_s^2(t)} = N^2 \frac{(w_1 - a)^2}{2(a\lambda_1^{\frac{1}{2}} + 1)^3\lambda_1^{\frac{1}{2}}} \tag{3-148}$$

If the amplitude of the saturation signal is to be less than the saturation level 99 per cent of the time, then (3-137) is used, where $\overline{q_s(t)} = 0$ and $\overline{q_s^2(t)}$ is given in (3-148). As a result, the Lagrange multiplier must satisfy

$$N^2 \frac{a(w_1 - a)^2}{2(a\lambda_1^{\frac{1}{2}} + 1)^3 a\lambda_1^{\frac{1}{2}}} = \frac{1}{9} Q_s^2 \tag{3-149}$$

The explicit solution for the Lagrange multiplier λ_1 is, in general, impossible to derive from (3-149) because a fourth-degree algebraic equation must be solved.

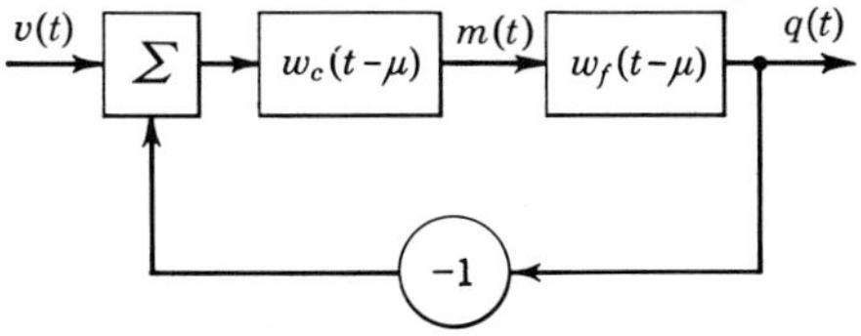

Fig. 3-12 Block diagram of control system with fixed elements.

In order to further investigate this design problem, the following relationship between the design parameters is assumed:

$$\frac{Q_s^2/N^2}{a(w_1 - a)^2} = \frac{9}{16} \tag{3-150}$$

For this relationship, the solution of (3-149) is found by inspection to be

$$\lambda_1 = \frac{1}{a^2} \tag{3-151}$$

Hence, the optimum transfer function of the control system is

$$W^*(s) = \frac{a(w_1 - a)}{2} \frac{1}{(s + w_1)(s + a)} \tag{3-152}$$

Suppose the control system shown in Fig. 3-12 is to be synthesized, where $w_f(t - \mu)$ and $w_c(t - \mu)$ are the impulse responses of the fixed-member and the compensation elements in the control system, respectively. The compensation elements are to be chosen so that the transfer function of the composite control system has the transfer function given in (3-152). In other words, this transfer function is expressed in terms of the transfer functions of the compensation and fixed member as

$$W^*(s) = \frac{W_c(s)W_f(s)}{1 + W_c(s)W_f(s)} \tag{3-153}$$

For the transfer function given in (3-152), the transfer function of the compensation elements is

$$W_c(s) = \frac{1}{W_f(s)} \frac{k}{s^2 + 2\zeta\omega_n s + \omega_n^2} \tag{3-154}$$

where $k = \dfrac{a(w_1 - a)}{2} \qquad \zeta = \left(\dfrac{w_1 + a}{2a}\right)^{\frac{1}{2}} > 1 \qquad \omega_n = \left[\dfrac{a(w_1 + a)}{2}\right]^{\frac{1}{2}}$ (3-155)

If the transfer function of the fixed member has three or more poles than zeros, then the transfer function of the compensation elements involves one or more differentiations at high frequency. In this case, where the input signal contains white gaussian noise, such differentiations will cause the mean-square value of the control signal $m(t)$ to be infinite. This situation, of course, is not physically realizable with a linear system, and the fixed member of the control system experiences saturation at its input. This difficulty in realizing compensation elements for control systems with complicated fixed members is inherent in the impulse-response optimization method and is a difficulty that very definitely limits the usefulness of the method. The addition of saturation constraints does ease the problem somewhat, but a more direct method of including the problem of realizing compensation elements in control systems designed by impulse-response optimization is needed.

3-8 Bandwidth minimization for stationary statistical signals

The problem of realizing compensation elements in optimum-impulse-response systems is that the bandwidth required for the optimum system may be larger than the bandwidth of the fixed member of the system. This suggests that an optimization procedure should be developed which deals directly with the bandwidth of the fixed elements of the system.[12] Therefore, the block diagram shown in Fig. 3-13 is assumed for the bandwidth-minimization problem. The signal $b(t)$ is assumed to be a signal that indicates the bandwidth of the optimum control system, and the signal $t(t)$ is a test signal introduced into the control system to determine its bandwidth. Minimizing the mean-square value of $b(t)$ is related to minimizing the bandwidth of the control system. However, this minimization of the mean-square value of the bandwidth signal is stated so that the mean-square value of the response error must be less than or equal to a given value. In other words, the minimum bandwidth control system must satisfy the constraint

$$\overline{[Q(t) - q(t)]^2} \leq c_1 \tag{3-156}$$

Therefore, the error index to be minimized is written as

$$e_c = \overline{b^2(t)} + \lambda_1\{\overline{[Q(t) - q(t)]^2} - c_1\} \tag{3-157}$$

where the Lagrange multiplier λ_1 is introduced in order to treat the constraint. Finally, the bandwidth signal for the stationary test signal $t(t)$ is

$$b(t) = \int_0^\infty w(\gamma)t(t - \gamma)\, d\gamma \tag{3-158}$$

When the minimization procedure indicated in (3-124) and (3-125) is applied, the condition for the optimum impulse response is

$$\int_0^\infty \eta(\beta)\left\{\lambda_1\overline{Q(t)v(t-\beta)} - \int_0^\infty w^*(\gamma)[\overline{t(t-\beta)t(t-\gamma)} + \lambda_1\overline{v(t-\beta)v(t-\gamma)}]\,d\gamma\right\}d\beta = 0 \quad (3\text{-}159)$$

As in the case of (3-126), this condition for the optimum impulse response is reduced to

$$\lambda_1\phi_{vQ}(\beta) = \int_0^\infty w^*(\gamma)[\phi_{tt}(\beta-\gamma) + \lambda_1\phi_{vv}(\beta-\gamma)]\,d\gamma \qquad \beta \geqq 0 \qquad (3\text{-}160)$$

If a modified correlation function due to the response error constraint is defined as

$$\phi_b(\beta-\gamma) = \phi_{tt}(\beta-\gamma) + \lambda_1\phi_{vv}(\beta-\gamma) \qquad (3\text{-}161)$$

then the optimum-control-system transfer function satisfies

$$W^*(s) = \lambda_1\frac{[\Phi_{vQ}(s)/\Phi_b^-(s)]_+}{\Phi_b^+(s)} \qquad (3\text{-}162)$$

where

$$\Phi_b(s) = \Phi_{tt}(s) + \lambda_1\Phi_{vv}(s) \qquad (3\text{-}163)$$

The physical interpretation of minimizing bandwidth subject to a mean-square error constraint can be stated by introducing the following arguments. If the

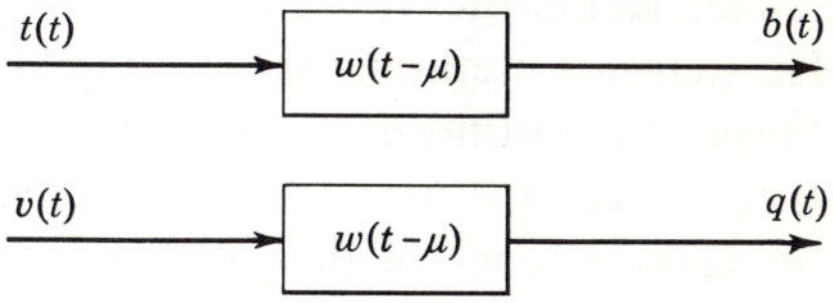

Fig. 3-13 Control-system block diagram for bandwidth minimization.

compensation elements shown in the diagram in Fig. 3-12 are assumed not to have differentiations at high frequencies and if the fixed member is assumed to have a larger number of poles than zeros in its transfer function, then the loop gain of the system shown in Fig. 3-12 becomes very small at high frequencies. In other words, if the number of poles minus the number of zeros in the transfer function of the fixed member in the system is n and if a cutoff frequency ω_c is defined as the frequency corresponding to the smallest time constant in this transfer function, then

$$W^*(s) \approx W_f(s) \approx \frac{1}{s^n} \qquad s \gg j\omega_c \qquad (3\text{-}164)$$

as found from (3-153). The relationship in (3-164) describes the high-frequency behavior of the optimum-control-system transfer function under the conditions that the compensation elements in the control system contain no differentiations at these frequencies. Also, the spectral density of the bandwidth signal must approach zero for large frequencies if this bandwidth signal is to have a finite mean-square value and hence a finite bandwidth. If this bandwidth signal is to have a bandwidth which is not in excess of the bandwidth of the fixed member of the control system, then

$$\Phi_{bb}(s) = W^*(s)W^*(-s)\Phi_{tt}(s) \approx \frac{1}{s^2} \qquad s \gg j\omega_c \qquad (3\text{-}165)$$

If the relationships given in (3-164) and (3-165) are used, then the test signal must have a spectral-density behavior

$$\Phi_{tt}(s) \approx s^{2(n-1)} \qquad s \gg j\omega_c \tag{3-166}$$

The interpretation of bandwidth minimization allows one to design an optimum control system that does not involve compensation elements that contain differentiations at high frequencies. The high-frequency behavior of the test signal is selected by utilizing (3-166), where n is the number of poles minus the number of zeros in the fixed member of the system; and the cutoff frequency is $\omega_c = 1/\tau_c$, where τ_c is defined as the smallest time constant in the fixed member of the system. However, the situation may arise where the value for the Lagrange multiplier λ_1 is very large for a given set of design conditions which includes the specifications of the fixed member of the system. In other words, it is difficult to satisfy the constraint on the mean-square response error of the system for the given input signals and fixed member of the control system. If the input signals are considered fixed, then the cutoff frequency of the fixed member should be larger. Therefore, bandwidth minimization can be considered as a method of selecting the fixed member for a given constraint on the mean-square response error. This fixed member has a minimum cutoff frequency and hence a minimum power requirement for the design problem at hand. The bandwidth-minimization procedure was first developed by Newton.[12]

In addition to the considerations discussed here, Newton has developed more refined techniques for selecting the test signal in the bandwidth-minimization problem. He suggests methods of selecting the low-frequency behavior of the test signal that shapes the optimum transfer function of the system to approximate the behavior of such filters as Butterworth. These shaping procedures give desirable characteristics for certain types of design problems. A more general interpretation of the bandwidth-minimization and the saturation-constraint design problems has been developed by Kaiser.[31] His procedures allow the designers to have more control over the characteristics of the optimum control system, and they improve the possibilities of realizing the compensating elements in the control system.

3-9 Wiener-Hopf condition for deterministic signals

As mentioned previously, minor modifications in the impulse-response optimization procedures presented here for statistical signals are required in design problems where deterministic signals are present. These minor modifications are discussed briefly here in terms of the saturation-constraint problem. The block diagram of the control system with a saturation constraint for deterministic signals is precisely the same as that given in Fig. 3-11 for statistical signals.

When $Q(t)$ and $v(t)$ are deterministic, the Wiener-Hopf condition derived for the mean-square error index reduces to

$$Q(t) = \int_0^\infty w^*(\gamma)v(t-\gamma)\,d\gamma \tag{3-167}$$

as is seen from (3-127) in the special case $\lambda_1 = 0$. Therefore, when no saturation constraint is present, the output of the optimum system is precisely the desired-response signal, and great difficulty generally is experienced in realizing the optimum system for this case. Unfortunately, imposing a constraint such as given in (3-120) for all t requires the introduction of a time-varying Lagrange multiplier and hence gives rise to an optimum time-varying system for deterministic signals. An optimization problem with time-varying Lagrange multipliers generally cannot be solved explicitly. In order to avoid this difficulty, the error index to be minimized for deterministic signals is of the integral squared type,

$$e = \int_t^\infty [Q(\sigma) - q(\sigma)]^2 \, d\sigma \tag{3-168}$$

subject to the constraint

$$\int_t^\infty q_s^2(\sigma) \, d\sigma \leqq c_1 \tag{3-169}$$

This form of constraint does provide some control over the instantaneous magnitude of the saturation signal $q_s(t)$, but it is not wholly adequate. Finally, the index to be minimized for the saturation-constraint problem stated in (3-168) and (3-169) for deterministic signals is expressed in terms of a Lagrange multiplier as

$$e_c = \int_t^\infty \{[Q(\sigma) - q(\sigma)]^2 + \lambda_1 q_s^2(\sigma)\} \, d\sigma - \lambda_1 c_1 \tag{3-170}$$

The system response and saturation signals are expressed in (3-122) and (3-123), respectively.

The minimization procedure for the error index given in (3-170) is formulated precisely in the manner defined in (3-124) and (3-125) for the statistical-signal case. When these procedures are applied, the condition for minimum error with the integral squared error index is

$$\begin{aligned}\int_0^\infty \eta(\beta)\Bigg(\int_0^\infty \Bigg\{Q(\sigma)v(\sigma - \beta) - \int_0^\infty w^*(\gamma)\Bigg[v(\sigma - \gamma)v(\sigma - \beta) \\ + \lambda_1 \int_0^\infty \int_0^\infty w_s(\eta)w_s(\mu)v(\sigma - \eta - \gamma)v(\sigma - \mu - \beta)\, d\eta\, d\mu\Bigg] d\gamma\Bigg\} d\sigma\Bigg) d\beta = 0\end{aligned} \tag{3-171}$$

In this condition for minimum error, the terms inside the parentheses now involve the integral of the product of two signals instead of the mean of the product of two signals. This is a difference between the statistical-signal and the deterministic-signal design problems. However, functions are defined that are analogous to cross-correlation and autocorrelation functions, so that

$$\psi_{vQ}(\beta) = \int_0^\infty Q(\alpha)v(\alpha - \beta) \, d\alpha \tag{3-172}$$

and

$$\psi_{vv}(\beta) = \int_0^\infty v(\alpha)v(\alpha - \beta) \, d\alpha \tag{3-173}$$

when the signals $v(t)$ and $Q(t)$ can be generated as the impulse responses of linear time-invariant systems. These functions are called the cross-translation and auto-translation functions, respectively. They appear in (3-171) when the change in variables $\sigma - t = \alpha$ is made.

If the terms within the parentheses of (3-171) are set equal to zero and the definitions for the translation functions are introduced, the Wiener-Hopf condition for integral squared error reduces to

$$\psi_{vQ}(\beta) = \int_0^\infty w^*(\gamma)\left[\psi_{vv}(\gamma - \beta) + \lambda_1 \int_0^\infty \int_0^\infty w_s(\eta)w_s(\mu)\psi_{vv}(\beta - \gamma + \eta - \mu)\, d\eta\, d\mu\right] d\gamma \qquad \beta \geqq 0 \quad (3\text{-}174)$$

This condition is precisely analogous to the condition given in (3-127) for statistical signals. Hence, the optimum transfer function of the control system is given in terms of spectral factorization as

$$W^*(s) = \frac{[\Psi_{vQ}(s)/\Psi_s^-(s)]_+}{\Psi_s^+(s)} \qquad (3\text{-}175)$$

where

$$\Psi_s(s) = [1 + \lambda_1 W_s(s)W_s(-s)]\Psi_{vv}(s) \qquad (3\text{-}176)$$

The spectral densities appearing in (3-175) and (3-176) are defined by

$$\Psi(s) = \frac{1}{2\pi}\int_{-\infty}^{\infty} \psi(\beta)e^{-s\beta}\, d\beta \qquad (3\text{-}177)$$

The introduction of translation functions instead of correlation functions makes the procedure for finding the optimum impulse response for deterministic signals

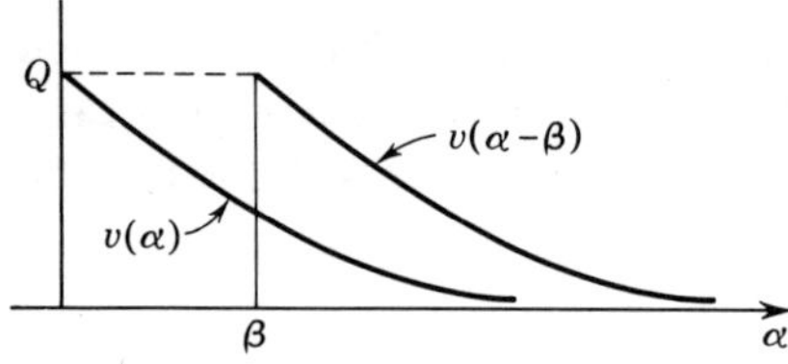

Fig. 3-14 Sketch of input signal for $\beta > 0$.

precisely analogous to that for finding the optimum impulse response for statistical signals.

In order to demonstrate another difference between the treatment of statistical signals and that of deterministic signals in impulse-response optimization, the following special example is considered. The desired-response and input signals are assumed to be

$$Q(t) = v(t) = \begin{cases} Qe^{-at} & t \geqq 0 \\ 0 & t < 0 \end{cases} \qquad (3\text{-}178)$$

These signals are sketched in Fig. 3-14 in order to aid in the evaluation of translation functions used in (3-175) and (3-176).

When the integration is performed, the cross-translation and autotranslation functions are

$$\psi_{vQ}(\beta) = \psi_{vv}(\beta) = \frac{Q^2}{2a}e^{-a|\beta|} \qquad (3\text{-}179)$$

for this example. The spectral densities of these translation functions are

$$\Psi_{vQ}(s) = \Psi_{vv}(s) = \frac{1}{2\pi}\frac{Q^2}{(s+a)(-s+a)} \qquad (3\text{-}180)$$

If the modified spectral density is evaluated for $W_s(s) = s$, then the spectral factorization indicated in (3-175) is performed, yielding

$$W^*(s) = \frac{1}{\lambda_1^{\frac{1}{2}}a + 1} \frac{1}{\lambda_1^{\frac{1}{2}}s + 1} \tag{3-181}$$

The evaluation of the Lagrange multiplier λ_1 is found by evaluating the integral squared value of the saturation signal by means of the tabulated integrals. The integral squared value of the saturation signal is equated to the constraint value specified in the statement of the design problem, and the value of the Lagrange multiplier is found from this equation.

The motivation for considering this example is to point out that if the input and desired-response signals are constants, that is, $a = 0$, then the cross-translation and autotranslation functions are unbounded. Also, the spectral densities of these translation functions have poles on the j axis, and the Fourier transform process used in arriving at (3-181) is not strictly applicable. However, a convergence factor can be introduced by treating the constant signals as a limiting case of exponential signals. Then the optimum transfer functions for constant desired and input signals are found with no difficulty by specializing the optimum transfer function given in (3-181) to $a = 0$.

From the considerations presented in this section, it is seen that impulse-response optimization for deterministic signals is performed in a way analogous to impulse-response optimization for statistical signals. However, the input and response signals in the design problem must be the impulse responses of linear time-invariant systems. Also, translation functions must be introduced for the deterministic signals. Finally, difficulty may arise in finding these translation functions and their spectral densities for some types of deterministic signals. But this difficulty can be overcome by considering a simple limiting process.

3-10 Summary

The concept of impulse-response optimization is helpful in introducing the concept of using mathematical optimization as a method of synthesizing control systems. Also, impulse-response optimization is used here to introduce the concepts of an optimum linear system and a linear optimum system. In the case of gaussian signals and deterministic signals which are generated as the impulse responses of linear systems, the optimum systems for mean-square and integral squared error indices, respectively, are linear. However, the approximation to the optimum system afforded by the optimum linear system is useful when the statistical signals are nongaussian. In particular, the additional conceptual and engineering difficulties arising in nonlinear optimum control systems may not be justified by the engineering compromise of increased performance versus increased system complexity.

However, the concept of impulse-response optimization suffers from two restrictions. First of all, the optimization of the system impulse response treats the control system as a composite system; hence the system structure is not retained. In other words, a very important concept in the optimization of control systems is the concept of the control-system configuration. The next chapter introduces concepts that relate to the optimum configuration of control systems. The second difficulty that arises in impulse-response optimization is that the methods apply only when linear models of the fixed member of the control system can be assumed. This is a severe limitation in design problems where the fixed member is in fact nonlinear, and useful operation is derived from these nonlinearities. For instance, in homing-missile control, high roll rates may be desirable when the missile has an aerodynamic configuration that requires roll in order to turn. High roll rates give rise to inertial cross-coupling in the missile; hence the response of the missile may be decidedly nonlinear. The mathematical developments in the next chapter also are not limited to the restriction that the fixed member of the control system be linear.

4

System Optimization Using the Calculus of Variations

In this chapter, the problem of control-system optimization is formulated so that the design-problem statement does not require initial assumptions concerning the configuration of the controller or the linearity of the overall system. Specifically, the optimum time function for the control signal is found directly. The optimum control signal is dependent not only on the design-problem parameters but also on the existing state of the dynamic process.

If the optimum control signal can be expressed as a function of the existing state of the dynamic process, this function constitutes a control equation. The dependence of the optimum control equation on the state of the dynamic process forms the basis for a control system with a feedback configuration. The feedback portion of the optimum control equation requires the measurement of the state signals of the dynamic process.

The system-optimization problem is developed in this chapter along the classical lines of the calculus of variations. This development is used to illustrate the physical, mathematical, and computational significances of the two-point boundary-value problem.

4-1 Elementary theory for a first-order dynamic process

In this section, the response and control signals are expressed in terms of the state signal of the dynamic process as

$$q(t) = G[x(t), x'(t), t] \tag{4-1}$$

and

$$m(t) = F[x(t), x'(t), t] \tag{4-2}$$

If the error index is taken to be

$$e(t) = \int_t^T h[q(\sigma), m(\sigma), \sigma]\, d\sigma \tag{4-3}$$

then the error measure h can be combined with the equations describing the dynamic process so that

$$e(t) = \int_t^T H[x(\sigma), x'(\sigma), \sigma]\, d\sigma \tag{4-4}$$

The problem of minimizing the error index as expressed in (4-4) is the most elementary and often treated problem in the calculus of variations.[23] Subsequent to the detailed treatment of this first-order case, suitable extensions for Nth-order dynamic processes, and hence more realistic control problems, are undertaken.

The minimization of (4-4) is performed by first assuming that the state signal is expressed as

$$x(\sigma) = x_*(\sigma) + \epsilon\eta(\sigma) \tag{4-5}$$

where $x_*(\sigma)$ is the state signal that minimizes (4-4). The parameter ϵ is considered to be an arbitrarily small parameter, and the function $\eta(\sigma)$ is considered to be an arbitrary and unrestricted perturbation. Also, the derivative of the state signal is written as

$$x'(\sigma) = x_*'(\sigma) + \epsilon\eta'(\sigma)$$

which results from differentiating (4-5). Because ϵ is taken to be arbitrarily small, the error measure is perturbed infinitesimally about the value $H[x_*(\sigma), x_*'(\sigma), \sigma]$. If H and its derivatives with respect to $x(\sigma)$ and $x'(\sigma)$ are continuous, the error measure is expanded in a Taylor series such that

$$\begin{aligned} H[x(\sigma), x'(\sigma), \sigma] = H[x_*(\sigma), x_*'(\sigma), \sigma] &+ \epsilon \frac{\partial H[x_*(\sigma), x_*'(\sigma), \sigma]}{\partial x_*(\sigma)} \eta(\sigma) \\ &+ \epsilon \frac{\partial H[x_*(\sigma), x_*'(\sigma), \sigma]}{\partial x_*'(\sigma)} \eta'(\sigma) + \epsilon^2 \cdots \end{aligned} \tag{4-6}$$

If this series converges uniformly, then the error index can be written as

$$\begin{aligned} e(t) = e_*(t) + \epsilon \int_t^T \Bigg\{ \eta(\sigma) & \frac{\partial H[x_*(\sigma), x_*'(\sigma), \sigma]}{\partial x_*(\sigma)} \\ &+ \eta'(\sigma) \frac{\partial H[x_*(\sigma), x_*'(\sigma), \sigma]}{\partial x_*'(\sigma)} \Bigg\}\, d\sigma + \epsilon^2 \cdots \end{aligned} \tag{4-7}$$

The term $e_*(t)$ is the value of the error index evaluated on the trajectory $x(\sigma) = x_*(\sigma)$, which presumably minimizes $e(t)$.

The minimization of the error index now can be treated as a problem in differential calculus, where ϵ is treated as an adjustable parameter tending to zero.

The first necessary condition for a minimum is

$$\left.\frac{\partial e(t)}{\partial \epsilon}\right|_{\epsilon=0} = 0 \tag{4-8}$$

when $\eta(\sigma)$ is treated as an arbitrary function. The application of (4-8) to (4-7) yields the condition §

$$\int_t^T \left\{ \eta(\sigma) \frac{\partial H[x_*(\sigma), x'_*(\sigma), \sigma]}{\partial x_*(\sigma)} + \eta'(\sigma) \frac{\partial H[x_*(\sigma), x'_*(\sigma), \sigma]}{\partial x'_*(\sigma)} \right\} d\sigma = 0 \tag{4-9}$$

However, (4-9) cannot be reduced directly to an explicit condition for minimum error because the integrand of (4-9) involves both $\eta(\sigma)$ and $\eta'(\sigma)$. Unlike the previous derivation for the optimum impulse response, this integrand has not yet been factored into a term not involving $\eta(\sigma)$.

This difficulty is resolved when the second term appearing in (4-9) is integrated by parts. This procedure yields

$$\int_t^T \eta'(\sigma) \frac{\partial H[x_*(\sigma), x'_*(\sigma), \sigma]}{\partial x'_*(\sigma)} d\sigma = -\int_t^T \eta(\sigma) \frac{d}{d\sigma} \left\{ \frac{\partial H[x_*(\sigma), x'_*(\sigma), \sigma]}{\partial x'_*(\sigma)} \right\} d\sigma + \eta(\sigma) \left. \frac{\partial H[x_*(\sigma), x'_*(\sigma), \sigma]}{\partial x'_*(\sigma)} \right|_{\sigma=t}^{\sigma=T} \tag{4-10}$$

Then (4-9) becomes

$$\int_t^T \eta(\sigma) \left(\frac{\partial H[x_*(\sigma), x'_*(\sigma), \sigma]}{\partial x_*(\sigma)} - \frac{d}{d\sigma} \left\{ \frac{\partial H[x_*(\sigma), x'_*(\sigma), \sigma]}{\partial x'_*(\sigma)} \right\} \right) d\sigma + \eta(\sigma) \left. \frac{\partial H[x_*(\sigma), x'_*(\sigma), \sigma]}{\partial x'_*(\sigma)} \right|_{\sigma=t}^{\sigma=T} = 0 \tag{4-11}$$

If the integrand appearing in this last equation is finite at $\sigma = t$ and $\sigma = T$, then the contribution to the left-hand side of (4-11) at these end points is due to only the second term. Therefore, this second term must satisfy

$$\eta(\sigma) \left. \frac{\partial H[x_*(\sigma), x'_*(\sigma), \sigma]}{\partial x'_*(\sigma)} \right|_{\sigma=t}^{\sigma=T} = 0 \tag{4-12}$$

This equation sometimes is called the *transversality condition*.¶ Furthermore, the remaining term in (4-11) must vanish for an arbitrary perturbation $\eta(\sigma)$ on the interval $t < \sigma < T$. Therefore, the integrand appearing in (4-11) must vanish independently of $\eta(\sigma)$ on this interval so that

$$\frac{\partial H[x_*(\sigma), x'_*(\sigma), \sigma]}{\partial x_*(\sigma)} - \frac{d}{d\sigma} \left\{ \frac{\partial H[x_*(\sigma), x'_*(\sigma), \sigma]}{\partial x'_*(\sigma)} \right\} = 0 \tag{4-13}$$

must hold. This differential equation is called the *Euler-Lagrange equation.*

The Euler-Lagrange equation is in general nonlinear and is second-order for a first-order dynamic process. This is easily demonstrated by expanding the total

§ More generally, this condition is necessary for a stationary value of the error index. For the present, the assumption is made that H is chosen properly so that only one stationary point exists and this point corresponds to a minimum.

¶ Given here for a restricted class of problems.

time derivative appearing in (4-13) in terms of partial derivatives. This procedure yields

$$\frac{\partial^2 H[x_*(\sigma), x'_*(\sigma), \sigma]}{\partial x'_*(\sigma)^2} x''_*(\sigma) + \frac{\partial^2 H[x_*(\sigma), x'_*(\sigma), \sigma]}{\partial x_*(\sigma)\, \partial x'_*(\sigma)} x'_*(\sigma) = \frac{\partial H[x_*(\sigma), x'_*(\sigma), \sigma]}{\partial x_*(\sigma)} - \frac{\partial^2 H[x_*(\sigma), x'_*(\sigma), \sigma]}{\partial \sigma\, \partial x'_*(\sigma)} \qquad \text{(4-14)}$$

The coefficients in this alternative form of the Euler-Lagrange equation clearly are dependent only upon $x_*(\sigma)$, $x'_*(\sigma)$, and σ. Also, conditions for which the Euler-Lagrange equation is linear are readily deduced from (4-14).

When explicit solutions of the Euler-Lagrange equation are to be found, appropriate boundary conditions are required. If these boundary conditions are specified as $x_*(t)$ and $x'_*(t)$ or as $x_*(T)$ and $x'_*(T)$, then the solution is found as the solution of the common initial-value or *one-point boundary-value problem.* Such problems often are solved on analog computers. However, if the boundary conditions are specified as $x_*(t)$ and $x_*(T)$ or as $x_*(t)$ and $x'_*(T)$, then the more complicated *two-point boundary-value problem* must be solved. Specifically, the Euler-Lagrange equation cannot be solved starting at either $\sigma = t$ or $\sigma = T$ without assuming an initial value for either $x_*(\sigma)$ or $x'_*(\sigma)$, as the case may be.

The boundary conditions for the solution of the Euler-Lagrange equation are defined by (4-12). First, the minimizing function $x_*(\sigma)$ at $\sigma = t$ must correspond to the existing state of the dynamic process. Therefore, one boundary condition is

$$x(t) = x_*(t) \qquad \text{(4-15)}$$

which requires that $\eta(t) = 0$ from the definition given in (4-5). The requirement that the minimizing function $x_*(\sigma)$ have a fixed end point is called a *fixed-point boundary condition.* Because $\eta(t) = 0$, (4-12) reduces to

$$\eta(T) \frac{\partial H[x_*(T), x'_*(T), T]}{\partial x'_*(T)} = 0 \qquad \text{(4-16)}$$

If the fixed-point terminal-boundary condition

$$x(T) = x_*(T) \qquad \text{(4-17)}$$

is imposed, then $\eta(T) = 0$ by definition, and (4-16) is satisfied automatically. On the other hand, if $x_*(T)$ is free to take on any finite value, then $\eta(T)$ is arbitrary. In this case, the so-called *free-point terminal-boundary condition* requires that

$$\frac{\partial H[x_*(T), x'_*(T), T]}{\partial x'_*(T)} = 0 \qquad \text{(4-18)}$$

in order to satisfy (4-16). The boundary conditions given in (4-15) and either (4-17) or (4-18) result in a two-point boundary-value problem and its accompanying difficulties.

Finally, the stationary value of the error index, defined by (4-9), must satisfy an additional condition in order to be considered a minimum value. As in the case of differential calculus, the condition

$$\left.\frac{\partial^2 e(t)}{\partial \epsilon^2}\right|_{\epsilon=0} > 0 \qquad \text{(4-19)}$$

in general must be satisfied if the stationary value corresponds to a minimum value. The application of this condition to (4-7) yields

$$\frac{1}{2}\int_t^T\left\{\eta^2(\sigma)\frac{\partial^2 H[x_*(\sigma), x'_*(\sigma), \sigma]}{\partial x_*(\sigma)^2} + 2\eta(\sigma)\eta'(\sigma)\frac{\partial^2 H[x_*(\sigma), x'_*(\sigma), \sigma]}{\partial x_*(\sigma)\,\partial x'_*(\sigma)} + [\eta'(\sigma)]^2\frac{\partial^2 H[x_*(\sigma), x'_*(\sigma), \sigma]}{\partial x'_*(\sigma)^2}\right\} d\sigma > 0 \qquad (4\text{-}20)$$

A sufficient condition for satisfying (4-20) is a positive integrand everywhere on the interval $t \leqq \sigma \leqq T$ for all perturbations $\eta(\sigma)$ and $\eta'(\sigma)$. A positive integrand is ensured when

$$\frac{\partial^2 H[x_*(\sigma), x'_*(\sigma), \sigma]}{\partial x_*(\sigma)^2} > 0$$

$$\frac{\partial^2 H[x_*(\sigma), x'_*(\sigma), \sigma]}{\partial x'_*(\sigma)^2} > 0 \qquad (4\text{-}21)$$

$$\sqrt{\frac{\partial^2 H[x_*(\sigma), x'_*(\sigma), \sigma]}{\partial x_*(\sigma)^2}\,\frac{\partial^2 H[x_*(\sigma), x'_*(\sigma), \sigma]}{\partial x'_*(\sigma)^2}} - \frac{\partial^2 H[x_*(\sigma), x'_*(\sigma), \sigma]}{\partial x_*(\sigma)\,\partial x'_*(\sigma)} > 0$$

The conditions given in (4-21) are equivalent to the requirement that the matrix of second derivatives of H be positive definite.

The sufficient condition for a minimum value of the error index given in (4-21) is very restrictive and can be weakened considerably. Unfortunately, sufficiency conditions in the calculus of variations are the subject of extensive and difficult mathematical treatments.[37] Furthermore, testing a given error index for sufficiency is difficult if not impossible. Therefore, most of the remaining material presented here is based on the tacit assumption that the Euler-Lagrange equation is not only a necessary condition but also a sufficient condition for a minimum error. The validity of this assumption is implied by requiring the use of error measures possessing the properties of a strictly convex function. Continuous functions that possess the properties given in (4-21) for all $x_*(\sigma)$ and $x'_*(\sigma)$ are *strictly convex functions.*

4-2 Example

The example introduced here is the simplest possible control problem. Such a simple example is chosen so that analytical solutions can be obtained from the minimization procedure. The dynamic process and the error measure are taken to be

$$q(t) = x(t) \qquad x'(t) = m(t)$$
$$h[q(\sigma), m(\sigma), \sigma] = [Q - q(\sigma)]^2 + \rho m^2(\sigma) \qquad (4\text{-}22)$$

For this example, the error measure H is found from appropriate substitutions into h. These substitutions yield

$$H[x_*(\sigma), x'_*(\sigma), \sigma] = [Q - x_*(\sigma)]^2 + \rho[x'_*(\sigma)]^2 \qquad (4\text{-}23)$$

The partial derivatives that appear in the Euler-Lagrange equation are

$$\frac{\partial H}{\partial x_*(\sigma)} = -2[Q - x_*(\sigma)] \qquad \frac{\partial H}{\partial x'_*(\sigma)} = 2\rho x'_*(\sigma)$$
$$\frac{d}{d\sigma}\left[\frac{\partial H}{\partial x'_*(\sigma)}\right] = 2\rho x''_*(\sigma) \tag{4-24}$$

Therefore, (4-13) reduces to

$$x''_*(\sigma) - \frac{1}{\rho}\, x_*(\sigma) = \frac{-Q}{\rho} \tag{4-25}$$

The sufficiency condition for a minimum given in (4-21) is checked easily for this example. The partial derivatives needed are

$$\frac{\partial^2 H}{\partial x_*(\sigma)^2} = 2 \qquad \frac{\partial^2 H}{\partial x'_*(\sigma)^2} = 2\rho \qquad \frac{\partial^2 H}{\partial x_*(\sigma)\, \partial x'_*(\sigma)} = 0 \tag{4-26}$$

Clearly, (4-21) is satisfied for $\rho > 0$.

For this example, the Euler-Lagrange equation is seen to be linear and hence can be solved analytically. A simple method of solving (4-25) is based on the introduction of the Laplace transform

$$x_*(s) = \int_0^\infty x_*(\sigma)e^{-s(\sigma - t)}\, d(\sigma - t) \tag{4-27}$$

When both sides of (4-25) are transformed with respect to $\sigma - t$, the transform of the optimum state signal is found to be

$$x_*(s) = \frac{-Q/\rho}{s(s^2 - 1/\rho)} + x(t)\frac{s}{s^2 - 1/\rho} + x'_*(t)\frac{1}{s^2 - 1/\rho} \tag{4-28}$$

From a partial fraction expansion of (4-28), the inverse transform of $x_*(s)$ is found to be

$$x_*(\sigma) = Q + \tfrac{1}{2}[-Q + x(t) - \rho^{\frac{1}{2}}x'_*(t)]e^{-(\sigma - t)/\rho^{\frac{1}{2}}} + \tfrac{1}{2}[-Q + x(t) + \rho^{\frac{1}{2}}x'_*(t)]e^{(\sigma - t)/\rho^{\frac{1}{2}}} \tag{4-29}$$

Hence, the optimum state signal $x_*(\sigma)$ is expressed analytically in terms of the initial condition $x(t)$ and the optimum initial derivative of this state signal $x'_*(t)$. Because of the unstable form of the Euler-Lagrange equation, the optimum state signal includes a component with an increasing exponential. The Euler-Lagrange equation is always unstable and poses some difficulties, which are discussed later.§

The complete solution for the optimum state signal cannot be obtained until the terminal-boundary condition is introduced. First, the fixed-point terminal-boundary condition

$$x_*(T) = Q \tag{4-30}$$

is assumed. When (4-29) is evaluated at $\sigma = T$ and the relationship given in (4-30) is inserted, the optimum initial value of the derivative of the state signal is found to be

$$x'_*(t) = \frac{Q - x(t)}{\rho^{\frac{1}{2}}} \coth \frac{T - t}{\rho^{\frac{1}{2}}} \tag{4-31}$$

§ An unstable Euler-Lagrange equation does not imply an unstable control equation. The significance of the stability of the Euler-Lagrange equation stems from numerical considerations.

Therefore, the optimum state signal $x_*(\sigma)$ becomes

$$x_*(\sigma) = Q - [Q - x(t)]\left(1 - \coth\frac{T-t}{\rho^{\frac{1}{2}}}\tanh\frac{\sigma - t}{\rho^{\frac{1}{2}}}\right)\cosh\frac{\sigma - t}{\rho^{\frac{1}{2}}} \tag{4-32}$$

and the optimum control signal $m_*(\sigma)$ is found, by differentiating (4-32), to be

$$m_*(\sigma) = \frac{1}{\rho^{\frac{1}{2}}}[Q - x(t)]\left(\coth\frac{T-t}{\rho^{\frac{1}{2}}} - \tanh\frac{\sigma - t}{\rho^{\frac{1}{2}}}\right)\cosh\frac{\sigma - t}{\rho^{\frac{1}{2}}} \tag{4-33}$$

Finally, the optimum value of the control signal in real time is found by specializing (4-34) to $\sigma = t$. This optimum control signal is written as

$$m_*(t) = k(t)[Q - q(t)] \tag{4-34}$$

where the optimum gain $k(t)$ is

$$k(t) = \frac{1}{\rho^{\frac{1}{2}}}\coth\frac{T-t}{\rho^{\frac{1}{2}}} \tag{4-35}$$

The block diagram of this control system is shown in Fig. 4-1. In the derivation of the Euler-Lagrange condition for minimum error, no assumptions are required

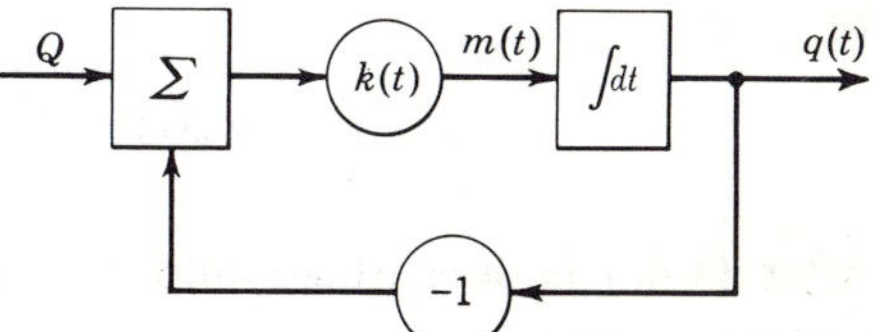

Fig. 4-1 Optimum feedback control for fixed terminal point.

concerning the linearity of the minimizing solution. Therefore, the system shown in Fig. 4-1 is the optimum system for the design problem specified in (4-22) and (4-30), and the system happens to be linear. Also, the solution is optimum for all initial conditions in the dynamic process and for all $t \leqq T$. Therefore, the control system shown in Fig. 4-1 is optimum for all initial conditions. The configuration of the control system shown in Fig. 4-1 is called the *optimum configuration* because the corresponding control equation is optimum for all initial conditions in the system. It should be stressed again that the solution of the minimization problem represented by the block diagram in Fig. 4-1 is not limited conceptually, as are parameter optimization and impulse-response optimization.

For the optimum control system in this example, the minimum value of the error index $e(t)$, now denoted as $e_*(t)$, is

$$e_*(t) = \int_t^T \{[Q - q_*(\sigma)]^2 + \rho m_*^2(\sigma)\}\, d\sigma \tag{4-36}$$

When (4-32) and (4-33) are substituted into (4-36) and the integration is performed, the minimum value of the error index is found to be

$$e_*(t) = \rho^{\frac{1}{2}}[Q - x(t)]^2 \coth\frac{T-t}{\rho^{\frac{1}{2}}} \tag{4-37}$$

If T, t, ρ, and Q are considered to be parameters in the minimization problem, then the minimum value of the error index is mathematically dependent on the square of the initial condition in the dynamic process. This quadratic dependence of the minimum error index on only the initial conditions in the dynamic process is very important and is exploited in the more general optimization problems developed in Chap. 7.

The example specified in (4-22) now is treated with the free-point terminal-boundary condition imposed. From (4-18), the free-point terminal-boundary condition is

$$\left.\frac{\partial H}{\partial x'_*}\right|_{\sigma=T} = 2\rho x'_*(T) = 0 \tag{4-38}$$

Because the free-point terminal-boundary condition is dependent upon the derivative of the optimum state signal, the optimum state signal given in (4-29) is differentiated, resulting in

$$x'_*(\sigma) = \frac{1}{2\rho^{\frac{1}{2}}}[Q - x(t) + \rho^{\frac{1}{2}}x'_*(t)]e^{-(\sigma-t)/\rho^{\frac{1}{2}}} + \frac{1}{2\rho^{\frac{1}{2}}}[-Q + x(t) + \rho^{\frac{1}{2}}x'_*(t)]e^{(\sigma-t)/\rho^{\frac{1}{2}}} \tag{4-39}$$

This relationship now is evaluated at $\sigma = T$. When the free-point terminal-boundary condition is satisfied, the derivative of the optimum state signal at $\sigma = t$ is found to be

$$x'_*(t) = \frac{Q - x(t)}{\rho^{\frac{1}{2}}} \tanh \frac{T-t}{\rho^{\frac{1}{2}}} \tag{4-40}$$

After (4-40) is substituted into (4-29) and (4-39), the optimum state and control signals become

$$x_*(\sigma) = Q - [Q - x(t)]\left(1 - \tanh\frac{T-t}{\rho^{\frac{1}{2}}}\tanh\frac{\sigma-t}{\rho^{\frac{1}{2}}}\right)\cosh\frac{\sigma-t}{\rho^{\frac{1}{2}}} \tag{4-41}$$

and

$$m_*(\sigma) = \frac{1}{\rho^{\frac{1}{2}}}[Q - x(t)]\left(\tanh\frac{T-t}{\rho^{\frac{1}{2}}} - \tanh\frac{\sigma-t}{\rho^{\frac{1}{2}}}\right)\cosh\frac{\sigma-t}{\rho^{\frac{1}{2}}} \tag{4-42}$$

From (4-42) when evaluated at $\sigma = t$, the optimum control signal is given by (4-34), where now the optimum gain is

$$k(t) = \frac{1}{\rho^{\frac{1}{2}}}\tanh\frac{T-t}{\rho^{\frac{1}{2}}} \tag{4-43}$$

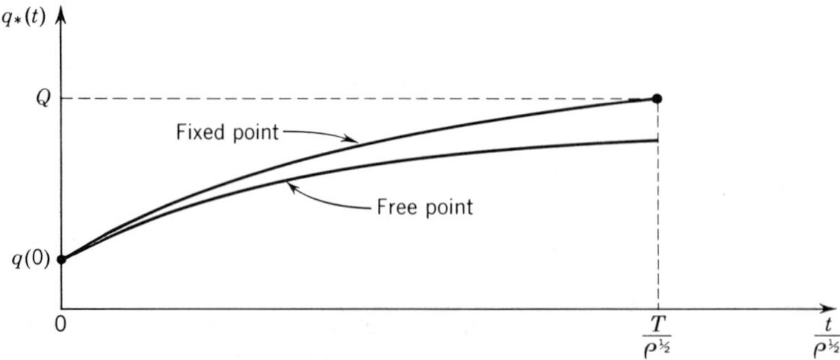

Fig. 4-2 Sketch of response signal for optimum feedback control system.

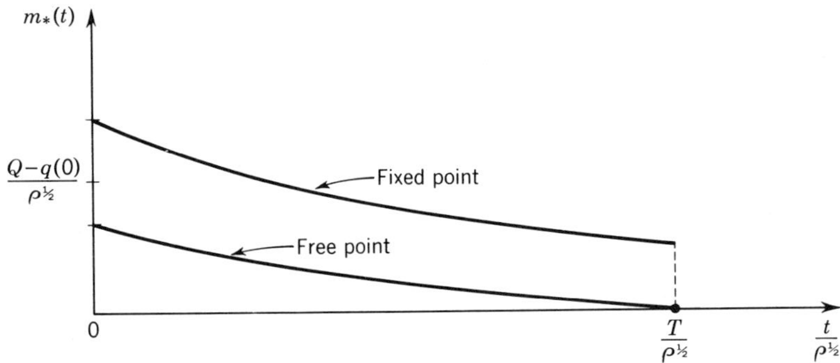

Fig. 4-3 Sketch of control signal for optimum feedback control system.

Finally, the substitutions of (4-41) and (4-42) into (4-36) yield the minimum value of the error index

$$e_*(t) = \rho^{\frac{1}{2}}[Q - x(t)]^2 \tanh \frac{T - t}{\rho^{\frac{1}{2}}} \tag{4-44}$$

A comparison of (4-37) and (4-44) shows that the mathematical form of the minimum values of the error index with respect to $x(t)$ is not dependent upon the terminal-boundary condition.

In order to develop a physical interpretation of the effects of the two types of terminal-boundary conditions, this example is pursued further. The response signal of the optimum control system shown in Fig. 4-1 is sketched in Fig. 4-2 for an initial condition $q(0)$. For the fixed-point terminal-boundary condition, the response trajectory is equal to the desired response at the terminal time, as indicated in Fig. 4-2. On the other hand, the trajectory for the free-point terminal-boundary condition has terminal error. The sketch of the optimum control signal for this example is shown in Fig. 4-3. The optimum control signal for the fixed-point terminal-boundary condition does not equal zero at the terminal point, but the optimum control signal for the free-point terminal-boundary condition does equal zero.

If the ratio R is defined as the ratio of the minimum value of the error index for the fixed-point terminal-boundary condition to the minimum value of the error index for the free-point terminal-boundary condition, then this ratio is

$$R = \coth^2 \frac{T - t}{\rho^{\frac{1}{2}}} \tag{4-45}$$

Equation (4-45) is found by dividing (4-37) by (4-44). This ratio is sketched in Fig. 4-4, and it is seen that the minimum value of the error index for the fixed-point

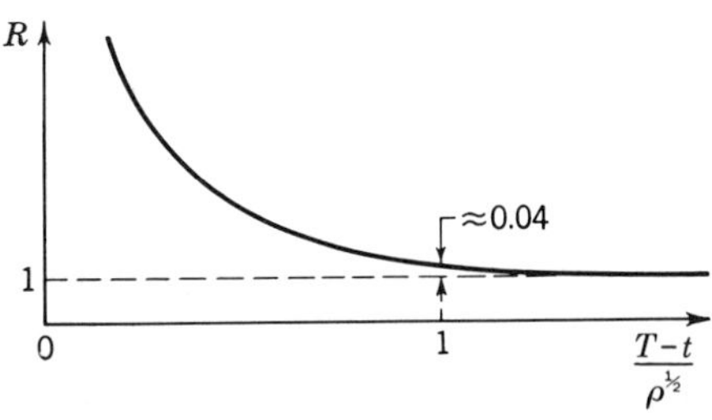

Fig. 4-4 Sketch of the ratio R.

terminal-boundary condition is always greater than the minimum value of the error index for the free-point terminal-boundary condition. Also, the ratio R is independent of the initial condition in the dynamic process. However, as the interval $T - t$ becomes arbitrarily large, the difference between these values of the error index becomes arbitrarily small. The physical interpretation of this asymptotic behavior of R is argued as follows. Imposing a requirement that the terminal value of $x(\sigma)$ be a specific value is a constraint on the behavior of the system response. The introduction of such a constraint necessarily causes a greater value of the error index. For an arbitrarily large time-to-go $T - t$, however, the system response approaches the value Q with little extra control effort required to satisfy $q(T) = Q$. For $T - t = \infty$, imposing this fixed-point terminal condition no longer is a constraint. The sketch of the ratio R is shown in terms of a normalized time-to-go

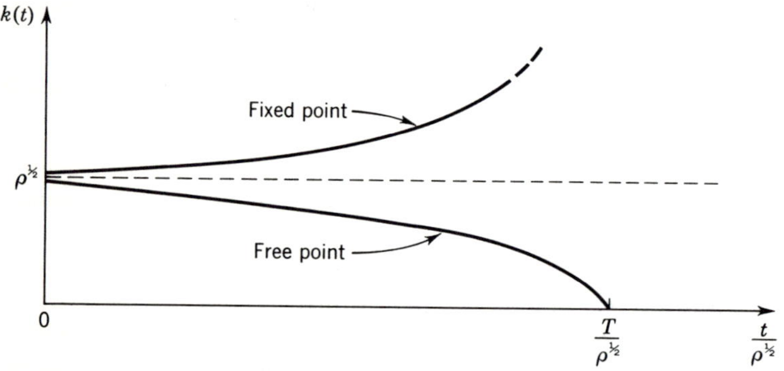

Fig. 4-5 Sketch of loop gain for optimum feedback control system.

which is defined as $(T - t)/\rho^{\frac{1}{2}}$. For a value of the normalized time-to-go approximately equal to 1, this sketch shows that very little difference is caused by the two types of boundary conditions. When the normalized time-to-go is greater than 1, the type of terminal-boundary condition imposed is of little or no consequence. It is significant to note that a normalized time-to-go can be defined which therefore assumes the usefulness of a design invariant. For large allowable control effort, as determined by a small value of ρ, the actual time-to-go can be small and yet the normalized time-to-go is greater than 1. On the other hand, if only small control effort is allowed, that is, ρ is large, then the actual time-to-go must be large in order to have a normalized time-to-go that is greater than 1. The concept of defining design invariants, such as normalized time-to-go in this particular example, is exploited in later work.

Finally, the optimum gain $k(t)$ is sketched in Fig. 4-5 for the two types of terminal-boundary conditions. For the fixed-point terminal-boundary condition, this gain approaches infinity as the interval $T - t$ approaches zero. Of course, if there is any response error when the system has reached the terminal point, this gain causes an infinite value for the control signal. Obviously, this situation may be very undesirable from a control-system standpoint, and saturation will result. Because of measurement noise and other disturbances, a nonzero error at the terminal point cannot be avoided and hence saturation will always occur. On the other hand, the free-point terminal-boundary condition gives rise to a gain $k(t)$ that decreases to zero as the terminal time is approached. The fact that the optimum gain for the free-point terminal-boundary condition goes to zero at the terminal point is reasonable from a physical standpoint. If the time-to-go $T - t$ becomes very small, then only extremely large values of the control signal can cause any appreciable change in the state of the dynamic process. Because the terminal state of the dynamic process is not required to be a particular value, the error index is minimized for small time-to-go by simply reducing the control signal to zero.

The final point of interest is that the optimum gain $k(t)$ for both types of boundary conditions is proportional to $\rho^{-\frac{1}{2}}$. If no weighting of the control signal had been included in the error measure, then the gain of the optimum system would be infinite and saturation would occur in the physical system. This phenomenon leads one to a tentative conclusion. If the control signal is not weighted in the error measure, then the optimum control system always is of the saturating variety. The validity of this conclusion is demonstrated in later chapters. Using an infinite gain in the control system and neglecting the possibility of saturation in the dynamic process are incompatible. Therefore, the control signal is always weighted in the error measure except in specific problems where a saturating controller is desired.

4-3 Extension of the elementary theory to an *N*th-order dynamic process

This section is a direct extension of the previous results for a first-order dynamic process. Here the response and control signals are expressed as

$$q(t) = G[x(t), x'(t), \ldots, x^{(N)}(t), t] \tag{4-46}$$

and

$$m(t) = F[x(t), x'(t), \ldots, x^{(N)}(t), t] \tag{4-47}$$

However, the error index used here is given by (4-3). For the present, the problem of reducing the differential equations defining the dynamic process to the form of (4-46) and (4-47) is not considered. The substitution of (4-46) and (4-47) into (4-3) yields

$$e(t) = \int_t^T H[x(\sigma), x'(\sigma), \ldots, x^{(N)}(\sigma), \sigma]\, d\sigma \tag{4-48}$$

which is the form of the error index treated here. The error measure H is assumed to be continuous and a strictly convex function of $x(\sigma)$ and its N derivatives.

The minimization is performed by first assuming

$$x^{(n)}(\sigma) = x_*^{(n)}(\sigma) + \epsilon\eta^{(n)}(\sigma) \qquad n = 0, 1, 2, \ldots, N \tag{4-49}$$

Then the error measure H is expanded in a Taylor series, and the application of (4-8) to this result yields

$$\int_t^T \sum_{n=0}^{N} \eta^{(n)}(\sigma) \frac{\partial H}{\partial x_*^{(n)}(\sigma)}\, d\sigma = 0 \tag{4-50}$$

Also, successive integrations by parts of the nth term in (4-50) gives

$$\int_t^T \eta^{(n)}(\sigma) \frac{\partial H}{\partial x_*^{(n)}(\sigma)}\, d\sigma = (-1)^n \int_t^T \eta(\sigma) \frac{d^n}{d\sigma^n} \frac{\partial H}{\partial x_*^{(n)}(\sigma)}\, d\sigma$$
$$+ \sum_{j=1}^{n} \left[(-1)^{j-1} \eta^{(n-j)}(\sigma) \frac{d^{j-1}}{d\sigma^{j-1}} \frac{\partial H}{\partial x_*^{(n)}(\sigma)} \right]_{\sigma=t}^{\sigma=T} \tag{4-51}$$

The condition for minimum error becomes

$$\int_t^T \eta(\sigma)\left[\sum_{n=0}^{N}(-1)^n \frac{d^n}{d\sigma^n}\frac{\partial H}{\partial x_*^{(n)}(\sigma)}\right] d\sigma + \sum_{n=1}^{N}\sum_{j=1}^{n}\left[(-1)^{j-1}\eta^{(n-j)}(\sigma)\frac{d^{j-1}}{d\sigma^{j-1}}\frac{\partial H}{\partial x_*^{(n)}(\sigma)}\right]_{\sigma=t}^{\sigma=T} = 0 \qquad (4\text{-}52)$$

when (4-51) is substituted into (4-50). Therefore, the transversality condition is

$$\sum_{n=1}^{N}\sum_{j=1}^{n}\left[(-1)^{j-1}\eta^{(n-j)}(\sigma)\left[\frac{d^{j-1}}{d\sigma^{j-1}}\frac{\partial H}{\partial x_*^{(n)}(\sigma)}\right]\right]_{\sigma=t}^{\sigma=T} = 0 \qquad (4\text{-}53)$$

which is rewritten as

$$\sum_{m=1}^{N}\left[\eta^{(m-1)}(\sigma)\sum_{n=m}^{N}(-1)^{n-m}\frac{d^{n-m}}{d\sigma^{n-m}}\frac{\partial H}{\partial x_*^{(n)}(\sigma)}\right]_{\sigma=t}^{\sigma=T} = 0 \qquad (4\text{-}54)$$

by collecting the coefficients of $\eta^{(m-1)}(\sigma)$. Finally, the Euler-Lagrange equation is found from (4-52) to be

$$\sum_{n=0}^{N}(-1)^n \frac{d^n}{d\sigma^n}\frac{\partial H}{\partial x_*^{(n)}(\sigma)} = 0 \qquad (4\text{-}55)$$

The boundary conditions on the solution of the Euler-Lagrange equation are found readily from the transversality condition as expressed in (4-54). The existing state of the dynamic process is defined by the state variables $x(t), x'(t), \ldots, x^{(N-1)}(t)$. Therefore, the fixed-point initial-boundary conditions are

$$x^{(n)}(t) = x_*^{(n)}(t) \qquad n = 0, 1, \ldots, N-1 \qquad (4\text{-}56)$$

and (4-54) is satisfied at $\sigma = t$ because $\eta^{(m-1)}(t) = 0$ for $m = 1, 2, \ldots, N$. At the terminal point $\sigma = T$, either $\eta^{(m-1)}(t) = 0$ for a fixed-point terminal-boundary condition or $\eta^{(m-1)}(T)$ is arbitrary for a free-point terminal-boundary condition. Hence, each term of the sum over m in (4-54) must vanish independently of other terms so that

$$\eta^{(m-1)}(T)\sum_{n=m}^{N}(-1)^{n-m}\frac{d^{n-m}}{dT^{n-m}}\frac{\partial H}{\partial x_*^{(n)}(T)} = 0 \qquad m = 1, 2, \ldots, N \qquad (4\text{-}57)$$

For the fixed-point terminal-boundary condition

$$x^{(m-1)}(T) = x_*^{(m-1)}(T) \qquad 1 \leq m \leq N \qquad (4\text{-}58)$$

(4-57) is satisfied because $\eta^{(m-1)}(T) = 0$. For the free-point terminal-boundary condition where $x_*^{(m-1)}(T)$ is arbitrary, (4-57) is satisfied if

$$\sum_{n=m}^{N}(-1)^{n-m}\frac{d^{n-m}}{dT^{n-m}}\frac{\partial H}{\partial x_*^{(n)}(T)} = 0 \qquad 1 \leq m \leq N \qquad (4\text{-}59)$$

The optimum control signal is found by replacing $x^{(N)}(t)$ by $x_*^{(N)}(t)$ in (4-47), thereby obtaining

$$m_*(t) = F[x(t), x'(t), \ldots, x^{(N-1)}(t), x_*^{(N)}(t), t] \qquad (4\text{-}60)$$

The initial value $x_*^{(N)}(t)$ corresponds to the solution of the Euler-Lagrange equation that satisfies the required initial- and terminal-boundary conditions of the design problem. If only the existing values of the state variables, namely $x(t), x'(t), \ldots,$

$x^{(N-1)}(t)$, plus real time t are treated as variables, then the solution of the Euler-Lagrange equation is written as

$$x_*(\sigma) = X_0[x(t), x'(t), \ldots, x^{(N-1)}(t); t, \sigma] \tag{4-61}$$

This function is written under the assumption that the coefficients of H, the terminal-boundary conditions, and the terminal time T are design-problem parameters and hence are fixed. Successive partial differentiations of (4-61) yield

$$x_*^{(N)}(\sigma) = X_N[x(t), x'(t), \ldots, x^{(N-1)}(t); t, \sigma] \tag{4-62}$$

where the definition

$$X_N = \frac{\partial^N X_0}{\partial \sigma^N} \tag{4-63}$$

has been introduced. By substituting (4-62), when evaluated at $\sigma = t$, into (4-60), the optimum control equation can be written as

$$m_*(t) = P[x(t), x'(t), \ldots, x^{(N-1)}(t), t] \tag{4-64}$$

At least in theory, the optimum control equation can be found for the optimum feedback control system constructed from the measurement of the process state signals. However, the steps from finding the time function $x_*^{(N)}(\sigma)$ to finding the control equation P are not straightforward, in general, and the steps require extensive further study. These steps constitute the *system-synthesis problem* in optimization theory. Furthermore, finding the numerical solution to the Euler-Lagrange equation that satisfies the required boundary conditions is no simple matter. These numerical considerations constitute the *two-point boundary-value problem* in optimization theory.

4-4 Linear optimum controls

In Sec. 4-3, the solution of the Euler-Lagrange equation subject to appropriate boundary conditions is indicated in implied form. Here, the feasibility of obtaining numerical solutions of the Euler-Lagrange equation for a class of design problems is investigated. This class of problems is characterized by error measures of the form

$$H[x_*(\sigma), x_*'(\sigma), \ldots, x_*^{(N)}(\sigma), \sigma] = f(\sigma) + \sum_{n=0}^{N} f_n(\sigma) x_*^{(n)}(\sigma) + \sum_{n=0}^{N} \sum_{m=0}^{N} f_{nm}(\sigma) x_*^{(n)}(\sigma) x_*^{(m)}(\sigma) \tag{4-65}$$

where $f_{nm}(\sigma) = f_{mn}(\sigma)$. This quadratic form always results when the dynamic process is linear and the error measure h, appearing in (4-3), is quadratic with respect to $q(\sigma)$ and $m(\sigma)$. The partial derivatives of (4-65) that are required in the Euler-Lagrange equation are

$$\frac{\partial H}{\partial x_*^{(n)}(\sigma)} = f_n(\sigma) + 2 \sum_{m=0}^{N} f_{nm}(\sigma) x_*^{(m)}(\sigma) \qquad n = 0, 1, \ldots, N \tag{4-66}$$

If the appropriate time derivatives of (4-66) are computed and if they are substituted into the Euler-Lagrange equation given in (4-55), the result is of the form

$$\sum_{n=0}^{2N} a_n(\sigma) x_*^{(n)}(\sigma) = M(\sigma) \tag{4-67}$$

The time-varying a coefficients appearing in (4-67) are combinations of the f functions which appear in (4-65) and their derivatives. Similarly, the driving function $M(\sigma)$ is found from these f functions and their derivatives. Therefore, when the error measure H is a quadratic with respect to the dynamic-process state variables and $x^{(N)}(\sigma)$, the Euler-Lagrange equation is a linear differential equation of order $2N$.

Now the feasibility of solving a linear differential equation when the boundary conditions are imposed at two points in time is examined. By the use of superposition, the variable $x_*(\sigma)$ can be expressed as a linear combination of the initial conditions occurring in the solution of (4-67). The solution of (4-67) is of the form

$$x_*(\sigma) = g(\sigma) + \sum_{n=0}^{N-1} g_n(\sigma)x^{(n)}(t) + \sum_{n=N}^{2N-1} g_n(\sigma)x_*^{(n)}(t) \tag{4-68}$$

In (4-68), the initial conditions appearing in the second sum are not known and must be determined from the terminal-boundary conditions. Equation (4-68) is differentiated successively, and the state signals are evaluated at $\sigma = T$. Therefore, the terminal-boundary conditions are expressed in terms of (4-68) as

$$x_*^{(m)}(T) = g^{(m)}(T) + \sum_{n=0}^{N-1} g_n^{(m)}(T)x^{(n)}(t) + \sum_{n=N}^{2N-1} g_n^{(m)}(T)x_*^{(n)}(t)$$
$$m = 0, 1, \ldots, N-1 \tag{4-69}$$

These N equations for the terminal-boundary conditions contain N unknown initial conditions.

Because the assumed form of the error measure H leads to a unique solution to the minimization problem, the solution for the N unknown initial conditions also is unique. Therefore, the properties of matrices can be used directly to solve (4-69) for the unknown initial conditions. If the matrix $[G]$ is defined as

$$[G] = \begin{bmatrix} g_N(T) & g_{N+1}(T) & \cdots & g_{2N-1}(T) \\ g'_N(T) & g'_{N+1}(T) & \cdots & g'_{2N-1}(T) \\ \cdot & \cdot & \cdot & \cdot \\ g_N^{(N-1)}(T) & g_{N+1}^{(N-1)}(T) & \cdots & g_{2N-1}^{(N-1)}(T) \end{bmatrix} \tag{4-70}$$

then the unknown initial condition $x_*^{(N)}(t)$ is found by a matrix inversion and is expressed as

$$x_*^{(N)}(t) = \sum_{m=0}^{N-1} \frac{|G|_m}{|G|}\left[x_*^{(m)}(T) - g^{(m)}(T) - \sum_{n=0}^{N-1} g_n^{(m)}(T)x^{(n)}(t)\right] \tag{4-71}$$

The notations $|G|$ and $|G|_m$ used in (4-71) denote the determinant of the $[G]$ matrix and the cofactor of the element $g_N^{(m)}(T)$ in the $[G]$ matrix, respectively. In this class of problems, the unknown initial condition $x_*^{(N)}(t)$ therefore can be expressed as a linear combination of the existing initial conditions in the dynamic process.

If the function F of the state equation given in (4-47) is also the linear function of the initial conditions and $x^{(N)}(t)$,

$$m(t) = l(t) + \sum_{n=0}^{N} l_n(t)x^{(n)}(t) \tag{4-72}$$

then the optimum control equation is found by substituting (4-71) into (4-72). This procedure results in the linear optimum control equation

$$m_*(t) = L(t) + \sum_{n=0}^{N-1} L_n(t)x^{(n)}(t) \tag{4-73}$$

Because the dynamic process is linear in this case as specified by (4-72), the optimum control system is linear with respect to all initial conditions. In this class of problems, the time functions $L(t)$ and $L_n(t)$, $n = 0, 1, \ldots, N - 1$, define the optimum configuration of the control system. Also, these time functions are optimum for all possible initial conditions in the dynamic process.

If the parameters of the optimum control equation given in (4-73) are to be found numerically, a sequence of computations for the g functions and their $N - 1$ derivatives appearing in (4-69) must be specified. Because the Euler-Lagrange equation is linear, superposition also is applied in the computation. From (4-69), the numerical values of the g functions and their derivatives are given by

$$g^{(m)}(T) = x_*{}^{(m)}(T) \qquad m = 0, 1, \ldots, N - 1; \text{ all } X^{(n)}(t) = 0 \tag{4-74}$$

$$g_l{}^{(m)}(T) = \frac{x_*{}^{(m)}(T)}{X^{(l)}(t)} \qquad m, l = 0, 1, \ldots, N - 1; \text{ all } X^{(n)}(t) = 0 \text{ except } n = l \tag{4-75}$$

where
$$X^{(l)}(t) = \begin{cases} x^{(l)}(t) & l = 0, 1, \ldots, N - 1 \\ x_*{}^{(l)}(t) & l = N, N + 1, \ldots, 2N - 1 \end{cases} \tag{4-76}$$

Equations (4-74) and (4-75) specify a set of $N + 1$ solutions of (4-67) that are required to compute the g functions and their derivatives. At the end of each computer solution, that is, at $\sigma = T$, this solution specifies one g function and its $N - 1$ derivatives. By this method, the linear two-point boundary-value problem is solved as a sequence of $N + 1$ computer solutions. However, any set of $N + 1$ computer solutions can be used for finding the g functions and their $N - 1$ derivatives provided that the initial conditions pertaining to (4-67) are chosen in sets that are linearly independent. The computer solutions specified by (4-74) and (4-75) are particularly convenient for (4-69).

An important difficulty arises in the solution of the linear two-point boundary-value problem in the special case $T = \infty$. The Euler-Lagrange equation is always unstable, a fact that can be proved from sufficiency conditions for a unique minimum. An example of this instability is seen in (4-25). Because of this instability, the solution of (4-67) over an infinite time interval generally causes infinite values for the optimum state signal and its $2N - 1$ derivatives. Therefore, the g functions appearing in (4-68) also become infinite at $\sigma = \infty$. This instability gives rise to an impossible situation in obtaining the exact solution for the case $T = \infty$. However, if T is made arbitrarily large, then in theory the solutions for the parameters appearing in (4-73), found by imposing boundary conditions at $\sigma = T$, approximate the exact optimum with an accuracy that increases with the increasing of T. The results of the simple example given in Sec. 4-2, as depicted in Fig. 4-4, suggest this method of approximation. However, the matrix inversion required in (4-71) becomes increasingly difficult with increasing values of T because small differences between large quantities occur. These difficulties can be eliminated with the alternative formulation of linear optimum controls presented in Chap. 7.

4-5 Nonlinear optimum controls for first-order dynamic processes

In Sec. 4-4, a method for solving linear Euler-Lagrange equations is developed. This method is based on the superposition principle, and a straightforward method for computing the time functions which appear in the optimum control law is obtained. However, in the case of nonlinear Euler-Lagrange equations, the principle of superposition cannot be utilized. Because boundary conditions occur at the initial point $\sigma = t$ and the terminal point $\sigma = T$, the solution of nonlinear Euler-Lagrange equations must be a trial-and-error process. When a computer is used for solutions, the initial conditions must be selected and then the equations are solved in order to check the terminal-boundary conditions. For a given set of assumed initial conditions, in general, the terminal-boundary conditions are not satisfied. Therefore, a new set of initial-boundary conditions must be specified and another computer solution obtained. This trial-and-error process, called *boundary-condition iteration*, may be very difficult when the order of the dynamic process is large. For these difficult cases, a search procedure is needed which has the property of rapid convergence to the selection of the correct initial conditions. However, search techniques are not introduced here. Instead, they are introduced when a more complete understanding of alternative forms of the condition for minimum error are developed.

However, analytical solutions can be obtained for a limited class of nonlinear problems when the dynamic process is first-order.[38] Because study of nonquadratic forms of H gives insight into the selection of appropriate error measures for particclar deșign problems, this limited class of nonlinear optimum controls is discussed here in detail. For a first-order dynamic process, $N = 1$, the Euler-Lagrange equation given in (4-55) becomes

$$\frac{d}{d\sigma}\left(\frac{\partial H}{\partial x'_*}\right) - \frac{\partial H}{\partial x_*} = 0 \tag{4-77}$$

In order to obtain an analytical solution for (4-77), the total time derivative of H is written in expanded form as

$$\frac{dH}{d\sigma} = \frac{\partial H}{\partial \sigma} + x'_* \frac{\partial H}{\partial x_*} + x''_* \frac{\partial H}{\partial x'_*} \tag{4-78}$$

The result of substituting (4-77) into (4-78) is written as

$$\frac{dH}{d\sigma} = \frac{\partial H}{\partial \sigma} + \frac{d}{d\sigma}\left(x'_* \frac{\partial H}{\partial x'_*}\right) \tag{4-79}$$

If the assumption is made that $\partial H/\partial \sigma = 0$, then (4-79) is rewritten as

$$\frac{d}{d\sigma}\left(H - x'_* \frac{\partial H}{\partial x'_*}\right) = 0 \tag{4-80}$$

Because (4-80) involves a total time derivative, the integration can be performed over the interval $[t, T]$, and the result is

$$x'_*(t) = \left.\frac{H - C}{\partial H/\partial x'_*}\right|_{\sigma=t} \tag{4-81}$$

where

$$C = \left(H - x'_* \frac{\partial H}{\partial x'_*}\right)_{\sigma=T} \tag{4-82}$$

If this constant of integration can be found without solving a nonlinear differential equation, then (4-81) directly gives a solution for the optimum initial value of the derivative of the state signal.

In order to investigate the solution of (4-81), the following special case is assumed:

$$H[x_*(\sigma), x'_*(\sigma), \sigma] = \left[\frac{Q - x_*(\sigma)}{E}\right]^{2n} + \frac{1}{2m-1}\left[\frac{x'_*(\sigma)}{L}\right]^{2m} \tag{4-83}$$

If the interval where control effort is applied is infinite, that is, $T = \infty$, then the terminal value for the optimum state signal is equal to the desired state signal Q. Also, the derivative of the optimum state signal is zero at the terminal point because the optimum state signal is a constant at $\sigma = \infty$. Therefore, the constant C vanishes, so that

$$C = 0 \qquad \text{for } T = \infty \tag{4-84}$$

and the initial derivative of the optimum state signal can be evaluated readily from (4-81). The partial derivative appearing in (4-81) is

$$\frac{\partial H}{\partial x'_*} = \frac{2m}{2m-1}\frac{1}{L}\left[\frac{x'_*(\sigma)}{L}\right]^{2m-1} \tag{4-85}$$

for this example. Finally, appropriate substitutions into (4-81) yield the initial derivative of the optimum state signal

$$x'_*(t) = \pm L\left|\frac{Q - x_*(t)}{E}\right|^{n/m} \tag{4-86}$$

The plus sign appearing in (4-86) is used for $[Q - x^*(t)] > 0$, and the minus sign is used for $[Q - x^*(t)] < 0$ in order to have decreasing errors. Also, the optimum derivative of the initial state signal is zero for $[Q - x^*(t)] = 0$ in order to maintain zero errors.

The implications of (4-86) are examined in terms of a control system for the following dynamic process:

$$m(t) = x'(t) \qquad q(t) = x(t) \tag{4-87}$$

Now the optimum control signal can be written as

$$m_*(t) = \mathscr{L}[e_q(t)] \tag{4-88}$$

where

$$e_q(t) = Q - q(t) \tag{4-89}$$

and

$$\mathscr{L}[e_q(t)] = \begin{cases} L\left|\dfrac{e_q(t)}{E}\right|^{n/m} & e_q(t) > 0 \\ 0 & e_q(t) = 0 \\ -L\left|\dfrac{e_q(t)}{E}\right|^{n/m} & e_q(t) < 0 \end{cases} \tag{4-90}$$

This form for the optimum control signal corresponds to (4-86). The block diagram of the optimum feedback control system for these design conditions is shown in Fig. 4-6, where the nonlinear gain $\mathscr{L}$ is given by (4-90). Limiting conditions for this gain, as determined by the ratio n/m, are shown in Fig. 4-7. For $n/m = 1$, the optimum feedback control system is linear. This value for the ratio n/m could correspond to an error measure where the weighting of both response errors and control effort is quadratic. On the other hand, this value for the ratio could correspond to quartic weighting of both response errors and control effort. In the limiting case $n/m \to \infty$, the optimum feedback control system approaches a dead-zone controller on the interval $[-E, E]$ of the system error. On the other hand, for the ratio $n/m \to 0$, the optimum feedback control system approaches a saturation control system.

The limiting cases $n/m = \infty$ and $n/m = 0$ have not been derived on a strictly rigorous mathematical basis. In particular, for $n/m = \infty$, the dead zone may cause no decrease in system response error, depending on the initial response error. In cases where the response error is not decreased to zero, the evaluation of C given in (4-84) is incorrect. Also, the condition $n/m = 0$ is difficult to treat because the error index is not strictly convex. However, these two limiting cases can be approximated with arbitrary accuracy by choosing the ratio n/m either arbitrarily large or arbitrarily small.

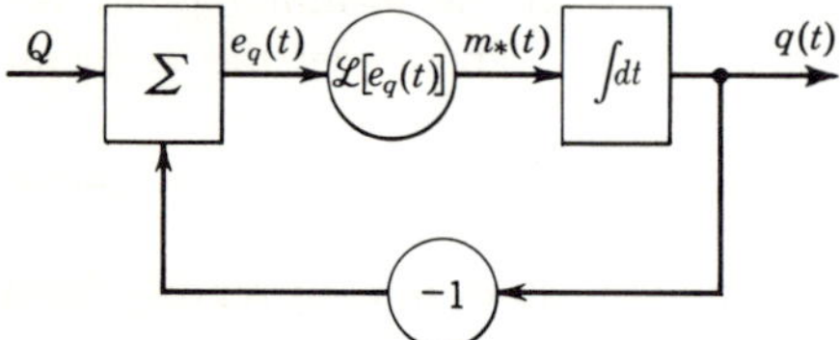

Fig. 4-6 Block diagram of optimum feedback control system.

The important point of this example is that the controller characteristics can be selected more or less at will by selecting the form of the error measure. The characteristics of the controller range from a dead-zone controller to a saturating-type controller. Also of importance is the fact that the relative weighting of response errors and control effort determines the characteristics of the optimum control system. For the mathematical form assumed in (4-83), the relative degree of the weighting of these two terms, as given by n/m, determines the characteristics of the controller. For instance, the limiting case $n/m \to \infty$ is achieved by making either n arbitrarily large or m arbitrarily small. Therefore, the desired characteristics of the controller generally are obtained by quadratic weighting of either response errors or control effort and by quartic or higher-degree weighting of the remaining terms in the error measure.

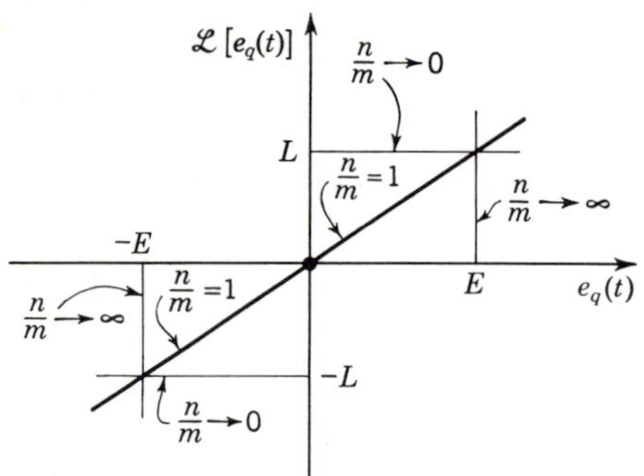

Fig. 4-7 Loop-gain characteristics of optimum feedback control system.

The reader should be careful not to extend the conclusions discussed here for a first-order dynamic process to dynamic processes of higher order without further considerations. In particular, the saturating-type controllers for higher-order systems are dependent specifically upon the form of function used for weighting response errors. These considerations are discussed in a subsequent chapter.

4-6 Hamiltonian formulation of the calculus of variations

The successful use of the variational-calculus method developed in Sec. 4-3 is based on the assumption that the error measure can be expressed explicitly in terms of the state signal and its N derivatives. This assumption is implied by the use of (4-46) and (4-47). Generally speaking, however, this mathematical description of the dynamic process cannot be obtained. As discussed in Chap. 1, the response of the dynamic process more generally is expressed as

$$\mathbf{q}(t) = \mathbf{g}[\mathbf{x}(t), t] \tag{4-91}$$

Also, the equation of state for the dynamic process generally is expressed as

$$\mathbf{x}'(t) = \mathbf{f}[\mathbf{x}(t), \mathbf{m}(t), t] \tag{4-92}$$

where the order N of the dynamic process is equal to the number of components of the state vector $\mathbf{x}(t)$. If the error index is defined as

$$e(t) = \int_t^T h[\mathbf{q}(\sigma), \mathbf{m}(\sigma), \sigma]\, d\sigma \tag{4-93}$$

then the error measure H cannot be found explicitly as assumed in (4-48). Therefore, this section is devoted to an alternative formulation of the minimizing condition for the more general design problem defined by (4-91) to (4-93).

First, the desired-response signals are eliminated from (4-93) by substituting (4-91) into this expression. Then, the error index minimized in this formulation is defined as

$$e(t) = \int_t^T H[\mathbf{x}(\sigma), \mathbf{m}(\sigma), \sigma]\, d\sigma \tag{4-94}$$

Because this error index cannot be written explicitly in terms of a single state signal and its N derivatives, (4-94) is minimized subject to the constraints imposed by (4-92). Specifically, the constrained error index

$$e_c(t) = \int_t^T H_c[\mathbf{x}(\sigma), \mathbf{m}(\sigma), \boldsymbol{\lambda}(\sigma), \sigma]\, d\sigma \tag{4-95}$$

is minimized, where the error measure H_c is written in terms of the Lagrange multipliers $\lambda_n(\sigma)$ as

$$H_c[\mathbf{x}(\sigma), \mathbf{m}(\sigma), \boldsymbol{\lambda}(\sigma), \sigma] = H[\mathbf{x}(\sigma), \mathbf{m}(\sigma), \sigma] + \sum_{n=1}^{N} \lambda_n(\sigma)\{f_n[\mathbf{x}(\sigma), \mathbf{m}(\sigma), \sigma] - x_n'(\sigma)\} \tag{4-96}$$

The minimization of (4-94) when (4-92) is satisfied is equivalent to minimizing (4-95) when the Lagrange multipliers are adjusted so that (4-92) is satisfied.[39] In other words, the error measure H_c is equal to the error measure H for values of the Lagrange multipliers that require the terms inside the braces of (4-96) to vanish.

The condition for minimum error now is developed by introducing definitions for the optimum state signals similar to those used in Sec. 4-3. The optimum state signals are defined in terms of

$$x_m(\sigma) = x_m^*(\sigma) + \epsilon_m \eta_m(\sigma) \qquad m = 1, 2, \ldots, N \tag{4-97}$$

and hence the derivatives of the optimum state signals are expressed in terms of

$$x_m'(\sigma) = x_m^{*\prime}(\sigma) + \epsilon_m \eta_m'(\sigma) \qquad m = 1, 2, \ldots, N \tag{4-98}$$

Similarly, the optimum control signals and the optimum Lagrange multipliers are defined as independent variables in terms of

$$m_m(\sigma) = m_m^*(\sigma) + \delta_m\mu_m(\sigma) \qquad m = 1, 2, \ldots, M \tag{4-99}$$

and

$$\lambda_m(\sigma) = \lambda_m^*(\sigma) + \zeta_m\xi_m(\sigma) \qquad m = 1, 2, \ldots, N \tag{4-100}$$

Now the error measure H_c is expanded in a Taylor series about the value H_c^*. This Taylor series expansion, with time-function notation dropped for brevity, is expressed as

$$H_c = H_c^* + \sum_{m=1}^{N}\left(\epsilon_m\eta_m\frac{\partial H_c^*}{\partial x_m^*} + \epsilon_m\eta_m'\frac{\partial H_c^*}{\partial x_m^{*\prime}} + \zeta_m\xi_m\frac{\partial H_c^*}{\partial \lambda_m^*}\right) + \sum_{m=1}^{M}\left(\delta_m\mu_m\frac{\partial H_c^*}{\partial m_m^*}\right) + \cdots \tag{4-101}$$

Because the perturbations $\eta_m(\sigma)$, $\mu_m(\sigma)$, and $\xi_m(\sigma)$ are treated as independent, the conditions for a minimum value of the error index e_c are

$$\left.\frac{\partial e_c(t)}{\partial \epsilon_k}\right|_{\substack{\epsilon=0\\ \delta=0\\ \zeta=0}} = 0 \qquad \left.\frac{\partial e_c(t)}{\partial \delta_k}\right|_{\substack{\epsilon=0\\ \delta=0\\ \zeta=0}} = 0 \qquad \left.\frac{\partial e_c(t)}{\partial \zeta_k}\right|_{\substack{\epsilon=0\\ \delta=0\\ \zeta=0}} = 0 \tag{4-102}$$

These conditions, when applied to (4-95) and (4-101), become

$$\int_t^T \left(\eta_k\frac{\partial H_c^*}{\partial x_k^*} + \eta_k'\frac{\partial H_c^*}{\partial x_k^{*\prime}}\right) d\sigma = 0 \qquad k = 1, 2, \ldots, N \tag{4-103}$$

$$\int_t^T \mu_k\frac{\partial H_c^*}{\partial m_k^*}\, d\sigma = 0 \qquad k = 1, 2, \ldots, M \tag{4-104}$$

and

$$\int_t^T \xi_k\frac{\partial H_c^*}{\partial \lambda_k^*}\, d\sigma = 0 \qquad k = 1, 2, \ldots, N \tag{4-105}$$

The first condition given in (4-103) involves both the perturbation and the derivative of this perturbation, and therefore the second term in (4-103) is integrated by parts. The result is

$$\int_t^T \eta_k\left[\frac{\partial H_c^*}{\partial x_k^*} - \frac{d}{d\sigma}\left(\frac{\partial H_c^*}{\partial x_k^{*\prime}}\right)\right]d\sigma + \left.\eta_k\frac{\partial H_c^*}{\partial x_k^{*\prime}}\right|_{\sigma=t}^{\sigma=T} = 0 \qquad k = 1, 2, \ldots N \tag{4-106}$$

Furthermore, the terms not inside the integral of (4-106) must vanish so that

$$\left.\eta_k\frac{\partial H_c^*}{\partial x_k^{*\prime}}\right|_{\sigma=t}^{\sigma=T} = 0 \qquad k = 1, 2, \ldots, N \tag{4-107}$$

Finally, (4-106) is satisfied for an arbitrary perturbation η_k if

$$\frac{\partial H_c^*}{\partial x_k^*} - \frac{d}{d\sigma}\left(\frac{\partial H_c^*}{\partial x_k^{*\prime}}\right) = 0 \qquad k = 1, 2, \ldots, N \tag{4-108}$$

Similarly, (4-104) and (4-105) are satisfied for arbitrary perturbations if

$$\frac{\partial H_c^*}{\partial m_k^*} = 0 \qquad k = 1, 2, \ldots, M \tag{4-109}$$

and

$$\frac{\partial H_c^*}{\partial \lambda_k^*} = 0 \qquad k = 1, 2, \ldots, N \tag{4-110}$$

Equations (4-108) to (4-110) are the first necessary condition for a minimum value of the error index, and they are analogous to the Euler-Lagrange equation developed in Sec. 4-3.

Equation (4-107) specifies the boundary conditions of the optimum solution for the particular design problem at hand. From the mathematical form of the constrained error measure H_c, (4-109) is seen to give algebraic relationships for the optimum control signals. If the values of the optimum control signals are inserted into (4-108) and (4-110), then these equations are solved as a set of $2N$ first-order differential equations.

In order to demonstrate the solution of this condition for a minimum, the following simple example is assumed:

$$q_1(\sigma) = x_1(\sigma) \qquad x_1'(\sigma) = m_1(\sigma)$$
$$h[\mathbf{q}(\sigma), \mathbf{m}(\sigma), \sigma] = [Q - q_1(\sigma)]^2 + m_1^2(\sigma) \tag{4-111}$$

Here, the error measure H is

$$H[\mathbf{x}(\sigma), \mathbf{m}(\sigma), \sigma] = [Q - x_1(\sigma)]^2 + m_1^2(\sigma) \tag{4-112}$$

and the minimizing value H_c^* of the constrained error measure is

$$H_c^* = [Q - x_1^*(\sigma)]^2 + [m_1^*(\sigma)]^2 + \lambda_1^*(\sigma)[m_1^*(\sigma) - x_1^{*\prime}(\sigma)] \tag{4-113}$$

When the differentiations indicated in (4-108) to (4-110) are performed, the conditions specifying the optimum system become

$$-2[Q - x_1^*(\sigma)] + \lambda_1^{*\prime}(\sigma) = 0 \tag{4-114}$$
$$2m_1^*(\sigma) + \lambda_1^*(\sigma) = 0 \tag{4-115}$$

and

$$m_1^*(\sigma) - x_1^{*\prime}(\sigma) = 0 \tag{4-116}$$

By manipulating the last three equations, the optimum state signal is found to satisfy the equation

$$x_1^{*\prime\prime}(\sigma) - x_1^*(\sigma) = -Q \tag{4-117}$$

Equation (4-117) is precisely the Euler-Lagrange equation for this particular example. Also, the boundary conditions specified in (4-107) are found to be

$$x_1^*(t) = x_1(t) \qquad x_1^*(T) = x_1(T) \tag{4-118}$$

for a fixed-point terminal-boundary condition or

$$x_1^*(t) = x_1(t) \qquad \lambda_1^*(T) = 0 \tag{4-119}$$

for the free-point terminal-boundary condition. These boundary conditions are equivalent to those found in the Euler-Lagrange formulation.

The development of the conditions for a minimum in this section have an interesting and useful analogy to hamiltonian dynamics.[40] First, the function $\mathscr{H}$ is defined as

$$\mathscr{H}[\mathbf{x}(\sigma), \mathbf{m}(\sigma), \boldsymbol{\lambda}(\sigma), \sigma] = \sum_{n=0}^{N} \lambda_n(\sigma) f_n[\mathbf{x}(\sigma), \mathbf{m}(\sigma), \sigma] \tag{4-120}$$

where

$$\lambda_0(\sigma) = 1 \qquad x_0'(\sigma) = f_0[\mathbf{x}(\sigma), \mathbf{m}(\sigma), \sigma] = H[\mathbf{x}(\sigma), \mathbf{m}(\sigma), \sigma] \tag{4-121}$$

This function $\mathscr{H}$ is nearly the error measure H_c defined in (4-96) but does not include the derivatives of the state signals. The conditions for a minimum given in (4-108) to (4-110) now can be rewritten in terms of this new function $\mathscr{H}$.

Specifically, (4-109) is rewritten as

$$\frac{\partial \mathscr{H}^*}{\partial m_k^*} = 0 \qquad k = 1, 2, \ldots, M \tag{4-122}$$

However, this expression is the condition that results from minimizing $\mathscr{H}$ with respect to the components of the control signal. Therefore, (4-122) is written as

$$\mathscr{H}^* = \min_{\mathbf{m}(\sigma)} \mathscr{H}[\mathbf{x}^*(\sigma), \mathbf{m}(\sigma), \boldsymbol{\lambda}^*(\sigma), \sigma] \tag{4-123}$$

The minimization with respect to the components of the control signal defined in (4-123) treats the components of the control-signal vector as adjustable parameters. In other words, this minimization is a problem in differential calculus.

In addition, (4-108) and (4-110) can be expressed in terms of $\mathscr{H}$. From the definitions of H_c^* and $\mathscr{H}^*$

$$\frac{\partial H_c^*}{\partial x_k^*} = \frac{\partial \mathscr{H}^*}{\partial x_k^*} \tag{4-124}$$

and

$$\frac{\partial H_c^*}{\partial x_k^{*\prime}} = -\lambda_k^* \tag{4-125}$$

are established readily. Therefore, (4-108) becomes

$$\frac{\partial \mathscr{H}^*}{\partial x_k^*} = -\lambda_k^{*\prime} \qquad k = 0, 1, \ldots, N \tag{4-126}$$

Also, the relationship

$$\frac{\partial H_c^*}{\partial \lambda_k^*} = \frac{\partial \mathscr{H}^*}{\partial \lambda_k^*} - x_k^{*\prime} \tag{4-127}$$

results from the definitions of H_c^* and $\mathscr{H}^*$ so that (4-110) becomes

$$\frac{\partial \mathscr{H}^*}{\partial \lambda_k^*} = x_k^{*\prime} \qquad k = 0, 1, \ldots, N \tag{4-128}$$

Equations (4-123), (4-126), and (4-128) are the first necessary condition for a minimum§ in terms of the function $\mathscr{H}^*$. This set of equations is a compact and elegant form of the first necessary condition for minimizing (4-94) subject to (4-92).

The symmetry appearing in (4-126) and (4-128) is recognized to be the symmetry which occurs in hamiltonian dynamics. Because of this symmetry, the function $\mathscr{H}^*$ is referred to as a *hamiltonian function*. In this analogy, the state variables x_k^* would be the generalized position coordinates, and the Lagrange multipliers λ_k^* would be the generalized momenta in the hamiltonian-dynamics formulation. Also, an important property of the hamiltonian $\mathscr{H}^*$ is established readily. The total time derivative of the hamiltonian can be expanded into the form

$$\frac{d\mathscr{H}^*}{d\sigma} = \frac{\partial \mathscr{H}^*}{\partial \sigma} + \sum_{k=0}^{N} \left(\lambda_k^{*\prime} \frac{\partial \mathscr{H}^*}{\partial \lambda_k^*} + x_k^{*\prime} \frac{\partial \mathscr{H}^*}{\partial x_k^*} \right) + \sum_{k=1}^{M} m_k^{*\prime} \frac{\partial \mathscr{H}^*}{\partial m_k^*} \tag{4-129}$$

§ In addition to the necessary conditions given in (4-123), (4-126), and (4-128), a sufficiency condition is tacitly assumed throughout. Specifically, the convexity condition given in (4-21) and the required continuity assumptions are applied to the function $\mathscr{H}^*$.

When the conditions given in (4-122), (4-126), and (4-128) are inserted into (4-129), the time derivative of the hamiltonian reduces to

$$\frac{d\mathscr{H}^*}{d\sigma} = \frac{\partial\mathscr{H}^*}{\partial\sigma} \tag{4-130}$$

Therefore, the hamiltonian of the optimum system only varies in time with respect to the independent time variable. If $\partial\mathscr{H}^*/\partial\sigma = 0$, then the hamiltonian of the optimum system is a constant.

The first condition for a minimum, as expressed in (4-123), (4-126), and (4-128), is equivalent to Pontryagin's equations for minimization problems where the control signals do not saturate.[41]

4-7 Control-signal saturation

An important limitation of the calculus of variations, when used for feedback-control-system optimization, is that the procedure as developed thus far cannot be applied directly to design problems where control signals are subjected to saturation. Because saturation often occurs in the physical system represented by the dynamic

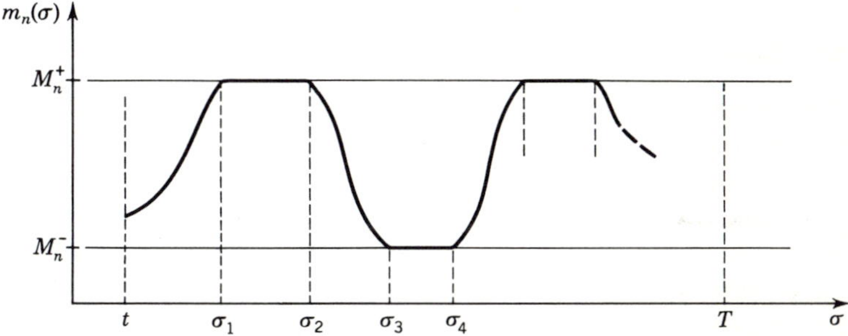

Fig. 4-8 Saturation of control signal $m_n(\sigma)$.

process, the need for a less restricted procedure is evident. The difficulty with saturation arises from the necessity of selecting a specific mathematical form for optimum control signals in the calculus of variations.

For design problems where saturation does not occur, the mathematical class of optimum control signals is selected in (4-99) by assuming unrestricted perturbations. Now the assumption is made that the control signal $m_n(\sigma)$ saturates, as depicted in Fig. 4-8. If the times $\sigma_1, \sigma_2, \ldots$ are known, then the variational process, such as expressed in (4-99), can be applied to $m_n(\sigma)$ on the time intervals $(\sigma_1 - t)$, $(\sigma_3 - \sigma_2)$, etc. Also, the assumption is made that on the intervals where the control signal saturates, such as $(\sigma_2 - \sigma_1)$, either $m_n^*(\sigma) = M_n{}^+$ or $m_n^*(\sigma) = M_n{}^-$ is specified as a segment of the optimum control signal. Unfortunately, the times $\sigma_1, \sigma_2, \sigma_3, \ldots$ are not known; hence the optimum control signal cannot be constructed explicitly by this process. Therefore, the procedures used in the traditional calculus of variations are not valid unless trial-and-error procedures are introduced.

Pontryagin extends the calculus of variations to a class of optimization problems that cannot be treated by the methods developed thus far. In particular, Pontryagin extends these methods to the case where the control-signal components experience saturation. Suppose that the control-signal vector must lie in a closed region of M-dimensional space which is defined as $\mathscr{M}$. Because the minimization process expressed in (4-123) treats the components of the control-signal vector as adjustable parameters, a reasonable assumption is that this minimization process can be extended so that only the values of the control signal which lie within or on the boundaries of $\mathscr{M}$ are taken to be allowable solutions. Supposedly, (4-123) could be rewritten as

$$\mathscr{H}^* = \min_{\mathbf{m}(\sigma) \in \mathscr{M}} \mathscr{H}[\mathbf{x}^*(\sigma), \mathbf{m}(\sigma), \boldsymbol{\lambda}^*(\sigma), \sigma] \tag{4-131}$$

These arguments suggest that (4-126), (4-128), and (4-131) express the condition for a minimum subject to the requirement that the control-signal components experience saturation at the boundaries of $\mathscr{M}$.

However, these arguments for control-signal saturation cannot be substantiated directly without mathematical considerations beyond the scope of this book. Instead, an indirect but formal development is given in Chap. 6. This development is based on the application of the elementary theory of partial differential equations to the continuous dynamic-programming condition for minimum error.

4-8 Summary

The concept of minimizing an error index by selecting the optimum control signal as opposed to the system parameters or the system impulse response greatly generalizes the concept of feedback-control-system optimization. In particular, the procedure presented in this chapter results in the specification of the optimum control signals not only in terms of the desired system response but also in terms of the existing state in the dynamic process. Also, no a priori specification that only linear feedback control systems are investigated is imposed. Therefore, depending on the form of the Euler-Lagrange equation, the optimum feedback control system is either nonlinear or linear.

The computational solutions for design problems where the optimum feedback control system is linear also are discussed. In this case, linear differential equations can be solved readily by using a computer. The appropriate boundary conditions also can be inserted into the solution of the Euler-Lagrange equation simply by utilizing the principle of superposition. However, in the case of nonlinear optimum feedback control systems, the search problem that is encountered when trying to satisfy the boundary conditions at two points in time may cause extreme difficulty. As mentioned previously, search techniques for solving the two-point boundary-value problem when the optimum control system is nonlinear are developed with the addition of new concepts in later chapters. If mathematical optimization theory is to be used as the basis for complex system design, numerical solutions must be obtained, and the success of the procedure is determined in part by the difficulty in obtaining these solutions. For nonlinear optimum control systems where the dynamic process is of first order, analytical solutions for the optimum control signal are obtained. The particular example given in this chapter demonstrates methods for selecting the form of the error measure as a method for selecting the characteristics of the optimum control equation. These characteristics range from dead-zone-type controllers to saturation-type controllers.

When saturation of the control signal occurs, the calculus of variations cannot be applied directly to the minimization problem. On the other hand, modifications due to Pontryagin can be introduced that provide a direct solution to the control-signal-saturation case. However, an explicit derivation of Pontryagin's method is not included in this chapter.

5

System Optimization Using Dynamic Programming

Dynamic programming is particularly important in feedback-control optimization because it is based on concepts that are very familiar to the control-system engineer. The development of the dynamic-programming equations depends upon the relationship of initial conditions to the solution of ordinary differential equations. These initial conditions, which define the existing state of the dynamic process, are the measured variables needed in the optimum control equation that leads to a control system with a feedback configuration.

The development of dynamic programming by Bellman[15] is in part motivated by two important problems that are mentioned in Chap. 4. In particular, dynamic programming is readily applicable to optimization problems where the control signals saturate. Also, the method of dynamic programming in theory eliminates the two-point boundary-value problem.

Dynamic programming applies to the control-signal-saturation case because it does not require the assumption of an explicit class of optimum control signals and unrestricted perturbations within this class. On the other hand, dynamic programming constructs the optimum control signal point by point. In terms of variational mathematics, each point of the control signal is treated as an adjustable parameter in determining the optimum sequence of points comprising the control signal.

The elimination of the two-point boundary-value problem is accomplished by a flooding procedure. This flooding procedure can be considered to be the embedding of the solution to the particular optimization problem desired for the system design into all possible optimization problems within the class specified in the system design. Specifically, dynamic programming embeds the solution to the optimization of the control system for a particular state of the dynamic process into the optimization of the control system for all possible states of the dynamic process. At the end of the computational process, the particular solution required is then selected from the solution for all possible initial states of the dynamic process. The computational process associated with this flooding procedure eliminates the two-point boundary-value problem but introduces new computational difficulties. These new computational difficulties are considered along with the computational difficulties of two-point boundary-value problems in a later chapter.

5-1 Elementary theory for a first-order dynamic process

As in Sec. 4-1, the dynamic-process equations

$$q(t) = G[x(t), x'(t), t] \tag{5-1}$$

and

$$m(t) = F[x(t), x'(t), t] \tag{5-2}$$

and the error index

$$e(t) = \int_t^T h[q(\sigma), m(\sigma), \sigma]\, d\sigma \tag{5-3}$$

are used to introduce the elementary concepts of the theory. Again the error measure h is combined with the equations of the dynamic process so that

$$e(t) = \int_t^T H[x(\sigma), x'(\sigma), \sigma]\, d\sigma \tag{5-4}$$

The condition for a minimum value of the error index, as expressed in (5-4), is developed by using the concepts of dynamic programming.

The initial concept in dynamic programming is the identification of the mathematical form of the minimum value of the error index. First, the viewpoint adopted here is that (5-4) is minimized by selecting $x'(\sigma)$. Then the variable $x(\sigma)$ is determined uniquely by $x'(\mu)$ on the interval $t \leqq \mu \leqq \sigma$ and by the initial condition. Of course, this dependence follows by definition and is written as

$$x(\sigma) = x(t) + \int_t^\sigma x'(\mu)\, d\mu \tag{5-5}$$

Furthermore, the minimizing value of $x'(\sigma)$, denoted as $x'_*(\sigma)$, is a function of the initial state $x(t)$, real time t, and σ assuming that T is a constant. In Chap. 4, this dependence is expressed as

$$x'_*(\sigma) = X_1[x(t); t, \sigma] \tag{5-6}$$

This dependence of $x'_*(\sigma)$ is used in conjunction with (5-5) to obtain

$$x_*(\sigma) = X_0[x(t); t, \sigma] \tag{5-7}$$

by definition. If (5-6) and (5-7) are substituted into (5-4), then a minimum value of $e(t)$ results, and this value is denoted as $e_*(t)$. When these substitutions are made, the integrand of (5-4) clearly is dependent on only $x(t)$, t, and σ, so that (5-4) is written as

$$e_*(t) = \int_t^T H_m[x(t); t, \sigma]\, d\sigma \tag{5-8}$$

Therefore, the minimum value of the error index can be defined as the function

$$e_*(t) = E[x(t), t] \tag{5-9}$$

This dependence of the minimum value of the error index on only $x(t)$ and t is demonstrated in Sec. 4-2. The function E is called the *minimum-error function*, and the identification of the arguments of E is the first conceptual step in dynamic programming.

The reasoning culminating in (5-9), where $x'(\sigma)$ is treated as the minimizing variable, is written compactly in the form

$$E[x(t), t] = \min_{\substack{x'(\sigma)\\ [t, T]}} \int_t^T H[x(\sigma), x'(\sigma), \sigma]\, d\sigma \tag{5-10}$$

The notation $[t, T]$ denotes that the minimization with respect to $x'(\sigma)$ is performed on the interval $t \leqq \sigma \leqq T$. If the minimization is restricted by

$$S^-(t) \leqq x'(t) \leqq S^+(t) \tag{5-11}$$

which usually is stated as

$$x'(t) \in S(t) \tag{5-12}$$

then $x_*'(\sigma)$ is still only a function of $x(t)$, t, and σ. This assumes that $S^-(t)$ and $S^+(t)$ are fixed parameters of the minimization problem. Therefore, the definition given in (5-9) is still valid, and the minimization of (5-4) subject to (5-12) is written as

$$E[x(t), t] = \min_{\substack{x'(\sigma) \in S(\sigma)\\ [t, T]}} \int_t^T H[x(\sigma), x'(\sigma), \sigma]\, d\sigma \tag{5-13}$$

In addition, a dummy time variable μ is introduced into (5-13) such that

$$E[x(\mu), \mu] = \min_{\substack{x'(\sigma) \in S(\sigma)\\ [\mu, T]}} \int_\mu^T H[x(\sigma), x'(\sigma), \sigma]\, d\sigma \tag{5-14}$$

This dummy time variable is defined on the interval $t \leqq \mu \leqq T$ and is introduced so that real time t can be treated as a constant during the minimization process. If the integrand appearing in (5-14) is finite at $\sigma = T$, then the integral appearing in (5-14) vanishes at $\mu = T$. Therefore, the boundary condition on the minimum-error function at $\mu = T$ is

$$E[x(T), T] = 0 \tag{5-15}$$

Because of the definition of the minimum-error function, the integral appearing in (5-4) satisfies

$$\int_\mu^T H[x(\sigma), x'(\sigma), \sigma]\, d\sigma - E[x(\mu), \mu] \geqq 0 \tag{5-16}$$

for all allowable functions $x'(\sigma)$. The equality occurring in (5-16) is satisfied for all functions $x'(\sigma)$ that correspond to the minimum value of the error index. The inequality occurring in (5-16) is satisfied for all other values of $x'(\sigma)$. If there is only one function $x'(\sigma)$ that satisfies the equality in (5-16), then the solution to the minimization problem is unique. By using the properties expressed in (5-16), the minimization problem now is expressed in the alternative form

$$\min_{\substack{x'(\sigma) \in S(\sigma) \\ [\mu, T]}} \left\{ \int_{\mu}^{T} H[x(\sigma), x'(\sigma), \sigma]\, d\sigma - E[x(\mu), \mu] \right\} = 0 \tag{5-17}$$

Furthermore, the integral appearing in (5-17) can be taken in two parts so that

$$\min_{\substack{x'(\sigma) \in S(\sigma) \\ [\mu, T]}} \left\{ \int_{\mu}^{\mu+\delta} H[x(\sigma), x'(\sigma), \sigma]\, d\sigma + \int_{\mu+\delta}^{T} H[x(\sigma), x'(\sigma), \sigma]\, d\sigma - E[x(\mu), \mu] \right\} = 0 \tag{5-18}$$

The second important concept in dynamic programming, namely, the principle of optimality, now is applied to (5-18).

The principle of optimality is a method used to replace the minimization on the interval $\mu \leqq \sigma \leqq T$ by a procedure of first minimizing on the interval $\mu + \delta \leqq \sigma \leqq T$ and then minimizing on the interval $\mu \leqq \sigma \leqq \mu + \delta$. Suppose that $x'(\sigma)$ is chosen arbitrarily on the interval $\mu \leqq \sigma \leqq \mu + \delta$. This choice would give rise to a state signal

$$x(\mu + \delta) = x(\mu) + \int_{\mu}^{\mu+\delta} x'(\sigma)\, d\sigma \tag{5-19}$$

Furthermore, all values of the state signal $x(\sigma)$ on the interval $\mu + \delta \leqq \sigma \leqq T$ are determined uniquely from $x(\mu + \delta)$ and $x'(\sigma)$ on the interval $\mu + \delta \leqq \sigma \leqq T$. Therefore, given any choice of $x'(\sigma)$ on $\mu \leqq \sigma \leqq \mu + \delta$ and hence the subsequent state signal $x(\mu + \delta)$, the error index can be minimized on the interval $\mu + \delta \leqq \sigma \leqq T$ by adjusting $x'(\sigma)$ on the interval $\mu + \delta \leqq \sigma \leqq T$ subject to whatever value of $x(\mu + \delta)$ is given. This cause-and-effect relationship is the principle of optimality and allows the replacing of (5-18) by

$$\min_{\substack{x'(\sigma) \in S(\sigma) \\ [\mu, \mu+\delta]}} \left(\min_{\substack{x'(\sigma) \in S(\sigma) \\ [\mu+\delta, T]}} \left\{ \int_{\mu}^{\mu+\delta} H[x(\sigma), x'(\sigma), \sigma]\, d\sigma + \int_{\mu+\delta}^{T} H[x(\sigma), x'(\sigma), \sigma]\, d\sigma - E[x(\mu), \mu] \right\} \right) = 0 \tag{5-20}$$

However, only the second integral appearing in (5-20) is dependent upon the choice of $x'(\sigma)$ on the interval $\mu + \delta \leqq \sigma \leqq T$, so that (5-20) becomes

$$\min_{\substack{x'(\sigma) \in S(\sigma) \\ [\mu, \mu+\delta]}} \left(\int_{\mu}^{\mu+\delta} H[x(\sigma), x'(\sigma), \sigma]\, d\sigma + \min_{\substack{x'(\sigma) \in S(\sigma) \\ [\mu+\delta, T]}} \left\{ \int_{\mu+\delta}^{T} H[x(\sigma), x'(\sigma), \sigma]\, d\sigma \right\} - E[x(\mu), \mu] \right) = 0 \tag{5-21}$$

Finally, the definition of the minimum-error function given in (5-14) is introduced into (5-21), thereby yielding

$$\min_{\substack{x'(\sigma) \in S(\sigma) \\ [\mu, \mu+\delta]}} \left\{ \int_{\mu}^{\mu+\delta} H[x(\sigma), x'(\sigma), \sigma]\, d\sigma + E[x(\mu + \delta), \mu + \delta] - E[x(\mu), \mu] \right\} = 0 \tag{5-22}$$

Equation (5-22) is the discrete form of the dynamic-programming condition for minimum error.

The continuous form of the dynamic-programming condition for minimum error is found from (5-22) by letting the parameter δ approach zero. For the parameter δ chosen to be arbitrarily small, the integral appearing in (5-22) becomes

$$\int_{\mu}^{\mu+\delta} H[x(\sigma), x'(\sigma), \sigma]\, d\sigma = \delta H[x(\mu), x'(\mu), \mu] + \delta^2 \cdots \tag{5-23}$$

Also, the difference appearing in (5-22) becomes

$$E[x(\mu+\delta), \mu+\delta] - E[x(\mu), \mu] = \delta \frac{dE[x(\mu), \mu]}{d\mu} + \delta^2 \cdots \tag{5-24}$$

If these expansions are substituted into (5-22), the terms within the braces are no longer dependent upon $x'(\sigma)$ on the interval $\mu \leqq \sigma \leqq \mu + \delta$ but are dependent upon $x'(\mu)$ only. Therefore, the minimization appearing in (5-22) simplifies to a minimization with respect to $x'(\mu)$. Also, the parameter δ does not enter into the minimization process and therefore can be taken outside. These steps yield

$$\delta\left(\min_{x'(\mu) \in S(\mu)} \left\{H[x(\mu), x'(\mu), \mu] + \frac{dE[x(\mu), \mu]}{d\mu} + \delta \cdots\right\}\right) = 0 \tag{5-25}$$

Because the parameter δ is arbitrary although small, (5-25) is satisfied if

$$\min_{x'(\mu) \in S(\mu)} \left\{H[x(\mu), x'(\mu), \mu] + \frac{dE[x(\mu), \mu]}{d\mu} + \delta \cdots\right\} = 0 \tag{5-26}$$

Finally, the limit $\delta \to 0$ yields the continuous form of the dynamic-programming condition for minimum error:

$$\min_{x'(\mu) \in S(\mu)} \left\{H[x(\mu), x'(\mu), \mu] + \frac{dE[x(\mu), \mu]}{d\mu}\right\} = 0 \tag{5-27}$$

A more traditional form of the continuous dynamic-programming condition for minimum error utilizes the expansion of the total time derivative in terms of partial derivatives. Specifically, the expansion used is

$$\frac{dE[x(\mu), \mu]}{d\mu} = \frac{\partial E[x(\mu), \mu]}{\partial \mu} + x'(\mu) \frac{\partial E[x(\mu), \mu]}{\partial x(\mu)} \tag{5-28}$$

The first term in this expansion is not dependent upon $x'(\mu)$ and hence can be taken outside the minimization process. Therefore, the substitution of (5-28) into (5-27) gives the alternative form

$$\min_{x'(\mu) \in S(\mu)} \left\{H[x(\mu), x'(\mu), \mu] + x'(\mu) \frac{\partial E[x(\mu), \mu]}{\partial x(\mu)}\right\} = -\frac{\partial E[x(\mu), \mu]}{\partial \mu} \tag{5-29}$$

The dynamic-programming condition for minimum error as expressed in either (5-27) or (5-29) actually implies two steps and hence two equations. The first step is the minimization of the sum of terms appearing in the braces of (5-29). The second step is equating the minimum value of this sum of terms to the right-hand side of (5-29). In minimizing the left-hand side of (5-29), the value $x'(\mu)$ is treated as an adjustable parameter; hence differential calculus is used. The error measure is assumed to be a continuous, strictly convex function of $x'(\mu)$ for the purposes of this discussion. This assumption ensures the existence of a unique minimizing value of

$x'(\mu)$. Specifically, the sum of terms appearing on the left-hand side of (5-29) also is strictly convex with respect to $x'(\mu)$, as is demonstrated by

$$\frac{\partial^2}{\partial x'(\mu)^2}\left\{H[x(\mu), x'(\mu), \mu] + x'(\mu)\frac{\partial E[x(\mu), \mu]}{\partial x(\mu)}\right\} = \frac{\partial^2 H[x(\mu), x'(\mu), \mu]}{\partial x'(\mu)^2} > 0 \quad (5\text{-}30)$$

Therefore, this sum possesses the structure indicated in Fig. 5-1. The minimum point on the curve sketched in Fig. 5-1 corresponds to the value of $x'(\mu)$ where the

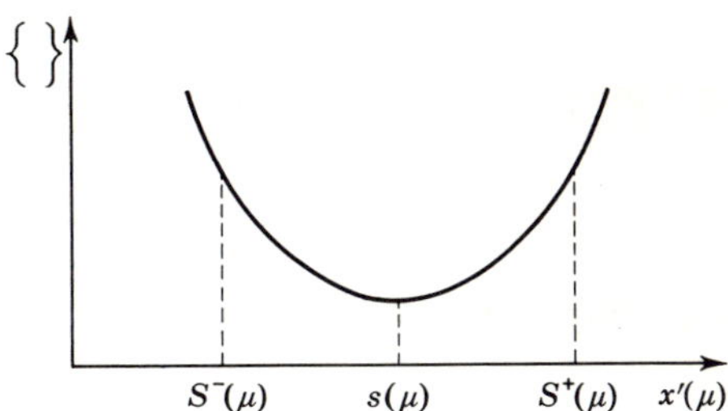

Fig. 5-1 Sketch of terms within the braces in Eq. (5-29).

partial derivative of the sum with respect to $x'(\mu)$ vanishes. Therefore, the switching signal $s(\mu)$ indicated in Fig. 5-1 satisfies

$$\frac{\partial H[x(\mu), s(\mu), \mu]}{\partial s(\mu)} + \frac{\partial E[x(\mu), \mu]}{\partial x(\mu)} = 0 \quad (5\text{-}31)$$

From simple geometrical considerations, the minimizing value of $x'(\mu)$ that satisfies (5-11) [denoted as $x'_*(\mu)$] is given by

$$x'_*(\mu) = \begin{cases} S^+(\mu) & S^+(\mu) \leq s(\mu) \\ s(\mu) & S^-(\mu) < s(\mu) < S^+(\mu) \\ S^-(\mu) & s(\mu) \leq S^-(\mu). \end{cases} \quad (5\text{-}32)$$

Finally, (5-29) reduces to

$$H[x(\mu), x'_*(\mu), \mu] + x'_*(\mu)\frac{\partial E[x(\mu), \mu]}{\partial x(\mu)} = -\frac{\partial E[x(\mu), \mu]}{\partial \mu} \quad (5\text{-}33)$$

Equations (5-31) to (5-33) are equivalent to the compact form given in (5-27) and are the equations defining the minimum-error function.

5-2 Relationships pertaining to the calculus of variations

Before methods for solving the partial differential equation appearing in (5-33) are discussed, some relationships between dynamic programming and the calculus of variations are established. These relationships are established for the case where the allowable region of the control signal is assumed to be arbitrarily large so that saturation does not occur. This assumption is needed, of course, because the elementary derivations in the calculus of variations do not apply to the saturation case.

Under this assumption, (5-31) and (5-32) reduce to

$$\frac{\partial H}{\partial x'} + \frac{\partial E}{\partial x} = 0 \quad (5\text{-}34)$$

where the switching signal $s(\mu)$ and the asterisk notation have been eliminated for the sake of simplicity. Also, (5-33) is rewritten as

$$H + \frac{\partial E}{\partial \mu} + x' \frac{\partial E}{\partial x} = 0 \tag{5-35}$$

The total time derivative of (5-34) is expanded to

$$\frac{d}{d\mu}\left(\frac{\partial H}{\partial x'}\right) + \frac{\partial^2 E}{\partial \mu\, \partial x} + x' \frac{\partial^2 E}{\partial x^2} = 0 \tag{5-36}$$

and the partial derivative with respect to x of (5-35) is

$$\frac{\partial H}{\partial x} + \frac{\partial^2 E}{\partial \mu\, \partial x} + x' \frac{\partial^2 E}{\partial x^2} = 0 \tag{5-37}$$

Equating (5-36) and (5-37) gives

$$\frac{d}{d\mu}\left(\frac{\partial H}{\partial x'}\right) - \frac{\partial H}{\partial x} = 0 \tag{5-38}$$

which is seen to be the Euler-Lagrange equation for a first-order dynamic process. Also, the boundary condition given in (5-15) for the minimum-error function is differentiated with respect to $x(T)$, resulting in the relationship

$$\frac{\partial E}{\partial x} = 0 \qquad \mu = T \tag{5-39}$$

Therefore, (5-34) reduces to

$$\frac{\partial H}{\partial x'} = 0 \qquad \mu = T \tag{5-40}$$

This is seen to be the free-point terminal-boundary condition arising in the calculus of variations.

In addition, two conditions pertaining to minima, which are traditional in the calculus of variations, can be derived directly from the dynamic-programming formulation. Without the a priori assumption that the terms in the braces appearing in (5-29) are a strictly convex function of $x'(\mu)$, the condition expressed in (5-34) specifies only that a stationary point on this function is found. A local minimum is achieved at this stationary point if the second partial derivative with respect to $x'(\mu)$ of the terms in the braces is greater than zero at the stationary point. Therefore, the condition for a relative minimum at the stationary point at x'_* is

$$\frac{\partial^2 H}{\partial x'^2} > 0 \qquad x' = x'_* \tag{5-41}$$

This condition for a relative minimum is the *Legendre condition* associated with the calculus of variations.[37]

If the stationary point defined by (5-34) is an absolute minimum, then the inequality

$$H[x(\mu), x'(\mu), \mu] + x'(\mu)\frac{\partial E[x(\mu), \mu]}{\partial x(\mu)} > H[x(\mu), x'_*(\mu), \mu] + x'_*(\mu)\frac{\partial E[x(\mu), \mu]}{\partial x(\mu)} \qquad x'(\mu) \neq x'_*(\mu) \tag{5-42}$$

is deduced geometrically from Fig. 5-1. Each side of the inequality appearing in (5-42) is the terms appearing in the braces of (5-29). This inequality is rewritten as

$$H[x(\mu), x'(\mu), \mu] - H[x(\mu), x'_*(\mu), \mu] + [x'(\mu) - x'_*(\mu)] \frac{\partial E[x(\mu), \mu]}{\partial x(\mu)} > 0 \qquad x'(\mu) \neq x'_*(\mu) \tag{5-43}$$

When the condition for a stationary point defined by (5-34) is substituted into (5-43), the result is

$$H[x(\mu), x'(\mu), \mu] - H[x(\mu), x'_*(\mu), \mu] - [x'(\mu) - x'_*(\mu)] \frac{\partial H[x(\mu), x'_*(\mu), \mu]}{\partial x'_*(\mu)} > 0 \qquad x'(\mu) \neq x'_*(\mu) \tag{5-44}$$

This condition for an absolute minimum is the *Weierstrass condition* associated with the calculus of variations.[37]

In Sec. 4-5, a simplification in the optimum control system is deduced for the special case where the error measure H is independent of σ and the interval of integration of the error index is infinite, $T = \infty$. Under these two conditions, the optimum control equation is time-invariant, which is to say that the optimum control equation does not depend explicitly on t. If H is independent of μ and if the minimum-error function has the form

$$E[x(\mu), \mu] = E_1[x(\mu)] + C\mu \tag{5-45}$$

then the solution of (5-34) and (5-35) is independent of μ and the optimum control equation is time-invariant. When (5-45) is substituted, (5-34) and (5-35) reduce to

$$\frac{\partial H}{\partial x'} + \frac{\partial E_1}{\partial x} = 0 \tag{5-46}$$

$$H + C + x' \frac{\partial E_1}{\partial x} = 0 \tag{5-47}$$

When the partials of E_1 with respect to x are eliminated from these last two equations, the optimum value of x' is given by

$$x' = \frac{H + C}{\partial H / \partial x'} \tag{5-48}$$

This is precisely the result obtained from the Euler-Lagrange equation. Therefore, the conclusion is drawn that when H is independent of σ and when $T = \infty$ in the error index, the minimum-error function must be of the form expressed in (5-45).

The results of this section demonstrate the equivalence of the equations of dynamic programming and those of the calculus of variations under the condition that control saturation does not occur. However, the method of dynamic programming offers a direct method for treating optimization problems with control-signal saturation.

5-3 Integral constraints

A useful tool in the calculus of variations is the introduction of a Lagrange multiplier in order to treat constraints. In this section, optimization problems subject to constraints are formulated in terms of dynamic programming. In this formulation, again only a first-order dynamic process is treated. The results easily can be extended to higher-order dynamic processes.

The optimization problem under consideration is the minimization of the error index

$$e(\mu) = \int_{\mu}^{T} H[x(\sigma), x'(\sigma), \sigma]\, d\sigma \tag{5-49}$$

subject to the integral-type constraint

$$y(\mu) = \int_{\mu}^{T} I[x(\sigma), x'(\sigma), \sigma]\, d\sigma \tag{5-50}$$

First, the form of the minimum-error function, such as given in (5-14) but now for the constraint problem, must be identified. For the first-order dynamic process, the existing state of the dynamic process is defined by $x(\mu)$. However, the minimum error is also dependent upon the value of the constraint $y(\mu)$. Therefore, the minimum-error function is defined as

$$E[x(\mu), y(\mu), \mu] = \min_{\substack{x'(\sigma) \in S(\sigma) \\ [\mu, T]}} \int_{\mu}^{T} H[x(\sigma), x'(\sigma), \sigma]\, d\sigma \tag{5-51}$$

Once the form of the minimum-error function is defined, the development of the condition for minimum error follows with precisely the steps used to derive (5-27). For the constraint problem, then, the condition for minimum error is

$$\min_{x'(\mu) \in S(\mu)} \left\{ H[x(\mu), x'(\mu), \mu] + \frac{dE[x(\mu), y(\mu), \mu]}{d\mu} \right\} = 0 \tag{5-52}$$

The total time derivative of the minimum-error function is expanded in a fashion similar to (5-28); hence the condition for minimum error becomes

$$\min_{x'(\mu) \in S(\mu)} \left\{ H[x(\mu), x'(\mu), \mu] + \frac{\partial E[x(\mu), y(\mu), \mu]}{\partial \mu} \right. \\ \left. + x'(\mu) \frac{\partial E[x(\mu), y(\mu), \mu]}{\partial x(\mu)} + y'(\mu) \frac{\partial E[x(\mu), y(\mu), \mu]}{\partial y(\mu)} \right\} = 0 \tag{5-53}$$

The derivative $y'(\mu)$ is found by differentiating (5-50). Under the condition that $x'(\mu)$ assumes the minimizing value $x'_*(\mu)$, the value of this derivative is

$$y'_*(\mu) = -I[x(\mu), x'_*(\mu), \mu] \tag{5-54}$$

Also, (5-53) reduces to

$$H[x(\mu), x'_*(\mu), \mu] + \frac{\partial E[x(\mu), y(\mu), \mu]}{\partial \mu} \\ + x'_*(\mu) \frac{\partial E[x(\mu), y(\mu), \mu]}{\partial x(\mu)} + y'_*(\mu) \frac{\partial E[x(\mu), y(\mu), \mu]}{\partial y(\mu)} = 0 \tag{5-55}$$

when the condition for minimum error is satisfied. The partial derivative of (5-55) with respect to $y(\mu)$ is

$$\frac{\partial^2 E[x(\mu), y(\mu), \mu]}{\partial \mu\, \partial y(\mu)} + x'_*(\mu) \frac{\partial^2 E[x(\mu), y(\mu), \mu]}{\partial x(\mu)\, \partial y(\mu)} + y'_*(\mu) \frac{\partial^2 E[x(\mu), y(\mu), \mu]}{\partial y(\mu)^2} = 0 \tag{5-56}$$

which is written in the alternative form

$$\frac{d}{d\mu} \left\{ \frac{\partial E[x(\mu), y(\mu), \mu]}{\partial y(\mu)} \right\} = 0 \tag{5-57}$$

Therefore, along the optimum trajectory as defined by $x'_*(\mu)$, the partial derivative of the minimum-error function with respect to the constraint variable is a constant. If (5-57) is integrated with respect to μ, then

$$\frac{\partial E[x(\mu), y(\mu), \mu]}{\partial y(\mu)} = -\lambda \tag{5-58}$$

where the constant of integration is defined as $-\lambda$.

Equations (5-54) and (5-58) now are substituted into (5-53), which becomes

$$\min_{x'(\mu) \in S(\mu)} \left\{ H[x(\mu), x'(\mu), \mu] + \lambda I[x(\mu), x'(\mu), \mu] + \frac{\partial E[x(\mu), y(\mu), \mu]}{\partial \mu} + x'(\mu) \frac{\partial E[x(\mu), y(\mu), \mu]}{\partial x(\mu)} \right\} = 0 \tag{5-59}$$

When (5-29) and (5-59) are compared, it is seen that in the constraint problem the error measure is replaced by the error measure plus a constant times the integrand of the constraint integral. In other words, the constraint problem can be treated by the introduction of a Lagrange multiplier λ. If the minimization problem is reconstructed as a minimization problem with a Lagrange multiplier, then the constrained error index is

$$e_c(\mu) = \int_\mu^T \left\{ H[x(\sigma), x'(\sigma), \sigma] + \lambda I[x(\sigma), x'(\sigma), \sigma] \right\} d\sigma \tag{5-60}$$

For the constrained error index defined in (5-60), the minimum-error function is defined as

$$E_c[x(\mu), \mu] = \min_{\substack{x'(\sigma) \in S(\sigma) \\ [\mu, T]}} \left(\int_\mu^T \left\{ H[x(\sigma), x'(\sigma), \sigma] + \lambda I[x(\sigma), x'(\sigma), \sigma] \right\} d\sigma \right) \tag{5-61}$$

The condition for minimum error corresponding to this definition of the minimum-error function results in (5-59). Therefore, minimization subject to an integral constraint of the type given in (5-50) can be treated simply by an extension of dynamic programming using Lagrange multipliers.

The type of problem discussed in this section is called an *isoperimetric problem.* One significant use of the isoperimetric problem is the introduction of fixed-point terminal-boundary conditions into the dynamic-programming formulation of the condition for minimum error. As observed previously, the dynamic-programming equation for minimum error leads to free-point terminal-boundary conditions. Therefore, if dynamic programming is used, a method for introducing fixed-point terminal-boundary conditions is needed for a number of engineering design problems.

5-4 Example

In order to demonstrate one of the advantages of the dynamic-programming formulation, a simple example is introduced here. For this example, the first-order dynamic process is described by

$$q(t) = x(t) \qquad x'(t) = m(t) \qquad -\infty < m(t) < \infty \tag{5-62}$$

and the error measure is defined as

$$h[q(\sigma), m(\sigma), \sigma] = [Q - q(\sigma)]^2 + \rho[m(\sigma)]^2 \tag{5-63}$$

Therefore, the error measure H becomes

$$H[x(\sigma), x'(\sigma), \sigma] = [Q - x(\sigma)]^2 + \rho[x'(\sigma)]^2 \tag{5-64}$$

The condition for minimum error expressed in (5-29) reduces to

$$\min_{x'(\mu)} \left\{[Q - x(\mu)]^2 + \rho[x'(\mu)]^2 + \frac{\partial E[x(\mu), \mu]}{\partial \mu} + x'(\mu)\frac{\partial E[x(\mu), \mu]}{\partial x(\mu)}\right\} = 0 \tag{5-65}$$

for this example. The notation indicating that the minimizing value of $x'(\mu)$ is restricted by the region $S(\mu)$ is dropped because the region now includes all finite values. The value of $x'(\mu)$ that minimizes the braces appearing in (5-65) is found by setting the partial of these braces with respect to $x'(\mu)$ equal to zero, thereby resulting in

$$2\rho x'(\mu) + \frac{\partial E[x(\mu), \mu]}{\partial x(\mu)} = 0 \tag{5-66}$$

When (5-66) is substituted into (5-65), the condition for minimum error becomes

$$[Q - x(\mu)]^2 + \frac{\partial E[x(\mu), \mu]}{\partial \mu} - \frac{1}{4\rho}\left\{\frac{\partial E[x(\mu), \mu]}{\partial x(\mu)}\right\}^2 = 0 \tag{5-67}$$

In Chap. 4, the minimum value of the error index is found to be quadratic with respect to the initial conditions in the dynamic process if the dynamic process is linear and the error measure is quadratic. With this information, a solution is assumed and then this solution is shown to satisfy the partial differential equation. This is an elementary procedure and is traditional in the solution of ordinary differential equations.

The solution for the minimum-error function is assumed to have the form

$$E[x(\mu), \mu] = \rho^{\frac{1}{2}}[Q^2k(\mu) - 2Qk_1(\mu)x(\mu) + k_{11}(\mu)x^2(\mu)] \tag{5-68}$$

The k parameters are assumed to be independent of the state signal $x(\mu)$. This solution, of course, is quadratic with respect to $x(\mu)$ and can be thought of as a power-series expansion of the minimum-error function where only the zeroth-, first-, and second-degree terms are retained. The partial derivatives that appear in (5-67) are found to be

$$\frac{\partial E[x(\mu), \mu]}{\partial \mu} = \rho^{\frac{1}{2}}[Q^2k'(\mu) - 2Qk_1'(\mu)x(\mu) + k_{11}'(\mu)x^2(\mu)] \tag{5-69}$$

and

$$\frac{\partial E[x(\mu), \mu]}{\partial x(\mu)} = -2\rho^{\frac{1}{2}}[Qk_1(\mu) - k_{11}(\mu)x(\mu)] \tag{5-70}$$

for this assumed solution. If these partial derivatives are substituted into (5-67) and if the terms involving the various powers of $x(\mu)$ are collected, the condition for minimum error reduces to

$$Q^2[1 + \rho^{\frac{1}{2}}k'(\mu) - k_1^2(\mu)] - 2Qx(\mu)[1 + \rho^{\frac{1}{2}}k_1'(\mu) - k_1(\mu)k_{11}(\mu)] + x^2(\mu)[1 + \rho^{\frac{1}{2}}k_{11}'(\mu) - k_{11}^2(\mu)] = 0 \tag{5-71}$$

The condition for minimum error expressed in (5-71) is satisfied for all finite values of $x(\mu)$, assuming the k parameters are independent of $x(\mu)$, only if each of the three bracketed terms occurring in (5-71) vanishes. Therefore, three independent conditions are imposed on the three k parameters. Because each of these bracketed terms must be equal to zero, the k parameters are described by the following three ordinary differential equations:

$$-\rho^{\frac{1}{2}}k'(\mu) = 1 - k_1{}^2(\mu) \tag{5-72}$$

$$-\rho^{\frac{1}{2}}k_1'(\mu) = 1 - k_1(\mu)k_{11}(\mu) \tag{5-73}$$

and

$$-\rho^{\frac{1}{2}}k_{11}'(\mu) = 1 - k_{11}{}^2(\mu) \tag{5-74}$$

This method of assuming a solution leads to the reduction of the problem of solving a partial differential equation to the problem of solving three ordinary first-order differential equations. The boundary conditions for the k parameters are deduced directly from the required boundary condition on the minimum-error function. Because (5-15) must be satisfied for all finite values of $x(T)$, the three k parameters appearing in (5-68) must vanish at $\mu = T$. Therefore, the boundary values of the solutions of (5-72) to (5-74) are

$$k(T) = k_1(T) = k_{11}(T) = 0 \tag{5-75}$$

For this simple example, the problem of finding the optimum control system is found to be a one-point boundary-value problem as opposed to the two-point boundary-value problem which would have resulted from a direct use of the calculus of variations. The optimum control signal is found by substituting (5-70) into (5-66), which gives

$$m(t) = \rho^{-\frac{1}{2}}[k_1(t)Q - k_{11}(t)x(t)] \tag{5-76}$$

at $\mu = t$. This control equation is optimum for all $x(t)$ because the k parameters are derived so that (5-71) is satisfied for all $x(t)$.

In order to examine the solutions of the k parameters, a transformation of variables is helpful and is chosen so that the boundary conditions occur at the beginning of the solution. In particular, if the transformation of variables

$$\eta = \rho^{-\frac{1}{2}}(T - \mu) \qquad k(\mu) = K(\eta) \qquad k_1(\mu) = K_1(\eta) \qquad k_{11}(\mu) = K_{11}(\eta) \tag{5-77}$$

is introduced, then the differential equations defining the k parameters reduce to

$$K'(\eta) = 1 - K_1{}^2(\eta) \tag{5-78}$$

$$K_1'(\eta) = 1 - K_1(\eta)K_{11}(\eta) \tag{5-79}$$

and

$$K_{11}'(\eta) = 1 - K_{11}{}^2(\eta) \tag{5-80}$$

The solutions for the K parameters are subject to the initial conditions

$$K(0) = K_1(0) = K_{11}(0) = 0 \tag{5-81}$$

These differential equations can be integrated by separation of variables, and the K parameters easily are shown to be

$$K(\eta) = K_1(\eta) = K_{11}(\eta) = \tanh \eta \tag{5-82}$$

From these solutions and the appropriate change of variables, the minimum-error function reduces to

$$E[x(t), t] = \rho^{\frac{1}{2}}[Q - x(t)]^2 \tanh \frac{T - t}{\rho^{\frac{1}{2}}} \tag{5-83}$$

and the optimum control equation reduces to

$$m(t) = \rho^{-\frac{1}{2}}[Q - q(t)] \tanh \frac{T - t}{\rho^{\frac{1}{2}}} \tag{5-84}$$

An important feature of the dynamic-programming formulation and the method of solution presented here is that the resulting ordinary differential equations are stable when written in terms of the time variable η. The solutions of these equations approach finite steady-state values as $T - t$ approaches infinity. In contrast, the solution of the Euler-Lagrange equation becomes infinite as the interval $T - t$ approaches infinity except for one set of initial conditions. The stability of these solutions is an important factor when more complicated design problems requiring the use of computers are formulated.

In order to impose the fixed-point terminal-boundary condition on the dynamic-programming formulation for this simple example, the methods associated with the isoperimetric problem are used. The constraint variable $y(\mu)$ imposed for the fixed-point terminal-boundary condition is

$$y(\mu) = \int_\mu^T q'(\sigma)\, d\sigma = q(T) - q(\mu) \tag{5-85}$$

and the fixed-point terminal-boundary condition is taken to be

$$q(T) = Q \tag{5-86}$$

For the minimization problem defined by (5-62) and (5-63) but subject to the constraint given in (5-85), the condition for minimum error expressed in (5-59) reduces to

$$\min_{x'(\mu)} \left\{ [Q - x(\mu)]^2 + \rho[x'(\mu)]^2 + \lambda x'(\mu) + \frac{\partial E_c[x(\mu), \mu]}{\partial \mu} + x'(\mu) \frac{\partial E_c[x(\mu), \mu]}{\partial x(\mu)} \right\} = 0 \tag{5-87}$$

The minimum of the terms appearing in the braces of (5-87) occurs when $x'(\mu)$ is defined by

$$2\rho x'(\mu) + \lambda + \frac{\partial E_c[x(\mu), \mu]}{\partial x(\mu)} = 0 \tag{5-88}$$

and therefore the condition for minimum error further reduces to

$$[Q - x(\mu)]^2 - \frac{\lambda^2}{4\rho} + \frac{\partial E_c[x(\mu), \mu]}{\partial \mu} - \frac{\lambda}{2\rho} \frac{\partial E_c[x(\mu), \mu]}{\partial x(\mu)} - \frac{1}{4\rho} \left\{ \frac{\partial E_c[x(\mu), \mu]}{\partial x(\mu)} \right\}^2 = 0 \tag{5-89}$$

The form of the minimum-error function in this isoperimetric problem is assumed to be

$$E_c[x(\mu), \mu] = \rho^{\frac{1}{2}}[Q^2 k(\mu) - 2Q k_1(\mu) x(\mu) + k_{11}(\mu) x^2(\mu)] \tag{5-90}$$

When operations similar to those given in (5-69) to (5-71) are performed, the ordinary differential equations describing the k parameters are found to be

$$-\rho^{\frac{1}{2}} k'(\mu) = 1 - \left(\frac{\lambda/Q}{2\rho^{\frac{1}{2}}}\right)^2 + \frac{\lambda/Q}{\rho^{\frac{1}{2}}} k_1(\mu) - k_1^2(\mu) \tag{5-91}$$

$$-\rho^{\frac{1}{2}} k_1'(\mu) = 1 - \frac{\lambda/Q}{2\rho^{\frac{1}{2}}} k_{11}(\mu) - k_1(\mu) k_{11}(\mu) \tag{5-92}$$

and

$$-\rho^{\frac{1}{2}} k_{11}'(\mu) = 1 - k_{11}^2(\mu) \tag{5-93}$$

The solutions of these k parameters again are subject to the boundary conditions

$$k(T) = k_1(T) = k_{11}(T) = 0 \tag{5-94}$$

and the optimum control equation is found from (5-88) to be

$$m(t) = -\frac{\lambda}{2\rho} + \rho^{-\frac{1}{2}}[k_1(t)Q - k_{11}(t)x(t)] \tag{5-95}$$

The solution of these differential equations proceeds in the manner used to arrive at (5-82). The two k parameters that appear in (5-95) are found to be

$$k_1(\mu) = \tanh\frac{T-\mu}{\rho^{\frac{1}{2}}} + \frac{\lambda/Q}{2\rho^{\frac{1}{2}}}\left(1 - \operatorname{sech}\frac{T-\mu}{\rho^{\frac{1}{2}}}\right) \tag{5-96}$$

and

$$k_{11}(\mu) = \tanh\frac{T-\mu}{\rho^{\frac{1}{2}}} \tag{5-97}$$

Finally, the value of the Lagrange multiplier λ must be specified from the response of the control system. The differential equation describing the state variable $x(\mu)$,

$$x'(\mu) + \left(\rho^{-\frac{1}{2}}\tanh\frac{T-\mu}{\rho^{\frac{1}{2}}}\right)x(\mu) = \rho^{-\frac{1}{2}}Q\tanh\frac{T-\mu}{\rho^{\frac{1}{2}}} - \frac{\lambda}{2\rho}\operatorname{sech}\frac{T-\mu}{\rho^{\frac{1}{2}}} \tag{5-98}$$

is found by combining the state equation of the dynamic process and (5-95). Equation (5-98) can be solved by finding an integrating factor; hence the equation defining the terminal value of the state variable $x(\mu)$ becomes

$$x(T) - x(t)\operatorname{sech}\frac{T-t}{\rho^{\frac{1}{2}}} = Q\left(1 - \operatorname{sech}\frac{T-t}{\rho^{\frac{1}{2}}}\right) - \frac{\lambda}{2\rho^{\frac{1}{2}}}\tanh\frac{T-t}{\rho^{\frac{1}{2}}} \tag{5-99}$$

If the fixed-point terminal-boundary condition given in (5-86) is imposed, then the Lagrange multiplier λ becomes

$$\frac{\lambda}{2\rho^{\frac{1}{2}}} = -[Q - x(t)]\frac{\operatorname{sech}(T-t)/\rho^{\frac{1}{2}}}{\tanh(T-t)/\rho^{\frac{1}{2}}} \tag{5-100}$$

This value of the Lagrange multiplier, along with the k parameters given in (5-96) and (5-97), results in the optimum control equation

$$m(t) = \rho^{-\frac{1}{2}}[Q - q(t)]\coth\frac{T-t}{\rho^{\frac{1}{2}}} \tag{5-101}$$

Of course, this optimum control equation is precisely the control equation found from the calculus of variations for this fixed-point terminal-boundary condition.

5-5 Extension of the elementary theory to an Nth-order dynamic process

This section is devoted to a direct extension of the results of Sec. 5-1 to higher-order dynamic processes. Here the response of an Nth-order dynamic process is expressed as

$$q(t) = G[x(t), x'(t), \ldots, x^{(N)}(t), t] \tag{5-102}$$

and the control signal is expressed as

$$m(t) = F[x(t), x'(t), \ldots, x^{(N)}(t), t] \tag{5-103}$$

Also, the dynamic process is assumed to possess saturation of the special form

$$S^-(t) \leqq x^{(N)}(t) \leqq S^+(t) \tag{5-104}$$

which is expressed alternatively as

$$x^{(N)}(t) \in S(t) \tag{5-105}$$

Finally, the error index is taken to be

$$e(t) = \int_t^T h[q(\sigma), m(\sigma), \sigma] \, d\sigma \tag{5-106}$$

If (5-102) and (5-103) are substituted into (5-106), then the error index treated here is written as

$$e(t) = \int_t^T H[x(\sigma), x'(\sigma), \ldots, x^{(N)}(\sigma), \sigma] \, d\sigma \tag{5-107}$$

The state variables of this dynamic process are $x(t), x'(t), \ldots, x^{(N-1)}(t)$. Hence, the error index given in (5-107) is minimized with respect to $x^{(N)}(\sigma)$, and the minimum-error function is defined as

$$E[x(\mu), \ldots, x^{(N-1)}(\mu), \mu] = \min_{\substack{x^{(N)}(\sigma) \in S(\sigma) \\ [\mu, T]}} \int_\mu^T H[x(\sigma), \ldots, x^{(N)}(\sigma), \sigma] \, d\sigma \tag{5-108}$$

where μ is a dummy time variable on the interval $t \leqq \mu \leqq T$. Furthermore, the free-point terminal-boundary condition is

$$E[x(T), \ldots, x^{(N-1)}(T), T] = 0 \tag{5-109}$$

The condition for minimum error is developed by treating

$$\min_{\substack{x^{(N)}(\sigma) \in S(\sigma) \\ [\mu, T]}} \left\{ \int_\mu^T H[x(\sigma), \ldots, x^{(N)}(\sigma), \sigma] \, d\sigma - E[x(\mu), \ldots, x^{(N-1)}(\mu), \mu] \right\} = 0 \tag{5-110}$$

with the procedures expressed in (5-18) through (5-27) used for the first-order dynamic process. The continuous condition for minimum error then is shown to be

$$\min_{x^{(N)}(\mu) \in S(\mu)} \left\{ H[x(\mu), \ldots, x^{(N)}(\mu), \mu] + \frac{dE[x(\mu), \ldots, x^{(N-1)}(\mu), \mu]}{d\mu} \right\} = 0 \tag{5-111}$$

Furthermore, an alternative form of (5-111),

$$\min_{x^{(N)}(\mu) \in S(\mu)} \left\{ H[x(\mu), \ldots, x^{(N)}(\mu), \mu] + \sum_{n=0}^{N-1} x^{(n+1)}(\mu) \frac{\partial E[x(\mu), \ldots, x^{(N-1)}(\mu), \mu]}{\partial x^{(n)}(\mu)} \right\} = -\frac{\partial E[x(\mu), \ldots, x^{(N-1)}(\mu), \mu]}{\partial \mu} \tag{5-112}$$

is found by the expansion of the total time derivative of the minimum-error function.

This section is included in order to indicate the relatively minor modifications required to extend the dynamic-programming condition for minimum error. In fact, the form of the results expressed in (5-111) is pleasingly simple, with only the error measure and the total time derivative of the minimum-error function appearing in the condition. The extension to the Nth-order dynamic process merely requires the introduction of additional variables in the functions H and E.

5-6 Generalized condition for minimum error

As suggested in Chap. 4, the form of the dynamic process assumed in (5-102) and (5-103) generally cannot be obtained from the physical description of the dynamic process. Therefore, the dynamic-programming condition for minimum error is presented here for the more general dynamic process described by

$$\mathbf{q}(t) = \mathbf{g}[\mathbf{x}(t), t] \tag{5-113}$$

$$\mathbf{x}'(t) = \mathbf{f}[\mathbf{x}(t), \mathbf{m}(t), t] \tag{5-114}$$

and

$$\mathbf{m}(t) \in \mathscr{M}(t) \tag{5-115}$$

The vector signals appearing in this description of the dynamic process are defined in (1-16). For this generalized case, the error index is taken to be

$$e(t) = \int_t^T h[\mathbf{q}(\sigma), \mathbf{m}(\sigma), \sigma]\, d\sigma \tag{5-116}$$

Because the assumed dynamic process is state-determined, the response vector can be represented by the functional

$$\mathbf{q}(\sigma) = \mathscr{G}[\mathbf{x}(t), \sigma, t; \mathbf{m}(\rho), t < \rho \leqq \sigma] \tag{5-117}$$

according to the arguments given in Appendix D. This functional is analogous to (5-5) for the initial derivation of dynamic programming. In this analogy, the vector $\mathbf{m}(\sigma)$ is equivalent to $x'(\sigma)$; hence $\mathbf{m}(\sigma)$ is taken here to be the minimizing set of variables. Furthermore, the state vector $\mathbf{x}(t)$ determines the minimizing values of $\mathbf{m}(\sigma)$. Therefore, the minimum-error function is defined as

$$E[\mathbf{x}(\mu), \mu] = \min_{\substack{\mathbf{m}(\sigma) \in \mathscr{M}(\sigma) \\ [\mu, T]}} \int_\mu^T h[\mathbf{q}(\sigma), \mathbf{m}(\sigma), \sigma]\, d\sigma \tag{5-118}$$

where

$$E[\mathbf{x}(T), T] = 0 \tag{5-119}$$

The condition for minimum error now can be developed in exactly the same fashion as developed in (5-18) through (5-27) of Sec. 5-1. The condition for minimum error for this generalized case becomes

$$\min_{\mathbf{m}(\mu) \in \mathscr{M}(\mu)} \left\{ h[\mathbf{q}(\mu), \mathbf{m}(\mu), \mu] + \frac{dE[\mathbf{x}(\mu), \mu]}{d\mu} \right\} = 0 \tag{5-120}$$

Furthermore, the total time derivative of the minimum-error function is expanded in terms of partial derivatives as

$$\frac{dE[\mathbf{x}(\mu), \mu]}{d\mu} = \frac{\partial E[\mathbf{x}(\mu), \mu]}{\partial \mu} + \sum_{n=1}^{N} x_n'(\mu) \frac{\partial E[\mathbf{x}(\mu), \mu]}{\partial x_n(\mu)} \tag{5-121}$$

When (5-113) and (5-114) are substituted into the condition for minimum error, the condition for minimum error (written explicitly in terms of the state- and control-signal vectors) becomes

$$\min_{\mathbf{m}(\mu) \in \mathscr{M}(\mu)} \Bigg(h\{\mathbf{g}[\mathbf{x}(\mu), \mu], \mathbf{m}(\mu), \mu\} + \sum_{n=1}^{N} f_n[\mathbf{x}(\mu), \mathbf{m}(\mu), \mu] \frac{\partial E[\mathbf{x}(\mu), \mu]}{\partial x_n(\mu)} \Bigg) = -\frac{\partial E[\mathbf{x}(\mu), \mu]}{\partial \mu} \tag{5-122}$$

If the minimizing control vector, defined as $\mathbf{m}^*(\mu)$ such that $\mathbf{m}^*(\mu) \in \mathscr{M}(\mu)$, is found from (5-122), then the minimum-error function is specified by the solution of

$$\frac{\partial E[\mathbf{x}(\mu), \mu]}{\partial \mu} + h\{\mathbf{g}[\mathbf{x}(\mu), \mu], \mathbf{m}^*(\mu), \mu\} + \sum_{n=1}^{N} f_n[\mathbf{x}(\mu), \mathbf{m}^*(\mu), \mu] \frac{\partial E[\mathbf{x}(\mu), \mu]}{\partial x_n(\mu)} = 0 \tag{5-123}$$

5-7 Condition for minimum error with statistical signals

The minimization problems discussed in Chap. 4 and in the preceding sections of this chapter pertain only to design problems with deterministic signals. Because statistical signals arise frequently in control systems, a treatment of statistical signals in optimization problems where the configuration of the system is established is needed. This section is devoted to a treatment of minimization problems in the dynamic-programming format with statistical signals.

The dynamic process assumed here is defined by

$$\mathbf{q}(t) = \mathbf{g}[\mathbf{x}(t), t] \qquad \mathbf{x}'(t) = \mathbf{f}\,[\mathbf{x}(t), \mathbf{m}(t), t] \qquad \mathbf{m}(t) \in \mathscr{M}(t) \tag{5-124}$$

However, the error measure is assumed to be dependent upon the vector signal $\mathbf{s}(\sigma)$, which is defined as

$$\mathbf{s}(\sigma) = \begin{bmatrix} s_1(\sigma) \\ s_2(\sigma) \\ \cdot \\ \cdot \\ \cdot \\ s_S(\sigma) \end{bmatrix} \tag{5-125}$$

If the vector $\mathbf{s}(\sigma)$ represents a set of deterministic signals, the error index is written as

$$e(t) = \int_t^T h[\mathbf{s}(\sigma), \mathbf{q}(\sigma), \mathbf{m}(\sigma), \sigma]\, d\sigma \tag{5-126}$$

Typical signals occurring in the vector $\mathbf{s}(\sigma)$ are desired-response and desired-control signals.

On the other hand, if the vector $\mathbf{s}(\sigma)$ represents a set of statistical signals, then the value of the error index expressed in (5-126) is dependent upon the particular ensemble member of this vector used in the integration. Specifically, this error index is written as

$${}_m e(t) = \int_t^T h[{}_m\mathbf{s}(\sigma), \mathbf{q}(\sigma), \mathbf{m}(\sigma), \sigma]\, d\sigma \tag{5-127}$$

where ${}_m\mathbf{s}(\sigma)$ is defined as the mth ensemble member of the vector $\mathbf{s}(\sigma)$. Clearly, the minimizing control vector for ${}_m e(t)$ depends upon the particular member of the ensemble chosen so that the optimum control signal must be represented as ${}_m\mathbf{m}^*(\sigma)$. Unfortunately, this minimizing control vector cannot be found, because the ensemble member ${}_m\mathbf{s}(\sigma)$ cannot be identified exactly on the future interval of time $t < \sigma \leqq T$. Therefore, an error index involving statistical averaging must be adopted.

The statistical averaging adopted here is the conditional mean. Therefore, the error index chosen for minimization is

$$\overline{e(t)}^{e|\mathbf{v}} = \int_{-\infty}^{\infty} e_t p(e_t \mid \mathbf{v}_\rho;\, t, -\infty < \rho \leqq t)\, de_t \tag{5-128}$$

using the notation defined in Appendix C. The vector $\mathbf{v}(\rho)$ is assumed to be a measured input vector and is defined as

$$\mathbf{v}(\rho) = \begin{bmatrix} v_1(\rho) \\ v_2(\rho) \\ \cdot \\ \cdot \\ \cdot \\ v_V(\rho) \end{bmatrix} \tag{5-129}$$

The conditional-probability density used in (5-128) depends on $\mathbf{v}(\rho)$ only over the interval of past and present time $-\infty < \rho \leqq t$, because this vector is assumed to be a measured set of signals. Finally, the error index to be minimized is expressed as

$$\overline{e(t)}^{e|\mathbf{v}} = \overline{\int_t^T h[\mathbf{s}(\sigma), \mathbf{q}(\sigma), \mathbf{m}(\sigma), \sigma]\, d\sigma}^{e|\mathbf{v}} \tag{5-130}$$

in terms of (5-126) and the abbreviated notation used for ensemble averages.

Before the condition for minimum error can be derived, certain assumptions about the dynamic process must be made. In particular, the dynamic process is assumed to be deterministic in the sense that the state vector $\mathbf{x}(\mu)$ is known deterministically when the control vector $\mathbf{m}(\sigma)$ is given as a set of deterministic functions on the interval $t \leqq \sigma \leqq \mu$. This assumption implies that the existing state $\mathbf{x}(t)$ is given deterministically and that the function $\mathbf{f}$ appearing in the state equation must be a deterministic function of its arguments including time.

Now, the minimization of (5-130) is developed under the assumption that the dynamic process is deterministic. In other words, the vector $\mathbf{s}(\sigma)$ does not contain either load-disturbance signals $\mathbf{u}(\sigma)$ or state signals $\mathbf{x}(\sigma)$. Suppose for the moment that real time t is considered a fixed number. At this instant of real time, the conditional-probability density contains complete statistical information and has parameters which are fixed by the measured $\mathbf{v}(\rho)$ on the interval $-\infty < \rho \leqq t$. With this conditional-probability density fixed, the control vector $\mathbf{m}(\sigma)$ must be chosen deterministically in order to specify an explicit control equation. Because the minimizing control vector must be treated as deterministic for a fixed value of real time, the state vector $\mathbf{x}(\sigma)$ and hence the response vector $\mathbf{q}(\sigma)$ are deterministic functions for a deterministic dynamic process. Therefore, the error index given in (5-130) is written as

$$\overline{e(t)}^{e|\mathbf{v}} = \int_t^T \overline{h[\mathbf{s}(\sigma), \mathbf{q}(\sigma), \mathbf{m}(\sigma), \sigma]}^{\mathbf{s}|\mathbf{v}}\, d\sigma \tag{5-131}$$

when the averaging is taken inside the integral.

As indicated in (5-131), the deterministic vectors $\mathbf{q}(\sigma)$ and $\mathbf{m}(\sigma)$ do not enter into the averaging process. The notation associated with the integrand of (5-131) implies the average

$$\overline{h[\mathbf{s}(\sigma), \mathbf{q}(\sigma), \mathbf{m}(\sigma), \sigma]}^{\mathbf{s}|\mathbf{v}} = \int_{-\infty}^{\infty} \cdots \int_{-\infty}^{\infty} h[\mathbf{s}(\sigma), \mathbf{q}(\sigma), \mathbf{m}(\sigma), \sigma] p[\mathbf{s}(\sigma) \mid \mathbf{v}(\rho);\, \sigma, -\infty < \rho \leqq t]\, ds_1(\sigma) \cdots ds_S(\sigma) \tag{5-132}$$

In theory, this average can be performed for any fixed value of real time; hence real time enters the resulting function as a parameter. The resulting function $\bar{h}$ is defined as

$$\overline{h[\mathbf{s}(\sigma), \mathbf{q}(\sigma), \mathbf{m}(\sigma), \sigma]}^{\mathbf{s}|\mathbf{v}} = \bar{h}[\mathbf{q}(\sigma), \mathbf{m}(\sigma), \sigma, t] \tag{5-133}$$

Finally, the error index to be minimized for every fixed value of real time t is

$$\overline{e(t)}^{e|\mathbf{v}} = \int_t^T \bar{h}[\mathbf{q}(\sigma), \mathbf{m}(\sigma), \sigma, t]\, d\sigma \tag{5-134}$$

The phrase *for every fixed value of real time* implies that the condition for minimizing (5-134) is an instantaneous condition and must be solved instantaneously as $\mathbf{v}(t)$ is measured during the operation of the control system. Only in special cases can this requirement for real-time solutions be circumvented.

The error index given in (5-134) contains no statistical variables; hence, as in the previous derivations, the minimum-error function is defined as

$$E[\mathbf{x}(t), t] = \min_{\substack{\mathbf{m}(\sigma) \in \mathscr{M}(\sigma) \\ [t,\, T]}} \int_t^T \bar{h}[\mathbf{q}(\sigma), \mathbf{m}(\sigma), \sigma, t]\, d\sigma \tag{5-135}$$

In this definition, real time t cannot be replaced by the dummy variable μ, because of the dependence of $\bar{h}$ on t. If t is replaced by μ in $\bar{h}$, then $\bar{h}$ is not known for $\mu > t$, because future values of $\mathbf{v}(\rho)$ cannot be measured and hence are not known. Therefore, a new function $\mathscr{E}$ is defined such that

$$\mathscr{E}[\mathbf{x}(\mu), \mu] = \min_{\substack{\mathbf{m}(\sigma) \in \mathscr{M}(\sigma) \\ [\mu,\, T]}} \int_\mu^T \bar{h}[\mathbf{q}(\sigma), \mathbf{m}(\sigma), \sigma, t]\, d\sigma \tag{5-136}$$

This function, called the *instantaneous minimum-error function*, possesses the properties

$$\mathscr{E}[\mathbf{x}(t), t] = E[\mathbf{x}(t), t]$$

and

$$\mathscr{E}[\mathbf{x}(\mu), \mu] \neq E[\mathbf{x}(\mu), \mu] \qquad t < \mu \leqq T \tag{5-137}$$

which are a direct result of the definitions.

The continuous dynamic-programming condition now is derived for $\mathscr{E}$ by the same methods used previously. Of course, real time t is treated as a fixed parameter in this derivation. Finally, the *instantaneous condition for minimum error* is written as

$$\min_{\mathbf{m}(\mu) \in \mathscr{M}(\mu)} \left\{ \bar{h}[\mathbf{q}(\mu), \mathbf{m}(\mu), \mu, t\,] + \frac{d\mathscr{E}[\mathbf{x}(\mu), \mu]}{d\mu} \right\} = 0 \tag{5-138}$$

subject to the boundary condition

$$\mathscr{E}[\mathbf{x}(T), T] = 0 \tag{5-139}$$

The solution of (5-138) on the interval $t \leqq \mu \leqq T$ for every value of t during system operation results in the optimum control vector $\mathbf{m}^*(t)$.

5-8 Example—statistical signals

In order to demonstrate the treatment of statistical signals in control-system optimization, a simple example is included here. The dynamic process is taken to be

$$q_1(t) = x_1(t) \qquad x_1'(t) = m_1(t) \qquad -\infty < m_1(t) < \infty \tag{5-140}$$

and the error measure is taken to be

$$h[\mathbf{s}(\sigma), \mathbf{q}(\sigma), \mathbf{m}(\sigma), \sigma] = \phi(\sigma)[Q_1(\sigma) - q_1(\sigma)]^2 + m_1^2(\sigma) \tag{5-141}$$

The statistical signals occurring in this example are defined by

$$s_1(\sigma) = Q_1(\sigma) \qquad v_1(\rho) = Q_1(\rho) \tag{5-142}$$

Therefore, the error measure $\bar{h}$ is

$$\bar{h}[\mathbf{q}(\sigma), \mathbf{m}(\sigma), \sigma, t] = \phi(\sigma)[\overline{Q_1^2(\sigma)}^{Q_1|Q_1} - 2q_1(\sigma)\overline{Q_1(\sigma)}^{Q_1|Q_1} + q_1^2(\sigma)] + m_1^2(\sigma) \tag{5-143}$$

For this example, the instantaneous condition for minimum-error given in (5-138) reduces to

$$\min_{m_1(\mu)} \left\{\phi(\mu)[\overline{Q_1^2(\mu)}^{Q_1|Q_1} - 2x_1(\mu)\overline{Q_1(\mu)}^{Q_1|Q_1} + x_1^2(\mu)] + m_1^2(\mu) + m_1(\mu)\frac{\partial\mathscr{E}[x_1(\mu), \mu]}{\partial x_1(\mu)}\right\} = -\frac{\partial\mathscr{E}[x_1(\mu), \mu]}{\partial\mu} \tag{5-144}$$

The value of the control signal $m_1(\mu)$ that minimizes the terms in the braces appearing in (5-144) is

$$m_1(\mu) = -\frac{1}{2}\frac{\partial\mathscr{E}[x_1(\mu), \mu]}{\partial x_1(\mu)} \tag{5-145}$$

Therefore, the instantaneous minimum-error function is specified by

$$\phi(\mu)[\overline{Q_1^2(\mu)}^{Q_1|Q_1} - 2x_1(\mu)\overline{Q_1(\mu)}^{Q_1|Q_1} + x_1^2(\mu)] - \frac{1}{4}\left\{\frac{\partial\mathscr{E}[x_1(\mu), \mu]}{\partial x_1(\mu)}\right\}^2 = -\frac{\partial\mathscr{E}[x_1(\mu), \mu]}{\partial\mu} \tag{5-146}$$

The solution of this last partial differential equation can be written as the solution of three ordinary differential equations by employing the same method described in Sec. 5-4 for deterministic signals. If the instantaneous minimum-error function for the first-order dynamic process is assumed to have the form

$$\mathscr{E}[x_1(\mu), \mu] = k(\mu) - 2k_1(\mu)x_1(\mu) + k_{11}(\mu)x_1^2(\mu) \tag{5-147}$$

then (5-146) reduces to

$$[\phi(\mu)\overline{Q_1^2(\mu)}^{Q_1|Q_1} + k'(\mu) - k_1^2(\mu)] - 2x_1(\mu)[\phi(\mu)\overline{Q_1(\mu)}^{Q_1|Q_1} + k_1'(\mu) - k_1(\mu)k_{11}(\mu)] + x_1^2(\mu)[\phi(\mu) + k_{11}'(\mu) - k_{11}^2(\mu)] = 0 \tag{5-148}$$

Therefore, the k parameters appearing in the instantaneous minimum-error function satisfy

$$-k'(\mu) = \phi(\mu)\overline{Q_1^2(\mu)}^{Q_1|Q_1} - k_1^2(\mu) \tag{5-149}$$

$$-k_1'(\mu) = \phi(\mu)\overline{Q_1(\mu)}^{Q_1|Q_1} - k_1(\mu)k_{11}(\mu) \tag{5-150}$$

and

$$-k_{11}'(\mu) = \phi(\mu) - k_{11}^2(\mu) \tag{5-151}$$

The boundary conditions for these k parameters are

$$k(T) = k_1(T) = k_{11}(T) = 0 \tag{5-152}$$

Finally, the optimum control signal, as expressed in terms of the k parameters, is

$$m_1^*(t) = k_1(t) - k_{11}(t)x_1(t) \tag{5-153}$$

In order to find an explicit solution to the differential equations describing the k parameters, the weighting factor $\phi(\mu)$ is specialized to

$$\phi(\mu) = \phi u_0(\mu - T) \tag{5-154}$$

where the function $u_0(x)$ is the unit-impulse function at $x = 0$. For this form of the weighting factor $\phi(\mu)$, which would be useful in a terminal-control system, the differential equations given in (5-149) to (5-151) reduce to

$$k'(\mu) = k_{11}{}^2(\mu) \qquad k_1'(\mu) = k_1(\mu)k_{11}(\mu) \qquad k_{11}'(\mu) = k_{11}{}^2(\mu) \tag{5-155}$$

where now the boundary conditions on the k parameters are

$$k(T) = \phi\overline{Q_1{}^2(T)}^{Q_1|Q_1} \qquad k_1(T) = \phi\overline{Q_1(T)}^{Q_1|Q_1} \qquad k_{11}(T) = \phi \tag{5-156}$$

These boundary conditions are found from evaluating the unit-impulse function in the integration of these equations.§ Finally, the solutions of the differential equations

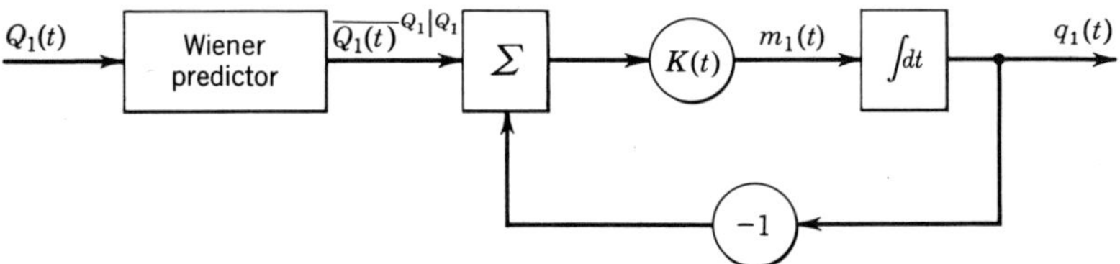

Fig. 5-2 Block diagram of optimum control system.

given in (5-155) subject to these boundary conditions readily are found to be

$$k(t) = \phi\overline{Q_1{}^2(T)}^{Q_1|Q_1} - \frac{(T-t)\phi^2}{(T-t)\phi + 1}[\overline{Q_1(T)}^{Q_1|Q_1}]^2 \tag{5-157}$$

$$k_1(t) = \frac{\phi}{(T-t)\,\phi + 1}\,\overline{Q_1(T)}^{Q_1|Q_1} \tag{5-158}$$

and

$$k_{11}(t) = \frac{\phi}{(T-t)\,\phi + 1} \tag{5-159}$$

Therefore, the optimum control equation is

$$m_1^*(t) = K(t)[\overline{Q_1(T)}^{Q_1|Q_1} - q_1(t)] \qquad K(t) = \frac{\phi}{(T-t)\phi + 1} \tag{5-160}$$

when the appropriate k parameters are substituted into (5-153). The optimum controller for this design problem is thus simply a time-varying gain $K(t)$ where the reference signal in the control system is the conditional mean of $Q_1(T)$.

Because the reference signal in the optimum control system is the conditional mean of the desired response at the future time T, this reference signal can be generated from a filter resulting from Wiener theory as developed in Chap. 3. Therefore, a block diagram of this optimum control system can be represented, as shown in Fig. 5-2, where the input signal is the measured signal $Q_1(t)$.

§ In this book, all impulse weighting factors introduced at the terminal point $\sigma = T$ are evaluated as the left-hand limit $\sigma = T^-$.

A point of interest is that the problems of prediction and control can be separated both conceptually and physically in the design of this optimum control system. The conceptual separation of prediction and control is indicated by the condition for minimum error in (5-138). The optimum control system is found from the solution of this partial differential equation when the conditional mean of the error measure is given. The conditional mean of the error measure relates to the problem of finding the conditional mean of the desired signal, which in turn is a problem in statistical prediction and can be treated separately on a conceptual basis. Also, in the case of the particular example discussed in this section, the control system and the prediction system are physically separable, as indicated in Fig. 5-2. This physical separation is developed in Chap. 7 for much more general design problems. In design problems where this physical separation occurs, the concepts of the instantaneous minimum-error function $\mathscr{E}$ and the real-time solution of (5-138) can be circumvented.

5-9 Summary

The concept of feedback-control-system optimization is reformulated here in the format of dynamic programming as opposed to the calculus of variations. These two formulations are equivalent when control-vector saturation does not occur.

The dynamic-programming formulation is based on a number of concepts that are extremely important and plausible to the control-systems engineer. Dynamic programming is formulated on the concept of a state associated with a dynamic process which is described by ordinary differential equations. In particular, the minimum-error function is shown to be a function only of the existing state in the dynamic process and the time at which the state is observed. Second, the selection of the optimum control vector is considered to be a sequence of optimal decisions, as is stated in Bellman's principle of optimality. This sequence of optimal decisions is conceptually useful in defining a class of adaptive controls in terms of optimization theory. In particular, the policy by which the optimum sequence of decisions is made alters with changing design specifications and characteristics of the dynamic process.

One of the advantages of the dynamic-programming formulation is that the traditional two-point boundary-value problem associated with the calculus of variations is eliminated, at least conceptually. This elimination results from the procedure of flooding, which is associated with the dynamic-programming formulation. In particular, the solution to the particular optimization problem, as determined by specific values for the state variables of the dynamic process, is contained in the problem of finding the optimum sequence of control decisions for all possible combinations of the intial state variables. This process of flooding therefore eliminates the computational difficulties associated with solving a two-point boundary-value problem. However, because this flooding process is exhaustive in nature in the sense that all possible solutions are generated, severe new computational difficulties arise. These new computational difficulties can be more restrictive than the computational difficulties associated with two-point boundary-value problems. In order to use mathematical optimization theory as a method for the design of control systems, numerical solutions must be obtainable for a wide variety of design problems; therefore computational considerations are of extreme importance in the context of this book.

6

Dynamic-programming Solutions Using Classical Methods

In order to determine the optimum control system using the dynamic-programming formulation, the partial differential equation given in (5-123) must be solved. When computers are used to determine the optimum control system, great simplifications would occur if this differential equation could be expressed alternatively as a set of ordinary differential equations. In the very simple example of Sec. 5-4, the optimum control system is expressed as the solution of a set of differential equations. This chapter is devoted to the classical method for solving partial differential equations of the type appearing in (5-123) with ordinary differential equations.

The type of partial differential equation of concern here is first-order because all partial derivatives are of first order. Furthermore, the partial differential equation is nonlinear because the partial derivatives $\partial E/\partial x_n$ occur in second or higher degrees. However, the dependent variable E does not occur in the partial differential equation, and the partial derivative $\partial E/\partial \mu$ always occurs in the first degree. An elementary example of this type of equation is seen in (5-67).

In the classical theory of partial differential equations, the type of equation described here is called the *Hamilton-Jacobi equation.*

Extensive literature is available on the solution of the Hamilton-Jacobi equation, but most of this literature is devoted to methods for obtaining closed-form analytical solutions. On the other hand, this chapter is devoted to only the portion of classical theory that is required for numerical solutions. In particular, Cauchy's method of characteristics is developed as a method of solving (5-123) by solving instead a set of ordinary differential equations.

6-1 Elementary theory for quasi-linear equations with two independent variables

The concepts of Cauchy's method are introduced by first considering the solution of the following quasi-linear partial differential equation of the first order:[21]

$$P(x, y, z)\frac{\partial z}{\partial x} + Q(x, y, z)\frac{\partial z}{\partial y} = R(x, y, z) \tag{6-1}$$

This equation is called *quasi-linear* because the partial derivatives of z with respect to x and y appear linearly in the equation. However, the occurrence of z in the functions P and Q implies that the equation is not linear with respect to z. The solution of this partial differential equation is of the mathematical form

$$u(x, y, z) = 0 \tag{6-2}$$

where x and y are considered to be independent variables determining the dependent variable z in terms of the function u. The function u defines a surface in xyz space which is called the *solution surface*.

Equation (6-2) now is differentiated with respect to the independent variables x and y, resulting in the partial differential equations

$$\frac{du}{dx} = \frac{\partial u}{\partial x} + \frac{\partial u}{\partial z}\frac{\partial z}{\partial x} = 0 \qquad y = \text{const} \tag{6-3}$$

and

$$\frac{du}{dy} = \frac{\partial u}{\partial y} + \frac{\partial u}{\partial z}\frac{\partial z}{\partial y} = 0 \qquad x = \text{const} \tag{6-4}$$

These last two equations are solved for the partial derivatives appearing in (6-1), thereby yielding

$$\frac{\partial z}{\partial x} = -\frac{\partial u/\partial x}{\partial u/\partial z} \qquad \frac{\partial z}{\partial y} = -\frac{\partial u/\partial y}{\partial u/\partial z} \qquad \text{for } \frac{\partial u}{\partial z} \neq 0 \tag{6-5}$$

If (6-5) is substituted into (6-1), then the original partial differential equation, now written in terms of the partial derivatives of the solution surface, becomes

$$P(x, y, z)\frac{\partial u}{\partial x} + Q(x, y, z)\frac{\partial u}{\partial y} + R(x, y, z)\frac{\partial u}{\partial z} = 0 \tag{6-6}$$

The form of the partial differential equation appearing in (6-6) is preferred to that of the form appearing in (6-1) because the partial derivatives play a symmetrical role in the equation.

If the vector $\mathbf{V}$ is defined as

$$\mathbf{V} = P(x, y, z)\mathbf{i}_x + Q(x, y, z)\mathbf{i}_y + R(x, y, z)\mathbf{i}_z \tag{6-7}$$

and if the gradient vector of the function u is defined as

$$\nabla u = \frac{\partial u}{\partial x}\mathbf{i}_x + \frac{\partial u}{\partial y}\mathbf{i}_y + \frac{\partial u}{\partial z}\mathbf{i}_z \tag{6-8}$$

then the partial differential equation appearing in (6-6) is written vectorially as

$$\mathbf{V} \cdot \nabla u = 0 \tag{6-9}$$

The gradient vector ∇u is a vector normal to the surface $u = 0$. Because of the relationship given in (6-9), the vector $\mathbf{V}$ is perpendicular to the gradient vector. Therefore, the vector $\mathbf{V}$ lies in the tangent plane of the solution surface, and this vector points in a direction on that surface for which u is equal to a constant. If a point (x, y, z) on the surface u that always travels in a direction defined by the direction of the vector $\mathbf{V}$ is considered, this point traces out a space curve. This particular space curve is called a *characteristic curve*. The surface u can be thought of as being constructed from a family of characteristic curves. This surface constructed from a family of characteristic curves has the property that any tangent plane to the surface contains the vector $\mathbf{V}$, and the normal vector to this surface is perpendicular to the vector $\mathbf{V}$. This surface constructed as a family of characteristic curves is called an *integral surface* of the partial differential equation.

A vector $\mathbf{r}$ now is defined from the origin of xyz space to the point on the integral surface. If the point (x, y, z) moves in space with an arc length along the trajectory defined as s, then this position vector is a function defined as $\mathbf{r}(s)$. Such a position vector has the property that the derivative of the position vector with respect to arc length is the unit tangent vector to the curve at that point. In other words, the rate of change of the position vector with respect to arc length is

$$\frac{d\mathbf{r}}{ds} = \frac{dx}{ds}\mathbf{i}_x + \frac{dy}{ds}\mathbf{i}_y + \frac{dz}{ds}\mathbf{i}_z \tag{6-10}$$

which has a direction that is tangent to the space curve at that point. If this tangent vector is to have the same direction as the vector $\mathbf{V}$, then the components of these two vectors must be proportional so that

$$P(x, y, z) = \mu(x, y, z)\frac{dx}{ds} \qquad Q(x, y, z) = \mu(x, y, z)\frac{dy}{ds} \qquad R(x, y, z) = \mu(x, y, z)\frac{dz}{ds} \tag{6-11}$$

The function $\mu(x, y, z)$ is called a *primitive* of the partial differential equation appearing in (6-6). Conceptually, a primitive possesses the same properties as an integrating factor in the theory of ordinary differential equations. Finally, Eq. (6-11) can be rewritten in the form

$$\frac{ds}{\mu(x, y, z)} = \frac{dx}{P(x, y, z)} = \frac{dy}{Q(x, y, z)} = \frac{dz}{R(x, y, z)} \tag{6-12}$$

The equation appearing in (6-12) is called the *subsidiary equation* of the original partial differential equation appearing in (6-1).

The second and third terms appearing in (6-12) can be used to obtain an ordinary differential equation. Also, the second and fourth terms in (6-12) can be used to find another ordinary differential equation. These two ordinary differential equations are

$$\frac{dy}{dx} = \frac{Q(x, y, z)}{P(x, y, z)} \qquad \frac{dz}{dx} = \frac{R(x, y, z)}{P(x, y, z)} \tag{6-13}$$

With this method of characteristics, the solution of the first-order quasi-linear partial differential equation appearing in (6-1) is reduced to the solution of two first-order ordinary differential equations given in (6-13). These two ordinary differential equations can be solved readily, for instance, on an analog computer if the variable x is identified with time. The solutions of these two ordinary differential equations can be thought of as defining the two-parameter surface

$$U(x, y, z; a, b) = 0 \tag{6-14}$$

Also, the parameters a and b can be thought of as the initial conditions for y and z assumed in the solution of the two differential equations appearing in (6-13). Therefore, the solution of the partial differential equation given in (6-1) defines a *two-parameter family of surfaces*. Conversely, a two-parameter family of surfaces implies a partial differential equation of first order with two independent variables.

The success in determining the two differential equations given in (6-13) for the solution of a partial differential equation is based on the property that the partial differential equation is quasi-linear. In other words, the functions P, Q, and R do not involve the partial derivatives of z with respect to x and y. However, this method can be extended to nonlinear partial differential equations of first order, which is of course the case dealt with in the solution of the condition for minimum error, by the introduction of additional ordinary differential equations.[42]

6-2 Characteristic equations for two independent variables

Now the partial differential equation

$$F(x, y, z, p, q) = 0 \tag{6-15}$$

is considered, where the partial derivatives are defined as

$$p = \frac{\partial z}{\partial x} \qquad q = \frac{\partial z}{\partial y} \tag{6-16}$$

for convenience. The function F appearing in (6-15) is arbitrary so that this equation may be a nonlinear first-order partial differential equation.

In order to develop a method possessing the same geometrical properties as the method of characteristics for quasi-linear equations, a plane is defined in xyz space as

$$P(x, y, z) = c_x(x - x_0) + c_y(y - y_0) + c_z(z - z_0) = 0 \tag{6-17}$$

The constants c_x, c_y, and c_z define the orientation of the plane. The vector normal to this plane has a direction given by the direction of

$$\nabla P = c_x\mathbf{i}_x + c_y\mathbf{i}_y + c_z\mathbf{i}_z \tag{6-18}$$

If ∇P has the same direction as ∇u at (x_0, y_0, z_0), then

$$\frac{\partial u/\partial x}{c_x} = \frac{\partial u/\partial y}{c_y} = \frac{\partial u/\partial z}{c_z} = \mu(x_0, y_0, z_0) \tag{6-19}$$

From the relationships between the partial derivatives of the solution surface given in (6-5) and from the definitions given in (6-16), the constants defining the plane P are found from (6-19) to be

$$\frac{c_x}{c_z} = -p \qquad \frac{c_y}{c_z} = -q \tag{6-20}$$

When (6-20) is substituted into (6-17), the plane having a normal parallel to ∇u is

$$z - z_0 = p(x - x_0) + q(y - y_0) \tag{6-21}$$

If z_0 is chosen so that $u(x_0, y_0, z_0) = 0$, then the plane given in (6-21) is the tangent plane to the solution surface $u(x, y, z) = 0$ at the point (x_0, y_0, z_0).

Finally, an additional relationship between p and q is needed in order to satisfy the partial differential equation given in (6-15). At the point (x_0, y_0, z_0), (6-15) gives a relationship between p and q; and this relationship is found by differentiating (6-15) so that

$$\frac{dF}{dp} = \frac{\partial F}{\partial p} + \frac{\partial F}{\partial q}\frac{dq}{dp} = 0 \tag{6-22}$$

Also, differentiation of (6-21), with respect to p so that only the orientation of the plane is varied, results in

$$0 = (x - x_0) + (y - y_0)\frac{dq}{dp} \tag{6-23}$$

The elimination of dq/dp from (6-22) and (6-23) yields

$$\frac{\partial F}{\partial p}(y - y_0) = \frac{\partial F}{\partial q}(x - x_0) \tag{6-24}$$

and the elimination of $(y - y_0)$ from (6-21) and (6-24) yields

$$\frac{\partial F}{\partial p}(z - z_0) = \left(p\frac{\partial F}{\partial p} + q\frac{\partial F}{\partial q}\right)(x - x_0) \tag{6-25}$$

If this tangent plane is reduced in size so that $x = x_0 + dx$, $y = y_0 + dy$, and $z = z_0 + dz$, then (6-24) and (6-25) reduce to

$$\frac{\partial F}{\partial p}dy = \frac{\partial F}{\partial q}dx \qquad \frac{\partial F}{\partial p}dz = \left(p\frac{\partial F}{\partial p} + q\frac{\partial F}{\partial q}\right)dx \tag{6-26}$$

Equation (6-26) is precisely the subsidiary equation given in (6-12) for the quasi-linear case. From (6-21), the tangent plane (reduced in size) is

$$dz = p\,dx + q\,dy \tag{6-27}$$

which is called the *elementary tangent plane* and is tangent to the solution surface $u(x, y, z) = 0$. The solutions of (6-26) therefore are characteristic curves.

As discussed previously, additional equations giving the increments dp and dq are needed in the subsidiary equation of a nonlinear partial differential equation because $\partial F/\partial p$ and $\partial F/\partial q$ involve p and q. These additional equations are found from expressing the incremental changes in p and q in terms of the incremental changes in the independent variables x and y. In terms of partial derivatives, these increments are

$$dp = \frac{\partial p}{\partial x}\,dx + \frac{\partial p}{\partial y}\,dy \tag{6-28}$$

and

$$dq = \frac{\partial q}{\partial x}\,dx + \frac{\partial q}{\partial y}\,dy \tag{6-29}$$

However, if p and q are continuous functions of both x and y, the order of partial differentiations can be interchanged so that

$$\frac{\partial p}{\partial y} = \frac{\partial q}{\partial x} \tag{6-30}$$

The substitution of the first equation in (6-26) and (6-30) into (6-28) and (6-29) yields the increment changes in p and q due to an incremental change in x. These increments are

$$dp = \left(\frac{\partial p}{\partial x}\frac{\partial F}{\partial p} + \frac{\partial q}{\partial x}\frac{\partial F}{\partial q}\right)\frac{dx}{\partial F/\partial p} \tag{6-31}$$

and

$$dq = \left(\frac{\partial p}{\partial y}\frac{\partial F}{\partial p} + \frac{\partial q}{\partial y}\frac{\partial F}{\partial q}\right)\frac{dx}{\partial F/\partial p} \tag{6-32}$$

Furthermore, the partials of p and q with respect to x and y are eliminated from these last two equations by taking the total derivatives of (6-15) with respect to x and y, thereby obtaining

$$\frac{\partial F}{\partial x} + p\,\frac{\partial F}{\partial z} + \frac{\partial p}{\partial x}\frac{\partial F}{\partial p} + \frac{\partial q}{\partial x}\frac{\partial F}{\partial q} = 0 \qquad y = \text{const} \tag{6-33}$$

and

$$\frac{\partial F}{\partial y} + q\,\frac{\partial F}{\partial z} + \frac{\partial p}{\partial y}\frac{\partial F}{\partial p} + \frac{\partial q}{\partial y}\frac{\partial F}{\partial q} = 0 \qquad x = \text{const} \tag{6-34}$$

The substitution of these last two equations into (6-31) and (6-32) results in the incremental changes of p and q with respect to x, expressed as

$$\frac{\partial F}{\partial p}\,dp = -\left(\frac{\partial F}{\partial x} + p\,\frac{\partial F}{\partial z}\right)dx \tag{6-35}$$

and

$$\frac{\partial F}{\partial p}\,dq = -\left(\frac{\partial F}{\partial y} + q\,\frac{\partial F}{\partial z}\right)dx \tag{6-36}$$

Finally, the subsidiary equation associated with (6-15) is found from (6-26), (6-35), and (6-36) to be

$$\frac{dx}{\partial F/\partial p} = \frac{dy}{\partial F/\partial q} = \frac{dz}{p\,\partial F/\partial p + q\,\partial F/\partial q} = \frac{dp}{-(\partial F/\partial x + p\,\partial F/\partial z)} = \frac{dq}{-(\partial F/\partial y + q\,\partial F/\partial z)} \tag{6-37}$$

From this subsidiary equation, the set of four ordinary differential equations

$$\frac{dy}{dx} = \frac{\partial F/\partial q}{\partial F/\partial p} \qquad \frac{dz}{dx} = \frac{p\,\partial F/\partial p + q\,\partial F/\partial q}{\partial F/\partial p}$$
$$\frac{dp}{dx} = \frac{-(\partial F/\partial x + p\,\partial F/\partial z)}{\partial F/\partial p} \qquad \frac{dq}{dx} = \frac{-(\partial F/\partial y + q\,\partial F/\partial z)}{\partial F/\partial p} \tag{6-38}$$

can be used to solve the partial equation appearing in (6-15). When the partial differential equation is nonlinear with respect to the partial derivatives of z, two extra ordinary differential equations are needed for obtaining a solution. In the special case where (6-15) is quasi-linear, the first two equations in the set given in (6-38) are independent of p and q and therefore are sufficient for obtaining a solution.

6-3 Characteristic equations for $N + 1$ independent variables

The concept of solving a first-order partial differential equation in terms of a set of ordinary differential equations must be generalized to partial differential equations with $N + 1$ independent variables.[43] This generalization leads directly to a solution for the condition for minimum error expressed in (5-123). The partial differential equation for $N + 1$ variables is taken to be

$$F(x_0, x_1, \ldots, x_N, y, p_0, p_1, \ldots, p_N) = 0 \tag{6-39}$$

where $x_0, x_1, \ldots, x_N$ are the independent variables, y is the dependent variable, and the partial derivatives are defined as

$$p_n = \frac{\partial y}{\partial x_n} \qquad n = 0, 1, \ldots, N \tag{6-40}$$

The derivation for this generalization is analogous to the previous derivation in all respects.

The elementary tangent hyperplane of the solution surface is

$$dy = \sum_{n=0}^{N} p_n\, dx_n \tag{6-41}$$

With the point $(x_0, x_1, \ldots, x_N)$ fixed in $N + 1$-dimensional space, the partial derivative of (6-41) with respect to p_m is

$$0 = dx_m + \frac{\partial p_N}{\partial p_m}\, dx_N \qquad m = 0, 1, \ldots, N - 1 \tag{6-42}$$

Similarly, the partial derivative of (6-39) with respect to p_m is

$$\frac{\partial F}{\partial p_m} + \frac{\partial p_N}{\partial p_m}\frac{\partial F}{\partial p_N} = 0 \qquad m = 0, 1, \ldots, N - 1 \tag{6-43}$$

In the derivation of both (6-42) and (6-43), p_N is treated as the dependent variable and $p_0, p_1, \ldots, p_{N-1}$ are treated as the independent variables in the differentiations. Equations (6-42) and (6-43) now are used to eliminate $\partial p_N/\partial p_m$, thereby resulting in

$$\frac{\partial F}{\partial p_N}\, dx_m = \frac{\partial F}{\partial p_m}\, dx_N \qquad m = 0, 1, \ldots, N - 1 \tag{6-44}$$

The substitution of (6-44) into (6-41) yields the additional incremental relationship

$$\frac{\partial F}{\partial p_N} dy = \left(\sum_{n=0}^{N} p_n \frac{\partial F}{\partial p_n} \right) dx_N \tag{6-45}$$

for the independent variable y. These last two equations correspond to (6-26) for the case of two independent variables.

The incremental changes in the partial derivatives that appear in (6-39) are found by partial differentiation to be

$$dp_m = \sum_{n=0}^{N} \frac{\partial p_m}{\partial x_n} dx_n \tag{6-46}$$

If the partial derivatives of p_m are continuous in all the independent variables, the order of partial differentiations can be interchanged so that

$$\frac{\partial p_m}{\partial x_n} = \frac{\partial p_n}{\partial x_m} \tag{6-47}$$

The substitution of (6-47) into (6-46) results in the expression for the incremental change in the partial derivative p_m as

$$dp_m = \sum_{n=0}^{N} \frac{\partial p_n}{\partial x_m} dx_n \qquad m = 0, 1, \ldots, N \tag{6-48}$$

Furthermore, the substitution of (6-44) into (6-48) yields

$$\frac{\partial F}{\partial p_N} dp_m = \left(\sum_{n=0}^{N} \frac{\partial p_n}{\partial x_m} \frac{\partial F}{\partial p_n} \right) dx_N \quad m = \quad 0, 1, \ldots, N \tag{6-49}$$

If (6-39) is differentiated with respect to x_m, holding the other independent variables constant, the equalities

$$\frac{\partial F}{\partial x_m} + p_m \frac{\partial F}{\partial y} + \sum_{n=0}^{N} \frac{\partial p_n}{\partial x_m} \frac{\partial F}{\partial p_n} = 0 \qquad m = 0, 1, \ldots, N \tag{6-50}$$

are obtained. Equations (6-49) and (6-50) now are combined to obtain the final incremental relationship

$$\frac{\partial F}{\partial p_N} dp_m = -\left(\frac{\partial F}{\partial x_m} + p_m \frac{\partial F}{\partial y} \right) dx_N \qquad m = 0, 1, \ldots, N \tag{6-51}$$

This last equation corresponds to (6-35) and (6-36) for the case of two independent variables.

Finally, the subsidiary equation for (6-39) is found by combining (6-44), (6-45), and (6-51). These three equations have the ratio

$$\frac{dx_N}{\partial F/\partial p_N}$$

in common so that the subsidiary equation becomes

$$\frac{dx_m}{\partial F/\partial p_m} = \frac{dy}{\sum_{n=0}^{N} p_n \, \partial F/\partial p_n} = \frac{dp_m}{-(\partial F/\partial x_m + p_m \, \partial F/\partial y)} \qquad m = 0, 1, \ldots, N \tag{6-52}$$

This equation contains $2(N+1)$ equalities for $N+1$ independent variables and is equivalent to (6-37) in the case of two independent variables. Furthermore, (6-52) can be used to obtain $2(N+1)$ ordinary differential equations in terms of the independent variable x_0. This set of differential equations is written as

$$\frac{dx_n}{dx_0} = \frac{\partial F/\partial p_n}{\partial F/\partial p_0} \qquad n = 1, 2, \ldots, N \tag{6-53}$$

$$\frac{dp_n}{dx_0} = -\frac{\partial F/\partial x_n + p_n\, \partial F/\partial y}{\partial F/\partial p_0} \qquad n = 1, 2, \ldots, N \tag{6-54}$$

$$\frac{dp_0}{dx_0} = -\frac{\partial F/\partial x_0 + p_0\, \partial F/\partial y}{\partial F/\partial p_0} \tag{6-55}$$

$$\frac{dy}{dx_0} = \frac{\sum_{n=0}^{N} p_n\, \partial F/\partial p_n}{\partial F/\partial p_0} \tag{6-56}$$

These four equations are the *characteristic equations* for (6-39) and are equivalent to the four equations appearing in (6-38) for the case of two independent variables.

6-4 Equations of Pontryagin

The characteristic equations for the partial differential equation resulting from dynamic programming now can be written directly by identifying variables. The following definitions are introduced:

$$x_0 = \mu \qquad y(\mu) = E[\mathbf{x}(\mu), \mu] \tag{6-57}$$

and the remaining independent variables $x_1, x_2, \ldots, x_N$ are taken to be the components of the state vector $\mathbf{x}(\mu)$. According to these definitions, the partial derivatives appearing in (6-39) are

$$p_0(\mu) = \frac{\partial E[\mathbf{x}(\mu), \mu]}{\partial \mu} \tag{6-58}$$

and

$$p_n(\mu) = \frac{\partial E[\mathbf{x}(\mu), \mu]}{\partial x_n(\mu)} \qquad n = 1, 2, \ldots, N \tag{6-59}$$

Time-function notation is adopted here for the dependent variables because the characteristic equations given in (6-53) through (6-56) treat x_0 as the independent variable.

Furthermore, the sum of terms appearing on the left-hand side of (5-122) is used to define the function

$$\mathscr{H}[\mathbf{x}(\mu), \mathbf{m}(\mu), \mathbf{p}(\mu), \mu] = h\{\mathbf{g}[\mathbf{x}(\mu), \mu], \mathbf{m}(\mu), \mu\} + \sum_{n=1}^{N} f_n[\mathbf{x}(\mu), \mathbf{m}(\mu), \mu]p_n(\mu) \tag{6-60}$$

in terms of (6-59). The vectors $\mathbf{x}(\mu)$ and $\mathbf{p}(\mu)$ are taken to be N-dimensional vectors in (6-60). Finally, the partial differential equation given in (5-123) becomes

$$p_0(\mu) + \mathscr{H}[\mathbf{x}(\mu), \mathbf{m}^*(\mu), \mathbf{p}(\mu), \mu] = 0 \tag{6-61}$$

according to the definitions introduced in this section. Equation (6-61) is the traditional form of the Hamilton-Jacobi equation occurring in nonconservative dynamics.[40] For the Hamilton-Jacobi equation, the partial derivatives needed in the characteristic equations are

$$\frac{\partial F}{\partial y} = 0 \qquad \frac{\partial F}{\partial p_0} = 1 \qquad \frac{\partial F}{\partial x_0} = \frac{\partial \mathscr{H}^*}{\partial \mu} + \sum_{m=1}^{M} m_m^{*\prime}(\mu) \frac{\partial \mathscr{H}^*}{\partial m_m^*(\mu)}$$
$$\frac{\partial F}{\partial x_n} = \frac{\partial \mathscr{H}^*}{\partial x_n(\mu)} + \sum_{m=1}^{M} \frac{\partial m_m^*(\mu)}{\partial x_n(\mu)} \frac{\partial \mathscr{H}^*}{\partial m_m^*(\mu)} \qquad n = 1, 2, \ldots, N$$
$$\frac{\partial F}{\partial p_n} = \frac{\partial \mathscr{H}^*}{\partial p_n(\mu)} + \sum_{m=1}^{M} \frac{\partial m_m^*(\mu)}{\partial p_n(\mu)} \frac{\partial \mathscr{H}^*}{\partial m_m^*(\mu)} \qquad n = 1, 2, \ldots, N \tag{6-62}$$

when the abbreviation $\mathscr{H}^* = \mathscr{H}[\mathbf{x}(\mu), \mathbf{m}^*(\mu), \mathbf{p}(\mu), \mu]$ is introduced. By definition, the hamiltonian $\mathscr{H}^*$ is given by

$$\mathscr{H}^* = \min_{\mathbf{m}(\mu) \in \mathscr{M}(\mu)} \mathscr{H}[\mathbf{x}(\mu), \mathbf{m}(\mu), \mathbf{p}(\mu), \mu] \tag{6-63}$$

which is equivalent to the minimization expressed in (5-122). As a result of this minimization, the property

$$\frac{\partial m_m^*(\mu)}{\partial x_n(\mu)} \frac{\partial \mathscr{H}^*}{\partial m_m^*(\mu)} = \frac{\partial m_m^*(\mu)}{\partial p_n(\mu)} \frac{\partial \mathscr{H}^*}{\partial m_m^*(\mu)} = 0 \tag{6-64}$$

is obtained because $\partial \mathscr{H}^*/\partial m_m^* = 0$ when m_m^* is not on the boundary of $\mathscr{M}(\mu)$, and $\partial m_m^*/\partial x_n = \partial m_m^*/\partial p_n = 0$ when m_m^* is on the boundary of $\mathscr{M}(\mu)$. In the latter case, m_m^* is equal to boundary value of $\mathscr{M}(\mu)$ and hence is equal to an independent time function.

The equations of Pontryagin are obtained directly from the previous relationships. The substitution of (6-62) and (6-64) into (6-53) and (6-54) yields

$$x_n'(\mu) = \frac{\partial \mathscr{H}^*}{\partial p_n(\mu)} \qquad n = 1, 2, \ldots, N \tag{6-65}$$

and

$$p_n'(\mu) = -\frac{\partial \mathscr{H}^*}{\partial x_n(\mu)} \qquad n = 1, 2, \ldots, N \tag{6-66}$$

Equations (6-63), (6-65), and (6-66) are the equations of Pontryagin, and they represent a condition for minimum error that is equivalent to the dynamic-programming condition given in (5-122). An important point to note is that these equations hold for control-signal saturation, which is not the case for (4-123), (4-126), and (4-128) derived from the calculus of variations. Also, the Lagrange multiplier λ_n in the calculus of variations, the partial derivative $\partial E/\partial x_n$ in dynamic programming, and the momentumlike variable p_n in the equations of Pontryagin are equivalent variables.

Although not needed in the solution of the equations of Pontryagin because the hamiltonian $\mathscr{H}^*$ is not dependent upon $p_0(\mu)$ and $y(\mu)$, the substitution of (6-62) into (6-55) and (6-56) yields the additional characteristic equations

$$p_0'(\mu) = -\frac{\partial \mathscr{H}^*}{\partial \mu} - \sum_{m=1}^{M} m_m^{*\prime}(\mu) \frac{\partial \mathscr{H}^*}{\partial m_m^*(\mu)} \tag{6-67}$$

and

$$y'(\mu) = p_0(\mu) + \sum_{n=1}^{N} p_n(\mu) \frac{\partial \mathscr{H}^*}{\partial p_n(\mu)} \tag{6-68}$$

Equation (6-67) provides an interesting interpretation of the condition for minimum error which often is referred to as Pontryagin's principle. Suppose the dynamic process is time-invariant so that

$$\mathbf{q}(\mu) = \mathbf{g}[\mathbf{x}(\mu)] \qquad \mathbf{x}'(\mu) = \mathbf{f}[\mathbf{x}(\mu), \mathbf{m}(\mu)] \tag{6-69}$$

and the control vector is confined to a closed space defined by constant boundaries such that

$$\mathbf{m}(\mu) \in \mathscr{M} \tag{6-70}$$

Also, suppose the error index is based on constant weighting of errors such that

$$e(t) = \int_t^T h[\mathbf{q}(\sigma), \mathbf{m}(\sigma)]\, d\sigma \tag{6-71}$$

For this class of problems, the hamiltonian is independent of time and is written as $\mathscr{H}^* = \mathscr{H}[\mathbf{x}(\mu), \mathbf{m}^*(\mu), \mathbf{p}(\mu)]$ so that

$$\frac{\partial \mathscr{H}^*}{\partial \mu} = 0 \tag{6-72}$$

Furthermore, if $m_m^*(\mu)$ is not on the boundary of $\mathscr{M}$, then $\partial\mathscr{H}^*/\partial m_m^*(\mu) = 0$ as a result of the minimization process implied by (6-63). On the other hand, if $m_m^*(\mu)$ is on the boundary of $\mathscr{M}$, then $m_m^{*\prime}(\mu) = 0$ because the boundary of $\mathscr{M}$ possesses constant dimensions. Therefore, the condition

$$m_m^{*\prime}(\mu)\frac{\partial \mathscr{H}^*}{\partial m_m^*(\mu)} = 0 \qquad m = 1, 2, \ldots, M \tag{6-73}$$

holds for this class of problems. From (6-72) and (6-73), the characteristic equation given in (6-67) results in $p_0'(\mu) = 0$ and hence $p_0(\mu) = -C$, where C is a constant of integration. Finally, the Hamilton-Jacobi equation gives

$$\mathscr{H}[\mathbf{x}(\mu), \mathbf{m}^*(\mu), \mathbf{p}(\mu)] = C \tag{6-74}$$

as seen from (6-61). This result states that the hamiltonian is a constant along the optimal trajectory specified by the equations of Pontryagin for the class of problems defined by (6-69) to (6-71). The final characteristic equation is shown to be

$$y'(\mu) = -h\{\mathbf{g}[\mathbf{x}(\mu), \mu], \mathbf{m}^*(\mu), \mu\} \tag{6-75}$$

by appropriate substitutions in (6-68).

The equations derived in this section—namely, (6-63), (6-65), and (6-66)—are referred to as the equations of Pontryagin. Apparently Katz[44] and Breakwell[45] also contributed these equations on an independent basis. However, Pontryagin has contributed an important additional result which pertains to the validity of using characteristic equations for the solution of the dynamic-programming equation. For the class of problems resulting in (6-74), the vector $\mathbf{p}(\mu)$ is continuous and finite. Hence, the reordering of partial differentiations required in (6-47) is permissible, and the characteristic equations give a valid solution to the dynamic-programming equation. In other problems, the continuity of $\mathbf{p}(\mu)$ and hence the validity of the characteristic equations must be checked from the numerical solutions of these equations.

6-5 Example

In order to demonstrate the use of Pontryagin's equations with control-signal saturation, the following simple example is chosen:

$$q_1(t) = x_1(t) \qquad x_1'(t) = m_1(t) \qquad -M_1 \leqq m_1(\sigma) \leqq M_1$$
$$h[\mathbf{q}(\sigma), \mathbf{m}(\sigma), \sigma] = f[Q - q_1(\sigma)] + \rho m_1^2(\sigma) \tag{6-76}$$

where

$$f(0) = 0 \qquad f'(y) = \begin{cases} > 0 & y > 0 \\ = 0 & y = 0 \\ < 0 & y < 0 \end{cases} \tag{6-77}$$

For this example, the function $\mathscr{H}$ defined in (6-60) is

$$\mathscr{H} = f[Q - x_1(\mu)] + \rho m_1^2(\mu) + p_1(\mu)m_1(\mu) \tag{6-78}$$

This function is strictly convex for $\rho > 0$, and a unique value of the control signal $m_1(\mu)$ that minimizes $\mathscr{H}$ subject to the magnitude constraint can be found.

The value of $m_1(\mu)$ that minimizes $\mathscr{H}$ neglecting the boundary constraint is denoted as $s_1(\mu)$. If $s_1(\mu)$ lies in the interval $-M_1 < s_1(\mu) < M_1$, then the optimum control signal is equal to $s_1(\mu)$. On the other hand, if $s_1(\mu)$ lies outside this interval, then the optimum control signal is equal to the nearest boundary of the region of space $\mathscr{M}$. Therefore, the optimum control signal is

$$m_1^*(\mu) = \begin{cases} M_1 & M_1 \leqq s_1(\mu) \\ s_1(\mu) & -M_1 < s_1(\mu) < M_1 \\ -M_1 & s_1(\mu) \leqq -M_1 \end{cases} \tag{6-79}$$

where

$$s_1(\mu) = -\frac{1}{2\rho} p_1(\mu) \tag{6-80}$$

The expression given in (6-80) is found from minimizing $\mathscr{H}$ with respect to $s_1(\mu)$. Furthermore, the conditions given in (6-65) and (6-66) for this example are

$$p_1'(\mu) = \frac{\partial f[Q - x_1(\mu)]}{\partial x_1(\mu)} \tag{6-81}$$

and

$$x_1'(\mu) = m_1^*(\mu) \tag{6-82}$$

For this example, the hamiltonian is independent of the variable μ; hence the hamiltonian $\mathscr{H}^*$ is a constant such that

$$f[Q - x_1(\mu)] + \rho[m_1^*(\mu)]^2 + p_1(\mu)m_1^*(\mu) = C \tag{6-83}$$

If the special case $T = \infty$ is considered, then

$$x_1(\mu) = Q \qquad x_1'(\mu) = m_1^*(\mu) = 0 \tag{6-84}$$

at $\mu = \infty$ because the desired response is a constant and the dynamic process is an integrator. Therefore, the constant of integration must satisfy $C = 0$, as seen by substituting (6-84) into (6-83). Then (6-83) is used to express the momentumlike variable as

$$p_1(\mu) = -\frac{f[Q - x_1(\mu)] + \rho[m_1^*(\mu)]^2}{m_1^*(\mu)} \tag{6-85}$$

Equation (6-85) then is substituted into (6-80); hence the optimum switching signal is given by

$$s_1(\mu) = \frac{1}{2\rho}\frac{f[Q - x_1(\mu)] + \rho[m_1^*(\mu)]^2}{m_1^*(\mu)} \tag{6-86}$$

Finally, the optimum control equation for this problem is found for $\mu = t$ to be

$$m_1^*(t) = \begin{cases} M_1 & M_1 \leqq \rho^{-\frac{1}{2}} f^{\frac{1}{2}}[Q - x_1(t)] \\ \rho^{-\frac{1}{2}} f^{\frac{1}{2}}[Q - x_1(t)] & -M_1 < \rho^{-\frac{1}{2}} f^{\frac{1}{2}}[Q - x_1(t)] < M_1 \\ -M_1 & \rho^{-\frac{1}{2}} f^{\frac{1}{2}}[Q - x_1(t)] \leqq -M_1 \end{cases} \tag{6-87}$$

by use of (6-79) and (6-86).

For the special case $f(y) = y^2$, the optimum control equation given in (6-87) is precisely the control equation depicted in Fig. 2-7. In other words, the optimum control system for this special case is a saturating system with a linear zone. In another special case $\rho = 0$, the control equation given in (6-87) reduces to

$$m_1^*(t) = \begin{cases} M_1 & Q - x_1(t) > 0 \\ 0 & Q - x_1(t) = 0 \\ -M_1 & Q - x_1(t) < 0 \end{cases} \tag{6-88}$$

Therefore, if the control effort is not weighted in the error measure, then the resulting optimum control system is a bang-bang control system. Bang-bang control systems generally result when control effort is not weighted.

The reader may conclude that the purely algebraic steps used in arriving at (6-87) could be employed to find the optimum control system for dynamic processes of higher order than 1. Unfortunately, this is not the case. For higher-order systems, differential equations must be solved in order to find the value of $\mathbf{p}(\mu)$ that appears in the hamiltonian.

6-6 Summary

The results of this chapter serve to illustrate the relationships between dynamic programming and modern extensions of the calculus of variations to problems where control-signal saturation is experienced. These relationships are established with the classical theory of first-order partial differential equations.

A comparison of the dynamic-programming equation and the ensuing characteristic equations leads to the conclusion that the extensions of the calculus of variations contributed by Pontryagin are more restrictive than dynamic programming.§ In particular, the characteristic equations require continuity in the vector $\mathbf{p}(\mu)$. Although such restrictions are important in some problems, the control engineer generally is not concerned with these difficulties. Instead, the successful application of optimization theory to the design of feedback control systems generally revolves around two considerations of a practical nature.

First, the specification of optimum control and response signals requires the numerical solutions of a condition defining minimum error. The dynamic-programming equation and the equations of Pontryagin represent the same condition but in

§ This statement is made with respect to problems that do not involve state-signal saturation. A direct development of the equations of Pontryagin does allow a countable number of discontinuities in the vector $\mathbf{p}(\mu)$.

a different computational format. On the one hand, the equations of Pontryagin suffer from the necessity of solving a two-point boundary-value problem. On the other hand, the dynamic-programming equation suffers from the necessity of solving a partial differential equation, usually by discrete approximations, and the dimensionality of the state vector becomes a critical factor.

Second, the construction of a control equation possessing feedback around the dynamic process poses an entirely different problem from the numerical problem. In particular, the solution to the condition for minimum error must be structured in a fashion that suggests physical components needed in the construction of the control equation. This structuring of the solution is really a control-system-synthesis problem. On the one hand, the equations of Pontryagin give solutions along a single trajectory and hence the control signals are found as functions of time and not as functions of the state vector and time. In other words, structure and hence a control equation are not obtained. On the other hand, the dynamic-programming equation expresses the condition for minimum error for all values of the state vector and hence provides hope for obtaining directly the structure of the control equation.

7

Synthesis of Linear Optimum Control Systems

The previous chapters are devoted to developing the variational mathematics pertinent to feedback-control-system optimization. This chapter is devoted to the subclass of control-system optimization problems where the optimum control equation is linear with respect to the state variables of the dynamic process.

Although there are many design problems where nonlinear control is desirable or even necessary, a large number of control problems can be resolved adequately with linear control equations. If a linear control equation is adequate, then the introduction of a nonlinear optimization problem would result in needless system complexity and expense. Hence, the investigation of linear optimum controls for a specific design problem could be regarded as an initial feasibility study of the simplest type of optimum system.

The class of design problems where such a feasibility study might be undertaken is characterized as follows: First, a linear dynamic process must be a valid representation of the fixed member of the system. Second, economic goals and other design considerations do not dictate a specific mathematical form of thc error measure. For this class of design problems, a quadratie error measure is chosen in conjunction with the linear dynamic process if the feasibility of linear optimum controls is to be investigated as a first step.

The specific goal of this chapter is the identification of the structural properties of linear optimum control systems. The structural properties of linear optimum control systems are important because they dictate the components required in the construction of the control equation. Furthermore, the component requirements differ widely in various classes of design problems. Finally, a thorough understanding of the linear optimization problem is necessary in order to appreciate the difficulties and limitations of the synthesis of nonlinear optimum control systems.

7-1 Linear optimum controls for deterministic dynamic processes

In Sec. 5-1, the condition for minimum error is derived under the conditions that a set of signals is statistical in nature and the error index is chosen to be the conditional mean. Because this set of statistical signals is not associated with the dynamic process and because the control signals are treated as adjustable deterministic functions, the state variables of the dynamic process are deterministic in the variational procedure. For deterministic state variables, the minimum-error function is defined as a function of the state vector in precisely the same fashion as for the wholly deterministic case. This is the class of variational problems treated in this section. On the other hand, when the load-disturbance vector is statistical in nature or when measurement noise is present in the measurement of the state vector, the state vector is no longer a deterministic quantity. The case of statistical state variables is treated in a later section of this chapter. When the state variables are statistical in nature, the definition of the minimum-error function must be based on additional concepts which are not obvious a priori.

For the subclass of linear optimum feedback control systems, the dynamic process must be linear. Therefore, the developments in this chapter are based on a dynamic process defined by the response vector expressed in terms of the state vector as

$$\mathbf{q}(t) = A\mathbf{x}(t) \tag{7-1}$$

where the matrix A is defined in terms of time-varying elements as

$$A = \begin{bmatrix} a_{11}(t) & a_{12}(t) & \cdots & a_{1N}(t) \\ a_{21}(t) & a_{22}(t) & \cdots & a_{2N}(t) \\ \cdot & \cdot & \cdot & \cdot \\ a_{Q1}(t) & a_{Q2}(t) & \cdots & a_{QN}(t) \end{bmatrix} \tag{7-2}$$

Also, the linear dynamic process is defined by the linear state equation

$$\mathbf{x}'(t) = B\mathbf{x}(t) + C\mathbf{m}(t) + \mathbf{u}(t) \tag{7-3}$$

where the matrices B and C are defined in terms of time-varying elements as

$$B = \begin{bmatrix} b_{11}(t) & b_{12}(t) & \cdots & b_{1N}(t) \\ b_{21}(t) & b_{22}(t) & \cdots & b_{2N}(t) \\ \cdot & \cdot & \cdot & \cdot \\ b_{N1}(t) & b_{N2}(t) & \cdots & b_{NN}(t) \end{bmatrix} \qquad C = \begin{bmatrix} c_{11}(t) & c_{12}(t) & \cdots & c_{1M}(t) \\ c_{21}(t) & c_{22}(t) & \cdots & c_{2M}(t) \\ \cdot & \cdot & \cdot & \cdot \\ c_{N1}(t) & c_{N2}(t) & \cdots & c_{NM}(t) \end{bmatrix} \tag{7-4}$$

Finally, if the dynamic process is linear, the optimum control signals must result in variations of the dynamic-process signals that remain in unsaturated regions. Therefore in this chapter, the allowable regions of the optimum control signals are not given and do not enter into the variational problem.

Also, for the subclass of linear optimum feedback control systems, the error measure is chosen to be quadratic in form. Therefore, the error index used in this section is taken to be

$$\overline{e(t)}^{\mathbf{s}|\mathbf{v}} = \sum_{n=1}^{Q} \int_t^T \{\chi_n(\sigma) + \phi_n(\sigma)\overline{[Q_n(\sigma) - q_n(\sigma)]}^t + \phi_{nn}(\sigma)\overline{[Q_n(\sigma) - q_n(\sigma)]^2}^t\}\, d\sigma$$
$$+ \sum_{n=1}^{M} \int_t^T \{\Omega_n(\sigma) + \psi_n(\sigma)\overline{[M_n(\sigma) - m_n(\sigma)]}^t + \psi_{nn}(\sigma)\overline{[M_n(\sigma) - m_n(\sigma)]^2}^t\}\, d\sigma \quad (7\text{-}5)$$

The statistical quantities in this error index are $Q_n(\sigma)$ and $M_n(\sigma)$—the desired-response and desired-control signals, respectively. The statistical averaging indicated in (7-5) is the conditional mean; which is denoted by an abbreviated form of the notation used in Sec. 5-7. In particular, the notation $Q_n \mid \mathbf{v}$ associated with the conditional mean is replaced merely by t in order to simplify the notation and yet indicate that the parameters of the conditional-probability densities are functions of real time t.

The terms not involving the response and control signals in the error measure associated with (7-5) are somewhat arbitrary but are defined here as

$$\chi_n(\sigma) = \frac{\phi_n^2(\sigma)}{4\phi_{nn}(\sigma)} \qquad \Omega_n(\sigma) = \frac{\psi_n^2(\sigma)}{4\psi_{nn}(\sigma)} \quad (7\text{-}6)$$

This definition and the conditions $\phi_{nn}(\sigma) \geqq 0$ and $\psi_{nn}(\sigma) > 0$ are imposed so that the integrand of (7-5) is positive and also strictly convex. The integrand of (7-5) is chosen to be positive because negative values of the error measure are not meaningful. When the integrand of (7-5) is strictly convex, the optimum control signals are unique. Also, the optimum control signals are bounded except in unrealistic situations where one or more of the desired signals are unbounded over a nonzero interval of time.

In (7-5), cross-product terms between two different response errors and cross-product terms between response and control errors are not included, because generally they do not occur in design problems. Also, the weighting factors between these cross-product terms are difficult to select from the concepts associated with the design problem. The results of this chapter could be readily generalized in these situations where cross-product terms could be identified in terms of the design problem at hand.

For the variational problem defined by the state equation and the error index given in (7-3) and (7-5), respectively, the instantaneous minimum-error function is defined as

$$\mathscr{E}[\mathbf{x}(\mu), \mu] = \min_{\substack{\mathbf{m}(\sigma) \\ [\mu, T]}} \overline{e(\mu)}^{\mathbf{s}|\mathbf{v}} \quad (7\text{-}7)$$

where t is replaced by μ in (7-5) everywhere except in the notation for the conditional mean. From (7-7) and the derivations given in Chap. 5, the condition for minimum error is written in the dynamic-programming format as

$$\min_{\mathbf{m}(\mu)}\left(\sum_{n=1}^{Q}\{\chi_n(\mu)+\phi_n(\mu)\overline{[Q_n(\mu)-q_n(\mu)]}^t+\phi_{nn}(\mu)\overline{[Q_n(\mu)-q_n(\mu)]^2}^t\}\right.$$
$$+\sum_{n=1}^{M}\{\Omega_n(\mu)+\psi_n(\mu)\overline{[M_n(\mu)-m_n(\mu)]}^t+\psi_{nn}(\mu)\overline{[M_n(\mu)-m_n(\mu)]^2}^t\}$$
$$\left.+\sum_{n=1}^{N}x_n'(\mu)\frac{\partial\mathscr{E}[\mathbf{x}(\mu),\mu]}{\partial x_n(\mu)}\right)=-\frac{\partial\mathscr{E}[\mathbf{x}(\mu),\mu]}{\partial\mu} \quad (7\text{-}8)$$

If the state equation is substituted into (7-8) and the terms involving the control-signal components are collected, then the optimum values of the control-signal components are defined by the variational process

$$\frac{\partial}{\partial m_i(\mu)}\left(\sum_{n=1}^{M}\{\psi_n(\mu)\overline{[M_n(\mu)-m_n(\mu)]}^t+\psi_{nn}(\mu)\overline{[M_n(\mu)-m_n(\mu)]^2}^t\}\right.$$
$$\left.+\sum_{n=1}^{N}\left\{\sum_{m=1}^{M}[c_{nm}(\mu)m_m(\mu)]\frac{\partial\mathscr{E}[\mathbf{x}(\mu),\mu]}{\partial x_n(\mu)}\right\}\right)=0 \quad (7\text{-}9)$$

The values of the control-signal components that minimize the sum appearing in (7-9) are found to be

$$m_i^*(\mu)=\frac{\psi_i(\mu)}{2\psi_{ii}(\mu)}+\overline{M_i(\mu)}^t-\frac{1}{2\psi_{ii}(\mu)}\sum_{n=1}^{N}c_{ni}(\mu)\frac{\partial\mathscr{E}[\mathbf{x}(\mu),\mu]}{\partial x_n(\mu)} \qquad i=1,2,\ldots,M \quad (7\text{-}10)$$

by setting the partial derivatives of the sum appearing in (7-9) with respect to the various components of the control vector equal to zero. For the optimum values of the control-vector components defined by (7-10), (7-8) becomes

$$\frac{\partial\mathscr{E}[\mathbf{x}(\mu),\mu]}{\partial\mu}+\sum_{n=1}^{Q}\{\chi_n(\mu)+\phi_n(\mu)\overline{[Q_n(\mu)-q_n(\mu)]}^t+\phi_{nn}(\mu)\overline{[Q_n(\mu)-q_n(\mu)]^2}^t\}$$
$$+\sum_{n=1}^{M}\{\Omega_n(\mu)+\psi_n(\mu)\overline{[M_n(\mu)-m_n^*(\mu)]}^t+\psi_{nn}(\mu)\overline{[M_n(\mu)-m_n^*(\mu)]^2}^t\}$$
$$+\sum_{n=1}^{N}x_n^{*\prime}(\mu)\frac{\partial\mathscr{E}[\mathbf{x}(\mu),\mu]}{\partial x_n(\mu)}=0 \quad (7\text{-}11)$$

The method used here to derive the optimum control equation is very elementary in concept.[22] First, the mathematical form of the solution is postulated by using a simple extension of the results presented in the examples of Chaps. 4 and 5. Then the postulated solution is substituted into the condition for minimum error, where the required mathematical operations are performed in conformity with the assumed form of the solution. Finally, the assumed solution is shown to be in fact the solution under certain mathematical conditions. These mathematical conditions uniquely specify the optimum control equation.

In the examples of Chap. 5 for linear dynamic processes and quadratic error measures, the minimum-error function for a first-order dynamic process is shown to be quadratic with respect to the state variable. For higher-order dynamic processes, the quadratic form of the minimum-error function is preserved. Therefore, the instantaneous minimum-error function associated with (7-11) is assumed to have the generalized quadratic form

$$\mathscr{E}[\mathbf{x}(\mu), \mu] = k(\mu) - 2\sum_{m=1}^{N} k_m(\mu)x_m(\mu) + \sum_{m=1}^{N}\sum_{k=1}^{N} k_{mk}(\mu)x_m(\mu)x_k(\mu) \tag{7-12}$$

where

$$k_{mk}(\mu) = k_{km}(\mu) \tag{7-13}$$

For this assumed form, the partial derivatives of the instantaneous minimum-error function appearing in (7-11) are

$$\frac{\partial\mathscr{E}[\mathbf{x}(\mu), \mu]}{\partial\mu} = k'(\mu) - 2\sum_{m=1}^{N} k'_m(\mu)x_m(\mu) + \sum_{m=1}^{N}\sum_{k=1}^{N} k'_{mk}(\mu)x_m(\mu)x_k(\mu) \tag{7-14}$$

and

$$\frac{\partial\mathscr{E}[\mathbf{x}(\mu), \mu]}{\partial x_n(\mu)} = -2k_n(\mu) + 2\sum_{m=1}^{N} k_{mn}(\mu)x_m(\mu) \tag{7-15}$$

Now if (7-1), (7-3), (7-10), (7-14), and (7-15) are substituted into the condition for the instantaneous minimum-error function given in (7-11), then algebraic manipulations show that the condition for minimum error can be written in the form

$$[k'(\mu) + \mathscr{F}(\mu)] - 2\sum_{m=1}^{N} [k'_m(\mu) + \mathscr{F}_m(\mu)]x_m(\mu) + \sum_{m=1}^{N}\sum_{k=1}^{N} [k'_{mk}(\mu) + \mathscr{F}_{mk}(\mu)]x_m(\mu)x_k(\mu) = 0 \tag{7-16}$$

This equation is found by collecting all the terms that multiply the various products of the state-vector components. In (7-16), the $\mathscr{F}$ functions depend upon the a, b, and c coefficients and load disturbances of the dynamic process, the ϕ and ψ weighting factors of the error measure, the conditional means of the desired-control and desired-response signals, and the k parameters appearing in the assumed form of the minimum-error function. However, the $\mathscr{F}$ functions do not depend upon the components of the state vector.

Because $1 + 2N + N!/2!(N - 2)!$ distinct k parameters are assumed in (7-12) as defining the instantaneous minimum-error function, an equal number of independent conditions defining the time derivatives of the k parameters must be imposed upon the form of the condition for minimum error given in (7-16). If the identity appearing in (7-13) is used, then the condition for minimum error given in (7-16) is rewritten as

$$[k'(\mu) + \mathscr{F}(\mu)] - 2\sum_{m=1}^{N} [k'_m(\mu) + \mathscr{F}_m(\mu)]x_m(\mu) + \sum_{m=1}^{N} [k'_{mm}(\mu) + \mathscr{F}_{mm}(\mu)]x_m{}^2(\mu) + 2\sum_{m=1}^{N}\sum_{k=1}^{m-1} [k'_{mk}(\mu) + \tfrac{1}{2}\mathscr{F}_{mk}(\mu) + \tfrac{1}{2}\mathscr{F}_{km}(\mu)]x_m(\mu)x_k(\mu) = 0 \tag{7-17}$$

Equation (7-17) combines the various products of the state-vector components that occur in the double sum appearing in (7-16). If the condition for minimum error, now written in the form given in (7-17), is to be satisfied for all possible values of the state-vector components, then the bracketed terms appearing in (7-17) must vanish. If each of the bracketed terms appearing in this equation is set equal to zero, the k parameters satisfy the ordinary differential equations

$$-k'(\mu) = \mathscr{F}(\mu) \tag{7-18}$$

$$-k'_m(\mu) = \mathscr{F}_m(\mu) \qquad m = 1, 2, \ldots, N \tag{7-19}$$

$$-k'_{mk}(\mu) = \tfrac{1}{2}[\mathscr{F}_{mk}(\mu) + \mathscr{F}_{km}(\mu)] \qquad k = 1, 2, \ldots, m;\ m = 1, 2, \ldots, N \tag{7-20}$$

The number of independent conditions imposed on the k parameter used in arriving at (7-18) to (7-20) is equal in number to the number of distinct k parameters assumed in the instantaneous minimum-error function.

From the algebraic manipulations required to arrive at (7-16), the $\mathscr{F}$ functions are found; therefore the ordinary differential equations defining the k parameters are

$$-k'(\mu) = \sum_{n=1}^{Q}\left[\frac{\phi_n}{4\phi_{nn}} + \phi_n\overline{Q_n}^t + \phi_{nn}\overline{Q_n^2}^t\right] - 2\sum_{n=1}^{N}[u_n k_n]$$
$$- \sum_{n=1}^{M}\left[-\psi_{nn}\overline{(M_n - \overline{M_n}^t)^2}^t + 2\left(\frac{\psi_n}{2\psi_{nn}} + \overline{M_n}^t\right)\sum_{i=1}^{N}\left(c_{in}k_i + \sum_{j=1}^{N}\frac{c_{jn}c_{in}}{\psi_{nn}}k_i k_j\right)\right] \quad (7\text{-}21)$$

$$-k'_m(\mu) = \sum_{n=1}^{Q}\left[\phi_{nn}\left(\frac{\phi_n}{2\phi_{nn}} + \overline{Q_n}^t\right)a_{nm}\right] + \sum_{n=1}^{N}[-u_n k_{nm} + b_{nm}k_n]$$
$$- \sum_{n=1}^{M}\left[\left(\frac{\psi_n}{2\psi_{nn}} + \overline{M_n}^t\right)\sum_{i=1}^{N}\left(c_{in}k_{im} + \sum_{j=1}^{N}\frac{c_{jn}c_{in}}{\psi_{nn}}k_i k_{jm}\right)\right] \quad (7\text{-}22)$$

and

$$-k'_{mk}(\mu) = \sum_{n=1}^{Q}[\phi_{nn}a_{nm}a_{nk}] + \sum_{n=1}^{N}[b_{nm}k_{nk} + b_{nk}k_{nm}]$$
$$- \sum_{n=1}^{M}\left[\sum_{i=1}^{N}\sum_{j=1}^{N}\left(\frac{c_{jn}c_{in}}{\psi_{nn}}k_{im}k_{jk}\right)\right] \quad (7\text{-}23)$$

The subscripts of the derivatives of the k parameters appearing in (7-22) and (7-23) are defined on the ranges indicated in (7-19) and (7-20). Finally, the boundary conditions for the k parameters are found from the boundary condition for the instantaneous minimum-error function. Because of the definition of this function, the instantaneous minimum-error function evaluated at $\mu = T$ is zero for all finite values of the state-vector components. Therefore from (7-12), the boundary conditions for the k parameters are

$$k(T) = k_m(T) = k_{mk}(T) = 0 \qquad 1 \leqq m, k \leqq N \quad (7\text{-}24)$$

The optimum control signals which appear in the real-time control system are found from (7-10) and (7-15) when evaluated at $\mu = t$. From appropriate substitutions, the optimum control signals are found to be

$$m_i^*(t) = \frac{\psi_i(t)}{2\psi_{ii}(t)} + \overline{M_i(t)}^t + \frac{1}{\psi_{ii}(t)}\sum_{m=1}^{N}c_{mi}(t)\left[k_m(t) - \sum_{k=1}^{N}k_{mk}(t)x_k(t)\right] \quad (7\text{-}25)$$

Furthermore, if terms are collected in (7-25), then the linear optimum control equation is written in the form

$$m_i^*(t) = R_i(t) - \sum_{k=1}^{N}K_{ik}(t)x_k(t) \qquad i = 1, 2, \ldots, M \quad (7\text{-}26)$$

where the reference signals are given by

$$R_i(t) = \frac{\psi_i(t)}{2\psi_{ii}(t)} + \overline{M_i(t)}^t + \frac{1}{\psi_{ii}(t)}\sum_{m=1}^{N}c_{mi}(t)k_m(t) \quad (7\text{-}27)$$

and where the feedback gains are given by

$$K_{ik}(t) = \frac{1}{\psi_{ii}(t)}\sum_{m=1}^{N}c_{mi}(t)k_{mk}(t) \quad (7\text{-}28)$$

Therefore, the optimum control equation for linear dynamic processes and quadratic error measures is constructed explicitly in terms of the k parameters that appear in the quadratic form of the minimum-error function. The optimum control system, as defined by the control equation given in (7-26), involves both a feedback part and a feedforward part. The feedback part of the control system is constructed with time-varying gains and hence is straightforward. On the other hand, the feedforward part of the control system has not been identified in terms of components. In other words, the reference signals are the outputs of a system of components that eventually must be structured for the purpose of design.

Because only the single- and double-subscripted k parameters define the optimum control signals, an unnecessary complication has been introduced in this initial derivation. In particular, the linear terms for weighting the response- and control-signal errors are unnecessary. From (7-22), the effect of ϕ_n and ψ_n is the introduction of biases in the conditional means of Q_n and M_n. Therefore, ϕ_n and ψ_n can be set equal to zero if the appropriate biases are added to the conditional means of the response and control signals. In other words, nothing is sacrificed conceptually if

$$\phi_n(\mu) = \psi_n(\mu) = 0 \tag{7-29}$$

Also, nothing is sacrificed conceptually if the assumption is made that not more than one control signal per equation of state occurs. Therefore, a less general dynamic process defined by

$$c_{nm}(\mu) = 0 \qquad n \neq m \tag{7-30}$$

is introduced. Finally, the less general error measure defined by

$$N = Q = M \tag{7-31}$$

is introduced at no sacrifice in concepts. For the simplifications afforded by (7-29) to (7-31), the differential equations defining the k parameters reduce to

$$-k'(\mu) = \sum_{n=1}^{N} \left[\phi_{nn}\overline{Q_n{}^2}^t + \psi_{nn}\overline{(M_n - \overline{M_n}^t)^2}^t - 2u_n k_n - 2c_{nn}\overline{M_n}^t k_n - \frac{c_{nn}{}^2}{\psi_{nn}} k_n{}^2 \right] \tag{7-32}$$

$$-k'_m(\mu) = \sum_{n=1}^{N} \left(\phi_{nn}\overline{Q_n}^t a_{nm} - u_n k_{nm} - c_{nn}\overline{M_n}^t k_{nm} + b_{nm}k_n - \frac{c_{nn}{}^2}{\psi_{nn}} k_n k_{nm} \right) \tag{7-33}$$

$$-k'_{mk}(\mu) = \sum_{n=1}^{N} \left(\phi_{nn}a_{nm}a_{nk} + b_{nm}k_{nk} + b_{nk}k_{nm} - \frac{c_{nn}{}^2}{\psi_{nn}} k_{nk}k_{nm} \right) \tag{7-34}$$

Also, the reference signals and feedback gains in the optimum control equation reduce to

$$R_i(t) = \overline{M_i(t)}^t + \frac{1}{\psi_{ii}(t)} c_{ii}(t)k_i(t) \tag{7-35}$$

and

$$K_{ik}(t) = \frac{1}{\psi_{ii}(t)} c_{ii}(t)k_{ik}(t) \tag{7-36}$$

Because these last equations are considerably simpler to write notationally and because no concepts are sacrificed, this more specialized case is discussed hereafter in this chapter.

7-2 General properties of linear optimum control systems

As expected in linear systems, a number of invariant and elegant properties of the systems can be deduced. In addition to the concepts resulting from superposition in linear systems, the optimum control systems possess some unexpected properties which are potentially very useful with respect to the adaptation of the optimum control equation to changing design conditions.

1. The optimum control system is linear with respect to the state signals of the dynamic process. This property results from the fact that the k parameters are independent of the state signals. Therefore, the optimum control equation given in (7-26) is linear with respect to the state signals. Because the assumed dynamic process also is linear with respect to the state signals, the optimum system, which includes both the dynamic process and the controls, is linear with respect to the state signals.

2. The feedback portion of the optimum control system is independent of the desired-response and desired-control signals and also is independent of the load-disturbance signals. The feedback portion of the optimum control system is defined by the double-subscripted k parameters. From (7-30), the double-subscripted k parameters are seen to be independent of the signals $\overline{Q_n}^t$, $\overline{M_n}^t$, and u_n. This independence results from the fact that the right-hand side of (7-34) does not involve these signals and also does not involve the single-subscripted k parameters. In other words, the feedback portion of the optimum control system does not adapt to changing desired-response and desired-control signals and also does not adapt to changing load-disturbance signals. The adaptation to these changing signals occurs only in the feedforward portion of the optimum control system, as specified by the single-subscripted k parameters.[48]

3. The feedforward portion of the optimum control system is linear with respect to the desired-response and desired-control signals and also the load-disturbance signals. Because the double-subscripted k parameters are independent of these signals, these parameters appear only as time-varying multipliers in (7-33), which defines the single-subscripted k parameters. Therefore, (7-33) represents a system of linear ordinary differential equations with respect to the single-subscripted k parameters. Also, these signals appear linearly on the right-hand side of (7-33). Therefore, the single-subscripted k parameters, which define the feedforward portion of the optimum control system, are linear with respect to the desired-response and desired-control signals and also the load-disturbance signals.

4. The instantaneous minimum-error function is positive and strictly convex with respect to the components of the state-signal vector. This property results from the assumption that the error measure is positive and strictly convex with respect to response- and control-signal errors. If the assumed form of the instantaneous minimum-error function given in (7-12) is strictly convex, then the matrix of second derivatives of $\mathscr{E}[\mathbf{x}(\mu), \mu]$ with respect to the components of $\mathbf{x}(\mu)$ is positive definite. This condition is equivalent to the requirement that the matrix D, defined as

$$D = [k_{nm}(t)] \qquad \text{for } t < T \tag{7-37}$$

be positive definite.[32] Furthermore, if the instantaneous minimum-error function is positive, then the matrix $\mathscr{D}$, defined as

$$\mathscr{D} = \left[\begin{array}{c|ccc} k(t) & k_1(t) & k_2(t) \cdots & k_N(t) \\ \hline k_1(t) & & & \\ k_2(t) & & & \\ \cdot & & D & \\ \cdot & & & \\ \cdot & & & \\ k_N(t) & & & \end{array}\right] \tag{7-38}$$

must be positive semidefinite.[47]

These requirements on the k parameters appearing in the instantaneous minimum-error function yield relationships between the feedback gains and reference signals in the optimum control equation. For a second-order dynamic process, for instance, the convexity requirement yields

$$k_{11}(t) > 0 \qquad k_{22}(t) > 0 \qquad |k_{12}(t)| < \sqrt{k_{11}(t)\,k_{22}(t)} \tag{7-39}$$

Furthermore, the positiveness requirement yields

$$k(t) \geqq 0 \qquad |k_1(t)| \leqq \sqrt{k(t)k_{11}(t)} \qquad |k_2(t)| \leqq \sqrt{k(t)k_{22}(t)} \tag{7-40}$$

The primary purpose of introducing these structural properties of the instantaneous minimum-error function is to point out the possibility of positive feedback in the optimum control equation. This possibility is verified by (7-39) because $k_{12}(t)$ can be negative. However, the incidence of this positive feedback does not imply that the overall feedback system, which includes both the dynamic process and the optimum control equation, results in instability of the system. The problem of stability is treated in a subsequent section.

7-3 Separable signals and the structure of linear optimum control systems

The structure of the feedforward portion of the optimum control equation given in (7-26) requires special considerations. Moreover, the structure of this portion of the control equation is dependent upon the class of desired-response and desired-control signals in addition to the class of load-disturbance signals that occur in the design problem. In this context, the most important class of signals is called separable.[22] The importance of separable signals is due to major simplications that result in the feedforward portion of the optimum control equation. Specifically, readily available components can be used in the construction of this portion of the system.

In the case of statistical signals, the signal $s(t)$ is called *separable* if the conditional mean $\overline{s(\mu)}^{t}$ can be written as

$$\overline{s(\mu)}^{t} = S(\mu) + \sum_{n=1}^{s} S_n(\mu, t)\overline{s_n(t)}^{t} \tag{7-41}$$

subject to the following interpretation. The *unseparated component* $S(\mu)$ is known a priori for all μ; hence the specification of $S(\mu)$ requires no measurements during the operation of the control system. Most commonly, the unseparated component of a statistical signal is the mean value of the signal. The *separated component* $S_n(\mu, t)\overline{s_n(t)}^t$ involves both a function $S_n(\mu, t)$ that is known a priori for all μ and t and a signal $\overline{s_n(t)}^t$ that is measured with a physical device during the operation of the system. The separated components arise when the impulse response of the Wiener predictor for gaussian signals is the homogeneous solution of a system of ordinary differential equations of order s.

As an example of separable statistical signals, suppose that the conditional mean of $s(\mu)$ is given by

$$\overline{s(\mu)}^t = \overline{s(\mu)} + \int_{-\infty}^{t} w^*(\mu, \sigma)v(\sigma)\, d\sigma \tag{7-42}$$

where

$$w^*(\mu, \sigma) = \sum_{n=1}^{s} F_n(\mu)G_n(\sigma) \tag{7-43}$$

The measured signals appearing in (7-41) are defined as

$$\overline{s_n(t)}^t = \int_{-\infty}^{t} F_n(t)G_n(\sigma)v(\sigma)\, d\sigma \tag{7-44}$$

for this example. Then, simple manipulations yield

$$S_n(\mu, t) = \frac{F_n(\mu)}{F_n(t)} \tag{7-45}$$

The question of separable deterministic signals arises naturally in conjunction with this discussion. If $s(\mu)$ is a deterministic signal, then the averages of signals appearing in (7-41) are replaced by the signals themselves. Therefore, the analogous form of (7-41) for separable deterministic signals is

$$s(\mu) = S(\mu) + \sum_{n=1}^{s} S_n(\mu, t)s_n(t) \tag{7-46}$$

The interpretation of (7-46) is similar to the statistical-signal case. The functions $S(\mu)$ and $S_n(\mu, t)$ are taken to be known a priori at the time of design, but the signal $s_n(t)$ is assumed to be measured during system operation. One form of (7-46), which is very useful as a method of approximating the future value of a signal, is a Taylor series expansion. In this case, (7-46) might be specialized to

$$S(\mu) = 0 \qquad S_n(\mu, t) = \frac{(\mu - t)^n}{n!} \qquad s_n(t) = s^{(n)}(t) \tag{7-47}$$

However, the case where S_n is taken to be a function of μ only and where s_n is taken to be a constant coefficient is a valid form of (7-46). This form of (7-46) arises when the signal $s(\mu)$ is a power series with coefficients unspecified at the time of system design.

Subject to these preliminary considerations, the desired-response and desired-control signals and the load-disturbance signals are assumed to be separable signals and are written in the form

$$\overline{Q_i(\mu)}^t = \mathcal{Q}_i(\mu) + \sum_{j=1}^{P_i} \mathcal{Q}_{ij}(\mu, t)\overline{Q_{ij}(t)}^t \tag{7-48}$$

$$\overline{M_i(\mu)}^t = \mathcal{M}_i(\mu) + \sum_{j=1}^{R_i} \mathcal{M}_{ij}(\mu, t)\overline{M_{ij}(t)}^t \tag{7-49}$$

and

$$u_i(\mu) = \mathcal{U}_i(\mu) + \sum_{j=1}^{S_i} \mathcal{U}_{ij}(\mu, t)U_{ij}(t) \tag{7-50}$$

The mathematical structure of the single-subscripted k parameters can be deduced directly for separable signals. Specifically, the linearity property of the single-subscripted k equations with respect to $\overline{Q_n(\mu)}^t$, $\overline{M_n(\mu)}^t$, and $u_n(\mu)$ is used in two forms. First, the solution to an additive set of components can be written as the sum of solutions found for each component separately. Second, the solution due to a single additive component can be divided by any coefficient not entering into the integration, thereby obtaining a modified solution. This modified solution is the solution found for this additive component divided by this same coefficient. For instance, the solution of the single-subscripted k equations is found for only the component $\mathcal{Q}_{ij}(\mu, t)\overline{Q_{ij}(t)}^t$ being treated as nonzero. If this solution is divided by $\overline{Q_{ij}(t)}^t$, then this modified solution is found as the solution of these same equations but with $\mathcal{Q}_{ij}(\mu, t)\overline{Q_{ij}(t)}^t$ being replaced by $\mathcal{Q}_{ij}(\mu, t)$.

Subject to these additional considerations, the single-subscripted k parameters can be written as

$$k_m(t) = l_m(t) + \sum_{i=1}^{N}\left[\sum_{j=1}^{P_i} p_{mij}(t)\,\overline{Q_{ij}(t)}^t\right] + \sum_{i=1}^{N}\left[\sum_{j=1}^{R_i} r_{mij}(t)\overline{M_{ij}(t)}^t\right] + \sum_{i=1}^{N}\left[\sum_{j=1}^{S_i} s_{mij}(t)U_{ij}(t)\right] \qquad (7\text{-}51)$$

for the assumed set of separable signals. When the single-subscripted k parameters satisfy (7-33), the functions $l_m(t)$, $p_{mij}(t)$, $r_{mij}(t)$, and $s_{mij}(t)$ are the solutions of the sets of differential equations

$$-l'_m(\mu) = \sum_{n=1}^{N}\left(\phi_{nn}\mathcal{Q}_n a_{nm} - \mathcal{U}_n k_{nm} - c_{nn}\mathcal{M}_n k_{nm} + b_{nm}l_n - \frac{c_{nn}^{\ 2}}{\psi_{nn}} l_n k_{nm}\right) \qquad (7\text{-}52)$$

$$-p'_{mij}(\mu) = \phi_{ii}\mathcal{Q}_{ij}a_{im} + \sum_{n=1}^{N}\left(b_{nm}p_{nij} - \frac{c_{nn}^{\ 2}}{\psi_{nn}} p_{nij}k_{nm}\right) \qquad (7\text{-}53)$$

$$-r'_{mij}(\mu) = -c_{ii}\mathcal{M}_{ij}k_{im} + \sum_{n=1}^{N}\left(b_{nm}r_{nij} - \frac{c_{nn}^{\ 2}}{\psi_{nn}} r_{nij}k_{nm}\right) \qquad (7\text{-}54)$$

$$-s'_{mij}(\mu) = -\mathcal{U}_{ij}k_{im} + \sum_{n=1}^{N}\left(b_{nm}s_{nij} - \frac{c_{nn}^{\ 2}}{\psi_{nn}} s_{nij}k_{nm}\right) \qquad (7\text{-}55)$$

Each of these last four equations represents a set of N simultaneous differential equations such that $m = 1, 2, \ldots, N$. Also, the solutions of these differential equations are subject to the boundary conditions

$$l_m(T) = p_{mij}(T) = r_{mij}(T) = s_{mij}(T) = 0 \qquad (7\text{-}56)$$

because $k_m(T) = 0$ must hold for all finite values of the components of the separable signals.

Before the optimum control equation for separable signals is introduced, the computational implications of rewriting (7-33) in terms of (7-52) through (7-55) should be discussed. Specifically, (7-52) through (7-55) involve only functions that are known a priori. Therefore, these equations can be solved at the time when the system is designed. On the other hand, the solution of (7-33) must be performed in real time because the conditional means $\overline{Q_n(\mu)}^t$ and $\overline{M_n(\mu)}^t$ are not known a priori. This result is of paramount importance because a very-high-speed on-line computer no longer is needed.

The optimum control equation now can be rewritten for separable signals by combining (7-26), (7-35), (7-36), and (7-51). The final form of the optimum control equation is

$$m_m^*(t) = L_m(t) + \sum_{i=1}^{N}\left[\sum_{j=1}^{P_i} P_{mij}(t)\overline{Q_{ij}(t)}^t\right] + \sum_{i=1}^{N}\left[\sum_{j=1}^{R_i} R_{mij}(t)\overline{M_{ij}(t)}^t\right] + \sum_{i=1}^{N}\left[\sum_{j=1}^{S_i} S_{mij}(t)U_{ij}(t)\right] - \sum_{n=1}^{N} K_{mn}(t)x_n(t) \quad (7\text{-}57)$$

The definitions of the new time functions appearing in (7-57) are

$$L_m(t) = \mathscr{M}_m(t) + \frac{c_{mm}(t)}{\psi_{mm}(t)}\, l_m(t) \quad (7\text{-}58)$$

$$P_{mij}(t) = \frac{c_{mm}(t)}{\psi_{mm}(t)}\, p_{mij}(t) \quad (7\text{-}59)$$

$$R_{mij}(t) = \delta_{mi}\mathscr{M}_{ij}(t, t) + \frac{c_{mm}(t)}{\psi_{mm}(t)}\, r_{mij}(t) \quad (7\text{-}60)$$

and

$$S_{mij}(t) = \frac{c_{mm}(t)}{\psi_{mm}(t)}\, s_{mij}(t) \quad (7\text{-}61)$$

where

$$\delta_{mi} = \begin{cases} 1 & m = i \\ 0 & m \neq i \end{cases} \quad (7\text{-}62)$$

The various terms in this form of the optimum control equation readily are identified in association with the use of components. In particular, the functions appearing in (7-57) are categorized as follows:

$\overline{Q_{ij}(t)}^t$, $\overline{M_{ij}(t)}^t$, $U_{ij}(t)$, $x_n(t)$ = measured signals

$L_m(t)$ = precomputed reference signal

$P_{mij}(t)$, $R_{mij}(t)$, $S_{mij}(t)$, $K_{mn}(t)$ = precomputed time-varying gains

Therefore, the optimum control equation can be constructed with readily available and inexpensive analog components. Of course, this statement is based on the assumption that the sensors required for signal measurements are readily available.

In this special subclass of optimization problems, adaptation to changing design conditions requires only the adjustment of the time-varying gains. In this sense, the linear optimum control systems with separable input signals are *linearly adaptive* with respect to all measured quantities in the optimum control system.[48] Even though a linearly adaptive control system as defined here is restrictive with respect to many concepts normally associated with adaptive control systems, this subclass of adaptive systems is important. This is one of the very few general classes of adaptive systems that can be structured precisely in terms of a finite number of available components. Also, the values of these components, as determined by the values of time-varying gains, are specified precisely as the solutions of the set of ordinary differential equations.

7-4 Linear time-invariant optimum control systems

In Sec. 7-3, the conditions for which the optimum control equation can be expressed wholly in terms of time-varying gains are determined. The conditions for which these gains become time-invariant also are important.

If the dynamic process is time-invariant, as specified by the time-invariant coefficients

$$a_{nm}(t) = a_{nm} \qquad b_{nm}(t) = b_{nm} \qquad c_{nn}(t) = c_{nn} \tag{7-63}$$

and if the weighting factors of the error measure are time-invariant, as specified by

$$\phi_{nn}(\sigma) = \phi_{nn} \qquad \psi_{nn}(\sigma) = \psi_{nn} \tag{7-64}$$

then the differential equation defining the double-subscripted k parameters is a nonlinear but time-invariant set of ordinary differential equations. If, in addition, the error measure is weighted over the semi-infinite interval

$$T = \infty \tag{7-65}$$

then the solutions of these nonlinear differential equations must assume steady-state values. Of course, steady-state values are achieved only when the differential equations are absolutely stable without limit cycles. This type of stability is ensured by the assumption of a strictly convex error measure. If the double-subscripted k parameters assume steady-state values, then the differential equation given in (7-34) reduces to the set of algebraic equations

$$0 = \sum_{n=1}^{N} \left(\phi_{nn} a_{nm} a_{nk} + b_{nm} k_{nk} + b_{nk} k_{nm} - \frac{c_{nn}^{\;2}}{\psi_{nn}} k_{nk} k_{nm} \right) \qquad \begin{matrix} k = 1, 2, \ldots, m; \\ m = 1, 2, \ldots, N \end{matrix} \tag{7-66}$$

The constant double-subscripted k parameters which satisfy (7-66) are denoted now by

$$k_{nm}(t) = k_{nm} \tag{7-67}$$

Therefore, when these three conditions are imposed, the feedback portion of the optimum control system is time-invariant, as specified by constant gains multiplying the measured state signals.

However, the feedforward portion of the optimum control system is time-varying unless additional conditions are imposed. The precomputed reference signal becomes a constant when the unseparated components of the separable signals are constants. That is, the conditions given in (7-63) to (7-65) and

$$\mathcal{Q}_n(t) = \mathcal{Q}_n \qquad \mathcal{M}_n(t) = \mathcal{M}_n \qquad \mathcal{U}_n(t) = \mathcal{U}_n \tag{7-68}$$

require the steady-state solution

$$0 = \sum_{n=1}^{N} \left(\phi_{nn} \mathcal{Q}_n a_{nm} - \mathcal{U}_n k_{nm} - c_{nn} \mathcal{M}_n k_{nm} + b_{nm} l_n - \frac{c_{nn}^{\;2}}{\psi_{nn}} l_n k_{nm} \right)$$
$$m = 1, 2, \ldots, N \tag{7-69}$$

of (7-52), where

$$l_m(t) = l_m \tag{7-70}$$

Furthermore, impulse responses can be defined for (7-53) to (7-55), which are the equations that are used to compute the feedforward gains. For (7-53), the value of p_{mij} at $\xi + \tau$ due to a unit impulse for $\mathcal{Q}_{ij}$ at ξ is defined as $w_{mi}(\xi + \tau, \xi)$. If the additional definition $\xi = T - \mu$ is introduced, then the superposition integral for $p_{mij}(t)$ becomes

$$p_{mij}(t) = \int_0^{T-t} w_{mi}(T - t, \xi)\mathcal{Q}_{ij}(T - \xi, t)\, d\xi \tag{7-71}$$

From a change in variables, (7-71) is rewritten as

$$p_{mij}(t) = \int_t^{T} w_{mi}(T - t, T - \mu)\mathcal{Q}_{ij}(\mu, t)\, d\mu \tag{7-72}$$

However, under the condition that the double-subscripted k parameters are constants, w_{mi} is the impulse response of a set of differential equations with constant coefficients and hence is a function of only time shift. Specifically, (7-72) reduces to

$$p_{mij}(t) = \int_t^{\infty} w_{mi}(\mu - t)\mathcal{Q}_{ij}(\mu, t)\, d\mu \tag{7-73}$$

when (7-63) to (7-65) hold. In the case of *stationary separable signals*, the function $\mathcal{Q}_{ij}$ is defined to be

$$\mathcal{Q}_{ij}(\mu, t) = \mathcal{Q}_{ij}(\mu - t) \tag{7-74}$$

Finally, the superposition integral given in (7-73) subject to (7-74) reduces to

$$p_{mij}(t) = \int_0^{\infty} w_{mi}(\xi)\mathcal{Q}_{ij}(\xi)\, d\xi = p_{mij} \tag{7-75}$$

with a change in variables. Under the conditions used to obtain (7-75), the feedforward gains in the optimum control equation are constants.

The optimum control equation is said to be time-invariant when all gains appearing in the control equation are constants and when the precomputed references also are constants.§ The feedback portion of this equation is invariant when the coefficients of the dynamic process are constants, the weighting factors are constants, and errors are weighted over all future time. Furthermore, the feedforward portion of this equation is invariant when all separable signals are stationary and all unseparated components are constants provided that the feedback portion also is invariant.

An interesting interpretation of (7-75) occurs when the time function $\mathcal{Q}_{ij}$ is of the special form

$$\mathcal{Q}_{ij}(\eta) = e^{-s\eta} \qquad s = \alpha + j\omega \tag{7-76}$$

This special form results for deterministic signals when the transform of the input signal is a rational polynomial in the complex-frequency variable. In the case of statistical signals, the special form given in (7-76) occurs when the spectral density of the input signal is rational in the complex-frequency variable. For this special form, (7-75) becomes

$$p_{mij} = \int_0^{\infty} w_{mi}(\xi)e^{-s\xi}\, d\xi \tag{7-77}$$

The integral appearing in this equation is noted to be a one-sided transform and is precisely the Laplace transform when s is a complex number. This transform is analogous to the spectral-factorization procedure of Wiener and is the transform of

§ Time invariance from a mathematical point of view does not require constant precomputed reference signals. However, time-varying reference signals require special components in the construction of the control equation.

the positive-time function defined in (3-73). Also, the solution of the algebraic equations given in (7-66) is analogous to the factorization process used to find the zeros of the spectral density defined in (3-131) for Wiener minimization subject to integral constraints.

7-5 Example with separable signal

In order to illustrate the physical realization of this subclass of optimum control systems, a simple example is assumed here. The dynamic process is taken to be first-order and of the special form

$$N = 1 \qquad a_{11}(t) = c_{11}(t) = 1 \qquad b_{11}(t) = 0 \qquad u_1(t) = 0 \tag{7-78}$$

Also, the error measure is specialized to

$$\phi_{nn}(\sigma) = \psi_{nn}(\sigma) = 1 \qquad M_1(\sigma) = 0 \tag{7-79}$$

The input signal $v_1(t)$ is taken to be

$$v_1(t) = Q_1(t) + n_1(t) \tag{7-80}$$

where the signal $n_1(t)$ is a noise signal, and these signals are assumed to be gaussian with the following correlation functions:

$$\phi_{Q_1Q_1}(\beta) = Q^2e^{-a|\beta|} \qquad \phi_{n_1n_1}(\beta) = N^2u_0(\beta) \qquad \phi_{n_1Q_1}(\beta) = 0 \tag{7-81}$$

This simple example is chosen so that a closed-form solution can be obtained.

For this example, the optimum control equation reduces to

$$m_1(t) = p_{111}(t)\overline{Q_{11}(t)}^{t} - k_{11}(t)x_1(t) \tag{7-82}$$

where the gains appearing in this equation are defined by

$$-p'_{111}(\mu) = \mathcal{Q}_{11}(\mu, t) - k_{11}(\mu)p_{111}(\mu) \tag{7-83}$$

and

$$-k'_{11}(\mu) = 1 - k_{11}{}^2(\mu) \tag{7-84}$$

These last three equations are specializations of (7-57), (7-53), and (7-34), respectively.

Because the signals Q_1 and n_1 are gaussian, the required signal $\overline{Q_1(\mu)}^{t}$ is the output of a Wiener predictor. For the correlation functions assumed here, this optimum predictor is derived in Sec. 3-4. The optimum impulse response for these signals is

$$w^*(\gamma) = (w_1 - a)e^{-a(\mu-t)}e^{-w_1\gamma} \tag{7-85}$$

where

$$w_1 = \sqrt{a^2 + \frac{2aQ^2}{N^2}} \tag{7-86}$$

When this impulse response is substituted into the superposition integral, the required signal $\overline{Q_1(\mu)}^{t}$ is written as

$$\overline{Q_1(\mu)}^{t} = e^{-a(\mu-t)}\int_{-\infty}^{t} (w_1 - a)e^{-w_1(t-\sigma)}v_1(\sigma)\, d\sigma \tag{7-87}$$

Therefore, the measured signal $\overline{Q_{11}(t)}^{t}$ appearing in the optimum control equation is

$$\overline{Q_{11}(t)}^{t} = \int_{-\infty}^{t} (w_1 - a)e^{-w_1(t-\sigma)}v_1(\sigma)\, d\sigma \tag{7-88}$$

and the time function appearing in the differential equation for the feedforward gain is

$$\mathscr{Q}_{11}(\mu, t) = e^{-a(\mu - t)} \tag{7-89}$$

For (7-89), the time-varying gains occurring in the control equation are found to be

$$p_{111}(t) = \frac{1}{1 - a^2}\{\tanh(T - t) - a[1 - e^{-a(T-t)}\operatorname{sech}(T - t)]\} \tag{7-90}$$

and
$$k_{11}(t) = \tanh(T - t) \tag{7-91}$$

from the appropriate integrations and boundary conditions. Of course, the optimum control system is time-varying because these gains are functions of time-to-go.

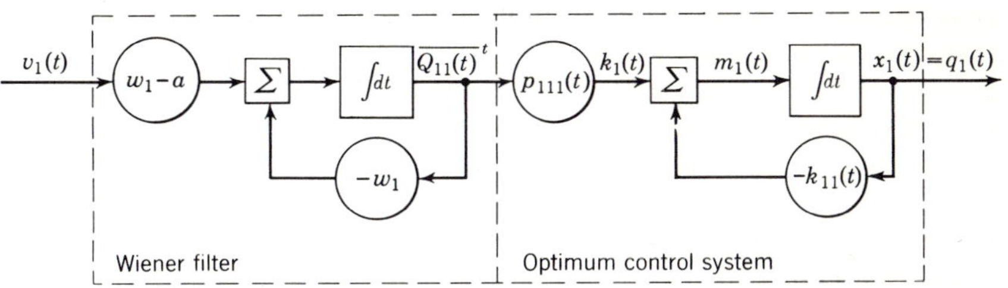

Fig. 7-1 Block diagram of linear optimum control system with a separable input signal.

In the special case where the terminal point is at infinity, the time-varying gains in the control equation reduce to the time-invariant gains

$$p_{111}(t) = \frac{1}{1 + a} \qquad k_{11}(t) = 1 \qquad T = \infty \tag{7-92}$$

The block diagram for this control system is shown in Fig. 7-1, where the time-varying gains are given by (7-90) and (7-91). From this block diagram, the functions of filtering and control are separable, which is a general result for separable signals and linear optimum control. This separation not only is important conceptually but also allows a simple realization of the optimum control system in terms of time-varying gains without the repetitive computation of the condition for minimum error as the input signals change with time.

Because the derivations in the previous sections pertain to a minimization problem with statistical input signals, one expects that a relationship with Wiener theory exists when integral constraints are placed on the optimization problem. In order to identify this relationship, this example is modified by

$$T = t + \tau \tag{7-93}$$

where the interval τ is considered to be a constant. This modification implies that the interval where the error measure is weighted is now a constant instead of shrinking with time. For this assumption, the gains appearing in the control equation become

$$K_1(\tau) = p_{111}(t) = \frac{1}{1 - a^2}[\tanh\tau - a(1 - e^{-a\tau}\operatorname{sech}\tau)] \tag{7-94}$$

and
$$K_{11}(\tau) = k_{11}(t) = \tanh\tau \tag{7-95}$$

Because the interval τ is considered to be a constant, the control system shown in Fig. 7-1 is now time-invariant. This example is specialized further by assuming that the noise originally appearing in the input signal $v_1(t)$ is now zero. This assumption yields

$$\overline{Q_{11}(t)}^{t} = Q_1(t) \qquad N^2 = 0 \tag{7-96}$$

by taking the limit of (7-88).

For these additional assumptions, the steady-state error in the control system appearing in Fig. 7-1 now is evaluated according to the measure

$$e_m(\tau) = \overline{[Q_1(t) - q_1(t)]^2} + \overline{m_1^2(t)} \tag{7-97}$$

This measure would be the measure used in Wiener theory and corresponds to the error measure used in the derivations leading to the results of this section.[49] However, the measure assumed in (7-97) pertains to only the steady-state error of the control system as opposed to initial-condition errors. The value of the measure $e_m(\tau)$ is written as a function of τ because the gains appearing in the control system are a function of τ, as specified in (7-94) and (7-95).

This measure now is evaluated by using the tabulated integrals discussed in Chap. 2. When the appropriate spectral densities and system transfer functions are found, the tabulated integral given in (2-33) results in the value of the steady-state error

$$e_m(\tau) = \frac{[K_{11}(\tau) - K_1(\tau)]^2 + aK_{11}(\tau)[1 + K_1^2(\tau)]}{K_{11}(\tau)[a + K_{11}(\tau)]} \tag{7-98}$$

For the special case $\tau = \infty$, the measure reduces to

$$e_m(\infty) = \frac{a(a+2)}{(a+1)^2} \tag{7-99}$$

The value for the error given in (7-99) is precisely the value that is found by minimizing (7-97) by using the procedures of Wiener for this particular design problem. In other words, the steady-state error of the control system indicated in Fig. 7-1 for $\tau = \infty$ is precisely the error obtained from Wiener theory.

Finally, the steady-state error of the control system indicated in Fig. 7-1 is studied for $\tau < \infty$. In the special case where the future values of the desired-response signal are not correlated with present values, the ratio $e_m(\tau)/e_m(\infty)$ is found to be

$$\frac{e_m(\tau)}{e_m(\infty)} = 1 \qquad a = \infty \tag{7-100}$$

On the other hand, when the desired response is a constant, this ratio reduces to

$$\frac{e_m(\tau)}{e_m(\infty)} = \coth 2\tau \qquad a = 0 \tag{7-101}$$

In Fig. 7-2, this ratio is plotted for various values of a. The sketch shows that the steady-state error in the control system indicated in Fig. 7-1 increases monotonically with decreasing τ except in the special case where the desired-response signal is similar to white noise. As the desired-response signal becomes more and more correlated, the steady-state error increases for a given τ.

The interpretation of these relative-error curves is related to the concept of prediction in control-system design. Because the interval τ is the interval of future time where response- and control-signal errors are weighted, increasing τ increases the amount of prediction used to find the control-system gains. Therefore, the more prediction included in the control system the smaller the system error, except when the desired-response signal is uncorrelated. This relationship between prediction and control provides a basis for important design and computational considerations discussed in the next chapter.

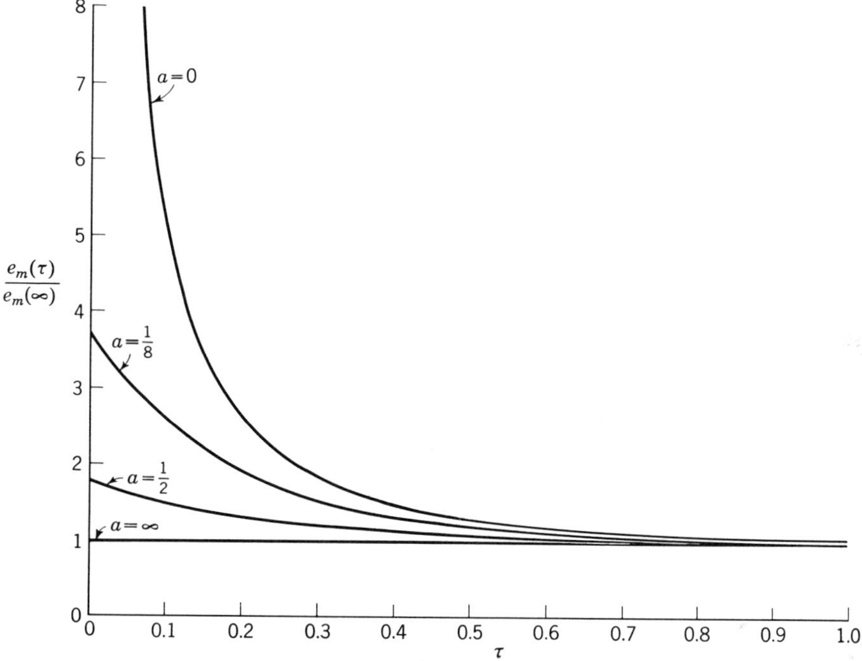

Fig. 7-2 Relative steady-state error.

7-6 Stability of linear optimum controls

Intuition may lead to the conclusion that a linear optimum control system is always a stable system, and indeed the problem of whether a linear optimum control system is stable is almost axiomatic. But the problem of designing a control system using optimization theory and with a particular equilibrium point in mind is somewhat more interesting. Furthermore, important relationships between optimization and stability theories exist and should be identified.

Suppose that a set of curves, denoted as $\mathbf{X}(t)$, is defined to be an equilibrium trajectory of the dynamic-process state vector $\mathbf{x}(t)$. In other words, the dynamic process is said to be in equilibrium when $\mathbf{x}(t) = \mathbf{X}(t)$. Also, suppose that a linear optimum control system is selected subject to the restriction that the dynamic process eventually reaches equilibrium as defined in terms of $\mathbf{X}(t)$.

If the state equation is given by

$$x_n'(t) = \sum_{m=1}^{N} [b_{nm}(t)x_m(t)] + c_{nn}(t)m_n(t) + u_n(t) \tag{7-102}$$

then the desired-control signals $M_n(t)$ are defined in terms of $\mathbf{X}(t)$ and $\mathbf{X}'(t)$ as

$$X_n'(t) = \sum_{m=1}^{N} [b_{nm}(t)X_m(t)] + c_{nn}(t)M_n(t) + u_n(t) \tag{7-103}$$

The selection of the equilibrium trajectory $\mathbf{X}(t)$ and the permissible load disturbances $\mathbf{u}(t)$ is restricted by the requirement that there exist a unique, finite desired-control vector $\mathbf{M}(t)$ which satisfies (7-103). If the response equation is given by

$$q_n(t) = \sum_{m=1}^{N} a_{nm}(t)x_m(t) \tag{7-104}$$

then the desired-response signals are defined in terms of $\mathbf{X}(t)$ as

$$Q_n(t) = \sum_{m=1}^{N} a_{nm}(t)X_m(t) \tag{7-105}$$

Finally, the error index to be minimized is taken to be

$$e(t) = \sum_{n=1}^{N} \int_t^T \{\phi_{nn}(\sigma)[Q_n(\sigma) - q_n(\sigma)]^2 + \psi_{nn}(\sigma)[M_n(\sigma) - m_n(\sigma)]^2\}\, d\sigma \tag{7-106}$$

where $T \to \infty$.

The minimization problem formulated here takes into account the desire to achieve an equilibrium state $\mathbf{X}(t)$. In particular, all errors are weighted relative to this equilibrium state so that zero error is assessed when the equilibrium state is achieved. These considerations are obvious when the vectors

$$\mathbf{z}(t) = \mathbf{x}(t) - \mathbf{X}(t) \tag{7-107}$$

and

$$\mathbf{y}(t) = \mathbf{m}(t) - \mathbf{M}(t) \tag{7-108}$$

are introduced. The state equation

$$z_n'(t) = \sum_{m=1}^{N} [b_{nm}(t)z_m(t)] + c_{nn}(t)y_n(t) \tag{7-109}$$

is found by combining (7-102) and (7-103). Also, the error index

$$e(t) = \sum_{n=1}^{N} \int_t^T \left\{\phi_{nn}(\sigma)\left[\sum_{m=1}^{N} a_{nm}(\sigma)z_m(\sigma)\right]^2 + \psi_{nn}(\sigma)[y_n(\sigma)]^2\right\} d\sigma \tag{7-110}$$

is found by combining (7-104) and (7-105).

For this optimization problem, the minimum-error function is defined as

$$E[\mathbf{z}(t), t] = \min_{\substack{\mathbf{y}(\sigma) \\ [t,\, T]}} e(t) \tag{7-111}$$

The techniques employed in Sec. 7-1 are used readily to show that the minimum-error function is given by

$$E[\mathbf{z}(t), t] = \sum_{m=1}^{N} \sum_{k=1}^{N} k_{mk}(t)z_m(t)z_k(t) \tag{7-112}$$

where

$$-k_{mk}'(\mu) = \sum_{n=1}^{N} \left[\phi_{nn}(\mu)a_{nm}(\mu)a_{nk}(\mu) + b_{nm}(\mu)k_{nk}(\mu) + b_{nk}(\mu)k_{nm}(\mu) - \frac{c_{nn}{}^2(\mu)}{\psi_{nn}(\mu)} k_{nm}(\mu)k_{nk}(\mu)\right] \tag{7-113}$$

and

$$k_{mk}(T) = 0 \tag{7-114}$$

Furthermore, the optimum control equation is given by

$$y_n^*(t) = -\frac{c_{nn}(t)}{\psi_{nn}(t)} \sum_{m=1}^{N} k_{nm}(t) z_m(t) \tag{7-115}$$

The final condition used here is

$$\frac{dE[\mathbf{z}(t), t]}{dt} = -\sum_{n=1}^{N} \left\{ \phi_{nn}(t) \left[\sum_{m=1}^{N} a_{nm}(t) z_m(t) \right]^2 + \psi_{nn}(t)[y_n^*(t)]^2 \right\} \tag{7-116}$$

which is found from the definition of the minimum-error function. The substitution of the optimum control equation into (7-116) yields

$$\frac{dE[\mathbf{z}(t), t]}{dt} = -\sum_{n=1}^{N} \sum_{k=1}^{N} \lambda_{mk}(t) z_m(t) z_k(t) \tag{7-117}$$

where

$$\lambda_{mk}(t) = \sum_{n=1}^{N} \left[\phi_{nn}(t) a_{nm}(t) + \frac{c_{nn}^2(t)}{\psi_{nn}(t)} k_{nm}(t) k_{nk}(t) \right] \tag{7-118}$$

Clearly, the matrix $[\lambda_{mk}]$ is positive definite because of the quadratic form indicated in (7-116).

The relationships developed thus far are sufficient conditions for the stability of the control system defined by

$$z_n'(t) = \sum_{m=1}^{N} \left[b_{nm}(t) - \frac{c_{nn}^2(t)}{\psi_{nn}(t)} k_{nm}(t) \right] z_m(t) \qquad n = 1, 2, \ldots, N \tag{7-119}$$

This set of equations is found by combining (7-109) and (7-115). Specifically, the minimum-error function is a Lyapunov function[50] satisfying the conditions

$$E[\mathbf{z}(t), t] > 0 \qquad \frac{dE[\mathbf{z}(t), t]}{dt} < 0 \qquad \text{for } \mathbf{z}(t) \neq \mathbf{0} \tag{7-120}$$

$$E[\mathbf{0}, t] = 0 \tag{7-121}$$

and

$$E[\mathbf{z}(t), t] \to \infty \qquad \text{for } \|\mathbf{z}(t)\| \to \infty \tag{7-122}$$

The condition given in (7-120) results from the requirement for a strictly convex error measure. The conditions given in (7-121) and (7-122) are seen directly from (7-112). The notation $\|\mathbf{z}(t)\|$ is used to denote the euclidean norm, which is

$$\|\mathbf{z}(t)\| = \sum_{n=1}^{N} z_n^2(t) \tag{7-123}$$

Because the system defined by (7-119) possesses a Lyapunov function and also satisfies $\mathbf{z}'(t) = \mathbf{0}$ for all t when $\mathbf{z}(t) = \mathbf{0}$, this system is asymptotically stable in the large. Asymptotic stability in the large requires that with any initial condition $\mathbf{z}(t_0) = \mathbf{z}_0$, the equilibrium condition $\mathbf{z}(t) \to \mathbf{0}$ when $t \to \infty$ must be achieved. The desire for asymptotic stability requires that $T = \infty$ be used in the error index.

This formulation of linear optimum controls, where the desired-response and desired-control signals are defined by (7-103) and (7-105), serves to illustrate that the minimum-error function is also a Lyapunov function in some instances. However, the minimum-error function is not a Lyapunov function in general.[51] Even in situations where a Lyapunov function is not found directly, the linear optimum control system is asymptotically stable provided that the error measure is strictly convex.§ This result is due to the fact that the feedback portion of the system is linear and is not dependent upon the desired-response and desired-control signals. Hence, the selection of the desired-response and desired-control signals does not affect the stability of the system but instead determines only the equilibrium curves $\mathbf{X}(t)$. However, the determination of the equilibrium curves for a given set of desired signals is a difficult problem of little practical importance.

7-7 Discrete approximation for statistical state signals

In a number of important design problems, statistical load disturbances and noise in the measurement of the state variables occur. When these types of statistical phenomena occur in system design, the state signals of the dynamic process can no longer be treated as deterministic. In these cases the procedures associated with dynamic programming cannot be used without additional concepts. The chief conceptual problem is that, because the state signals are no longer deterministic, the definition of a function for minimum error is nebulous. In other words, there is no immediately obvious analogy to the function $E[\mathbf{x}(\mu), \mu]$. This section is devoted to developing additional concepts which allow the appropriate definition of the mathematical form for the minimum error in a number of special cases for statistical state signals.

If the functional-equation approach is abandoned altogether, then an alternative is to approximate the integral occurring in the error index by a finite sum. Also, the differential equations defining the changes of state in the dynamic process can be approximated by difference equations. This procedure leads to the necessity of solving a set of simultaneous algebraic equations. Because the number of these equations equals the number of terms in the approximating sum for the integral, the solution becomes arbitrarily complex for arbitrary accuracy. This solution has a further complexity caused by the statistical phenomena. In order to circumvent the major complexity of dimensionality associated with accuracy, the identification of any cases for which the functional-equation approach applies is very important.

This discrete approximation of the minimization problem can be used to ascertain the mathematical form of the minimum-error function in a number of cases.[52] The discrete approximation is an unsophisticated way to bridge this conceptual gap because recourse to dynamic programming is not required. In the case of a quadratic error measure and a linear dynamic process, the algebraic equations that must be solved are linear. Therefore, the properties of matrices are useful in establishing the mathematical form of the optimum control equation and hence the mathematical form of the minimum-error function.

In this section, a first-order dynamic process is assumed to be

$$q(t) = a(t)x(t) \qquad x'(t) = b(t)x(t) + c(t)m(t) + u(t) \tag{7-124}$$

and also the error index is assumed to have the special form

$$e(t) = \int_t^T \left[\lambda(\sigma)\overline{[Q(\sigma) - q(\sigma)]^2}^{\,t} + \overline{[M(\sigma) - m(\sigma)]^2}^{\,t} \right] d\sigma \tag{7-125}$$

§ This result is always valid when the feedback portion of the optimum control system is time-invariant but may not always be valid when the feedback portion of the optimum control system is time-varying.

where the abbreviated notation introduced in Sec. 5-7 for conditional means is used. This error index is approximated by the sum

$$e(t) \approx \epsilon \sum_{n=1}^{R} \left[\lambda_n \overline{(Q_n - q_n)^2}^t + \overline{(M_n - m_n)^2}^t \right] \tag{7-126}$$

where the subscript notation appearing in this equation is defined as

$$f_n = f(t + n\epsilon) \qquad R = \frac{T - t}{\epsilon} \tag{7-127}$$

With the use of this subscript notation, the response signal is given by

$$q_n = a_n x_n \tag{7-128}$$

and the state equation is given by

$$x_n = x_{n-1} + \epsilon(b_{n-1}x_{n-1} + c_{n-1}m_{n-1} + u_{n-1}) \tag{7-129}$$

where the derivative of the state signal is approximated by a first difference. Finally, the value of the state variable x_n can be written in terms of the initial state signal x_1 by solving a set of difference equations recursively. From this procedure, the nth value of the state signal is found to be

$$x_n = \left[\prod_{m=1}^{n-1} (1 + b_m) \right] x_1 + \epsilon \sum_{m=1}^{n-1} \left[(c_m m_m + u_m) \prod_{r=m+1}^{n-1} (1 + \epsilon b_r) \right] \tag{7-130}$$

which is an approximation to the superposition integral. The notation introduced in (7-130) is defined as

$$\prod_{m=1}^{M} f_m = f_1 f_2 \cdots f_M \tag{7-131}$$

The condition for minimizing the approximated error index is found by considering the variation of the error index with respect to each of the R segments of the control signal. In other words, the error index is a function of each segment of the control signal; hence the condition for minimum error is

$$\frac{\partial e(t)}{\partial m_\mu} = 0 \qquad \mu = 1, 2, \ldots, R \tag{7-132}$$

When this condition for minimum error is applied and the appropriate equations for the segments of the control and response signals are substituted, the following set of linear algebraic equations is obtained:

$$m_\mu^* = \overline{M_\mu}^t + \epsilon \sum_{n=\mu+1}^{R} \overline{\lambda_n a_n c_\mu Q_n \prod_{r=\mu+1}^{n-1} (1 + \epsilon b_r)}^t - \epsilon \sum_{n=\mu+1}^{R} \overline{\lambda_n a_n^2 c_\mu x_n \prod_{r=\mu+1}^{n-1} (1 + \epsilon b_r)}^t \qquad \mu = 1, 2, \ldots, R \tag{7-133}$$

This set of R equations, when (7-130) is substituted into them, is used to find the R segments of the control signals.

A. Measurement noise and statistical load disturbances

Here, the a, b, and c coefficients of the dynamic process are assumed to be known deterministically, but the load-disturbance signal and the initial state x_1 are assumed to be statistical variables. When (7-133) is expanded under these conditions, the set of algebraic equations which specify the segments of the optimum control signal reduce to the form

$$\sum_{n=1}^{R} \alpha_{\mu n} m_n^* = \gamma_\mu - \delta_\mu \overline{x_1}^t \qquad \mu = 1, 2, \ldots, R \tag{7-134}$$

The coefficients appearing in this equation are found as combinations of the coefficients of the dynamic process, the weighting factors in the error measure, and the conditional means of the desired-response, desired-control, and load-disturbance signals. The properties of matrices now can be used to solve (7-134) simultaneously, thereby yielding the optimum segments of the control signal

$$m_n^* = \Gamma_n - \Delta_n \overline{x_1}^t \qquad n = 1, 2, \ldots, R \tag{7-135}$$

The limit as ϵ approaches zero gives

$$\lim_{\epsilon \to 0} \overline{x_1}^t = \overline{x(t)}^t \tag{7-136}$$

As a result of this limit, the optimum control signal is seen to be a linear function of the conditional mean of the initial state of the dynamic process such that

$$m^*(\sigma) = \Gamma(\sigma) - \Delta(\sigma)\overline{x(t)}^t \tag{7-137}$$

This mathematical relationship holds for all time on the interval $t \leqq \sigma \leqq T$. If the optimum control signal is substituted into the error index and the integration is performed, then the minimum value of the error index, $e_*(t)$, is a function of the conditional mean of the initial state signal. Because of this dependence of the minimum value of the error index, the minimum-error function is defined as the function

$$e_*(t) = E[\overline{x(t)}^t, t] \tag{7-138}$$

This definition holds when the state signal of the dynamic process is statistical because of random load disturbances and measurement noise.

B. Measurement noise, statistical load disturbances, and c parameter with random coefficients

Here, the mathematical form of the minimum-error function is investigated under the condition that the c parameter of the dynamic process is also a statistical variable. Under this condition, the algebraic equations defining the optimum control signal are expanded into the form

$$\sum_{n=1}^{R} \alpha_{\mu n} m_n^* = \gamma_\mu - \phi_\mu \overline{c_\mu x_1}^t \qquad \mu = 1, 2, \ldots, R \tag{7-139}$$

When this last equation is solved simultaneously for the R segments of the optimum control signal, the properties of matrices show that the optimum control signal becomes

$$m_n^* = \Gamma_n - \sum_{\alpha=1}^{R} \Phi_{n\alpha} \overline{c_\alpha x_1}^t \qquad n = 1, 2, \ldots, R \tag{7-140}$$

The optimum control equation derived here shows that the control signal is linearly dependent upon the correlations of the c parameter at various points in time with the initial value of the state signal of the dynamic process. In this case, the optimum control equation is linearly dependent upon an infinite number of these correlations when $\epsilon \to 0$. Clearly, this is an impractical situation in the limit because an infinite number of terms would be required in the construction of the control system. Also, the minimum-error function becomes a function of an infinite number of variables. Therefore, the special case is considered where this c parameter is of the form

$$c(\sigma) = \sum_{x=1}^{X} C_x c_x(\sigma) \tag{7-141}$$

and $c_x(\sigma)$ is a known function of time but C_x is a randomly distributed coefficient. For this special case, the optimum control signal given in (7-140) reduces to

$$m_n^* = \Gamma_n - \sum_{x=1}^{X} \Psi_{nx} \overline{C_x x_1}^t \tag{7-142}$$

Now the optimum control equation is linearly dependent upon only X correlation functions, and this number is invariant with respect to the limiting process $\epsilon \to 0$. Therefore, this case would result in a practical control system. Also, the minimum-error function would then be dependent only upon X variables which are these correlation functions. Because of the limit

$$\lim_{\epsilon \to 0} \overline{C_x x_1}^t = \overline{C_x x(t)}^t \tag{7-143}$$

the minimum value of the error index possesses the mathematical form

$$e_*(t) = E[\overline{\mathbf{C}x(t)}^t, t] \tag{7-144}$$

where

$$\overline{\mathbf{C}x(t)}^t = \begin{bmatrix} \overline{C_1 x(t)}^t \\ \overline{C_2 x(t)}^t \\ \cdot \\ \cdot \\ \cdot \\ \overline{C_X x(t)}^t \end{bmatrix} \tag{7-145}$$

This form is due to precisely the same reasoning used in arriving at (7-138).

The results of this derivation can be generalized to the condition where the a parameter of the dynamic process is also a random variable. However, the practical applications for statistically determined coefficients of the dynamic process are very limited. The purpose of including this derivation is merely to illustrate the concepts and difficulties.

C. Statistical b parameters

Here, the parameter b is assumed to be a random variable, but for the sake of simplicity all other variables are assumed to be deterministic. Then, even though arguments similar to those previously presented can be brought forth, this case appears to be academic because the required statistical information is not available. More specifically, terms appearing in (7-133) (such as $\overline{b_1 b_2 \cdots b_M}^t$ for $1 \leq M \leq R$) imply that an infinite amount of statistical information is needed for the solution when $\epsilon \to 0$. At this time, these difficulties are considered to be a fundamental barrier for practical applications, and therefore this case is pursued no further.

7-8 Linear optimum controls for statistical load disturbances and measurement noise

The results of Sec. 7-7 for statistical load disturbances and measurement noise can be generalized simply to the case of an Nth-order dynamic process. This generalization shows that the minimum-error function is dependent upon the conditional means of the N state signals of the dynamic process. Therefore, for the error index given in (7-5), the minimum-error function is defined notationally as

$$E[\overline{\mathbf{x}(t)}^t, t] = \min_{\substack{\mathbf{m}(\sigma) \\ [t, T]}} \overline{e(t)}^t \tag{7-146}$$

where

$$\overline{\mathbf{x}(t)}^t = \begin{bmatrix} \overline{x_1(t)}^t \\ \overline{x_2(t)}^t \\ \cdot \\ \cdot \\ \cdot \\ \overline{x_N(t)}^t \end{bmatrix} \tag{7-147}$$

Furthermore, the instantaneous minimum-error function is defined by

$$\mathscr{E}[\overline{\mathbf{x}(\mu)}^t, \mu] = \min_{\substack{\mathbf{m}(\sigma) \\ [\mu, T]}} \overline{e(\mu)}^t \tag{7-148}$$

subject to the boundary condition

$$\mathscr{E}[\overline{\mathbf{x}(T)}^t, T] = 0 \tag{7-149}$$

The application of the principle of optimality yields the discrete form of the condition for minimum error

$$\min_{\mathbf{m}(\mu)} \left(\sum_{n=1}^{N} \left\{ \phi_{nn}(\mu)\overline{[Q_n(\mu) - q_n(\mu)]^2}^t + \psi_{nn}(\mu)\overline{[M_n(\mu) - m_n(\mu)]^2}^t \right\} + \frac{\mathscr{E}[\overline{\mathbf{x}(\mu+\delta)}^t, \mu+\delta] - \mathscr{E}[\overline{\mathbf{x}(\mu)}^t, \mu]}{\delta} \right) = 0 \tag{7-150}$$

when the error measure given in (7-5) is specialized by the conditions given in (7-29) and (7-31). In order to take the limit $\delta \to 0$, the conditional mean of the state signal $x_n(\mu + \delta)$ is expanded as

$$\overline{x_n(\mu+\delta)}^t \approx \overline{x_n(\mu)}^t + \delta \overline{x_n'(\mu)}^t \tag{7-151}$$

If (7-151) is substituted into the minimum-error function for the N state signals, then the instantaneous minimum-error function can be expanded in a Taylor series where only the zeroth- and first-degree terms in the expansion are retained. Therefore, the condition for minimum error in continuous form reduces to

$$\min_{\mathbf{m}(\mu)} \left(\frac{\partial \mathscr{E}[\overline{\mathbf{x}(\mu)}^t, \mu]}{\partial \mu} + \sum_{n=1}^{N} \left\{ \overline{\phi_{nn}(\mu)[Q_n(\mu) - q_n(\mu)]^2}^t \right.\right.$$
$$\left.\left. + \overline{\psi_{nn}(\mu)[M_n(\mu) - m_n(\mu)]^2}^t + \overline{x_n'(\mu)}^t \frac{\partial \mathscr{E}[\overline{\mathbf{x}(\mu)}^t, \mu]}{\partial \overline{x_n(\mu)}^t} \right\} \right) = 0 \quad (7\text{-}152)$$

The solution to the optimization problem now can be expressed in terms of a set of ordinary differential equations in almost precisely the same manner as carried out earlier in this chapter for deterministic state signals.

The means of the derivatives of the state signals, which are needed in (7-152), are found by averaging both sides of the state equation given in (7-3). This averaging process results in

$$\overline{x_n'(\mu)}^t = \sum_{m=1}^{N} [b_{nm}(\mu)\overline{x_m(\mu)}^t] + c_{nn}(\mu)m_n(\mu) + \overline{u_n(\mu)}^t \qquad n = 1, 2, \ldots, N \quad (7\text{-}153)$$

when the dynamic process is specialized by (7-30). When (7-153) is substituted into (7-152), the optimum control signals are found from

$$\frac{\partial}{\partial m_i(\mu)} \left(\sum_{n=1}^{N} \left\{ \overline{\psi_{nn}(\mu)[M_n(\mu) - m_n(\mu)]^2}^t + m_n(\mu)c_{nn}(\mu) \frac{\partial \mathscr{E}[\overline{\mathbf{x}(\mu)}^t, \mu]}{\partial \overline{x_n(\mu)}^t} \right\} \right) = 0 \quad (7\text{-}154)$$

This equation is analogous to (7-9) for deterministic state signals. When the minimization indicated in (7-154) is performed, the optimum control signals are found to be

$$m_i^*(\mu) = \overline{M_i(\mu)}^t - \frac{c_{ii}(\mu)}{2\psi_{ii}(\mu)} \frac{\partial \mathscr{E}[\overline{\mathbf{x}(\mu)}^t, \mu]}{\partial \overline{x_i(\mu)}^t} \qquad i = 1, 2, \ldots, N \quad (7\text{-}155)$$

Under this condition for the optimum control signals, (7-152) reduces to

$$\frac{\partial \mathscr{E}[\overline{\mathbf{x}(\mu)}^t, \mu]}{\partial \mu} + \sum_{n=1}^{N} \left\{ \overline{\phi_{nn}(\mu)[Q_n(\mu) - q_n(\mu)]^2}^t \right.$$
$$\left. + \overline{\psi_{nn}(\mu)[M_n(\mu) - m_n^*(\mu)]^2}^t + \overline{x_n^{*\prime}(\mu)}^t \frac{\partial \mathscr{E}[\overline{\mathbf{x}(\mu)}^t, \mu]}{\partial \overline{x_n(\mu)}^t} \right\} = 0 \quad (7\text{-}156)$$

The minimum-error function can be shown to be a quadratic function with respect to the conditional means of the state signals from the derivations in Sec. 7-7 when the coefficients of the dynamic process are deterministic quantities. Therefore, the instantaneous minimum-error function is taken to be

$$\mathscr{E}[\overline{\mathbf{x}(\mu)}^t, \mu] = k(\mu) - 2\sum_{m=1}^{N} k_m(\mu)\overline{x_m(\mu)}^t + \sum_{m=1}^{N}\sum_{k=1}^{N} k_{mk}(\mu)\overline{x_m(\mu)}^t\,\overline{x_k(\mu)}^t \quad (7\text{-}157)$$

where
$$k_{mk}(\mu) = k_{km}(\mu) \quad (7\text{-}158)$$

When manipulations analogous to (7-14) through (7-20) are performed, the ordinary differential equations specifying the k parameters which appear in the instantaneous minimum-error function are found to be

$$-k'(\mu) = \sum_{n=1}^{N} \left\{ \phi_{nn} \overline{\left[Q_n - \sum_{j=1}^{N} a_{nj}(x_j - \overline{x_j}^t) \right]^2}^t + \psi_{nn} \overline{[M_n - \overline{M_n}^t]^2}^t - 2\overline{u_n}^t k_n - 2c_{nn}\overline{M_n}^t k_n - \frac{c_{nn}^{\ 2}}{\psi_{nn}} k_n^{\ 2} \right\} \tag{7-159}$$

$$-k'_m(\mu) = \sum_{n=1}^{N} \left(\phi_{nn} \overline{Q_n}^t a_{nm} - \overline{u_n}^t k_{nm} - c_{nn} \overline{M_n}^t k_{nm} + b_{nm} k_n - \frac{c_{nn}^{\ 2}}{\psi_{nn}} k_n k_{nm} \right) \tag{7-160}$$

$$-k'_{mk}(\mu) = \sum_{n=1}^{N} \left(\phi_{nn} a_{nm} a_{nk} + b_{nm} k_{nk} + b_{nk} k_{nm} - \frac{c_{nn}^{\ 2}}{\psi_{nn}} k_{nm} k_{nk} \right) \tag{7-161}$$

A comparison of (7-33) and (7-160) shows that the only change in the computation of the k parameters that appear in the optimum control equation involves the replacement of the load disturbances by their conditional means.

However, the unsubscripted k parameter, which does not enter into the optimum control equation, is seen to be somewhat different from that arising in the case of deterministic state signals due to the first term appearing in (7-159). This change results from the use of the relationship

$$\overline{\left(Q_n - \sum_{j=1}^{N} a_{nj} x_j \right)^2}^t = \overline{\left[Q_n - \sum_{j=1}^{N} a_{nj}(x_j - \overline{x_j}^t) \right]^2}^t - 2\overline{Q_n}^t \sum_{j=1}^{N} a_{nj} \overline{x_j}^t + \sum_{j=1}^{N} \sum_{i=1}^{N} a_{nj} a_{ni} \overline{x_i}^t \overline{x_j}^t \tag{7-162}$$

The use of this relationship is necessary in order to collect all similar terms involving the conditional means of the state signals so that the k parameters are truly independent of these variables. The first term appearing in (7-159) and also appearing in (7-162), is independent of the conditional means of the state signals because this term involves just the variances of the state signals.

Finally, the optimum control equation is found to be

$$m_i^*(t) = \overline{M_i(t)}^t + \frac{c_{ii}(t)}{\psi_{ii}(t)} \left[k_i(t) - \sum_{m=1}^{N} k_{im}(t) \overline{x_m(t)}^t \right] \qquad i = 1, 2, \ldots, N \tag{7-163}$$

Therefore, the construction of the optimum control equation requires the use of Wiener filters to generate the conditional means $\overline{x_m(t)}^t$. However, in the case of zero measurement noise, $\overline{x_m(t)}^t = x_m(t)$, and no energy-storage components are required in the feedback portion of the optimum control equation. Furthermore, the method developed in Sec. 7-3 for separable signals can be used here for statistical load disturbances.

7-9 Example with a statistical load disturbance

This section is devoted to an example of statistical state signals for a simple first-order system with a statistical load-disturbance signal. The dynamic process is specialized to

$$N = 1 \qquad a_{11}(t) = c_{11}(t) = 1 \qquad b_{11}(t) = 0 \tag{7-164}$$

and the error measure also is specialized to

$$\phi_{nn}(\sigma) = \psi_{nn}(\sigma) = 1 \qquad M_1(\sigma) = Q_1(\sigma) = 0 \tag{7-165}$$

The statistical load-disturbance signal is measured in terms of the input signal

$$v_1(\sigma) = u_1(\sigma) + n_1(\sigma) \tag{7-166}$$

where the signal $n_1(\sigma)$ is a measurement noise. The correlation functions assumed for these signals are

$$\phi_{u_1u_1}(\beta) = U^2e^{-a|\beta|} \qquad \phi_{n_1n_1}(\beta) = N^2u_0(\beta) \qquad \phi_{u_1n_1}(\beta) = 0 \tag{7-167}$$

and the signals are assumed to be gaussian.

For this example, the optimum control equation is given by

$$m_1(t) = s_{111}(t)\overline{U_{11}(t)}^t - k_{11}(t)x_1(t) \tag{7-168}$$

which is similar to the example given in Sec. 7-5. The conditional mean $\overline{x_1(t)}^t$ reduces to $x_1(t)$ because zero measurement noise for this state signal is assumed. Also, the feedforward gain is defined by the differential equation

$$-s'_{111}(\mu) = -\mathscr{U}_{11}(\mu, t)k_{11}(\mu) - k_{11}(\mu)s_{111}(\mu) \tag{7-169}$$

Because the correlation functions of the statistical input signal are precisely the same as those assumed in Sec. 7-5, the reference signal to the control system is given by

$$\overline{U_{11}(t)}^t = \int_{-\infty}^{t} (w_1 - a)e^{-w_1(t-\sigma)}v_1(\sigma)\, d\sigma \tag{7-170}$$

and the time function used in the solution of (7-169) is

$$\mathscr{U}_{11}(\mu, t) = e^{-a(\mu-t)} \tag{7-171}$$

The solution of (7-169) subject to the appropriate terminal-boundary condition is found to be

$$s_{111}(t) = -\frac{1}{1-a^2}\,[1 - e^{-a(T-t)}\,\text{sech}\,(T-t) - a\tanh\,(T-t)] \tag{7-172}$$

When (7-172) is used as the feedforward gain in the optimum control equation, two special cases are considered. First, the assumption is made that the mean-square value of the measurement noise is zero. In this case, the impulse response appearing in (7-170) becomes a unit impulse, and (7-168) reduces to

$$m_1(t) = s_{111}(t)u_1(t) - k_{11}(t)x_1(t) \qquad N^2 = 0 \tag{7-173}$$

Hence, the reference signal in the optimum control equation is simply the load-disturbance signal itself when no measurement noise is present. The other special case occurs when the mean-square value of the noise signal becomes infinite. In this case, the impulse response appearing in (7-170) limits to zero; hence the optimum control equation reduces to

$$m_1(t) = -k_{11}(t)x_1(t) \qquad N^2 = \infty \tag{7-174}$$

These two limiting cases can be interpreted in a slightly different sense. The first case, defined by (7-173), is simply the measurement of the load-disturbance signal with no measurement noise. However, the second case, defined by (7-174), can be considered to be the case where no measurement of the load-disturbance signal is possible. This is a very practical case because generally load disturbances are unmeasurable in most control-system design problems, and the engineer relies on the feedback configuration of the system to counteract load-disturbance signals. Of course, if the load-disturbance signal is not measurable, then the conditional mean is equal to the mean value, which happens to be zero for the statistical characteristics assumed in (7-167). If the conditional mean of the load disturbance is zero, then the single-subscripted k parameter appearing in (7-163) is zero because (7-160) has no driving function.

Now the optimum control equation is rewritten for these two special cases where the load disturbance is measurable with no measurement noise and where the load

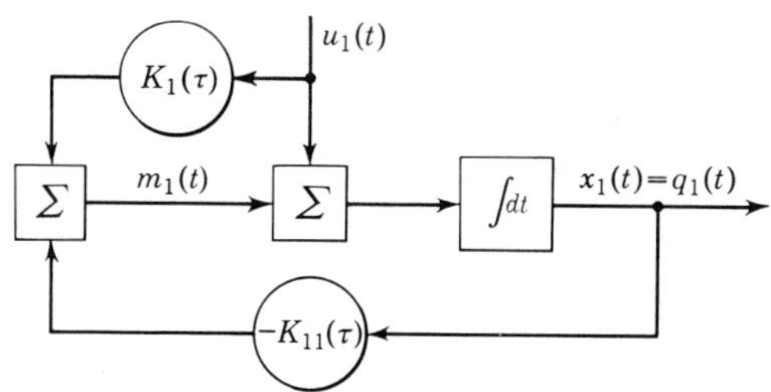

Fig. 7-3 Block diagram of linear optimum control system with separable load-disturbance signal.

disturbance is not measurable because instrumentation is not available. Specifically, the control equation is written as

$$m_1(t) = K_1(\tau)u_1(t) - K_{11}(\tau)x_1(t) \tag{7-175}$$

where the K parameter appearing in the feedforward portion of the optimum control system is given by

$$K_1(\tau) = \begin{cases} -\dfrac{1}{1-a^2}(1 - e^{-a\tau}\operatorname{sech}\tau - a\tanh\tau) & \text{disturbance measurable} \\ 0 & \text{disturbance not measurable} \end{cases} \tag{7-176}$$

In (7-176), the interval where response-signal error and control effort are weighted is taken to be a fixed interval $\tau = T - t$. The block diagram of the optimum control system for the statistical-load-disturbance-signal case is shown in Fig. 7-3. The two cases, where the load-disturbance signal is measurable and where it is not measurable, are defined by the K parameter given in (7-176).

The steady-state error of the optimum control system due to statistical load disturbances now is considered As in Sec. 7-5, the measure of steady-state error must correspond to the error measure used in the system design and therefore is again chosen to be (7-97). When this steady-state error is evaluated by using tabulated integrals presented in Chap. 2, the result is

$$e_m(\tau) = \frac{[1 + K_1(\tau)]^2 + K_{11}^2(\tau) + aK_1^2(\tau)K_{11}(\tau)}{K_{11}(\tau)[a + K_{11}(\tau)]} \tag{7-177}$$

The notation $e_m(\tau)$ is used to indicate the value of this error when the load disturbance is measurable, and $e_{nm}(\tau)$ is introduced to indicate the value of the error when the load-disturbance signal is not measured.

The special case $\tau = \infty$ for the measurable load disturbance gives the limiting value for this steady-state error

$$e_m(\infty) = \frac{1 + 2a}{(1 + a)^2} \tag{7-178}$$

For the special cases where the load disturbance is uncorrelated ($a = \infty$) and the

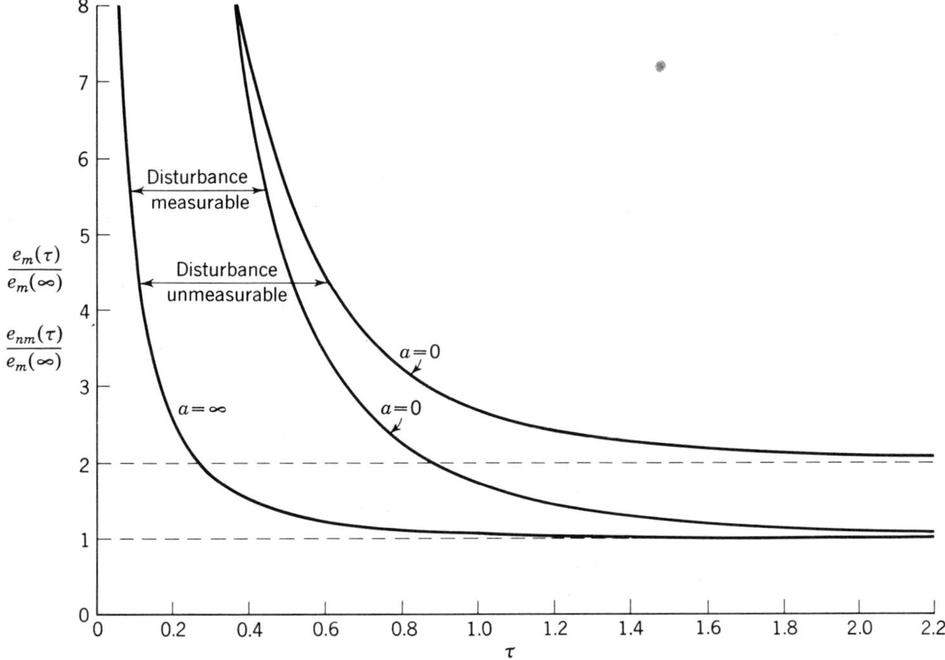

Fig. 7-4 Relative steady-state error.

load disturbance is a constant ($a = 0$), the relative steady-state error for a measurable load-disturbance signal is given by

$$\frac{e_m(\tau)}{e_m(\infty)} = \begin{cases} \coth 2\tau & a = \infty \\ \coth^2 \tau & a = 0 \end{cases} \tag{7-179}$$

Also, the relative steady-state error when the load-disturbance signal cannot be measured is given by

$$\frac{e_{nm}(\tau)}{e_m(\infty)} = \begin{cases} \coth 2\tau & a = \infty \\ 2(\coth \tau)(\coth 2\tau) & a = 0 \end{cases} \tag{7-180}$$

In (7-179) and (7-180), the steady-state indices are normalized by $e_m(\infty)$.

The curves in Fig. 7-4 show that the value of the error when the load disturbance is not measurable is always greater than or equal to the value of the error when the load disturbance is measurable. Figure 7-4 also shows that when the load-disturbance signal is uncorrelated, the results are the same for both measurable and unmeasurable load-disturbance signals. This is, of course, due to the fact that when the load-disturbance signal is uncorrelated, the conditional mean of the load disturbance is zero for all future time and hence corresponds to the unmeasurable case.

A comparison of Figs. 7-2 and 7-4 shows that load-disturbance signals, in general, require more prediction than desired-response signals. This is due, of course, to the fact that the load-disturbance signal enters directly into the energy storage associated with the dynamic process, and therefore additional time is required in order to counteract perturbations in the response due to the disturbances.

7-10 Linear optimum controls for dynamic processes with pure time delay

When the dynamic process is described by a finite set of ordinary first-order differential equations, the optimum control system is constructed with a finite number of measuring devices. Specifically, each state signal in the dynamic process must be measured in order to construct the feedback portion of the optimum control system. Therefore, the number of measuring devices is equal to the order of the differential equations describing the dynamic process. However, many dynamic processes of interest involve pure time delay due to distributed parameter effects in the physical system. Because pure time delay can be considered to result from a physical system which is described by an ordinary differential equation of infinite order, the presence of pure time delay would require an infinite number of measuring devices. Clearly, then, the optimization techniques presented thus far would not be practical for constructing an optimum control system when the dynamic process includes pure time delay. In fact, such dynamic processes are not state-determined by definition. Fortunately, new concepts can be added to the previous optimization theory so that some design problems involving pure time delay can be treated.[48]

In this section, a special form of dynamic process with pure time delay is considered in order to illustrate these concepts. In particular, the response variables of the dynamic process are assumed to be

$$q_n(t+\tau) = \sum_{m=1}^{N} a_{nm}(t+\tau)x_m(t+\tau) \qquad n = 1, 2, \ldots, N \tag{7-181}$$

Also, the state equation is assumed to be

$$x_n'(t+\tau) = \sum_{m=1}^{N} [b_{nm}(t+\tau)x_m(t+\tau)] + c_{nn}(t+\tau)m_n(t) + u_n(t+\tau) \qquad n = 1, 2, \ldots, N \tag{7-182}$$

These equations describe a dynamic process for which the control signals are delayed by τ sec. In other words, the dynamic process can be thought of in block-diagram form as the control signals entering a delay device of τ sec, and these delayed control signals then enter the remainder of the dynamic process, which is described by N ordinary first-order differential equations. The choice of this particular form for the dynamic process with pure time delay is somewhat arbitrary but is introduced here for the purposes of illustrating the treatment of pure time delay. However, this form of dynamic process possesses a number of useful applications. One example is the control of fuel flow to a boiler, where the fuel experiences a transport lag between the valve control and the burner.

Some of the new concepts which must be added to the optimization theory concern the selection of the error index. These new concepts are facilitated by the sketches of particular control and response signals in the dynamic process indicated in Fig. 7-5. Normally, control effort and response errors are weighted over the time interval $t \leqq \sigma \leqq T$. However, for a pure time delay of τ sec, the segment of the control signal on the interval $T - \tau \leqq \sigma \leqq T$ does not affect the response signal on the interval $t \leqq \sigma \leqq T$. Therefore, this segment of the control signal is not weighted in the error index, because this segment is arbitrary with respect to achieving a reduction in response error. On the other hand, the segment of the response signal on the interval $t \leqq \sigma \leqq t + \tau$ is determined completely by the input signals to the dynamic process over a past interval of time. Of course, only present and future values of the control signals in the dynamic process can be selected. Hence,

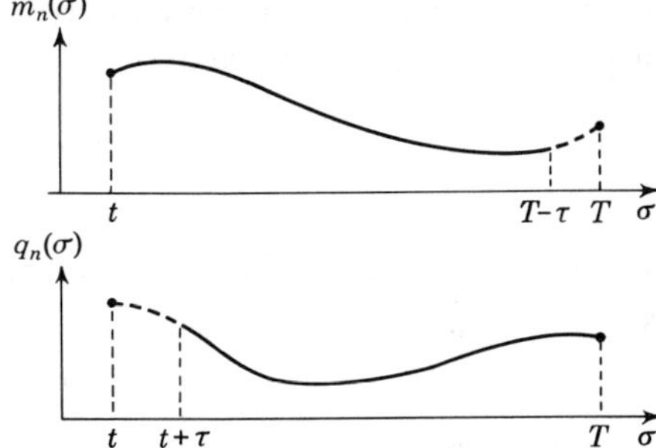

Fig. 7-5 Weighting intervals for dynamic process with pure-time delay τ.

only the segment of the response signal on the interval $t + \tau \leqq \sigma \leqq T$ is weighted in the error index. Based on these considerations of cause and effect, the modified error index

$$e(t) = \sum_{n=1}^{N} \left\{ \int_{t+\tau}^{T} \phi_{nn}(\sigma)[Q_n(\sigma) - q_n(\sigma)]^2 \, d\sigma + \int_{t}^{T-\tau} \psi_{nn}(\sigma)[M_n(\sigma) - m_n(\sigma)]^2 \, d\sigma \right\} \tag{7-183}$$

is used in design problems where a pure time delay of τ sec occurs in the dynamic process defined by (7-181) and (7-182). A change in variables for the error index given in (7-183) reduces this index to

$$e(t) = \sum_{n=1}^{N} \int_{t}^{T-\tau} \{\phi_{nn}(\sigma + \tau)[Q_n(\sigma + \tau) - q_n(\sigma + \tau)]^2 + \psi_{nn}(\sigma)[M_n(\sigma) - m_n(\sigma)]^2\} \, d\sigma \tag{7-184}$$

This is the error index that is minimized by using considerations pertinent to dynamic programming.

For this dynamic process with pure time delay, the state of the dynamic process now must be defined. First, the response vector is written in terms of the functional

$$\mathbf{q}(\sigma) = \mathscr{W}\,[\mathbf{m}(\rho), -\infty < \rho \leqq \sigma - \tau] \tag{7-185}$$

according to the reasoning used in establishing an appropriate error index. Then, in order to define the state of the dynamic process, the distribution of stored energy must be assessed. The stored energy in the portion of the dynamic process not involving the pure time delay is defined by $\mathbf{x}(t)$. Furthermore, the stored energy in the pure-time-delay devices is defined by the control vector $\mathbf{m}(\rho)$ on the past segment of time $t - \tau < \rho \leqq t$. Therefore, the representation of this dynamic process in terms of a transition functional is

$$\mathbf{q}(\sigma) = \mathscr{G}[\mathbf{x}(t), \sigma, t; \mathbf{m}(\rho), t - \tau < \rho \leqq \sigma - \tau] \tag{7-186}$$

However, if t is replaced by $t + \tau$ in (7-186), the response vector is expressed as

$$\mathbf{q}(\sigma) = \mathscr{G}[\mathbf{x}(t + \tau), \sigma, t + \tau; \mathbf{m}(\rho), t < \rho \leqq \sigma - \tau] \tag{7-187}$$

This functional involves only the present and future segment of the control vector, which is the segment selected in the minimization of the error index. Also, the state of this dynamic process is defined completely by the vector $\mathbf{x}(t + \tau)$.

For the purposes of dynamic programming, the state of the dynamic process is defined by the N components of the vector $\mathbf{x}(t + \tau)$. Therefore, the minimum value of the error index is a function of these components and real time t. In terms of a dummy variable μ, the minimum-error function for a dynamic process with this form of pure time delay is defined as

$$E[\mathbf{x}(\mu + \tau), \mu] = \min_{\substack{\mathbf{m}(\sigma) \\ [\mu, T-\tau]}} e(\mu) \tag{7-188}$$

According to this definition, the boundary condition on the minimum-error function is

$$E[\mathbf{x}(T), T - \tau] = 0 \tag{7-189}$$

With an appropriately defined error index and minimum-error function, the discrete dynamic-programming condition for minimum error is developed in precisely the same manner as for dynamic processes without pure time delay. This development results in the discrete condition

$$\min_{\mathbf{m}(\mu)} \left(\sum_{n=1}^{N} \{\phi_{nn}(\mu + \tau)[Q_n(\mu + \tau) - q_n(\mu + \tau)]^2 + \psi_{nn}(\mu)[M_n(\mu) - m_n(\mu)]^2\} + \frac{E[\mathbf{x}(\mu + \tau + \delta), \mu + \delta] - E[\mathbf{x}(\mu + \tau), \mu]}{\delta} \right) = 0 \tag{7-190}$$

Because the increment δ is arbitrarily small, a first-order approximation to the state variable at $\mu + \tau + \delta$ is

$$x_n(\mu + \tau + \delta) = x_n(\mu + \tau) + \delta x_n'(\mu + \tau) \tag{7-191}$$

Therefore, under the limiting condition $\delta \to 0$, the continuous condition for minimum error is derived, using the expansion rules of differentiation, to be

$$\min_{\mathbf{m}(\mu)} \left(\frac{\partial E[\mathbf{x}(\mu + \tau), \mu]}{\partial \mu} + \sum_{n=1}^{N} \left\{ \phi_{nn}(\mu + \tau)[Q_n(\mu + \tau) - q_n(\mu + \tau)]^2 + \psi_{nn}(\mu)[M_n(\mu) - m_n(\mu)]^2 + x_n'(\mu + \tau) \frac{\partial E[\mathbf{x}(\mu + \tau), \mu]}{\partial x_n(\mu + \tau)} \right\} \right) = 0 \tag{7-192}$$

This continuous condition for minimum error is in exactly the same form as the condition for dynamic processes without pure time delay, and therefore familiar techniques are used for reducing this partial differential equation to a set of ordinary differential equations.

The minimum-error function, for this linear dynamic process with pure time delay and for the error index defined in (7-184), is quadratic with respect to the components of $\mathbf{x}(\mu + \tau)$. Hence, the minimum-error function is written in the form

$$E[\mathbf{x}(\mu + \tau), \mu] = k(\mu) - 2 \sum_{m=1}^{N} k_m(\mu)x_m(\mu + \tau) + \sum_{m=1}^{N} \sum_{k=1}^{N} k_{mk}(\mu)x_m(\mu + \tau)x_k(\mu + \tau) \tag{7-193}$$

where

$$k_{mk}(\mu) = k_{km}(\mu) \tag{7-194}$$

If the manipulations performed for dynamic processes without pure time delay are performed now for (7-192) and (7-193), the optimum control equation is found to be

$$m_i^*(t) = M_i(t) + \frac{c_{ii}(t + \tau)}{\psi_{ii}(t)}\left[k_i(t) - \sum_{m=1}^{N} k_{im}(t)x_m(t+ \tau)\right] \qquad i = 1, 2, \ldots, N \tag{7-195}$$

Also, the ordinary differential equations defining the k parameters can be found, and these differential equations result in a form similar to (7-32) to (7-34). However, a

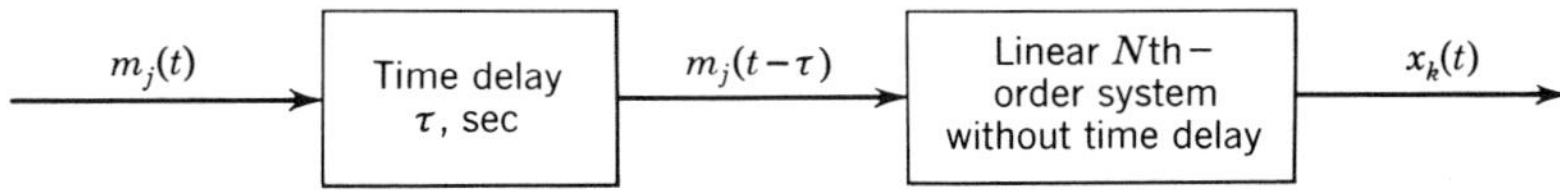

Fig. 7-6 Block diagram of dynamic process for zero load disturbances.

number of variables appearing in these differential equations are not functions of μ but now are functions of $\mu + \tau$. These variables are the a, b, and c coefficients of the dynamic process, the ϕ weighting factors, and also the load disturbances u and the desired responses Q. No other changes are necessary.

The construction of the optimum control equation given in (7-195) requires the measurement of $\mathbf{x}(t + \tau)$. This vector, of course, is the state vector τ sec in the

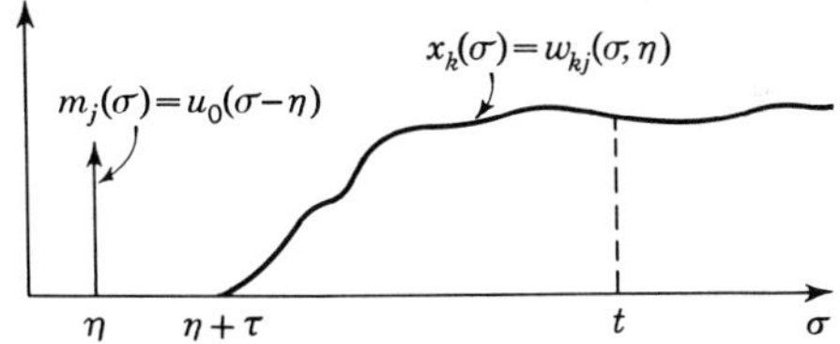

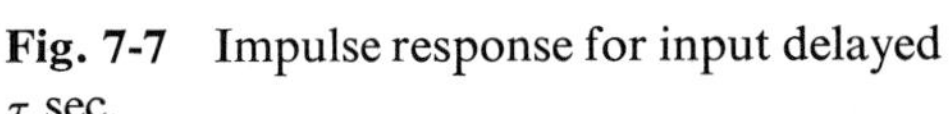
Fig. 7-7 Impulse response for input delayed τ sec.

future and hence cannot be measured directly. However, this vector can be computed in terms of the measured state vector $\mathbf{x}(t)$ and the measured control vector $\mathbf{m}(\sigma)$ on the past interval of time $t - \tau \leq \sigma \leq t$. In order to ascertain the required computations, a block diagram of the dynamic process is shown in Fig. 7-6 and now is considered. The mathematical relationship indicated in Fig. 7-6 is based upon the linearity property of the dynamic process and also is based upon the assumption that the load-disturbance signals are zero.

Under these assumptions, the impulse response from the jth control signal to the kth state signal is found. If an impulse input occurring at $\sigma = \eta$ is postulated, then the response of the system, shown in Fig. 7-6, results as indicated in Fig. 7-7. The curve shown in the last figure is the impulse response $w_{kj}(t, \eta)$ between the jth control signal and the kth state signal. Time t is defined as the time of observance of this response. Because of the pure time delay of τ sec, this impulse response satisfies

$$w_{kj}(t, \eta) = 0 \qquad t < \eta + \tau \tag{7-196}$$

which is indicated in Fig. 7-7. Finally, the kth state signal is written in terms of a sum of superposition integrals as

$$x_k(t) = \sum_{j=1}^{N} \int_{-\infty}^{t-\tau} w_{kj}(t, \eta) m_j(\eta)\, d\eta \qquad u_j(\eta) = 0 \tag{7-197}$$

when the load-disturbance signals are assumed to be zero.

When this superposition integral is used for these dynamic processes which include pure time delay, the future value of the kth state signal can be deduced from further considerations. In particular, the contributions to the kth state signal at time $t + \tau$ come from three sources. The first source of contribution is due to the load disturbances in the dynamic process. Because the load disturbances are assumed to be known deterministically except in statistical cases, they need not be measured and their contributions are simply functions of t and τ. The second contribution is due to the state signals at time t. The contributions of the state signals at time t appear in the kth state signal at time $t + \tau$ simply as the values of these existing state signals multiplied by the appropriate functions of t and τ. The third contribution to this future value of the state signal is due to the control-signal segment lying on the past interval of time $t - \tau \leqq \sigma \leqq t$. This segment of the control signals appears in the superposition integral where the control signals are weighted only over $t - \tau \leqq \eta \leqq t$. Therefore, the kth state signal at the future time $t + \tau$ can be written as

$$x_k(t + \tau) = f_k(t, \tau) + \sum_{j=1}^{N} \left[g_{kj}(t, \tau) x_j(t) + \int_{t-\tau}^{t} w_{kj}(t + \tau, \eta) m_j(\eta)\, d\eta \right] \qquad k = 1, 2, \ldots, N \tag{7-198}$$

The functions $g_{kj}(t, \tau)$ and $w_{kj}(t + \tau, \eta)$ are found from an analysis of the portion of the dynamic process that does not involve the pure time delay. However, the function $f_k(t, \tau)$ depends upon both an analysis of the dynamic process not involving the pure time delay and the known load-disturbance signals on the interval $t \leqq \sigma < t + \tau$.

From (7-198), the future values of the required state signals which appear in the optimum control equation are computed from measurements of both the existing state signals $\mathbf{x}(t)$ and the past segments of the control signals on the interval $t - \tau \leqq \sigma \leqq t$. Therefore, the optimum control system for dynamic processes with pure time delay can be constructed in a feedback configuration similar to the cases without pure time delay. Specifically, the optimum control equation can be rewritten in the form

$$m_i^*(t) = R_i(t) - \sum_{j=1}^{N} \int_{t-\tau}^{t} W_{ij}(t + \tau, \eta) m_j(\eta)\, d\eta - \sum_{j=1}^{N} K_{ij}(t) x_j(t) \tag{7-199}$$

where

$$R_i(t) = M_i(t) + \frac{c_{ii}(t + \tau)}{\psi_{ii}(t)} \left[k_i(t) - \sum_{m=1}^{N} k_{im}(t) f_m(t, \tau) \right] \tag{7-200}$$

$$W_{ij}(t + \tau, \eta) = \frac{c_{ii}(t + \tau)}{\psi_{ii}(t)} \sum_{m=1}^{N} k_{im}(t) w_{mj}(t + \tau, \eta) \tag{7-201}$$

$$K_{ij}(t) = \frac{c_{ii}(t + \tau)}{\psi_{ii}(t)} \sum_{m=1}^{N} k_{im}(t) g_{mj}(t, \tau) \tag{7-202}$$

However, the additional measurement of the finite-interval segment of the control signal requires a rather complicated filtering process because of the finite interval of the superposition integral appearing in (7-199). In particular, this filtering process requires a delay line or a lumped-parameter approximation of the delay line.

In order to illustrate the derivations of the time functions indicated in (7-198), the following special dynamic process is assumed:

$$x_1'(t + \tau) = m_1(t) \qquad x_2'(t + \tau) = x_1(t + \tau) \tag{7-203}$$

Because the load disturbances are assumed to be zero in this specialized case, the first term in (7-198) is

$$f_1(t, \tau) = f_2(t, \tau) = 0 \tag{7-204}$$

However, the g functions which appear in (7-198) are found to be

$$g_{11}(t, \tau) = 1 \qquad g_{12}(t, \tau) = 0 \qquad g_{21}(t, \tau) = \tau \qquad g_{22}(t, \tau) = 1 \tag{7-205}$$

by simple block-diagram considerations. Finally, the impulse responses appearing in (7-198) for the special dynamic process assumed here are found to be

$$w_{12}(t + \tau, \eta) = w_{22}(t + \tau, \eta) = 0 \tag{7-206}$$

$$w_{11}(t + \tau, \eta) = \begin{cases} 1 & \eta \leqq t \\ 0 & \eta > t \end{cases} \tag{7-207}$$

$$w_{21}(t + \tau, \eta) = \begin{cases} t - \eta & \eta \leqq t \\ 0 & \eta > t \end{cases} \tag{7-208}$$

The concepts introduced here for pure time delay avoid the necessity of introducing an infinite number of feedback loops because of the requirement of measuring the state of the dynamic process. Although only a special form of dynamic process with pure time delay is considered in this section, the procedures developed here can be extended to a number of alternative forms of dynamic processes with pure time delay. Because the number of possible forms is excessive, the derivation presented in this chapter suffices. Also, statistical signals and state variables are not treated here for the dynamic process with pure time delay, because no new concepts are needed and the results can be deduced from Secs. 7-1 and 7-8 by direct analogy.

7-11 Summary

This chapter is devoted to a detailed study of the theory and properties of the subclass of optimization problems where the dynamic process is linear and the error measure is quadratic. The solution to this class of problems is formulated by an elementary procedure of expanding the minimum-error function in terms of a set of parameters. These parameters also are the coefficients of the optimum control equation. The solution to this class of optimization problems is found by solving a set of ordinary differential equations that define the k parameters. These parameters are defined by the so-called matrix Riccati equations, often treated in the classical theory of ordinary differential equations.[53]

An advantage of expressing the solutions to these optimization problems in terms of the k parameters is that the traditional two-point boundary-value problem is eliminated. In particular, the two-point boundary-value problem is replaced by a sequence of two one-point boundary problems. The first one-point boundary-value problem is the system design, which requires the solution of the differential equations defining the k parameters. The second one-point boundary-value problem is the system evaluation or simulation, which requires the solution of the state equation in conjunction with the optimum control equation. Furthermore, these solutions are always found from stable differential equations, thereby greatly reducing numerical difficulties. Finally, the coefficients of the equations defining the k parameters do not involve the derivatives of signals and coefficients of the dynamic process, thereby greatly decreasing the required design information and system instrumentation.

Another advantage of expressing the solutions to these optimization problems in terms of the k parameters is that the structure and properties of the optimum control equation are obtained directly. When separable signals occur, the optimum control equation is structured completely in terms of linear time-varying or time-invariant gains and precomputed reference signals. Furthermore, no important conceptual difficulties are required for the treatment of statistical desired-response, desired-control, and load-disturbance signals. Also, no important practical difficulties are introduced if these statistical signals are separable and the Wiener prediction theory applies.

A class of dynamic processes which includes pure time delay is considered in terms of the solution of optimization problems. Again, with the introduction of additional concepts, the solution to the optimization problem plus the construction of the optimum control system essentially is not different from that for dynamic processes without pure time delay.

The considerations presented here for optimum control systems where the dynamic process is linear and the error measure is quadratic provide a broad basis for a generalized design procedure of linear control systems. Not only does this procedure apply to dynamic processes which are time-varying, but also it applies to design problems where multiple inputs and outputs must be considered. Because the numerical solutions of the optimization problems can be obtained and also because the optimum control equation is constructed readily in terms of components, optimization theory as presented here for control-system design is extremely useful not only from a conceptual standpoint but also from a practical standpoint.

8

Design Considerations for Linear Optimum Control Systems

If optimization theory is to be used as the basis for feedback-control-system design, then a number of important and practical details must be considered. First, the advisability and the need for using optimization theory must be investigated for the design problem at hand. Primarily, the need for using optimization theory is evaluated in terms of the dynamic requirements and difficulties of the design problem. Clearly, a method of categorizing design problems in terms of the degree of dynamic difficulty, which does not require recourse to large computational facilities, is needed.

The evaluation of design problems in terms of dynamic difficulties requires an approximate method for selecting the magnitudes of the weighting factors appearing in the error measure. Again, such a method should not require the use of extensive computations. Furthermore, precise information concerning the dynamic process generally is not available in practical problems. Hence, an understanding of the sensitivity of control-system performance to assumption errors is desirable.

A number of design invariants are presented in this chapter which can be used for categorizing problems. These design invariants are selected by a method very similar to those used for scaling analog-computer solutions of differential equations. Also, other design considerations such as numerical approximations to certain types of design problems, simplifications in the computation of the k parameters, and incomplete state-vector measurement are included.

8-1 Normalization

In order to determine the important design invariants associated with optimization theory, a dimensional analysis of the k parameters which appear in the optimum control equation is necessary. When the dimensions of these control-system parameters are determined, normalization of the differential equations is carried out. The normalized computations are much less sensitive to the important design-problem parameters than the computations that would be performed with the unnormalized differential equations.[49] The ideas and procedures associated with normalization are considered here for a first-order dynamic process. Also, dynamic processes without pure time delay are considered, and only deterministic signals are treated. Under these conditions, the k parameters that appear in the optimum control equation reduce, from (7-33) and (7-34), to

$$-k_1'(\mu) = \phi_{11}(\mu)Q_1(\mu)a_{11}(\mu) - u_1(\mu)k_{11}(\mu) - c_{11}(\mu)M_1(\mu)k_{11}(\mu) + b_{11}(\mu)k_1(\mu) - \frac{c_{11}^2(\mu)}{\psi_{11}(\mu)}k_1(\mu)k_{11}(\mu) \tag{8-1}$$

$$-k_{11}'(\mu) = \phi_{11}(\mu)a_{11}^2(\mu) + 2b_{11}(\mu)k_{11}(\mu) - \frac{c_{11}^2(\mu)}{\psi_{11}(\mu)}k_{11}^2(\mu) \tag{8-2}$$

where

$$k_1(T) = k_{11}(T) = 0 \tag{8-3}$$

The normalization procedure starts by finding the dimensions of k_1 and k_{11} in terms of parameters of the dynamic process and error measure. The dimensions of these two k parameters are found from the property that each of the additive terms appearing in (8-1) or (8-2) must have the same dimensions. For instance, the first and third terms appearing on the right-hand side of (8-2) must be equal dimensionally. From these two terms, the dimensions of k_{11} are found to be

$$\{k_{11}\} = \{\phi_{11}\}^{\frac{1}{2}}\{\psi_{11}\}^{\frac{1}{2}}\{a_{11}\}\{c_{11}\}^{-1} \tag{8-4}$$

The notation $\{x\}$ is defined as the dimensions of the variable x. The dimensions of k_{11}, as given in (8-4), are expressed wholly in terms of the parameters of the dynamic process and error measure. The dimensions of the parameter k_1 now can be found from the first and fifth terms appearing on the right-hand side of (8-1). From these two terms, the dimensions of k_1 are found to be

$$\{k_1\} = \{Q_1\}\{\phi_{11}\}^{\frac{1}{2}}\{\psi_{11}\}^{\frac{1}{2}}\{c_{11}\}^{-1} \tag{8-5}$$

Another important variable that appears in the optimum control equation is the independent variable time. Time is an important variable because the k parameters are a function of time-to-go $(T - t)$. The dimension of time now is expressed in terms of the parameters of the dynamic process and error index. The left-hand side of (8-2) must have the dimensions of k_{11} divided by the dimension of time because of the time differentiation of this term. When the dimensions of the left-hand side of (8-2) are equated to the dimensions of the first term on the right-hand side of (8-2), the dimension of time is found to be

$$\{\mu\} = \{\phi_{11}\}^{-\frac{1}{2}}\{\psi_{11}\}^{\frac{1}{2}}\{a_{11}\}^{-1}\{c_{11}\}^{-1} \tag{8-6}$$

The dimensions of k_1, k_{11}, and μ are not expressed in terms of the dimensions of the dynamic-process parameter b_{11}, the load-disturbance signal u_1, or the desired-response signal M_1. Therefore, the dimensions of this parameter and these two signals can be found in terms of the dimensions of the parameters appearing in (8-4) to (8-6). When the dimensions of the first and second terms appearing on the right-hand side of (8-2) are equated, the dimensions of the parameter b_{11} are found to be

$$\{b_{11}\} = \{\phi_{11}\}^{\frac{1}{2}}\{\psi_{11}\}^{-\frac{1}{2}}\{a_{11}\}\{c_{11}\} \tag{8-7}$$

By similar reasoning associated with (8-1), the dimensions of the load-disturbance signal u_1 and the desired-control signal M_1 are found to be

$$\{u_1\} = \{Q_1\}\{\phi_{11}\}^{\frac{1}{2}}\{\psi_{11}\}^{-\frac{1}{2}}\{c_{11}\} \tag{8-8}$$

and

$$\{M_1\} = \{Q_1\}\{\phi_{11}\}^{\frac{1}{2}}\{\psi_{11}\}^{-\frac{1}{2}} \tag{8-9}$$

From the previous dimensional analysis, normalized functions and parameters are introduced into the computations of the optimum control equation. First, normalized functions of the parameters and signal appearing on the right-hand side of the dimensional equalities (8-4) through (8-9) are introduced so that

$$\begin{gathered}\phi_{11}(\mu) = \phi_{11}(t)\Phi_{11}(\eta) \qquad \psi_{11}(\mu) = \psi_{11}(t)\Psi_{11}(\eta) \qquad a_{11}(\mu) = a_{11}(t)A_{11}(\eta) \\ c_{11}(\mu) = c_{11}(t)C_{11}(\eta) \qquad Q_1(\mu) = Q_1(t)\mathcal{Q}_1(\eta)\end{gathered} \tag{8-10}$$

For instance, the dimensions and amplitude of $\phi_{11}(\mu)$ are contained in the factor $\phi_{11}(t)$. Therefore, the function $\Phi_{11}(\eta)$ is a dimensionless function and is normalized by the amplitude of ϕ_{11} at $\mu = t$. The other equalities appearing in (8-10) are formulated on the same basis. The normalized time variable η, which appears in (8-10), is formulated according to the dimensions of time given in (8-6). In particular, the normalized time-to-go η is defined as

$$\eta = |a_{11}(t)c_{11}(t)|\sqrt{\frac{\phi_{11}(t)}{\psi_{11}(t)}}\,(T - \mu) \tag{8-11}$$

Similarly, normalized k parameters are defined from the dimensional equalities given in (8-4) and (8-5) as

$$K_1(\eta) = \frac{|c_{11}(t)|\,\text{sgn}\,a_{11}(t)}{Q_1(t)\sqrt{\phi_{11}(t)\psi_{11}(t)}}\,k_1(\mu) \qquad K_{11}(\eta) = \frac{|c_{11}(t)|}{|a_{11}(t)|\,\sqrt{\phi_{11}(t)\psi_{11}(t)}}\,k_{11}(\mu) \tag{8-12}$$

The function sgn x is defined as

$$\text{sgn}\,x = \begin{cases} 1 & x \geqq 0 \\ -1 & x < 0 \end{cases} \tag{8-13}$$

Finally, the dynamic-process parameter b_{11} and the signals u_1 and M_1 are normalized according to (8-7), (8-8), and (8-9), respectively, so that

$$B_{11}(\eta) = \frac{\sqrt{\psi_{11}(t)/\phi_{11}(t)}}{|a_{11}(t)c_{11}(t)|}\, b_{11}(\mu) \qquad U_1(\eta) = \frac{\sqrt{\psi_{11}(t)/\phi_{11}(t)}}{Q_1(t)\,|c_{11}(t)|}\,[\text{sgn}\ a_{11}(t)]u_1(\mu)$$
$$\mathscr{M}_1(\eta) = \frac{\sqrt{\psi_{11}(t)/\phi_{11}(t)}}{Q_1(t)}\,[\text{sgn}\ a_{11}(t)c_{11}(t)]M_1(\mu) \tag{8-14}$$

The original differential equations given in (8-1) and (8-2) now are rewritten in terms of the normalized functions defined in (8-10) through (8-14) by the appropriate substitutions. When these substitutions are made, (8-1) and (8-2), respectively, reduce to

$$K_1'(\eta) = \Phi_{11}(\eta)\mathscr{Q}_1(\eta)A_{11}(\eta) - U_1(\eta)K_{11}(\eta) - C_{11}(\eta)\mathscr{M}_1(\eta)K_{11}(\eta) + B_{11}(\eta)K_1(\eta) - \frac{C_{11}{}^2(\eta)}{\Psi_{11}(\eta)}K_1(\eta)K_{11}(\eta) \tag{8-15}$$

$$K_{11}'(\eta) = \Phi_{11}(\eta)A_{11}{}^2(\eta) + 2B_{11}(\eta)K_{11}(\eta) - \frac{C_{11}{}^2(\eta)}{\Psi_{11}(\eta)}K_{11}{}^2(\eta) \tag{8-16}$$

The boundary conditions on the K parameters are

$$K_1(0) = K_{11}(0) = 0 \tag{8-17}$$

The optimum control equation for this first-order dynamic process is

$$m_1(t) = M_1(t) + \frac{c_{11}(t)}{\psi_{11}(t)}\,[k_1(t) - k_{11}(t)x_1(t)] \tag{8-18}$$

If a normalized time-to-go from real time t is defined as

$$T_n = |a_{11}(t)c_{11}(t)|\sqrt{\frac{\phi_{11}(t)}{\psi_{11}(t)}}\,(T - t) \tag{8-19}$$

then the optimum control equation is rewritten in terms of the K parameters so that

$$m_1(t) = M_1(t) + [\text{sgn}\ a_{11}(t)c_{11}(t)]\sqrt{\frac{\phi_{11}(t)}{\psi_{11}(t)}}\,[K_1(T_n)Q_1(t) - K_{11}(T_n)q_1(t)] \tag{8-20}$$

The differential equations given in (8-15) and (8-16) plus the optimum control equation given in (8-20) completely define the optimum control system in terms of normalized gains. These normalized gains are dimensionless and generally are positive numbers.

At this stage in the development, the conceptual and practical advantages of normalization are not necessarily obvious. In particular, (8-15) and (8-16) appear to be just as complicated as (8-1) and (8-2), respectively. In fact, they appear to be exactly the same as (8-1) and (8-2) except for a change in notation. However, the equations given in (8-15) and (8-16) simplify considerably in certain important special cases, whereas (8-1) and (8-2) do not simplify. In particular, the case where all the parameters and signals of the dynamic process and error measure are constants is considered so that

$$\phi_{11}(\mu) = \phi_{11} \qquad \psi_{11}(\mu) = \psi_{11} \qquad a_{11}(\mu) = a_{11} \qquad c_{11}(\mu) = c_{11}$$
$$b_{11}(\mu) = b_{11} \qquad Q_1(\mu) = Q_1 \qquad M_1(\mu) = M_1 \qquad u_1(\mu) = u_1 \tag{8-21}$$

Under this simplifying assumption, (8-15) and (8-16) reduce to

$$K_1'(\eta) = 1 - (U_1 + \mathscr{M}_1)K_{11}(\eta) + B_{11}K_1(\eta) - K_1(\eta)K_{11}(\eta) \tag{8-22}$$

and

$$K_{11}'(\eta) = 1 + 2B_{11}K_{11}(\eta) - K_{11}{}^2(\eta) \tag{8-23}$$

These last two equations defining the K parameters are considerably simpler than the original differential equations defining the k parameters. In addition to the simplification sometimes afforded by this process of normalization, some important and interesting properties of the optimum system become apparent.

The block diagrams of the optimum control equation, both normalized and not normalized, are shown in Fig. 8-1. First, the normalized form of the control equation, represented in block-diagram form, is more conventional with a reference signal input, etc. Also, important quantities such as loop gain are identified immediately. The dimensional portion of control-equation feedback gain is $\sqrt{\phi_{11}(t)/\psi_{11}(t)}$.

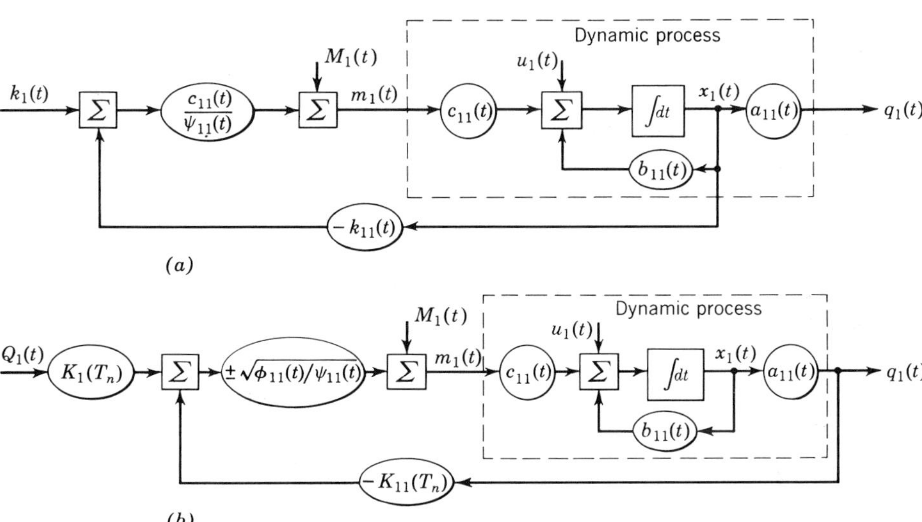

Fig. 8-1 (*a*) Block diagram of optimum control equation not normalized; (*b*) block diagram of optimum control equation normalized.

In addition, the magnitude of the gain internal to the dynamic process is $|a_{11}(t)c_{11}(t)|$ if the effects of $b_{11}(t)$ are neglected. Therefore, the dimensions and approximate magnitude of the major loop gain are given by $|a_{11}(t)c_{11}(t)|\sqrt{\phi_{11}(t)/\psi_{11}(t)}$. The reciprocal of this quantity has the dimensions of time and appears in the normalization of time-to-go. Finally, the sign of the major loop gain is always negative because $K_{11}(T_n)$ is positive. The gain of the feedback portion of the control equation and the gain of the dynamic process are always opposite in sign.

This major loop gain plays a major role in the normalization procedures presented in this section. In particular, the normalized time-to-go is given by the loop gain times the actual time-to-go, as expressed in (8-19). Therefore, if the loop gain is very large, the normalized time-to-go becomes very large also. Because the K parameters appearing in the optimum control equation become the steady-state solutions of (8-22) and (8-23) under the condition that the normalized time-to-go approaches infinity, the actual time-to-go behaves as if it approaches infinity under this condition. Also, the normalized B parameter of the dynamic process is given by b_{11} divided by the loop gain, as expressed in (8-14). Under the condition that the loop gain becomes very large, the normalized parameter B_{11} approaches zero. Under the condition that B_{11} approaches zero, the solutions of (8-22) and (8-23) become independent of the time constant of the dynamic process, which is given by the reciprocal of b_{11}. In other words, when the loop gain becomes very large, the optimum control system can be constructed without a knowledge of both the actual time-to-go $(T - t)$ and the dynamic-process time constant $1/b_{11}$. Stated in even another way, the optimum control system is insensitive to time-to-go and the time constant of the dynamic process under the conditions that the loop gain becomes arbitrarily large. Of course, this property of insensitivity is precisely the question raised in many practical situations. The problem of sensitivity is the motivation for pursuing further the normalizations presented in this section.

8-2 Sensitivity analysis

In order to evaluate quantitatively the sensitivity of the optimum control system to changes in time-to-go and the time constant of the dynamic process, the solutions of (8-22) and (8-23) are carried out under the simplifying assumption

$$M_1 = u_1 = 0 \tag{8-24}$$

When these two differential equations are integrated subject to the appropriate boundary conditions, the K parameters appearing in the optimum control equation are found to be

$$K_1(T_n) = \frac{1}{\sqrt{B_{11}^2 + 1}} \frac{\sinh \tau_n + \beta_{11}(1 - \cosh \tau_n)}{\cosh \tau_n - \beta_{11} \sinh \tau_n} \tag{8-25}$$

and

$$K_{11}(T_n) = \frac{1}{\sqrt{B_{11}^2 + 1}} \frac{\sinh \tau_n}{\cosh \tau_n - \beta_{11} \sinh \tau_n} \tag{8-26}$$

where the definitions $\tau_n = \sqrt{B_{11}^2 + 1}\,T_n$ and $\beta_{11} = B_{11}/\sqrt{B_{11}^2 + 1}$ have been introduced. These parameters are plotted in Figs. 8-2 and 8-3 in order to demonstrate their behavior as functions of normalized time-to-go.

Because the K parameters are found in closed mathematical form, sensitivity functions due to perturbations in normalized time-to-go and normalized time constant are introduced which are similar to those used by Bode[54] in sensitivity studies associated with feedback amplifiers. In particular, a *sensitivity function* is defined as

$$S_y^x(z) = \frac{\delta y(z)/y(z)}{\delta x/x} \tag{8-27}$$

The variable x appearing in (8-27) is the variable which is perturbed by an amount δx. In considerations presented here, the variable x is associated with normalized time-to-go and the normalized b_{11} parameter of the dynamic process. The variable y appearing in (8-27) is associated with the K parameters K_1 and K_{11}. The independent variable z appearing in (8-27) is associated with normalized time-to-go because these sensitivity functions are to be plotted as functions of the normalized time-to-go.

When the mathematical operations indicated in (8-27) are performed with respect to the K parameters expressed in (8-25) and (8-26), these sensitivity functions are found to be

$$S_{K_1}{}^{B_{11}}(T_n) = \beta_{11}\left[\frac{(1 - B_{11}T_n)(\beta_{11}\sinh\tau_n - \cosh\tau_n) + 1}{\sinh\tau_n + \beta_{11}(1 - \cosh\tau_n)} + \frac{(1 + \beta_{11}{}^2 - \beta_{11}\tau_n)\sinh\tau_n - \beta_{11}(2 - B_{11}T_n)\cosh\tau_n}{\cosh\tau_n - \beta_{11}\sinh\tau_n}\right] \tag{8-28}$$

$$S_{K_1}{}^{T_n}(T_n) = \tau_n\left[\frac{\cosh\tau_n - \beta_{11}\sinh\tau_n}{\sinh\tau_n + \beta_{11}(1 - \cosh\tau_n)} + \frac{\beta_{11}\cosh\tau_n - \sinh\tau_n}{\cosh\tau_n - \beta_{11}\sinh\tau_n}\right] \tag{8-29}$$

$$S_{K_{11}}^{B_{11}}(T_n) = \beta_{11}\frac{\sinh\tau_n - \beta_{11}\cosh\tau_n + B_{11}T_n\operatorname{csch}\tau_n}{\cosh\tau_n - \beta_{11}\sinh\tau_n} \tag{8-30}$$

$$S_{K_{11}}^{T_n}(T_n) = \tau_n\frac{\operatorname{csch}\tau_n}{\cosh\tau_n - \beta_{11}\sinh\tau_n} \tag{8-31}$$

Although the mathematical expressions for these sensitivity functions are complex functions of B_{11} and T_n, their behavior as functions of normalized time-to-go could be deduced directly from Figs. 8-2 and 8-3. These sensitivity functions are plotted in Figs. 8-4 through 8-7.

The sensitivity functions associated with K_{11} relate to the changes in stability of the control system due to computational errors, because K_{11} is the feedback gain of the control system. When the dynamic process is stable, $B_{11} \leqq 0$, the magnitude of $S_{K_{11}}^{B_{11}}$ increases monotonically with increasing normalized time-to-go T_n. Also, for any particular T_n, the magnitude of this sensitivity function increases monotonically with decreasing B_{11} for stable dynamic processes. For unstable dynamic processes, $B_{11} > 0$, the magnitude of this sensitivity function no longer is monotonic with respect to T_n.

The magnitude of the sensitivity function $S_{K_{11}}^{T_n}$ decreases monotonically for decreasing B_{11} when the dynamic process is stable. Also, this sensitivity function is less than that for the case $B_{11} = 0$ under the condition of very large normalized time-to-go, and $S_{K_{11}}^{T_n} \to 0$ for $T_n \to \infty$. This limiting behavior arises because steady-state solutions occur for very large normalized time-to-go.

The behavior of the sensitivity functions associated with the feedforward gain K_1 is somewhat more complicated except under the limiting condition $T_n \to \infty$. Under this limiting condition, $S_{K_1}{}^{T_n}$ approaches zero, and $|S_{K_1}{}^{B_{11}}|$ approaches a constant which increases monotonically with increasing $|B_{11}|$.

The behavior of these sensitivity functions leads to fairly general conclusions. When normalized time-to-go is large, the time constant of the dynamic process should be known accurately. But on the other hand, the time-to-go need not be known accurately. For small normalized time-to-go, the time-to-go should be known accurately, but the time constant of the dynamic process need not be known accurately.

Another interpretation of these sensitivity functions arises in the situation that the actual time-to-go is equal to infinity, and approximate computations of the k parameters are required. In the situation that the actual time-to-go is equal to infinity, the original differential equations given in (8-1) and (8-2) theoretically must

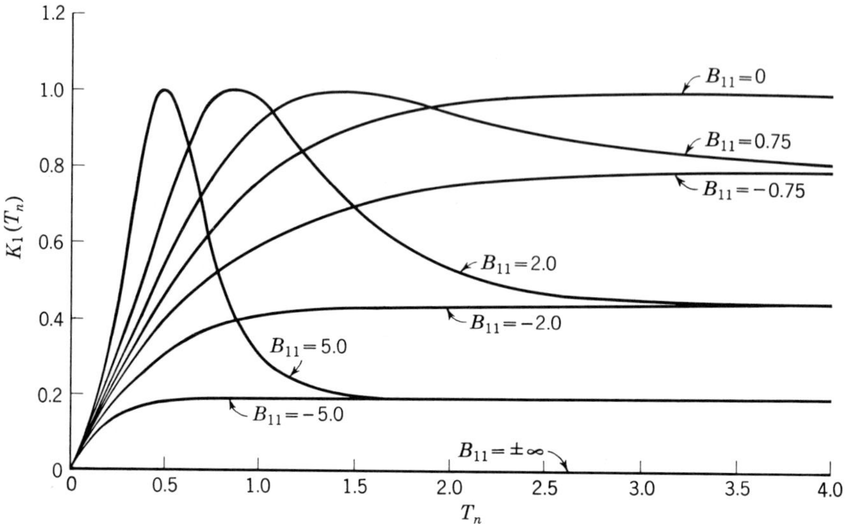

Fig. 8-2 Normalized feedforward gain.

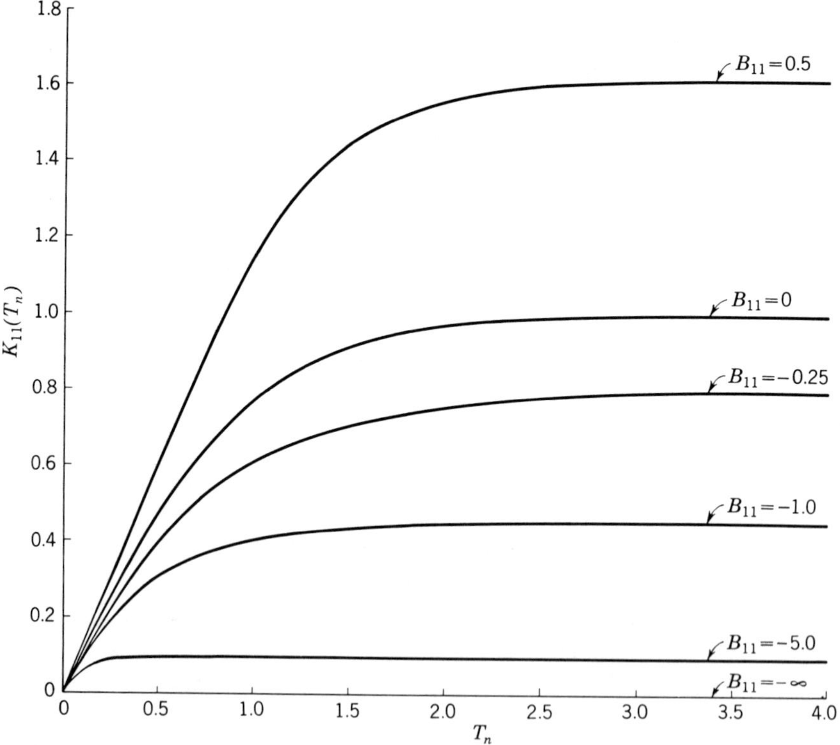

Fig. 8-3 Normalized feedback gain.

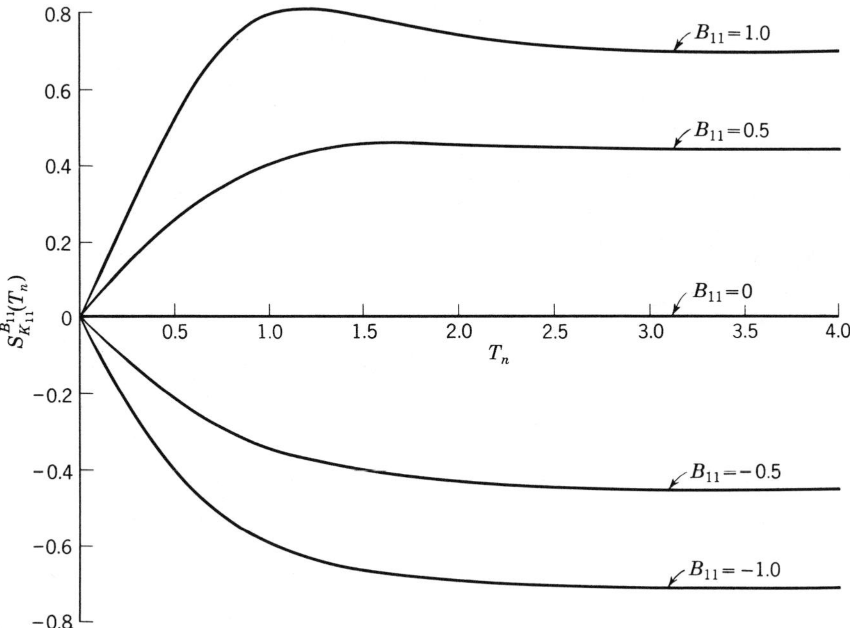

Fig. 8-4 Sensitivity of normalized feedback gain due to B_{11}.

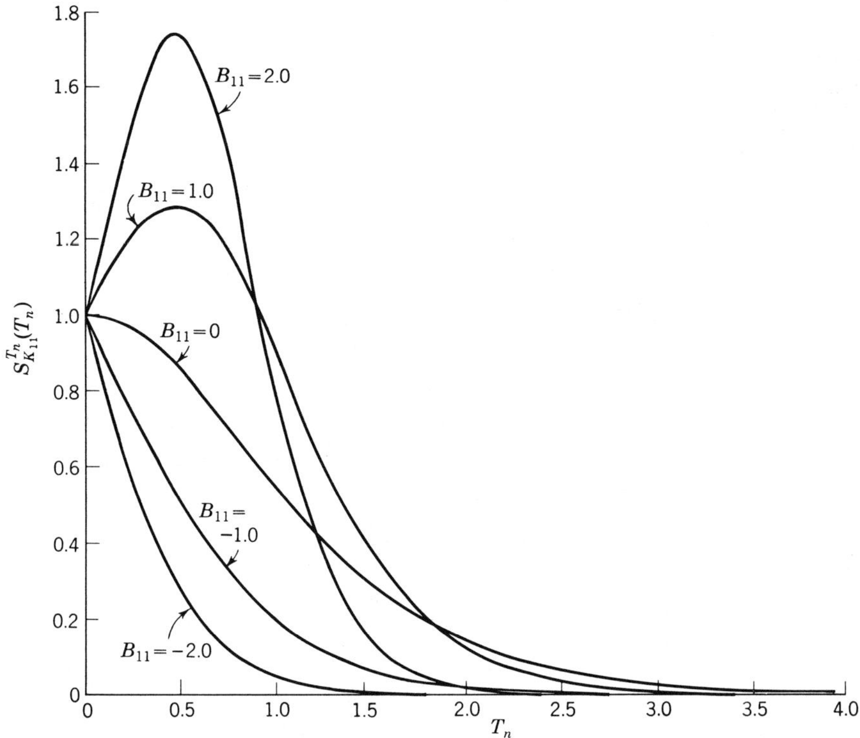

Fig. 8-5 Sensitivity of normalized feedback gain due to T_n.

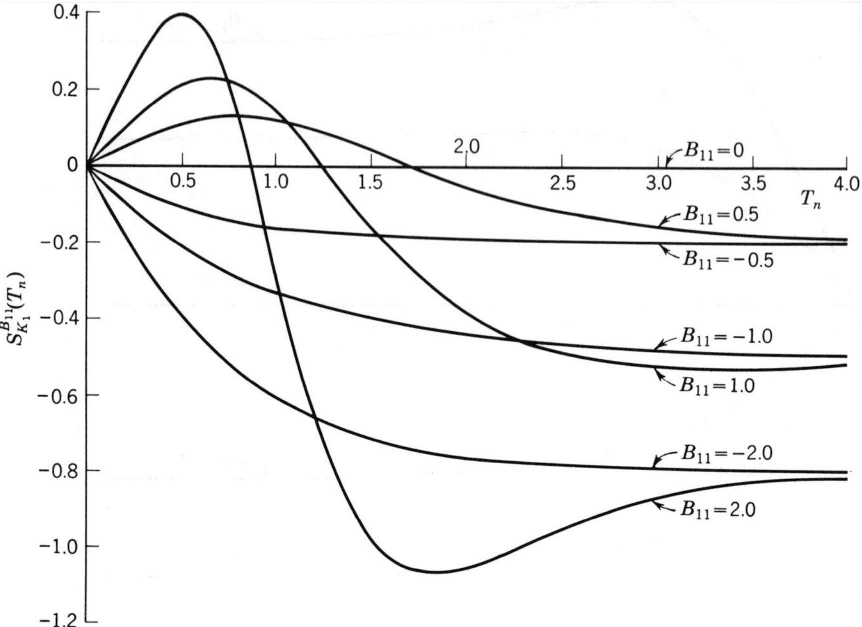

Fig. 8-6 Sensitivity of normalized feedforward gain due to B_{11}.

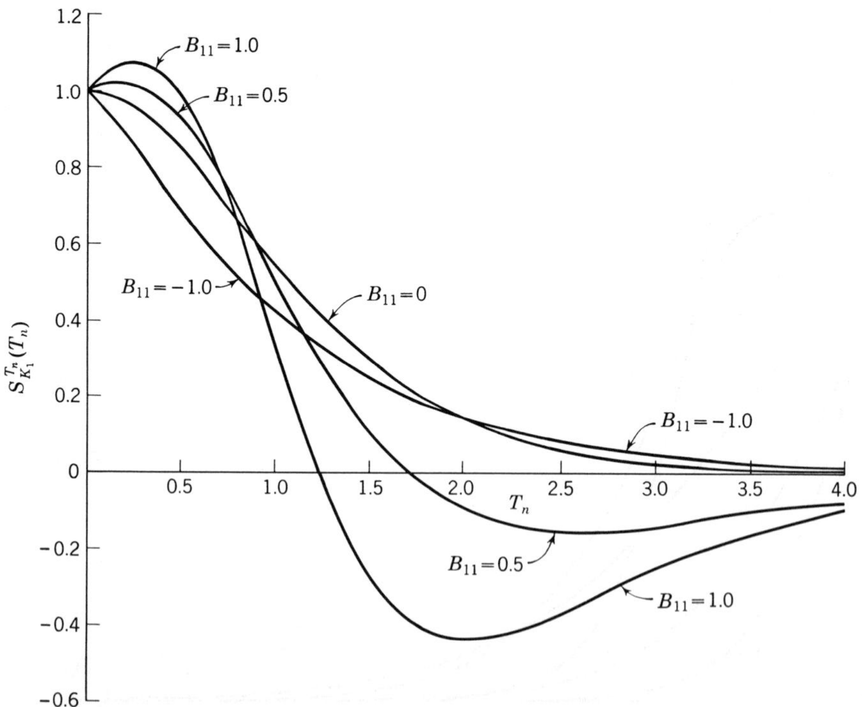

Fig. 8-7 Sensitivity of normalized feedforward gain due to T_n.

be solved over an infinite time interval. Of course, this is impractical from the standpoint of computations. Therefore, the sensitivity function with respect to perturbations in normalized time-to-go can be used to estimate a finite time-to-go which, when used instead of the infinite-interval computation, does not give rise to appreciable error in the computation of the k parameters. From these sensitivity functions, the worst case arises in the situation when $B_{11} = 0$. If a normalized time-to-go is chosen to be equal to or greater than 4, then the error due to approximating the infinite-interval computation with a finite-interval computation is always very small for any time constant of the dynamic process. This particular numerical value of normalized time-to-go is called a *pseudo time-to-go* and is chosen from Fig. 8-5 for use in the approximation of the infinite time-to-go computations. This pseudo time-to-go is important and can be developed for higher-order dynamic processes. Therefore, the normalization procedures developed in this section now are extended to the case where the dynamic process is of order N.

8-3 Dimensionless parameters of the optimum control system

For illustrative purposes, the error index assumed here is

$$e(t) = \sum_{n=1}^{N} \int_t^T \{\phi_{nn}[Q_n - q_n(\sigma)]^2 + \psi_{nn}[m_n(\sigma)]^2\}\, d\sigma \tag{8-32}$$

Also, the dynamic process is defined by the response signals

$$q_n(t) = \sum_{m=1}^{N} a_{nm} x_m(t) \qquad n = 1, 2, \ldots, N \tag{8-33}$$

and the state equation

$$x_n'(t) = \sum_{m=1}^{N} [b_{nm} x_m(t)] + c_{nn} m_n(t) \qquad n = 1, 2, \ldots, N \tag{8-34}$$

In Sec. 8-2, the important design invariant is shown to be the major loop gain of the dynamic process and optimum control system. Therefore, the dimensional analysis is developed here for the Nth-order dynamic process by finding the appropriate grouping of design parameters that define the major loop gain.

For the Nth-order dynamic process, the feedforward portion of the major loop of the system is defined from the first control signal m_1 to the Nth response signal q_N. This definition of the major loop is arbitrary but is convenient from the standpoints of notation and block diagrams. Generally speaking, the signal flow in the system is drawn conveniently on a block diagram from left to right, and the signals are labeled conveniently according to increasing indices from left to right. On the block diagram, the major loop is intended to pass through the maximum number of connected integrators in the dynamic process. The feedforward portion of the major loop is illustrated for a second-order system in Fig. 8-8. This block diagram clearly indicates the ambiguous situation that would arise without an agreed-upon convention.

With this convention for defining the major loop, the dimensional analysis proceeds in a way analogous to that of Sec. 8-1. The dimensions of the error index are found from (8-32) to be

$$\{e\} = \{\phi_{NN}\}\{q_N\}^2\{t\} = \{\psi_{11}\}\{m_1\}^2\{t\} \tag{8-35}$$

where only the terms involving the response error of the Nth response signal and the control effort of the first control signal are retained. The second equality appearing in (8-35) is used to obtain the dimensional relationship

$$\{q_N\}\{m_1\}^{-1} = \{\phi_{NN}\}^{-\frac{1}{2}}\{\psi_{11}\}^{\frac{1}{2}} \tag{8-36}$$

Further dimensional relationships now can be found from the dynamic process. In particular, the dimensions of the Nth response signal are found from (8-33) to be

$$\{q_N\} = \{a_{NN}\}\{x_N\} \tag{8-37}$$

Also, the state equation expressed in (8-34) gives the dimensional relationships

$$\{x_n'\} = \{b_{n,n-1}\}\{x_{n-1}\} \qquad n = 2, 3, \ldots, N \tag{8-38}$$

This last dimensional equality now is expressed in terms of the dimensions of time as

$$\{x_n\} = \{b_{n,n-1}\}\{x_{n-1}\}\{t\} \qquad n = 2, 3, \ldots, N \tag{8-39}$$

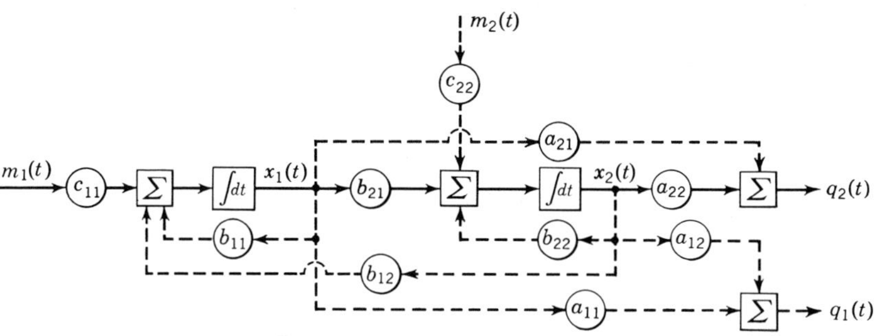

Fig. 8-8 Feedforward portion of the major loop for a second-order dynamic process.

This dimensional relationship is a recursion formula, and the dimensions of x_N can be expressed in terms of the dimensions of x_1 as

$$\{x_N\} = \left(\prod_{n=2}^{N} \{b_{n,n-1}\}\right)\{x_1\}\{t\}^{N-1} \tag{8-40}$$

The new notation appearing in (8-40) is defined as

$$\prod_{m=r}^{s} x_m = \begin{cases} x_r x_{r+1} \cdots x_s & s \geqq r \\ 1 & s < r \end{cases} \tag{8-41}$$

Again, the state equation given in (8-34) can be used to find the dimensions of x_1 in terms of m_1 so that

$$\{x_1'\} = \{c_{11}\}\{m_1\} \tag{8-42}$$

or

$$\{x_1\} = \{c_{11}\}\{m_1\}\{t\} \tag{8-43}$$

Finally, (8-37), (8-40), and (8-43) are combined to obtain the dimensional relationship

$$\{q_N\}\{m_1\}^{-1} = \{a_{NN}\}\{c_{11}\}\{t\}^N \prod_{n=2}^{N}\{b_{n,n-1}\} \tag{8-44}$$

If (8-36) is divided by (8-44) and the Nth root of the result is taken, the inverse dimensions of time are found to be

$$\{t\}^{-1} = \left(\{a_{NN}\}\{c_{11}\}\prod_{n=2}^{N}\{b_{n,n-1}\}\right)^{1/N}\left(\{\phi_{NN}\}\{\psi_{11}\}^{-1}\right)^{1/2N} \tag{8-45}$$

This last dimensional equality plays the same important role as the dimensional equality given in (8-6) of Sec. 8-1. In particular, $\{t\}^{-N}$ has the dimensions of the major loop gain, as is seen in the subsequent development.

When the dimensions of time are expressed in terms of design parameters associated with the major loop gain, the normalization factors associated with the k parameters appearing in the optimum control equation can be obtained from further dimensional analysis. The minimum-error function for the dynamic process and error index assumed here has the form

$$E[\mathbf{x}(t), t] = k(t) - 2\sum_{m=1}^{N} k_m(t)x_m(t) + \sum_{m=1}^{N}\sum_{k=1}^{N} k_{mk}(t)x_m(t)x_k(t) \tag{8-46}$$

Because this function is the minimum value of the error index, this function possesses the same dimensions as e. Therefore, the dimensions of the unsubscripted k parameter are given by

$$\{k\} = \{e\} = \{\phi_{NN}\}\{Q_N\}^2\{t\} \tag{8-47}$$

When the dimensions of the first and second terms in (8-46) are equated, the dimensions of the single-subscripted k parameters are

$$\{k_m\} = \{k\}\{x_m\}^{-1} \tag{8-48}$$

Now returning to the dynamic process, the dimensions of x_m are found to be

$$\{x_m\} = \{c_{11}\}\{\phi_{NN}\}^{\frac{1}{2}}\{\psi_{11}\}^{-\frac{1}{2}}\{t\}^m\{Q_N\}\prod_{n=2}^{m}\{b_{n,n-1}\} \tag{8-49}$$

by the same manipulations used to arrive at (8-40). Therefore, the dimensions of k_m are

$$\{k_m\} = \{c_{11}\}^{-1}\{Q_N\}\{\phi_{NN}\}^{\frac{1}{2}}\{\psi_{11}\}^{\frac{1}{2}}\{t\}^{1-m}\prod_{n=2}^{m}\{b_{n,n-1}\}^{-1} \tag{8-50}$$

Finally, when the dimensions of the second and third terms appearing in (8-46) are equated, the dimensions of the double-subscripted k parameters are

$$\{k_{mk}\} = \{k_m\}\{x_k\}^{-1} \tag{8-51}$$

and hence

$$\{k_{mk}\} = \{c_{11}\}^{-2}\{\psi_{11}\}\{t\}^{1-k-m}\left(\prod_{n=2}^{m}\{b_{n,n-1}\}^{-1}\right)\left(\prod_{n=2}^{k}\{b_{n,n-1}\}^{-1}\right) \tag{8-52}$$

Similar to the definition of the normalized time variable given in (8-11) for a first-order dynamic process, the definition for the normalized time variable is given here as

$$\eta = \mathcal{N}(T - \mu) \tag{8-53}$$

According to the dimensions of time expressed in (8-45), the normalizing factor appearing in (8-53) is chosen to be

$$\mathcal{N}^N = \left|a_{NN}c_{11}\prod_{n=2}^{N} b_{n,n-1}\right|\sqrt{\frac{\phi_{NN}}{\psi_{11}}} \tag{8-54}$$

Also, the normalized k parameters appearing in the optimum control equation are defined as

$$k_m(\mu) = \mathcal{N}_m K_m(\eta) \tag{8-55}$$

and

$$k_{mk}(\mu) = \mathcal{N}_{mk} K_{mk}(\eta) \tag{8-56}$$

The normalizing factors appearing in (8-55) and (8-56) are chosen to be

$$\mathcal{N}_m = (\operatorname{sgn} a_{NN}) Q_N \frac{\mathcal{N}^{m-1}\sqrt{\phi_{NN}\psi_{11}}}{\left| c_{11} \prod_{n=2}^{m} b_{n,n-1} \right|} \tag{8-57}$$

and

$$\mathcal{N}_{mk} = \frac{\mathcal{N}^{k+m-1}\psi_{11}}{c_{11}^2 \left| \prod_{n=2}^{m} b_{n,n-1} \right| \left| \prod_{n=2}^{k} b_{n,n-1} \right|} \tag{8-58}$$

The completion of this normalization process requires the definition of normalized design parameters associated with the dynamic process and error measure. Unfortunately, the definitions of these normalized design parameters are not unique, because the dimensional relationship between the Nth and the nth response variables is arbitrary until the design problem at hand is specified. A similar point of ambiguity arises with the dimensional relationship between the first and the nth control variables. If normalized design parameters cannot be defined, then in general the differential equations defining the normalized k parameters cannot be written in a compact and meaningful form. However, the normalized form of the k parameters is not needed for the purposes of obtaining a numerical solution unless the differential equations are to be solved on an analog computer or on a digital computer in fixed-point arithmetic. The large number of multiplications appearing in these differential equations usually requires the use of a digital computer, and floating-point arithmetic generally is available for such machines. In spite of the difficulties in completing the normalization, the normalizing factors given in (8-54), (8-57), and (8-58) provide an algebraic means for checking the approximate magnitudes of the effective time-to-go and of the k parameters.

The focal point of the normalization procedure presented in this section is the development of the effective or normalized time-to-go, which is given by

$$T_n = \mathcal{N}(T - t) \tag{8-59}$$

Normalized time-to-go is the primary design invariant associated with the dynamic difficulties of the design problem. Also, this invariant is used in a number of computational approximations. Furthermore, the normalizing factor $\mathcal{N}$ bears a relationship to the importance of loop gain in feedback systems. Specifically, $\mathcal{N}^N$ is equal to the major loop gain, as defined in this section. This relationship is verified easily by expressing the optimum control equation for $m_1(t)$ in terms of the normalized k parameters. This optimum control equation for (8-32) to (8-34) is

$$m_1(t) = \frac{c_{11}}{\psi_{11}} \left[k_1(t) - \sum_{m=1}^{N} k_{1m}(t) x_m(t) \right] \tag{8-60}$$

In terms of the normalized k parameters defined in (8-57) and (8-58), this control equation is rewritten as

$$m_1(t) = (\text{sgn } a_{NN}c_{11})\sqrt{\frac{\phi_{NN}}{\psi_{11}}}\left[K_1(T_n)Q_N - \sum_{m=1}^{N} \mathscr{K}_{1m}(T_n)a_{NN}x_m(t)\right] \tag{8-61}$$

where

$$\mathscr{K}_{1m}(T_n) = \frac{\prod_{n=m+1}^{N} |b_{n,n-1}|}{\mathscr{N}^{N-m}} K_{1m}(T_n) \tag{8-62}$$

Because $\mathscr{K}_{1N}(T_n) = K_{1N}(T_n)$, the feedback portion of the major loop gain is given by $\sqrt{\phi_{NN}/\psi_{11}}\,|a_{NN}|$. The remaining terms appearing in (8-54) are the absolute value of the major loop gain between $m_1(t)$ and $x_N(t)$ internal to the dynamic process.

The sensitivity of the k parameters to the coefficients of the dynamic process is related intimately to the major loop gain, as might be suspected from conventional feedback theory. This sensitivity is investigated easily in terms of the normalized k parameters. For the purposes of discussion, the error index is specialized to

$$e(t) = \int_t^T \{\phi_{NN}[Q_N - x_N(\sigma)]^2 + \psi_{11}m_1^2(\sigma)\}\,d\sigma \tag{8-63}$$

and the dynamic process is specialized to

$$x_n'(t) = x_{n-1}(t) \qquad n = 2, 3, \ldots, N \tag{8-64}$$

and

$$x_1'(t) = \sum_{m=1}^{N} [b_{1m}x_m(t)] + c_{11}m_1(t) \tag{8-65}$$

For convenience, the coefficient c_{11} is taken to be positive. The optimum control equation for this special case reduces to

$$m_1(t) = \frac{\mathscr{N}^N}{c_{11}}\left[K_1(T_n)Q_N - \sum_{m=1}^{N} \frac{K_{1m}(T_n)}{\mathscr{N}^{N-m}} x_m(t)\right] \tag{8-66}$$

The block diagram of both the dynamic process and the optimum control equation is shown in Fig. 8-9.

For this specialized dynamic process and error index, the differential equations defining the normalized k parameters can be written in a convenient form. When the definitions of the K parameters are introduced into (7-33) and (7-34), the result is

$$K_m'(\eta) = \delta_{mN} + \frac{b_{1m}}{\mathscr{N}^m} K_1(\eta) + \Delta_m K_{m+1}(\eta) - K_1(\eta)K_{1m}(\eta) \tag{8-67}$$

and

$$K_{mk}'(\eta) = \delta_{mN}\delta_{mk} + \frac{b_{1k}}{\mathscr{N}^k} K_{1m}(\eta) + \frac{b_{1m}}{\mathscr{N}^m} K_{1k}(\eta) + \Delta_m K_{m+1,k}(\eta) + \Delta_k K_{k+1,m}(\eta) - K_{1m}(\eta)K_{1k}(\eta) \tag{8-68}$$

where the functions δ_{mn} and Δ_m are defined as

$$\delta_{mn} = \begin{cases} 1 & m = n \\ 0 & m \neq n \end{cases} \tag{8-69}$$

and

$$\Delta_m = \begin{cases} 1 & m = 1, 2, \ldots, N-1 \\ 0 & m = N \end{cases} \tag{8-70}$$

These equations are useful in determining the dependence of the K parameters on the normalizing factor $\mathcal{N}$.

The case with no feedback internal to the dynamic process, that is,

$$b_{1m} = 0 \qquad m = 1, 2, \ldots, N \tag{8-71}$$

is considered. For this case, (8-67) and (8-68) reduce to

$$K'_m(\eta) = \delta_{mN} + \Delta_m K_{m+1}(\eta) - K_1(\eta)K_{1m}(\eta) \tag{8-72}$$

and

$$K'_{mk}(\eta) = \delta_{mN}\delta_{mk} + \Delta_m K_{m+1,k}(\eta) + \Delta_k K_{k+1,m}(\eta) - K_{1m}(\eta)K_{1k}(\eta) \tag{8-73}$$

Neither of these differential equations is dependent on the coefficients of the dynamic

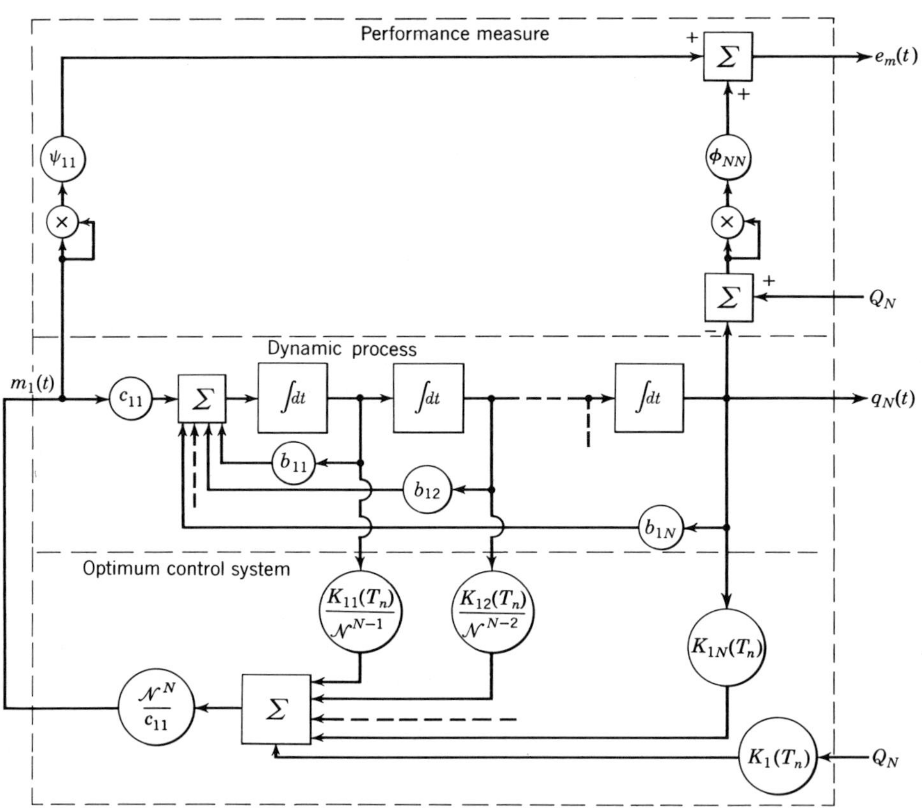

Fig. 8-9 Block diagram of optimum control system.

process; hence $K_m(\eta)$ and $K_{mk}(\eta)$ are also independent of the dynamic process. However, the values of the K parameters used in the optimum control equation are $K_1[\mathcal{N}(T-t)]$ and $K_{1m}[\mathcal{N}(T-t)]$. If the major loop gain becomes arbitrarily large, then the values of the K parameters correspond to the steady-state solutions of (8-72) and (8-73). Of course, the steady-state solutions $K_1(\infty)$ and $K_{1m}(\infty)$ are completely independent of $\mathcal{N}$.

In the more general situation where the dynamic process possesses internal feedback, the solutions $K_1[\mathcal{N}(T-t)]$ and $K_{1m}[\mathcal{N}(T-t)]$ are dependent upon the coefficients of the dynamic process. This dependency is due to the terms involving $b_{1m}\mathcal{N}^{-m}$ in (8-67) and (8-68). However, as the major loop gain $\mathcal{N}^N$ becomes arbitrarily large, these terms become vanishingly small. In other words, the internal feedback due to the dynamic-process coefficients is of little or no consequence for large major loop gain. Therefore, the time constants of the dynamic process need not be known accurately in situations where this gain is very large. This result also arises in sensitivity analyses associated with conventional feedback theory.

The sensitivity considerations presented here also indicate the relative importance of the various b_{1m} coefficients. For $\mathcal{N} > 1$, the relative importance of b_{1m} decreases for increasing m. For $\mathcal{N} < 1$, the reverse sequence of relative importance applies. Furthermore, if feedforward coefficients in the dynamic process (such as b_{m1} for $m = 3, 4, \ldots, N$) are not zero, then terms involving $b_{m1}\mathcal{N}^m$ appear in differential equations defining the K parameters. Of course, these feedforward terms are accentuated in situations with large loop gain and conversely are attenuated for small loop gain. When the dynamic process has no feedforward terms, the dynamic limitations for large loop gain are due primarily to the order N of the dynamic process. In these situations, the high-frequency behavior of the system fixed member therefore cannot be neglected altogether in selecting the dynamic-process equations. Also, even though the major loop gain is very large, the normalized time-to-go may still be small for dynamic processes of high order. Normalized time-to-go is proportional to $\mathcal{N}$, the Nth root of the major loop gain.

8-4 Infinite-interval problem approximation

A number of design problems arise where the design considerations correspond to $T = \infty$. That is, response errors are weighted over all future time. Because the case $T = \infty$ would require the solution of differential equations over an infinite time interval, approximations are necessary. This approximation problem arbitrarily is called the *infinite-interval problem.* For dynamic processes of order greater than $N = 2$, closed-form solutions of the k equations generally are not possible and computed solutions must be obtained.

The theoretical considerations presented in previous chapters demonstrate that as $T \to \infty$ the terminal-boundary conditions have a diminishing effect upon the optimum control equation and hence upon the k parameters. Therefore, a method of approximating the infinite-interval problem is to compute the k parameters with a finite but large T. Notationally, this approximation utilizes the property

$$k_m(t) \equiv F_m(t, T)\big|_{T\to\infty} = \mathscr{F}_m(t)$$
$$k_{mn}(t) \equiv F_{mn}(t, T)\big|_{T\to\infty} = \mathscr{F}_{mn}(t) \qquad (8\text{-}74)$$

Two questions arise concerning this property: How large must T be for the approximation to be accurate, and is the convergence to the case $T = \infty$ uniform? The exact selection of T for an accurate approximation unfortunately is empirical, although the normalization techniques developed in Sec. 8-3 greatly simplify the selection problem. Also, if $T - t = \tau_s$ is defined as the largest interval length for which the closed-loop system is unstable, then the performance error decreases monotonically with increasing $T - t$ so that $T - t > \tau_s$. Even though the performance error is monotonic, the asymptotic behavior of the k parameters, as expressed in (8-74), is not uniform. Therefore, if a system simulation, used to compute the performance error, is to be circumvented, then the k-parameter computations must be performed for a large enough T so that all the k parameters, not just the k parameters appearing in the control equation, are in an asymptotic region. The value of T required for the asymptotic behavior of all the k parameters is in general larger than that required from a performance-error standpoint.

In order to illustrate these considerations, an example is introduced here. The example chosen is normalized with a dynamic process consisting of just integrators

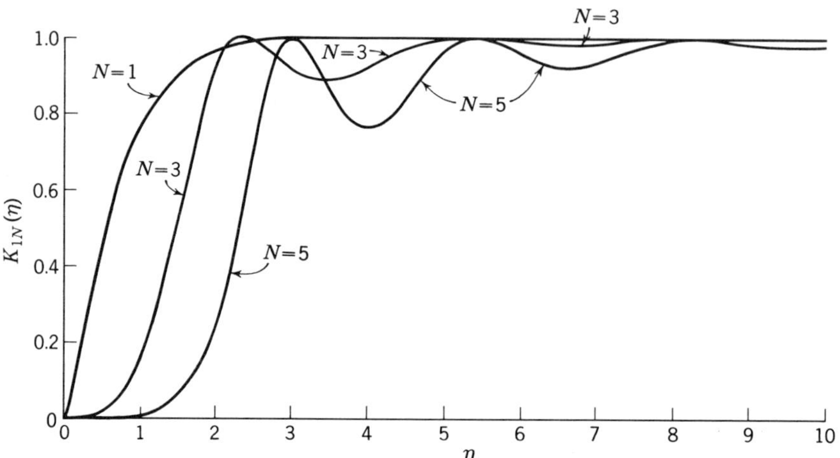

Fig. 8-10 Position feedback gain for integrator dynamic process.

and a weighting of only the control effort and position error. For N integrators, the dynamic process is

$$x_n'(t) = x_{n-1}(t) \qquad n = 2, 3, \ldots, N$$
$$x_1'(t) = m_1(t) \tag{8-75}$$

and the error index is chosen to be

$$e(t) = \int_t^T [x_N^2(\sigma) + m_1^2(\sigma)]\, d\sigma \tag{8-76}$$

The optimum control equation for this dynamic process and error index is

$$m_1(t) = -\sum_{n=1}^{N} K_{1n}(\eta)x_n(t) \tag{8-77}$$

where

$$\eta = T - t \tag{8-78}$$

For this specialized class of design problems, Fig. 8-10 is a plot of the position feedback gains for first-, third-, and fifth-order dynamic processes. Because of the oscillatory behavior of these curves, the convergence of the gains is not uniform except for $N = 1$. Therefore, care must be exercised when attempting to determine a small value of η for which the infinite-interval problem is approximated accurately. Figure 8-11, for a third-order dynamic process, shows that the existence of steady-state conditions can be identified more reliably when all the feedback gains are observed simultaneously. Also, Fig. 8-11 indicates that the position loop gain, as opposed to rate and acceleration, is the last gain to reach steady state for increasing η. Hence, the gain $K_{1N}(\eta)$ is the best single indicator for determining when steady-state conditions are achieved.

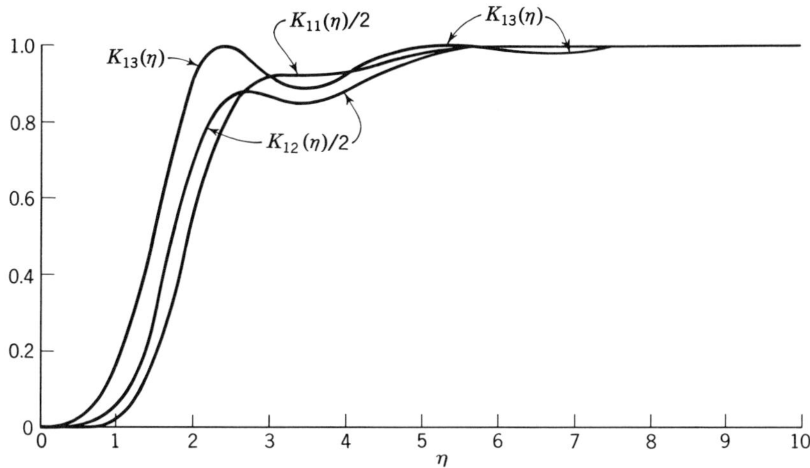

Fig. 8-11 Feedback gains for integrator dynamic process: $N = 3$.

As a matter of interest, Table 8-1 is included in order to indicate the steady-state values of the feedback gains for a dynamic process consisting of integrators only. These gains possess an interesting symmetry, and they correspond to the Butterworth filter.

In the approximation of the infinite-interval problem, the heart of the approximation problem is associated with the increase in system performance error caused by a finite T. If the steady-state error index

$$e(\eta) = \int_t^\infty [x_N^2(\sigma) + m_1^2(\sigma)]\, d\sigma \tag{8-79}$$

Table 8-1 *Steady-state values of controller feedback gains*

N	$K_{11}(\infty)$	$K_{12}(\infty)$	$K_{13}(\infty)$	$K_{14}(\infty)$	$K_{15}(\infty)$	$K_{16}(\infty)$
1	1.000					
2	1.414	1.000				
3	2.000	2.000	1.000			
4	2.605	3.400	2.605	1.000		
5	3.235	5.234	5.234	3.235	1.000	
6	3.813	7.461	9.137	7.461	3.863	1.000

is chosen, then the performance of the controllers constructed for different η can be compared. Figure 8-12 shows the steady-state performance for dynamic processes of different order. These curves are computed from tabulated integrals under the condition of zero initial conditions. These performance-error curves are infinite for $\eta \leqq \tau_s$ because the closed-loop system is unstable in this region. For $\eta > \tau_s$, the performance-error curves decrease monotonically with increasing η. The points where instability occurs and the asymptotic values of the performance error are summarized in Tables 8-2 and 8-3.

A comparison of the steady-state performance error and position-feedback-gain curves shows that the approximation to the infinite-interval problem is accurate for a value of η that is about one-half the value expected from Fig. 8-10. Of course, this one-half factor pertains only to this example. However, the infinite-interval approximation becomes accurate with increasing η, in general, before the controller gains reach steady state. In this example, steady-state controller gains are constant gains. With time-varying coefficients in the dynamic process or with time-varying weighting factors, the steady-state gains are functions of time. In such situations, steady state implies that the time functions are insensitive to incremental changes in T.

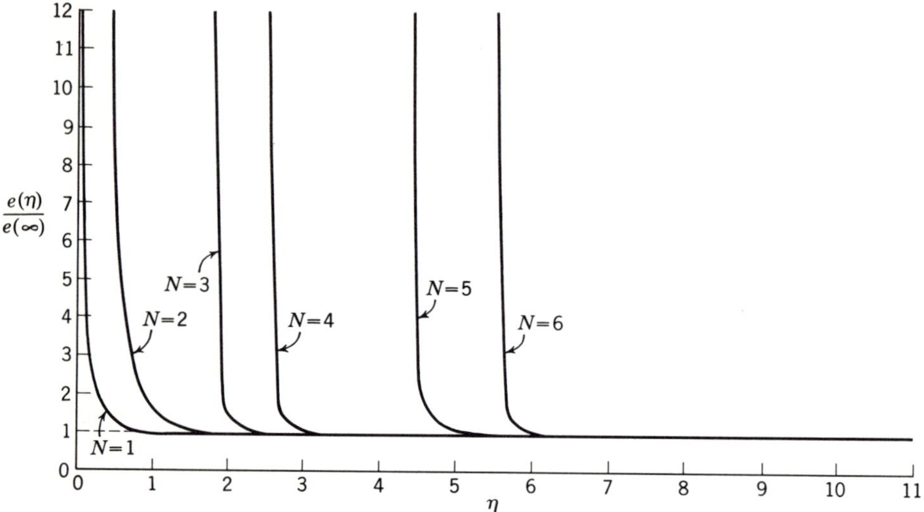

Fig. 8-12 Normalized steady-state performance error.

Table 8-2 *Stability abscissa for steady-state performance error*

N	1	2	3	4	5	6
τ_s	0.0	0.0+	1.80	2.55	4.37	5.55

Table 8-3 *Asymptotic values of steady-state performance error*

N	1	2	3	4	5	6
$e(\infty)$	1.000	1.500	2.000	2.613	3.236	3.864

The relative spacing of the curves in Fig. 8-12 merits some mention. In particular, the increasing order of the dynamic process, when the order becomes odd, gives rise to a disproportionate increase in the stability abscissa. No firm explanation of this occurrence is known. However, if the order is odd, there must be at least one real pole in the transfer function of the closed-loop system.

The methods presented here for approximating the infinite-interval problem are at best trial-and-error procedures, which are performed with the aid of a computer for design problems of interest. However, the normalization procedures in Sec. 8-3 are useful in extrapolating the results of one problem to the next. For instance, if the example had been defined by

$$\begin{aligned} x_n'(t) &= b_{n,n-1}x_{n-1}(t) \qquad n = 2, 3, \ldots, N \\ x_1'(t) &= c_{11}m_1(t) \end{aligned} \tag{8-80}$$

$$q_N(t) = a_{NN}x_N(t) \tag{8-81}$$

and

$$e(t) = \int_t^T [\phi_{NN}q_N^2(\sigma) + \psi_{11}m_1^2(\sigma)]\, d\sigma \tag{8-82}$$

then the optimum control equation could be written as

$$m_1(t) = -\frac{a_{NN}}{c_{11}} \sum_{n=1}^{N} \left| c_{11}\sqrt{\frac{\phi_{NN}}{\psi_{11}}} \right|^{n/N} K_{1n}(\eta)x_n(t) \tag{8-83}$$

where

$$\eta = (T - t)\left| a_{NN}c_{11}\sqrt{\frac{\phi_{NN}}{\psi_{11}}} \prod_{n=2}^{N} b_{n,n-1} \right|^{1/N} \tag{8-84}$$

In conjunction with the notation defined in (8-56) and (8-84) for $K_{1N}(\eta)$ and η, respectively, Figs. 8-10 to 8-12 apply to this modified design problem. Therefore, the use of normalizing factors makes the accuracy of the approximations less sensitive to changes in weighting factors and in coefficients of the dynamic process. In other words, the use of normalizing factors may significantly reduce the amount of trial and error.

In practice, the steady-state solutions of the k parameters most likely are found by the use of computers. If a digital computer is used, programmed means for finding steady-state solutions are needed.

The initial selection of the terminal time T in the general time-varying case cannot be specified explicitly, but the normalizing factor $\mathcal{N}$ can be used to some advantage. Suppose that the control system is operated over an interval of real time $t_0 \leqq t \leqq t_f$. Then the terminal time must be chosen so that $T > t_f$. Furthermore, the limiting case of large loop gain requires a normalized time-to-go approximately equal to the order N of the dynamic process. This result for steady-state performance error is indicated in Fig. 8-12. Therefore, an initial choice for terminal time could be made by the "rule of thumb"

$$T_0 = \frac{N}{\mathcal{N}} + t_f \tag{8-85}$$

Inevitably, this value of terminal time is not large enough, and also the suitability of this value must be checked from the computed solutions of the k parameters at the point $\mu = t_f$. Suppose that the k parameters are computed for two terminal times T_n and T_{n+1}. Then the condition for achieving the steady-state values of the k parameters could be written in terms of (8-74) as

$$\begin{aligned} \left| \frac{F_m(t_f, T_{n+1}) - F_m(t_f, T_n)}{F_m(t_f, T_{n+1})} \right| &< \delta \\ \left| \frac{F_{mk}(t_f, T_{n+1}) - F_{mk}(t_f, T_n)}{F_{mk}(t_f, T_{n+1})} \right| &< \delta \end{aligned} \tag{8-86}$$

If (8-86) is satisfied for all m and k, then T_{n+1} is an acceptable pseudo terminal time for approximating the infinite-interval case. When (8-86) is satisfied, the segments of the k parameters needed in the optimum control equation are found by continuing the computations, for the solution corresponding to T_{n+1}, over the interval $t_0 \leqq \mu \leqq t_f$. The increase in terminal time between successive solutions should correspond roughly to the value of normalized time-to-go required for steady-state performance error. Therefore, a "rule of thumb" for increasing terminal time between successive solutions is chosen to be

$$T_{n+1} = \frac{N}{\mathcal{N}} + T_n \tag{8-87}$$

This method of successive approximations to the infinite-interval case for time-varying systems is programmed easily for use on a digital computer. Also, the test given in (8-86) is chosen so that the accuracy of the required computations can be specified directly in terms of the accuracy of the components used to construct the control system. However, if all or part of the optimum control system is time-invariant, then the procedure for approximating the steady-state values of the k parameters simplifies significantly.

Suppose that the weighting factors in the error measure and the coefficients of the dynamic-process equations are constants. Then the double-subscripted k parameters simplify to

$$k_{mk}(t) = F_{mk}(T - t) \tag{8-88}$$

Of course, these parameters become constants as the time-to-go increases without limit, and these constants are the solutions to a set of algebraic equations. Unfortunately, the algebraic equations are nonlinear, and their solution is difficult and requires special computing techniques.[55] An alternative and straightforward procedure is the use of the differential equations as an algorithm for solving the algebraic equations. In other words, the steady-state values of the k parameters are found as the approximate steady-state solutions of the differential equations. The condition for the steady-state solution of a set of differential equations with constant coefficients is that all derivatives vanish simultaneously. If the increment of time used in an algorithm for numerical integration is defined as Δ, then the condition for an approximate steady-state solution can be chosen as

$$\left| \frac{k_{mk}(\mu - \Delta) - k_{mk}(\mu)}{k_{mk}(\mu - \Delta)} \right| < \delta \tag{8-89}$$

for all m and k. This criterion is chosen to correspond to the accuracy of components eventually to be used in the system construction, and (8-89) is written in a form which indicates that the numerical integration is performed backward in time from known terminal-boundary conditions.

Normally, the free-point terminal-boundary condition is associated with the infinite-interval problem. However, this does not mean that the computation of the steady-state values of the double-subscripted k parameters must begin with the boundary conditions $k_{mk}(T) = 0$. In fact, the computations can be initiated with any set of boundary conditions so that the matrix $[k_{mk}(T)]$ is positive definite. In other words, these boundary conditions can be taken to be the solutions of any other design problems with dynamic processes of equal order. This property is due to the fact that the matrix Riccati equations, although nonlinear, are asymptotically stable for all initial conditions forming a positive-definite or a positive-semidefinite set. This process of initializing the solutions can be thought of as being due to a properly chosen set of impulse-function weighting factors at $\mu = T$. The selection of nonzero values of $k_{mk}(T)$ also can be used as an initial estimate of the steady-state values of k_{mk}. As a "rule of thumb," the normalizing factors § defined in (8-58) can be used as the initial estimate such that

$$k_{mk}(T) = \mathcal{N}_{mk}$$

However, if steady-state solutions to design problems similar to the one at hand are available, these solutions generally are better initial estimates. Accurate initial estimates greatly reduce the amount of computing time required to find the steady-state solutions. Once the steady-state values of the double-subscripted k parameters are found, these values can be used in the computations for the steady-state values of the single-subscripted k parameters.

8-5 Floating-interval problem approximation

Another class of engineering design problems gives rise to an additional computational problem, which is called the *floating-interval* problem. The infinite-interval problem can be thought of in terms of a constant interval $T - t$, but this interval is usually

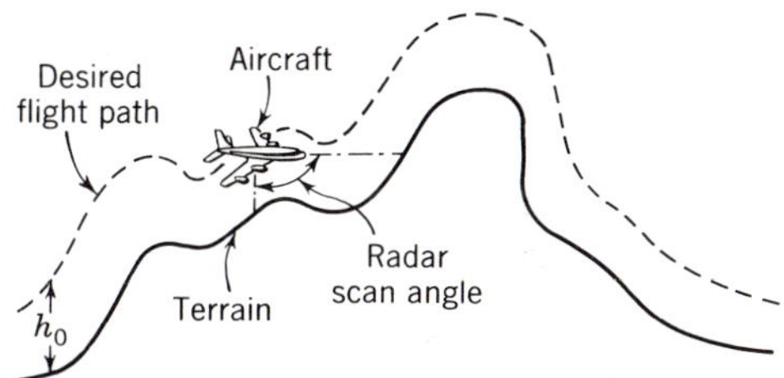

Fig. 8-13 Terrain Following System.

much greater than the time interval where system operation occurs. On the other hand, the floating-interval problem is characterized by a constant-interval length $\tau = T - t$, where τ is less than or equal to the real-time interval where system operation occurs. Note, however, that the constant interval τ must be greater than the stability abscissa τ_s, if the control system is operated indefinitely, in order to achieve stable operation. The stability abscissa is of the order of magnitude given by $N/\mathcal{N}$.

In order to clarify the engineering origin of floating-interval problems, an example is discussed before the computational problems are discussed. The example is a Terrain Following System for aircraft. This system would be used for avoiding terrain obstructions in a long-range, high-speed, and very-low-altitude bombing mission. The exact terrain would be unknown prior to the actual flight, so that the terrain immediately in front of the aircraft must be continuously scanned by radar during the flight. Figure 8-13 depicts the system design problem. Because the radar scanning can "look at" only a limited segment of the terrain, only a segment of the desired aircraft altitude is known at any time during the system operation. Furthermore, no prior statistics can be assigned to the terrain locale. Therefore, a

§ This set is positive semidefinite.

finite and constant interval where response errors can be weighted must be used in the error index.

In light of this example, the computational difficulties of the floating-interval problem become obvious. As the aircraft proceeds over the terrain, the k parameters of the optimum controller must be recomputed continuously. The minimal rate at which these computations must be repeated in order not to degrade system performance is generally unknown and must be found experimentally. However, the degrading of system performance can be expressed functionally in terms of accuracy of the computed gains. First, suppose that Δt sec is required to compute a solution of the k equations over a time increment τ. If the assumption is made that a solution is begun at $t - \Delta t$, then the terminal point used in this solution is $t + \tau - \Delta t$ and the solution is available at time t. According to (8-74), this solution is expressed as

$$k_m(t) = F_m(t, t + \tau - \Delta t)$$

and

$$k_{mn}(t) = F_{mn}(t, t + \tau - \Delta t) \tag{8-90}$$

However, this solution must be used for the next Δt sec before the next solution is available for use in the control system. Therefore, the solution expressed in (8-90) becomes

$$\begin{aligned} k_m(t + \Delta t) &= F_m(t + \Delta t, t + \tau - \Delta t) \\ k_{mn}(t + \Delta t) &= F_{mn}(t + \Delta t, t + \tau - \Delta t) \end{aligned} \tag{8-91}$$

at the time that the next solution is available. The second solution, according to the reasoning used to arrive at (8-90), is

$$k_m(t + \Delta t) = F_m(t + \Delta t, t + \tau)$$

and

$$k_{mn}(t + \Delta t) = F_{mn}(t + \Delta t, t + \tau) \tag{8-92}$$

If the errors between these two solutions are small for all t, then the system performance is not degraded because of the solution time Δt. In other words, a condition for acceptable system performance is

$$F_m(t + \Delta t, t + \tau) \approx F_m(t + \Delta t, t + \tau - \Delta t)$$

and

$$F_{mn}(t + \Delta t, t + \tau) \approx F_m(t + \Delta t, t + \tau - \Delta t) \tag{8-93}$$

An alternative condition for acceptable system performance with small Δt is found from a Taylor series expansion of (8-91) and (8-92). This condition is written as

$$\begin{aligned} \Delta t &\ll \left| \frac{F_m(t, t + \tau)}{\partial F_m(t, t + \tau)/\partial \tau} \right| \\ \Delta t &\ll \left| \frac{F_{mn}(t, t + \tau)}{\partial F_{mn}(t, t + \tau)/\partial \tau} \right| \end{aligned} \tag{8-94}$$

From the previous discussion on the infinite-interval problem, the k-parameter solutions become independent of τ as τ becomes very large, and the solution time Δt is not important.

A simple example of this condition can be carried out analytically for the case

$$N = a_{11} = c_{11} = \phi_{11} = \psi_{11} = 1 \qquad \text{and} \qquad b_{11} = Q_1 = M_1 = 0 \tag{8-95}$$

For this case, the k parameters are

$$k_1(t) = k_{11}(t) = \tanh(\tau - \Delta t) \tag{8-96}$$

and (8-94) becomes

$$\Delta t \ll \tfrac{1}{2} \sinh 2\tau \tag{8-97}$$

If the limiting case $\tau \to 0$ is considered, (8-97) becomes

$$\Delta t \ll \tau \tag{8-98}$$

Even in floating-interval problems, recomputation of the k parameters may not be necessary. In other words, the floating-interval case does not have to be solved in real time as a sequence of shrinking-interval problems. These cases can be identified from results obtained in Chap. 7 on separable signals. If the coefficients of the dynamic process, a_{nm}, b_{nm}, and c_{nm}, are constants and if the weighting factors ϕ_{nn} and ψ_{nn} are constants, then the k_{nm} parameters are dependent upon only $T - t$. Under these conditions for the floating-interval problem, these parameters are constants and are expressed as

$$k_{nm}(t) = F_{nm}(\tau) \tag{8-99}$$

Therefore, these k parameters do not have to be recomputed in real time. If there is a further restriction that the unseparated components of the desired signals Q_n and M_n and the load disturbance u_n are constants, then the precomputed reference signals are constants such that

$$l_m(t) = L_m(\tau) \tag{8-100}$$

If there are no separable components and these conditions occur, then no real-time computation of the k parameters is necessary. However, if separable components occur, then further restrictions must be imposed in order to eliminate the need for real-time computations. The separable-component gain can be expressed in terms of the superposition integral given in (7-71) as

$$p(t) = \int_0^\tau w(\tau, \xi)\mathcal{Q}(t + \tau - \xi, t)\, d\xi \tag{8-101}$$

In this equation, the subscripts are omitted for the sake of convenience. If the separable driving component $\mathcal{Q}$ is stationary and hence is restricted to the form

$$\mathcal{Q}(\mu, t) = \mathcal{Q}(\mu - t) \tag{8-102}$$

then the separable-component gain becomes

$$p(t) = P(\tau) = \int_0^\tau w(\tau, \xi)\mathcal{Q}(\tau - \xi)\, d\xi \tag{8-103}$$

For this condition, the separable-component gain also does not have to be recomputed.

Approximations that allow circumventing the need for real-time computations in floating-interval problems are important to consider before the control system is constructed. The requirements associated with the need for recomputations may make the construction of the optimum control system unfeasible. Therefore, approximations that allow circumventing the need for real-time computations in floating-interval problems, especially for the double-subscripted k parameters, are important and should be made whenever possible.

8-6 Selection of weighting factors

When optimization theory is used as the basis for the design of linear control systems, the error measure must be taken to be quadratic in form. There remains, however, the problem of selecting the appropriate weighting factors appearing in the error measure for the design problem under consideration. The selection of weighting factors, although somewhat a matter of experience and ingenuity, is suggested generally by the performance specifications of the design problem. This selection seldom is dictated precisely by the performance specifications, which is fortunate because simplifications can be introduced. The allowable latitude in this selection is very much a function of the dynamic difficulty of the design problem. In particular, more difficult design problems require more accurately selected weighting factors if the performance specifications are to be met. Many times the weighting factors can be chosen as simple functions of time, and a number of performance specifications can be neglected altogether. These general considerations are illustrated in the case study presented in Appendix E.

An approximate procedure for selecting the weighting factors can be developed using very elementary concepts. If the error measure at any point in real time is

$$e_m(t) = \sum_{n=1}^{Q} \phi_{nn}(t)[Q_n(t) - q_n(t)]^2 + \sum_{n=1}^{M} \psi_{nn}(t)[M_n(t) - m_n(t)]^2 \tag{8-104}$$

then the weighting factors $\phi_{nn}(t)$ and $\psi_{nn}(t)$ can be selected at each point in real time by considering the allowable errors. Philosophically, the maximum allowable response errors at any point in time should contribute equally to the error measure because the control system is designed to minimize the integrated sum of the errors. According to this concept, the relationships

$$\phi_{nn}(t) = \left[\frac{\epsilon_Q(t)}{\epsilon_n(t)}\right]^2 \phi_{QQ}(t) \qquad n = 1, 2, \ldots, Q - 1 \tag{8-105}$$

are found by equating terms in (8-104). The variable $\epsilon_n(t)$ is defined as

$$\epsilon_n(t) = [Q_n(t) - q_n(t)]_{\text{max allowable}} \qquad n = 1, 2, \ldots, Q \tag{8-106}$$

If the same reasoning is applied to the amount of control effort used, then

$$\psi_{nn}(t) = \left[\frac{\delta_1(t)}{\delta_n(t)}\right]^2 \psi_{11}(t) \qquad n = 2, 3, \ldots, M \tag{8-107}$$

is obtained, where

$$\delta_n(t) = [M_n(t) - m_n(t)]_{\text{max available}} \qquad n = 1, 2, \ldots, M \tag{8-108}$$

Presumably, $\epsilon_n(t)$ and $\delta_n(t)$ can be specified from the performance specifications, and then the relationships between the various ϕ's and the relationships between the various ψ's are found numerically.

The problem remaining, however, is to find the relationship between $\phi_{QQ}(t)$ and $\psi_{11}(t)$. This problem is resolved from arguments similar to those used to find (8-105) and (8-107). Specifically, the total contribution due to the maximum allowable response errors should roughly equal the total contribution due to the maximum available control effort in the error measure. From this argument, the additional relationship

$$\sum_{n=1}^{Q} \phi_{nn}(t)\epsilon_n^2(t) = \sum_{n=1}^{M} \psi_{nn}(t)\, \delta_n^2(t) \tag{8-109}$$

is obtained. When (8-105) and (8-107) are substituted into (8-109), the final relationship

$$\phi_{QQ}(t) = \frac{M}{Q}\left[\frac{\delta_1(t)}{\epsilon_Q(t)}\right]^2 \psi_{11}(t) \tag{8-110}$$

is obtained. From the normalization procedures presented earlier in this chapter, the optimum control equation is found to be dependent upon the ratios of the weighting factors. Therefore, no generality is sacrificed by choosing $\psi_{11}(t) = 1$. With this choice for $\psi_{11}(t)$, (8-105), (8-107), and (8-110) are sufficient to specify the numerical values of the weighting factors.

The weighting factors found by this selection procedure should be considered as initial estimates. In some instances, the maximum allowable response errors do not occur simultaneously in time, thereby invalidating the equalities. A value of $\phi_{QQ}(t)$ which is larger than that found from (8-110) generally can be used in these situations. In other instances, the value of $\phi_{QQ}(t)$ must be decreased because the actual amount of control effort that would occur in system operation exceeds the maximum available control effort. These occurrences are due to the particular dynamic process and desired signals used in the design problem. In this situation, either larger response errors must be tolerated or the fixed member of the system must be redesigned for greater response capabilities.

On the other hand, the values of the weighting factors found by this procedure are sufficiently accurate to perform preliminary analyses in conjunction with the normalization procedures discussed previously in this chapter but without recourse to a digital computer. In fact, the procedure usually can be made more approximate in many design problems. For instance, some design specifications may be neglected completely. The most widely used approximation is to choose a constant weighting factor if an error is weighted over the whole interval of time. In this approximation, the value of maximum allowable error used in the selection might be the maximum error allowable for the whole interval of real-time operation of the system.

8-7 Test for zero k parameters

In large systems, possible simplifications may be difficult to identify without an organized method of analysis. In particular, significant savings in computer time can be made if zero k parameters are eliminated from the k equations before computations are started. With a large set of k equations, many times it is difficult to determine by inspection which k parameters vanish, if any. The test for zero k parameters developed in this section is easy to carry out by hand and also can be used to form the basis of a computer program for deriving k equations. This test is sufficient, in a mathematical sense, but is not necessary for a k parameter to be zero. In other words, this test may not identify all the zero k parameters. However, cases where zero k parameters are missed correspond to very specialized cases.

The basis for this test can be stated simply in words. The double-subscripted k parameters are considered first because they are dependent only upon the dynamic-process coefficients and weighting factors. Specifically, they are not dependent upon the single-subscripted k parameters and the separable-component gains. The parameter $k_{mk}(\mu)$ is zero over the interval $t \leq \mu \leq T$ if $k'_{mk}(\mu)$ is zero over this entire interval, because $k_{mk}(T) = 0$ as the boundary condition. The desired test is for $k_{mk}(\mu) = 0$ for all μ on this interval.

First, the assumption is made that these derivatives are defined by the differential equations

$$-k'_{mk}(\mu) = \sum_{n=1}^{N} \left[\phi_{nn}(\mu)a_{nm}(\mu)a_{nk}(\mu) + b_{nm}(\mu)k_{nk}(\mu) + b_{nk}(\mu)k_{nm}(\mu) - \frac{c_{nn}{}^2(\mu)}{\psi_{nn}(\mu)} k_{nm}(\mu)k_{nk}(\mu) \right] \tag{8-111}$$

The test involves determining whether the right-hand side of (8-111) is zero or not zero and hence whether $k_{mk}(\mu)$ is zero or not zero. In other words, the test involves a binary result, and a nonzero value is repesented by 1 and a zero value is represented by 0. Therefore, a logical equivalent of $k_{mk}(\mu)$ is defined so that

$$\begin{aligned} k_{mk}{}^P &= 1 \qquad k_{mk}(\mu) \neq 0 \text{ anywhere on } t \leqq \mu \leqq T \\ k_{mk}{}^P &= 0 \qquad k_{mk}(\mu) = 0 \text{ everywhere on } t \leqq \mu \leqq T \end{aligned} \tag{8-112}$$

The superscript p of the logical variable $k_{km}{}^p$ is used to denote the transition number in a sequence of transitions from zero to nonzero values. The initial condition on this sequence of transitions is

$$k_{mk}{}^0 = 0 \tag{8-113}$$

because of the boundary condition $k_{mk}(T) = 0$.

At this boundary condition, the first transition results in $k_{mk}{}^1 = 1$ if the right-hand side of (8-111) is nonzero and $k_{mk}{}^1 = 0$ if the right-hand side of (8-111) is zero. If any term on the right-hand side of (8-111) is nonzero, the condition $k_{mk}{}^1 = 1$ results by definition; conversely, if all terms on the right-hand side of (8-111) are zero, the condition $k_{mk}{}^1 = 0$ results by definition. According to this definition, the right-hand side of (8-111) is tested according to binary arithmetic. Therefore, logical equivalents are defined such that

$$\begin{aligned} \phi_{nn}{}^0 &= 1 \qquad \phi_{nn}(\mu) \neq 0 \text{ anywhere on } t \leqq \mu \leqq T \\ a_{nm}{}^0 &= 1 \qquad a_{nm}(\mu) \neq 0 \text{ anywhere on } t \leqq \mu \leqq T \\ b_{nm}{}^0 &= 1 \qquad b_{nm}(\mu) \neq 0 \text{ anywhere on } t \leqq \mu \leqq T \\ c_{nn}{}^0 &= 1 \qquad c_{nn}(\mu) \neq 0 \text{ anywhere on } t \leqq \mu \leqq T \\ \phi_{nn}{}^0 &= 0 \qquad \phi_{nn}(\mu) = 0 \text{ everywhere on } t \leqq \mu \leqq T \\ a_{nm}{}^0 &= 0 \qquad a_{nm}(\mu) = 0 \text{ everywhere on } t \leqq \mu \leqq T \\ b_{nm}{}^0 &= 0 \qquad b_{nm}(\mu) = 0 \text{ everywhere on } t \leqq \mu \leqq T \\ c_{nn}{}^0 &= 0 \qquad c_{nn}(\mu) = 0 \text{ everywhere on } t \leqq \mu \leqq T \\ \psi_{nn}{}^0 &= 1 \qquad \text{all conditions} \end{aligned} \tag{8-114}$$

Furthermore, the two algebraic operations of addition and multiplication are defined for the logical variables $z_1{}^0$ and $z_2{}^0$ in Tables 8-4 and 8-5. The logical equivalent of

Table 8-4 *Addition:* $z_1{}^0 + z_2{}^0$

	$z_1{}^0$	
$z_2{}^0$	0	1
0	0	1
1	1	1

Table 8-5 *Multiplication:* $z_1{}^0 z_2{}^0$

	$z_1{}^0$	
$z_2{}^0$	0	1
0	0	0
1	0	1

the right-hand side of (8-111) at the first transition reduces to the right-hand side of

$$k_{mk}{}^1 = \sum_{n=1}^{N} \phi_{nn}{}^0 a_{nm}{}^0 a_{nk}{}^0 \tag{8-115}$$

From (8-115), the first transitions can be computed readily. If any of the $k_{mk}(\mu)$'s are found to be nonzero as indicated by $k_{mk}{}^1 = 1$, then additional terms on the right-hand side of (8-111) may be nonzero and hence additional $k_{mk}(\mu)$'s may be nonzero. The logical equivalent of the right-hand side of (8-111) for the second transition can be used to determine the nonzero $k_{mk}{}^2$'s. This sequence of transitions can be computed for $k_{mk}{}^3$, $k_{mk}{}^4$, . . . until all the nonzero $k_{mk}(\mu)$'s are found. Obviously this sequence of computations terminates when no new nonzero $k_{mk}(\mu)$'s are found as the result of a transition or when all the $k_{mk}(\mu)$'s are nonzero. This test for zero double-subscripted k parameters now can be stated formally.

1. Construct the logical arrays $[a_{nm}{}^0]$, $[b_{nm}{}^0]$, $[c_{nn}{}^0]$, and $[\phi_{nn}{}^0]$ according to the rules of (8-114).

2. Construct the sequence of logical arrays $[k_{mk}{}^p]$, $p = 1, 2, \ldots, P$, from the relationship

$$k_{mk}{}^{p+1} = \sum_{n=1}^{N} (\phi_{nn}{}^0 a_{nm}{}^0 a_{nk}{}^0 + b_{nm}{}^0 k_{nk}{}^p + b_{nk}{}^0 k_{nm}{}^p + c_{nn}{}^0 k_{nm}{}^p k_{nk}{}^p) \tag{8-116}$$

where $k_{mk}{}^0$ is defined by (8-113). The last array, corresponding to $p = P$, is defined by either

$$\left.\begin{aligned} k_{mk}{}^P &= 1 \\ k_{mk}{}^P &= k_{mk}{}^{P-1} \end{aligned}\right. \qquad \text{for all } m, k \tag{8-117}$$

3. The set $\{m, k\}$, for which zeros occur in $[k_{mk}{}^P]$, corresponds to the set $\{k_{mk}(\mu)\}$, for which the double-subscripted k parameters are zero everywhere on the interval $t \leqq \mu \leqq T$.

A similar test can be used for determining zero single-subscripted k parameters. The logical equivalents of $k_m(\mu)$ are defined similarly to (8-112) and (8-113). Also, the logical equivalents of $Q_n(\mu)$, $M_n(\mu)$, and $u_n(\mu)$ are defined similarly to $\phi_{nn}{}^0$ in (8-114). Then the logical equivalent of

$$-k'_m(\mu) = \sum_{n=1}^{N} \Bigg[\phi_{nn}(\mu) Q_n(\mu) a_{nm}(\mu) - u_n(\mu) k_{nm}(\mu) - c_{nn}(\mu) M_n(\mu) k_{nm}(\mu) + b_{nm}(\mu) k_n(\mu) - \frac{c_{nn}{}^2(\mu)}{\psi_{nn}(\mu)} k_n(\mu) k_{nm}(\mu) \Bigg] \tag{8-118}$$

is

$$k_m{}^{r+1} = \sum_{n=1}^{N} (\phi_{nn}{}^0 Q_n{}^0 a_{nm}{}^0 + u_n{}^0 k_{nm}{}^P + c_{nn}{}^0 M_n{}^0 k_{nm}{}^P + b_{nm}{}^0 k_n{}^r + c_{nn}{}^0 k_n{}^r k_{nm}{}^P) \tag{8-119}$$

The sequence of arrays, starting with $k_m{}^0 = 0$, terminates at $r = R$ under the same conditions as the test for $k_{mk}{}^P$. The use of $k_{mk}{}^P$ in (8-119) is required because $k_m(\mu)$ is dependent upon $k_{mk}(\mu)$; hence the test for $k_m{}^R$ is performed after the test for $k_{mk}{}^P$. Of course, tests for zero separable-component gains can be constructed on a similar basis.

In order to illustrate this test, the following example is chosen:

$$[a_{nm}^0] = \begin{bmatrix} 1 & 0 & 0 \\ 0 & 1 & 0 \\ 0 & 0 & 1 \end{bmatrix} \qquad [b_{nm}^0] = \begin{bmatrix} 1 & 0 & 0 \\ 0 & 1 & 0 \\ 1 & 1 & 1 \end{bmatrix}$$
$$[c_{nm}^0] = \begin{bmatrix} 1 & 0 & 0 \\ 0 & 1 & 0 \\ 0 & 0 & 0 \end{bmatrix} \tag{8-120}$$

For this example, (8-116) becomes

$$\begin{aligned} k_{11}^{p+1} &= \phi_{11}^0 + k_{11}^p + k_{12}^p + k_{13}^p \\ k_{12}^{p+1} &= k_{12}^p + k_{13}^p + k_{23}^p + k_{11}^p k_{12}^p + k_{12}^p k_{22}^p \\ k_{13}^{p+1} &= k_{13}^p + k_{33}^p + k_{11}^p k_{13}^p + k_{12}^p k_{23}^p \\ k_{22}^{p+1} &= \phi_{22}^0 + k_{12}^p + k_{22}^p + k_{23}^p \\ k_{23}^{p+1} &= k_{23}^p + k_{33}^p + k_{12}^p k_{13}^p + k_{22}^p k_{23}^p \\ k_{33}^{p+1} &= \phi_{33}^0 + k_{13}^p + k_{23}^p + k_{33}^p \end{aligned} \tag{8-121}$$

If the case $\phi_{11}^0 = \phi_{22}^0 = \phi_{33}^0 = 1$ is carried out, then $P = 3$ and the sequence of arrays is

$$[k_{mk}^1] = \begin{bmatrix} 1 & 0 & 0 \\ & 1 & 0 \\ & & 1 \end{bmatrix} \qquad [k_{mk}^2] = \begin{bmatrix} 1 & 0 & 1 \\ & 1 & 1 \\ & & 1 \end{bmatrix}$$
$$[k_{mk}^3] = \begin{bmatrix} 1 & 1 & 1 \\ & 1 & 1 \\ & & 1 \end{bmatrix} \tag{8-122}$$

On the other hand, if $\phi_{11}^0 = \phi_{22}^0 = 1$ and $\phi_{33}^0 = 0$, then $P = 2$ and

$$[k_{mk}^1] = \begin{bmatrix} 1 & 0 & 0 \\ & 1 & 0 \\ & & 0 \end{bmatrix} \qquad [k_{mk}^2] = \begin{bmatrix} 1 & 0 & 0 \\ & 1 & 0 \\ & & 0 \end{bmatrix} \tag{8-123}$$

The second case results in a reduction of system order from $N = 3$ to $N = 2$ and also results in a dynamic decoupling of the optimum system to two first-order systems because $k_{12}(\mu)$ is zero. The occurrence of this decoupling is important because noninteracting control results and because one numerical problem degenerates into two separate numerical problems.

A test for noninteracting optimum control systems is a simple interpretation of the test for zero k parameters. Suppose that noninteracting system 1 has the minimum-error function

$$E_1[\mathbf{x}(\mu), \mu] = {}_1k(\mu) - 2 \sum_{n=1}^{N_1} k_n(\mu) x_n(\mu) + \sum_{n=1}^{N_1} \sum_{m=1}^{N_1} k_{nm}(\mu) x_n(\mu) x_m(\mu) \tag{8-124}$$

Also suppose that the state variables of a second noninteracting system (system 2) are labeled so that this second system has the minimum-error function

$$E_2[\mathbf{x}(\mu), \mu] = {}_2k(\mu) - 2\sum_{n=N_1+1}^{N} k_n(\mu)x_n(\mu) + \sum_{n=N_1+1}^{N}\sum_{m=N_1+1}^{N} k_{nm}(\mu)x_n(\mu)x_m(\mu) \tag{8-125}$$

Furthermore, the minimum-error function of a composite system made up of both these independent systems is the sum of E_1 and E_2. Therefore, the minimum-error function of the composite system is

$$E_{1,2}[\mathbf{x}(\mu), \mu] = {}_1k(\mu) + {}_2k(\mu) - 2\sum_{n=1}^{N} k_n(\mu)x_n(\mu) + \sum_{n=1}^{N_1}\sum_{m=1}^{N_1} k_{nm}(\mu)x_n(\mu)x_m(\mu) + \sum_{n=N_1+1}^{N}\sum_{m=N_1+1}^{N} k_{nm}(\mu)x_n(\mu)x_m(\mu) \tag{8-126}$$

Equation (8-126) is immediately seen to be a specialization of the general quadratic form of the minimum-error function for an Nth order system. This specialization of the general form requires that

$$k_{mk}(\mu) = 0 \qquad 1 \leqq m \leqq N_1 \qquad N_1 + 1 \leqq k \leqq N \tag{8-127}$$

Conversely, if (8-127) results, then the optimum system decomposes into two noninteracting subsystems, one containing state variables $x_1, x_2, \ldots, x_{N_1}$ and the other containing state variables $x_{N_1+1}, x_{N_1+2}, \ldots, x_N$. In terms of the test for zero k parameters, the final array appears as

$$[k_{mk}{}^P] = \left[\begin{array}{c|c} \ddots\,1 & 0 \\ \hline & 1\,\ddots \end{array}\right] \begin{array}{l} \}N_1 \text{ rows} \\ \\ \end{array} \tag{8-128}$$

N_1 columns

The problem of designing an optimum noninteracting controller which also satisfies a set of design specifications by the appropriate selection of weighting factors cannot be solved in general. Unfortunately, the selection required here for noninteraction generally involves the cancellation of terms on the right-hand side of (8-111). The test for zero k parameters derived in this section neglects cancellations, which is why the test does not necessarily indicate all zero k parameters. Also, these cancellations generally require the addition of cross-product terms in the error measure.

8-8 Incomplete measurement of state

In design problems where the dynamic process must be of high order for an accurate description of the system fixed member, the measurement of the complete set of state signals invariably cannot be accomplished. This difficulty is due to the expense or unavailability of appropriate sensors. One possible solution to this problem is the approximation of the dynamic process with a pure time delay followed by a low-order system of ordinary differential equations. This type of approximation is used often in conventional control-system design. However, the theory presented in Chap. 7 demonstrates that the optimum control system for dynamic processes of this type requires the construction of compensation filters with finite memory. The expense and difficulties of finite-memory filters should be avoided if possible.

An alternative procedure for the construction of the optimum control equation with incomplete measurement of state is the on-line computation of the state signals that cannot be measured. This computation is accomplished by constructing a simulator or model of all or part of the system fixed member as part of the optimum control system.[56] Figure 8-14 depicts the situation where the state signal is measured directly with a sensor. This is the ideal case, where the exact value of $x_n(t)$ is found and used in the optimum control equation. Furthermore, direct feedback is

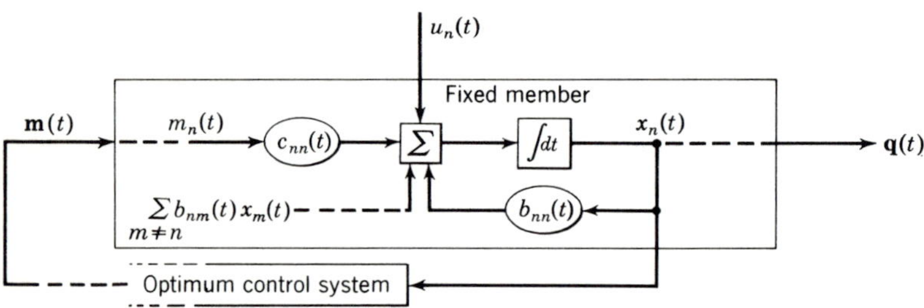

Fig. 8-14 Block diagram of state-signal measurement.

established around the system fixed member, thereby giving the maximum insensitivity of the system performance to possible unpredictabilities of the load disturbance $u_n(t)$ and perturbations of the parameters of the system fixed member.

Suppose, however, that the state signal $x_n(t)$ cannot be measured directly but that all other state signals can be measured. In this case, the state signal must be computed; this computation is diagrammed in Fig. 8-15. The input to the model $x_n(t_0)$ is the value of this state signal at time $t = t_0$ where control-system operation is initiated. If this model is exact, that is, the inputs and coefficients of the model are identical to those of the fixed member, then there is no degradation of system performance. In other words, the responses of the systems shown in Figs. 8-14 and 8-15 are identical under ideal conditions. However, ideal conditions never are achieved.

The most likely unknowns are the load-disturbance signal $u_n(t)$ and the initial condition $x_n(t_0)$. Load disturbances generally are statistical and cannot be measured directly. Also, the initial condition cannot be measured, because of the assumption that $x_n(t)$ cannot be measured, and seldom is known a priori. Sometimes, however, this initial condition is known exactly, such as the flight-path angle of a missile at blast-off. If the load disturbance and initial condition are specified statistically, then the required state signal, corresponding to the conditional-mean error index, also can be computed from a model.

For the model shown in Fig. 8-15, a superposition integral of a first-order system can be written so that

$$x_n(t) = x_n(t_0)w_n(t, t_0) + \int_{t_0}^{t} w_n(t, \sigma)\left\{\sum_{m \neq n} [b_{nm}(\sigma)x_m(\sigma)] + c_{nn}(\sigma)m_n(\sigma) + u_n(\sigma)\right\} d\sigma \tag{8-129}$$

where the impulse response is

$$w_n(t, \sigma) = \exp\left[\int_\sigma^t b_{nn}(\xi)\, d\xi\right] \tag{8-130}$$

The model is treated to be first-order, even though the fixed member is Nth-order, because the state signals $x_m(\sigma)$ for $m \neq n$, the optimum control signal $m_n(\sigma)$, and

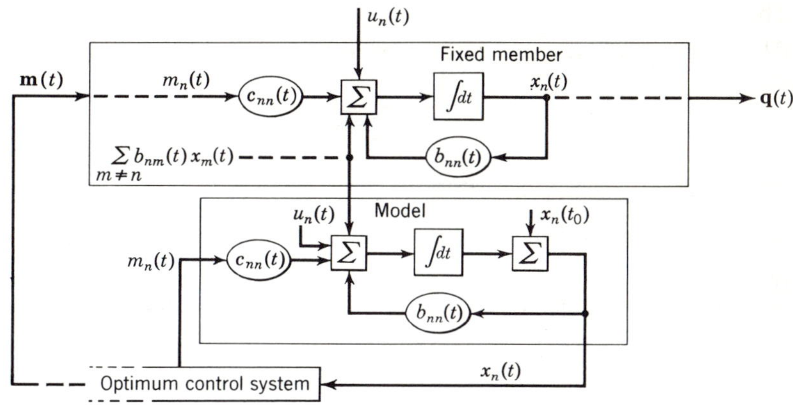

Fig. 8-15 Block diagram of state-signal computation.

the load disturbance $u_n(\sigma)$ are inputs to the model. The conditional mean is found by averaging both sides, thereby obtaining

$$\overline{x_n(t)}^t = \overline{x_n(t_0)}w_n(t, t_0) + \int_{t_0}^t w_n(t, \sigma)\left\{\sum_{m \neq n}[b_{nm}(\sigma)x_m(\sigma)] + c_{nn}(\sigma)m_n(\sigma) + \overline{u_n(\sigma)}\right\} d\sigma \tag{8-131}$$

because of the following assumptions. The initial state $x_n(t_0)$ cannot be estimated from correlated signals, so that the conditional mean of $x_n(t_0)$ becomes the mean value. The same assumption is made for the load disturbance $u_n(\sigma)$. Also, the state signals $x_m(\sigma)$ for $m \neq n$ are measured precisely on the past interval $t_0 \leqq \sigma \leqq t$ of system operation so that the conditional mean of $x_m(t)$ becomes the state signal itself. The same reasoning is applied to the optimum control signal $m_n(\sigma)$ which is generated, by virtue of the optimum control equation, from $x_m(t)$ for $m \neq n$ and $\overline{x_n(t)}^t$. With the use of (8-130) and (8-131), the equation

$$\frac{d}{dt}\left[\overline{x_n(t)}^t - \overline{x_n(t_0)}\right] = b_{nn}(t)\overline{x_n(t)}^t + \sum_{m \neq n}[b_{nm}(t)x_m(t)] + c_{nn}(t)m_n(t) + \overline{u_n(t)} \tag{8-132}$$

is obtained by differentiation. This equation describes the model shown in Fig. 8-16 for a statistical, but not measurable, initial state and load disturbance. In other words, the system indicated in Fig. 8-16 is the optimum system subject to the constraint, imposed by the design problem, that $x_n(t)$ cannot be measured and that the only statistics available for computing $\overline{x_n(t)}^t$ are $\overline{x_n(t_0)}$ and $\overline{u_n(t)}$. This type of constraint is typical of many design problems.

In a number of design problems, the computations required for a state signal that cannot be measured are equivalent to feedforward compensation networks.

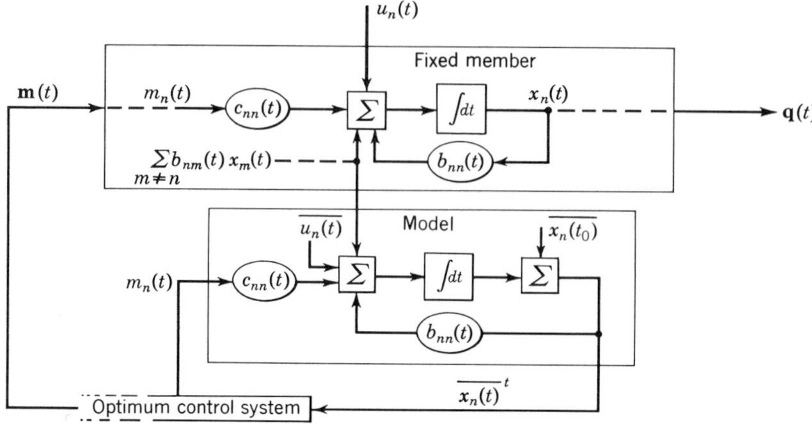

Fig. 8-16 Block diagram of computation for conditional mean of state signal.

This situation occurs when $\overline{u_n(t)} = 0$, $\overline{x_n(t_0)} = 0$, and the double-subscripted k parameters are independent of time. The transfer functions of these optimum compensation networks are found by defining

$$\epsilon_i(t) = M_i(t) + \frac{c_{ii}}{\psi_{ii}}\left[k_i(t) - \sum_{m \neq n} k_{im}x_m(t)\right] \tag{8-133}$$

so that the optimum control equation is given by

$$m_i(t) + \frac{c_{ii}}{\psi_{ii}} k_{in}x_n(t) = \epsilon_i(t) \tag{8-134}$$

If the transfer function of the dynamic process from the ith control signal to the nth state signal is defined as

$$\frac{x_n(s)}{m_i(s)} = W_{ni}(s) \tag{8-135}$$

then the transform of (8-134) results in

$$\frac{m_i(s)}{\epsilon_i(s)} = W_{ci}(s) = \frac{1}{1 + (c_{ii}/\psi_{ii})k_{in}W_{ni}(s)} \tag{8-136}$$

The function $W_{ci}(s)$ defined in (8-136) is the transfer function of the optimum compensation network. The optimun control equation with this compensation network, in terms of the corresponding impulse response $w_{ci}(t - \sigma)$, is shown in Fig. 8-17.

This problem of incomplete measurement of state is probably the most frequently occurring constraint placed on the design of control systems. In the case of linear optimum controls, no conceptual difficulties are introduced by this problem,

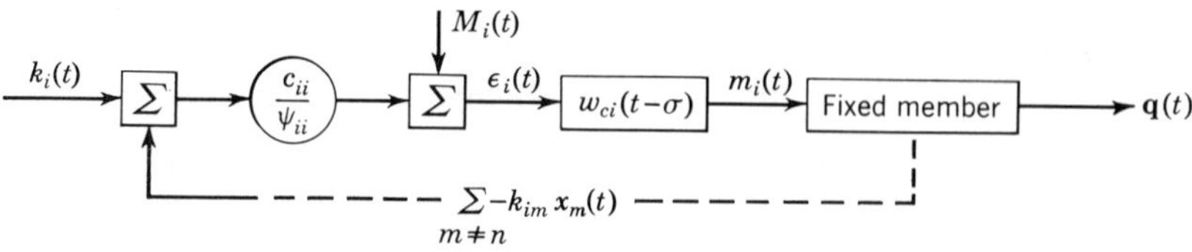

Fig. 8-17 Block diagram for control equation with optimum compensation.

and the results presented here extend directly to design problems where more than one state signal cannot be measured.§

8-9 Summary

The material presented in this chapter is intended to provide the engineer with a number of considerations required when the results and procedures of optimization theory are used as the basis for feedback-control-system design. Although this material is restricted to linear control equations, an important class of design problems can be resolved efficiently and effectively by using optimization theory. The design problem discussed in Appendix E is typical of this class.

This class of design problems is characterized roughly by the following points:

1. A linear dynamic process is a suitable description of the system fixed member over the region of state space where the control system must operate.

2. The performance requirements do not include minimum time, minimum energy, or specified economic modes of operation. Instead, the performance requirements involve a number of design conditions that are difficult to satisfy and cannot be violated in successful operation of the system. A weighted quadratic error measure then can be chosen.

The desirability of applying optimization theory to this class of design problems is primarily due to the following considerations:

1. Multiple response and control variables are treated with no difficulty. Also, statistical desired-response, desired-control, and load-disturbance signals can be introduced without greatly increasing the complexity of the optimum system. Finally, the incomplete measurement of state resulting from the unavailability of sensors is resolved with the introduction of compensating computations or filters.

2. The weighting factors appearing in the error measure are selected readily and essentially without regard to the stability of the resulting system. The time-function forms of these weighting factors usually are specified directly from the design-problem performance requirements. Also, the initial estimates of the magnitudes of these time functions are found from simple algebraic equations. Any adjustments in these initial estimates, required in order to satisfy all the performance requirements simultaneously, are made readily because of the monotonic relationship between the magnitudes of the weighting factor and the resulting response error.

3. The numerical solutions of the k parameters are found with relative ease on automatic computing machines. These solutions are found from a one-point boundary-value problem involving stable differential equations. A computer program constructed for obtaining these solutions is outlined in Appendix F.

4. The optimum control equation can be constructed with standard, inexpensive analog-computer components.

In spite of the many applications of optimum linear feedback control, the nonlinearities associated with many important design problems cannot be avoided. Furthermore, many applications require the use of nonquadratic error measures. Therefore, the difficulties of nonlinear optimum control ultimately cannot be circumvented. The remainder of this book is devoted to design problems of these types.

§ However, practical difficulties may arise when an unstable portion of the fixed member cannot be observed.[14] Exact cancellation of the effects of this unstable portion is impossible.

9

Synthesis of Nonlinear Optimum Control Systems for Small Regions of State Space

The purpose of this chapter is to investigate the structure of the optimum control equation in design problems where the error measure is nonquadratic and the dynamic process is nonlinear. Specifically, the class of design problems under consideration is defined by the error index

$$e(t) = \int_t^T h[\mathbf{q}(\sigma), \mathbf{m}(\sigma), \sigma]\, d\sigma \tag{9-1}$$

and the dynamic-process response equation

$$\mathbf{q}(t) = \mathbf{g}[\mathbf{x}(t), t] \tag{9-2}$$

state equation

$$\mathbf{x}'(t) = \mathbf{f}[\mathbf{x}(t), \mathbf{m}(t), t] \tag{9-3}$$

and control-saturation equation

$$\mathbf{m}(t) \in \mathscr{M}(t) \tag{9-4}$$

Throughout this chapter, the response vector is eliminated from (9-1) by the use of (9-2) so that the error index is written as

$$e(t) = \int_t^T H[\mathbf{x}(\sigma), \mathbf{m}(\sigma), \sigma]\, d\sigma \tag{9-5}$$

and (9-3) to (9-5) define the variational problem undertaken here.

As discussed in Chaps. 4 and 5, a major difficulty in nonlinear optimum control arises when obtaining the numerical solution to the variational problem. On the one hand, the two-point boundary-value nature of Pontryagin's equations poses one set of difficulties. On the other hand, discrete approximations to the dynamic-programming equation and the ensuing computations pose another set of difficulties. However, the synthesis problem for nonlinear optimum control is treated before these computational problems are considered in the next chapter because, in part, the computations required for system design are dictated by the mathematical form of the optimum control equation. The mathematical form of the optimum control equation is in turn the synthesis problem associated with the optimum feedback control.

Unfortunately, closed-form mathematical expressions for the optimum control equation in nonlinear problems cannot be obtained analytically except in very special and simple cases. Therefore, the synthesis problem, by necessity, must be treated as an approximation problem. The emphasis of this chapter is directed to the form and complexity of these approximations to the optimum control equation.§ The theoretical development of this approximation problem is facilitated by making a distinction between two classes of design problems.

The first class represents design problems where the system operates in a "small region" of state space about an optimum trajectory associated with a nominal set of design conditions. The term *small region* is used to indicate that system operation is limited to a suitably small region of state space. The size of this region, of course, is determined both by the particular design problem at hand and by the form and complexity of the approximations used. A typical example of this class of design problems is the aircraft landing system discussed in Appendix E.

The second class represents design problems where the system operates in a "large region" of state space. A region is called *large* when the boundaries are specified and when this region cannot be associated with a single optimum trajectory. The approximations used for small regions of state space do not give sufficient accuracy, by definition, for this second class of problems. Hence, recourse to different methods of approximation is required for large regions of state space, and these methods are discussed in Chap. 11.

9-1 The synthesis problem for small regions of state space

The synthesis problem for small regions of state space is concerned with the approximation of the optimum control equation

$$\mathbf{m}(t) = \mathbf{P}[\mathbf{x}(t), t] \tag{9-6}$$

This equation implies feedback as a result of the assumed measurement of the state

§ The reader should note that the validity of the approximations discussed here is presupposed. Such a supposition is in fact invalid for certain isolated and unusual design problems. The mathematical reasons for these difficulties are delicate and are considered to be beyond the scope of this book. In addition, not all the necessary conditions for the validity of the approximations can be checked analytically for specific practical design problems. These conditions, however, can be identified in terms of certain computational difficulties and hence are reconsidered in Chap. 10.

vector $\mathbf{x}(t)$. First, the results of the theory presented in Chaps. 5 and 6 are reexamined in terms of the information supplied about the form of the optimum control equation.

If the characteristic equations are solved for a given initial state, defined here as $\hat{\mathbf{x}}(t_0)$, then the solution specifies a particular optimum trajectory, defined here as $\hat{\mathbf{x}}(t)$, which corresponds to this initial state. Furthermore, if the optimum control vector is recorded along this optimum trajectory, then a set of signals $\hat{\mathbf{m}}(t)$ is obtained. In terms of (9-6), this set of signals is written as

$$\hat{\mathbf{m}}(t) = \mathbf{P}[\hat{\mathbf{x}}(t), t] \tag{9-7}$$

With this single solution, no information concerning the structure of $\mathbf{P}$ has been obtained. In other words, a single solution of the characteristic equations yields a programmed control system possessing no feedback for counteracting disturbances and other departures from the ideal conditions assumed in the design. Of course, a number of solutions could be obtained for different values of $\hat{\mathbf{x}}(t_0)$, thereby establishing a number of points on the function $\mathbf{P}$. Then interpolation could be used to approximate $\mathbf{P}$ from these solutions. However, the cost due to computer time per solution generally is large so that this procedure generally is inefficient.

An alternative procedure for constructing $\mathbf{P}$ is based on the solution of the discrete form of the dynamic-programming equation. Such a solution requires that both time and each component of the state vector be treated as discrete variables. At each point in this discrete approximation, values of the control signals are computed. Therefore, the optimum control equation given in (9-6) can be thought of as a table of numbers in this case, and the control system can be considered as a table-lookup device requiring a digital memory. Of course, such a system could include interpolation between points or perhaps be based on a mathematical approximation of the table. In any event, the chief difficulty in a scheme involving table lookup is due to dimensionality and the number of points. Suppose that the dimensions of the design problem are

M = number of control signals
N = number of state signals
Q = number of points required for each state-signal amplitude
R = number of points required for time

Then the number of points S required in the table is

$$S = MR(Q)^N \tag{9-8}$$

Even for relatively simple design problems, this number may be unreasonably large, which is primarily due to the dependence of S on the order of the dynamic process.

In situations where the control system can be confined to operate in a small region of state space about the optimum trajectory $\hat{\mathbf{x}}(t)$, an obvious synthesis method is to approximate $\mathbf{P}$ with time-varying gains. Specifically, a Taylor series expansion of (9-6) about $\hat{\mathbf{x}}(t)$ is written as

$$m_n(t) \approx \hat{g}_n(t) + \sum_{m=1}^{N} \hat{g}_{nm}(t)[x_m(t) - \hat{x}_m(t)] \tag{9-9}$$

where only the zeroth- and first-degree terms are retained. The time-varying parameters $\hat{g}_n(t)$ and $\hat{g}_{nm}(t)$ generally are dependent upon the optimum trajectory chosen and are given by

$$\hat{g}_n(t) = P_n[\hat{\mathbf{x}}(t), t] \tag{9-10}$$

and

$$\hat{g}_{nm}(t) = \frac{\partial P_n[\hat{\mathbf{x}}(t), t]}{\partial \hat{x}_m(t)} \tag{9-11}$$

If the system is on the optimum trajectory so that $\mathbf{x}(t) = \hat{\mathbf{x}}(t)$, then the control signal is given by (9-7) and the system remains on the optimum trajectory. On the other hand, if $\mathbf{x}(t) \approx \hat{\mathbf{x}}(t)$, then (9-9) gives an approximation to the optimum control signal where the errors are of second degree in the various components of the state vector. Therefore, the control equation given in (9-9) provides a basis for constructing a quasi-optimum feedback control system for small perturbations due to initial-condition errors and other disturbances. Because $\hat{\mathbf{x}}(t)$ is considered to be a known function of time by definition, the Taylor series approximation generally is rewritten as a power series such that

$$m_n(t) \approx \hat{G}_n(t) - \sum_{m=1}^{N} \hat{G}_{nm}(t) x_m(t) \tag{9-12}$$

where

$$\hat{G}_n(t) = \hat{g}_n(t) - \sum_{m=1}^{N} \hat{g}_{nm}(t) \hat{x}_m(t) \tag{9-13}$$

and

$$\hat{G}_{nm}(t) = -\hat{g}_{nm}(t) \tag{9-14}$$

The form of the quasi-optimum control equation given in (9-12) generally is preferable to (9-9) because system construction is simplified. Specifically, the measurement or computation of the errors in the state-vector components requires extra equipment and is sensitive to noise.

Even though the power-series expansion of the optimum control equation is a logical synthesis method for small regions of state space, a slightly different method is pursued here. In particular, approximation methods are applied to the minimum-error function instead of to the optimum control equation. For a first-order dynamic process, this procedure is illustrated in Fig. 9-1. The procedure involves the construction of a surface $S[x, \{\hat{p}(y)\}]$ which approximates the minimum-error function $E(x, y)$ in some neighborhood $\hat{\mathcal{N}}(y)$, involving $x(y)$, about the optimum trajectory $\hat{x}(y)$. The set $\{\hat{p}(y)\}$ are parameters appearing in the mathematical form taken for the approximating surface S. Presumably, these parameters are computed along the optimum trajectory $\hat{x}(t)$ and hence are functions of time.

This procedure of approximating the minimum-error function possesses the conceptual advantage that the general shape of the minimum-error function is known a priori. This information, such as convexity, may be helpful in judiciously selecting S to give the required accuracy in the approximation with a minimum number of parameters to compute. Furthermore, this procedure may yield greater accuracy than the direct expansion of the optimum control equation.

For a first-order dynamic process, the optimum control equation can be written in the alternative form

$$m(t) = P\left\{x(t), \frac{\partial E[x(t), t]}{\partial x(t)}, t\right\} \tag{9-15}$$

where the direct dependence of $m(t)$ on the minimum-error function is indicated. For the situation where

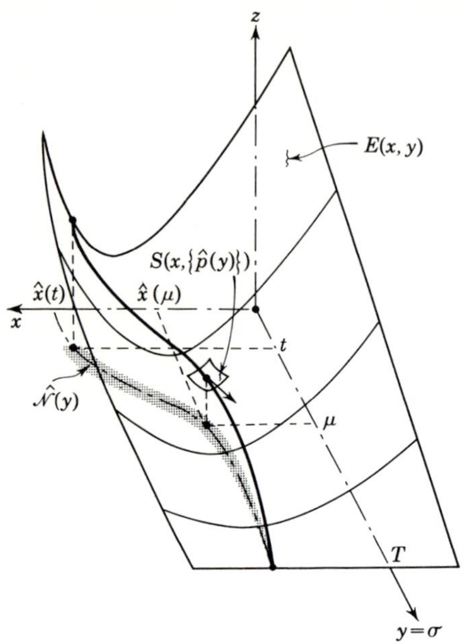

Fig. 9-1 Construction of the approximating surface S.

$$\frac{\partial E[x(t), t]}{\partial x(t)} \approx \frac{\partial S[x(t), \{\hat{p}(t)\}]}{\partial x(t)} \tag{9-16}$$

in the region $x(t) \in \hat{\mathcal{N}}(t)$, then the quasi-optimum control system is defined by

$$m(t) \approx P\left\{x(t), \frac{\partial S[x(t), \{\hat{p}(t)\}]}{\partial x(t)}, t\right\} \tag{9-17}$$

When this method of approximation is used, there is no need to approximate the function P, which would introduce additional errors. These additional errors can be very substantial in some design problems. In any event, (9-9) can be obtained from (9-17) if desired by expanding the function P in a Taylor series.

9-2 Review of the characteristic equations and the first-degree expansion of the minimum-error function

The characteristic equations derived in Chap. 6 play an important role in the synthesis procedure proposed here for small regions of state space. First, the parameters of the surface S are to be computed along an optimum trajectory $\hat{\mathbf{x}}(\mu)$ which is also a characteristic curve. Furthermore, the derivation of the characteristic equations along classical lines requires the construction of an elementary plane which also is the simplest form for S. The purpose of this section is to rederive the characteristic equations in a manner that can be extended readily to more complicated approximations of the minimum-error function.

The dynamic-programming equation for the minimization problem defined by (9-3) through (9-5) is written as

$$\frac{\partial E[\mathbf{x}(\mu), \mu]}{\partial \mu} + \min_{\mathbf{m}(\mu) \in \mathcal{M}(\mu)} \left\{H[\mathbf{x}(\mu), \mathbf{m}(\mu), \mu] + \sum_{n=1}^{N} f_n[\mathbf{x}(\mu), \mathbf{m}(\mu), \mu] \frac{\partial E[\mathbf{x}(\mu), \mu]}{\partial x_n(\mu)}\right\} = 0 \tag{9-18}$$

Here, the hamiltonian function is defined for convenience as

$$\mathscr{H}[\mathbf{x}(\mu), \mathbf{m}(\mu), \mu] = H[\mathbf{x}(\mu), \mathbf{m}(\mu), \mu] + \sum_{n=1}^{N} f_n[\mathbf{x}(\mu), \mathbf{m}(\mu), \mu] \frac{\partial E[\mathbf{x}(\mu), \mu]}{\partial x_n(\mu)} \tag{9-19}$$

which is somewhat different from the definition used in (6-60). Also, the notation

$$\mathscr{H}^*[\mathbf{x}(\mu), \mu] = \min_{\mathbf{m}(\mu) \in \mathscr{M}(\mu)} \mathscr{H}[\mathbf{x}(\mu), \mathbf{m}(\mu), \mu] \tag{9-20}$$

is used here as opposed to (6-63). By definition, this function can be written as

$$\mathscr{H}^*[\mathbf{x}(\mu), \mu] = H[\mathbf{x}(\mu), \mathbf{m}^*(\mu), \mu] + \sum_{n=1}^{N} f_n[\mathbf{x}(\mu), \mathbf{m}^*(\mu), \mu] \frac{\partial E[\mathbf{x}(\mu), \mu]}{\partial x_n(\mu)} \tag{9-21}$$

The purpose of treating $\mathscr{H}^*$ as dependent upon only $\mathbf{x}(\mu)$ and μ, as opposed to also including the dependence on $\mathbf{m}^*(\mu)$ and $\mathbf{p}(\mu)$, is to eliminate the control vector $\mathbf{m}(\mu)$ from the characteristic equations. The desire is eventually to determine the dependence of the characteristic equations on $\hat{\mathbf{x}}(\mu)$, μ, and the parameters associated with the surface S. In (9-21), the control vector is treated as a specific function of $\mathbf{x}(\mu)$ and μ. From these definitions, the dynamic-programming equation given in (9-18) becomes

$$\frac{\partial E[\mathbf{x}(\mu), \mu]}{\partial \mu} + \mathscr{H}^*[\mathbf{x}(\mu), \mu] = 0 \tag{9-22}$$

which is the form of the Hamilton-Jacobi equation treated in this chapter.

The characteristic equations of (9-22), with the notation defined in (6-58) and (6-59) for the required partial derivatives, now are derived readily. Specifically,

$$x'_n(\mu) = f_n[\mathbf{x}(\mu), \mathbf{m}^*(\mu), \mu] \qquad n = 1, 2, \ldots, N \tag{9-23}$$

and

$$p'_n(\mu) = -\frac{\partial \mathscr{H}^*[\mathbf{x}(\mu), \mu]}{\partial x_n(\mu)} \qquad n = 1, 2, \ldots, N \tag{9-24}$$

are obtained from (6-53) and (6-54). Equations (6-65) and (9-23) are identical. Furthermore, (6-66) and (9-24) are shown to be identical by expanding the partial derivative appearing in (9-24) so that

$$\begin{aligned}\frac{\partial \mathscr{H}^*[\mathbf{x}(\mu), \mu]}{\partial x_n(\mu)} &= \frac{\partial H[\mathbf{x}(\mu), \mathbf{m}^*(\mu), \mu]}{\partial x_n(\mu)} + \sum_{m=1}^{N} \frac{\partial f_m[\mathbf{x}(\mu), \mathbf{m}^*(\mu), \mu]}{\partial x_n(\mu)} p_m(\mu) \\ &\quad + \sum_{j=1}^{M} \left\{ \frac{\partial H[\mathbf{x}(\mu), \mathbf{m}^*(\mu), \mu]}{\partial m_j^*(\mu)} + \sum_{m=1}^{N} \frac{\partial f_m[\mathbf{x}(\mu), \mathbf{m}^*(\mu), \mu]}{\partial m_j^*(\mu)} p_m(\mu) \right\} \frac{\partial m_j^*(\mu)}{\partial x_n(\mu)}\end{aligned} \tag{9-25}$$

However, the minimization process implied by (9-20) yields

$$\begin{aligned}\frac{\partial H[\mathbf{x}(\mu), \mathbf{m}^*(\mu), \mu]}{\partial m_j^*(\mu)} + \sum_{m=1}^{N} \frac{\partial f_m[\mathbf{x}(\mu), \mathbf{m}^*(\mu), \mu]}{\partial m_j^*(\mu)} p_m(\mu) = 0 &\qquad m_j^*(\mu) \text{ not on boundary of } \mathscr{M}(\mu) \\ \frac{\partial m_j^*(\mu)}{\partial x_n(\mu)} = 0 &\qquad m_j^*(\mu) \text{ on boundary of } \mathscr{M}(\mu)\end{aligned} \tag{9-26}$$

Therefore, (9-24) reduces to

$$p'_n(\mu) = -\left\{ \frac{\partial H[\mathbf{x}(\mu), \mathbf{m}^*(\mu), \mu]}{\partial x_n(\mu)} + \sum_{m=1}^{N} \frac{\partial f_m[\mathbf{x}(\mu), \mathbf{m}^*(\mu), \mu]}{\partial x_n(\mu)} p_m(\mu) \right\} \tag{9-27}$$

which is precisely the partial derivative used in (6-66).

In this section, the approximating properties of the characteristic equations to the minimum-error function are investigated along a particular characteristic curve $\hat{\mathbf{x}}(\mu)$. The equations defining this characteristic curve are found by evaluating (9-23) and (9-24) along this trajectory so that

$$\hat{x}_n'(\mu) = f_n[\hat{\mathbf{x}}(\mu), \hat{\mathbf{m}}(\mu), \mu] \qquad n = 1, 2, \ldots, N \tag{9-28}$$

and

$$\hat{p}_n'(\mu) = -\left.\frac{\partial \mathscr{H}^*[\mathbf{x}(\mu), \mu]}{\partial x_n(\mu)}\right|_{\mathbf{x}(\mu)=\hat{\mathbf{x}}(\mu)} \qquad n = 1, 2, \ldots, N \tag{9-29}$$

Suppose that an elementary plane E_1 is defined as

$$E_1[\mathbf{x}(\mu), \mu] = \hat{p}_0(\mu) + \sum_{n=1}^{N} \hat{p}_n(\mu)[x_n(\mu) - \hat{x}_n(\mu)] \tag{9-30}$$

where

$$\hat{p}_0(\mu) = E[\hat{\mathbf{x}}(\mu), \mu] \tag{9-31}$$

and

$$\hat{p}_n(\mu) = \left.\frac{\partial E[\mathbf{x}(\mu), \mu]}{\partial x_n(\mu)}\right|_{\mathbf{x}(\mu)=\hat{\mathbf{x}}(\mu)} \tag{9-32}$$

Then the minimum-error function can be written as

$$E[\mathbf{x}(\mu), \mu] = E_1[\mathbf{x}(\mu), \mu] + \sum_{n=1}^{N}\sum_{m=1}^{N} \hat{P}_{nm}(\mu)[x_n(\mu) - \hat{x}_n(\mu)][x_m(\mu) - \hat{x}_m(\mu)] \tag{9-33}$$

by use of Taylor's remainder formula,[65] where

$$\hat{P}_{nm}(\mu) = \hat{P}_{mn}(\mu) = \frac{1}{2}\frac{\partial^2 E[\boldsymbol{\xi}(\mu), \mu]}{\partial \xi_n(\mu)\,\partial \xi_m(\mu)} \tag{9-34}$$

and

$$\hat{x}_n(\mu) < \xi_n(\mu) < x_n(\mu) \tag{9-35}$$

The elementary plane E_1 is the form of the approximating surface S used in this section and is a first-degree Taylor series expansion of the minimum-error function about the point $\hat{\mathbf{x}}(\mu)$. Throughout this section, the tacit assumption is made that $\mathbf{x}(\mu) \in \hat{\mathscr{N}}_1(\mu)$ so that $\hat{\mathscr{N}}_1(\mu)$ is a neighborhood of $\hat{\mathbf{x}}(\mu)$ where the required derivatives exist and the required expansions converge uniformly. In terms of E_1, the partial derivatives of the minimum-error function required in the Hamilton-Jacobi equation are§

$$\frac{\partial E[\mathbf{x}(\mu), \mu]}{\partial \mu} = \frac{\partial E_1[\mathbf{x}(\mu), \mu]}{\partial \mu} - 2\sum_{n=1}^{N}\sum_{m=1}^{N} \hat{P}_{nm}(\mu)\hat{x}_n'(\mu)[x_m(\mu) - \hat{x}_m(\mu)]$$
$$+ \sum_{n=1}^{N}\sum_{m=1}^{N} \hat{P}_{nm}'(\mu)[x_n(\mu) - \hat{x}_n(\mu)][x_m(\mu) - \hat{x}_m(\mu)] \tag{9-36}$$

and

$$\frac{\partial E[\mathbf{x}(\mu), \mu]}{\partial x_n(\mu)} = \frac{\partial E_1[\mathbf{x}(\mu), \mu]}{\partial x_n(\mu)} + 2\sum_{m=1}^{N} \hat{P}_{nm}(\mu)[x_m(\mu) - \hat{x}_m(\mu)]$$
$$+ \sum_{m=1}^{N}\sum_{k=1}^{N} \hat{P}_{nmk}(\mu)[x_m(\mu) - \hat{x}_m(\mu)][x_k(\mu) - \hat{x}_k(\mu)] \tag{9-37}$$

§ The functions $\hat{P}'_{nm}(\mu)$ and $\hat{P}_{nmk}(\mu)$ are defined as

$$\hat{P}'_{nm}(\mu) = \frac{\partial \hat{P}_{nm}(\mu)}{\partial \mu} = \frac{1}{2}\frac{\partial^3 E[\boldsymbol{\xi}(\mu), \mu]}{\partial \mu\,\partial \xi_n(\mu)\,\partial \xi_m(\mu)} + \sum_{i=1}^{N}\sum_{j=1}^{N} \frac{1}{2}\frac{\partial^3 E[\boldsymbol{\xi}(\mu), \mu]}{\partial \xi_i(\mu)\,\partial \xi_n(\mu)\,\partial \xi_m(\mu)}\frac{\partial \xi_i(\mu)}{\partial \hat{x}_j(\mu)}\hat{x}_j'(\mu)$$

$$\hat{P}_{nmk}(\mu) = \frac{\partial \hat{P}_{mk}(\mu)}{\partial x_n(\mu)} = \sum_{i=1}^{N} \frac{1}{2}\frac{\partial^3 E[\boldsymbol{\xi}(\mu), \mu]}{\partial \xi_i(\mu)\,\partial \xi_m(\mu)\,\partial \xi_k(\mu)}\frac{\partial \xi_i(\mu)}{\partial x_n(\mu)}$$

Now suppose that a *truncated hamiltonian function* is defined as

$$\mathscr{H}_1[\mathbf{x}(\mu), \mathbf{m}(\mu), \mu] = H[\mathbf{x}(\mu), \mathbf{m}(\mu), \mu] + \sum_{n=1}^{N} f_n[\mathbf{x}(\mu), \mathbf{m}(\mu), \mu] \frac{\partial E_1[\mathbf{x}(\mu), \mu]}{\partial x_n(\mu)} \tag{9-38}$$

In terms of (9-38), the approximate value of the optimum control vector $\mathbf{m}^\dagger(\mu)$ is defined in terms of the minimization process

$$\mathscr{H}_1^\dagger[\mathbf{x}(\mu), \mu] = \min_{\mathbf{m}(\mu) \in \mathscr{M}(\mu)} \mathscr{H}_1[\mathbf{x}(\mu), \mathbf{m}(\mu), \mu] \tag{9-39}$$

Furthermore, the function $\mathscr{H}_1^\dagger$ is written in terms of $\mathbf{m}^\dagger(\mu)$ as

$$\mathscr{H}_1^\dagger[\mathbf{x}(\mu), \mu] = H[\mathbf{x}(\mu), \mathbf{m}^\dagger(\mu), \mu] + \sum_{n=1}^{N} f_n[\mathbf{x}(\mu), \mathbf{m}^\dagger(\mu), \mu] \frac{\partial E_1[\mathbf{x}(\mu), \mu]}{\partial x_n(\mu)} \tag{9-40}$$

from the previous definitions. In order to determine the relationship between $\mathscr{H}_1^\dagger$ and E_1 that is imposed by the Hamilton-Jacobi equation, the function $\mathscr{H}^*$ must be expressed in terms of $\mathscr{H}_1^\dagger$. This relationship is derived in Appendix G by the use of appropriate Taylor series expansions in the neighborhood $\hat{\mathscr{N}}_1(\mu)$ of the optimum trajectory $\hat{\mathbf{x}}(\mu)$. Specifically, the relationship

$$\begin{aligned}\mathscr{H}^*[\mathbf{x}(\mu), \mu] = \mathscr{H}_1^\dagger[\mathbf{x}(\mu), \mu] &+ 2\sum_{n=1}^{N} \sum_{m=1}^{N} \hat{P}_{nm}(\mu)\hat{x}_n'(\mu)[x_m(\mu) - \hat{x}_m(\mu)] \\ &+ \sum_{n=1}^{N} \sum_{m=1}^{N} \hat{Q}_{nm}(\mu)[x_n(\mu) - \hat{x}_n(\mu)][x_m(\mu) - \hat{x}_m(\mu)]\end{aligned} \tag{9-41}$$

is obtained.

Equations (9-36) and (9-41) now are substituted into the Hamilton-Jacobi equation. From this substitution, (9-22) becomes§

$$\begin{aligned}\frac{\partial E_1[\mathbf{x}(\mu), \mu]}{\partial \mu} &+ \mathscr{H}_1^\dagger[\mathbf{x}(\mu), \mu] \\ &+ \sum_{n=1}^{N} \sum_{m=1}^{N} [\hat{P}_{nm}'(\mu) + \hat{Q}_{nm}(\mu)][x_n(\mu) - \hat{x}_n(\mu)][x_m(\mu) - \hat{x}_m(\mu)] = 0\end{aligned} \tag{9-42}$$

From the definition given in (9-30), the partial derivative $\partial E_1/\partial \mu$ involves only zeroth- and first-degree terms in $[x_n(\mu) - \hat{x}_n(\mu)]$. In order to collect all the zeroth- and first-degree terms in (9-42), the function $H_1^\dagger$ is defined as

$$H_1^\dagger[\mathbf{x}(\mu), \mu] = \hat{H}_0(\mu) + \sum_{n=1}^{N} \hat{H}_n(\mu)[x_n(\mu) - \hat{x}_n(\mu)] \tag{9-43}$$

where

$$\hat{H}_0(\mu) = \mathscr{H}_1^\dagger[\hat{\mathbf{x}}(\mu), \mu] \tag{9-44}$$

and

$$\hat{H}_n(\mu) = \left.\frac{\partial \mathscr{H}_1^\dagger[\mathbf{x}(\mu), \mu]}{\partial x_n(\mu)}\right|_{\mathbf{x}(\mu)=\hat{\mathbf{x}}(\mu)} \qquad n = 1, 2, \ldots, N \tag{9-45}$$

Then the function $\mathscr{H}_1^\dagger$ is written as

$$\mathscr{H}_1^\dagger[\mathbf{x}(\mu), \mu] = H_1^\dagger[\mathbf{x}(\mu), \mu] + \sum_{n=1}^{N} \sum_{m=1}^{N} \hat{O}_{nm}(\mu)[x_n(\mu) - \hat{x}_n(\mu)][x_m(\mu) - \hat{x}_m(\mu)] \tag{9-46}$$

§ The key to this derivation is the cancellation of the second terms in (9-36) and (9-41).

in the neighborhood $\hat{\mathscr{N}}_1(\mu)$, again using Taylor's remainder formula. Finally, the Hamilton-Jacobi equation, when written in terms of E_1 and $H_1^\dagger$, becomes

$$\frac{\partial E_1[\mathbf{x}(\mu), \mu]}{\partial \mu} + H_1[\mathbf{x}(\mu), \mu] + \sum_{n=1}^{N} \sum_{m=1}^{N} [\hat{O}_{nm}(\mu) + \hat{P}'_{nm}(\mu) + \hat{Q}_{nm}(\mu)][x_n(\mu) - \hat{x}_n(\mu)][x_m(\mu) - \hat{x}_m(\mu)] = 0 \quad (9\text{-}47)$$

The first two terms in (9-47) involve only zeroth- and first-degree terms in $[x_n(\mu) - \hat{x}_n(\mu)]$; hence the neighborhood $\hat{\mathscr{N}}_1(\mu)$ can be chosen to be suitably small so that the second-degree remainder term can be neglected. Under this condition, the *first-degree expansion of the Hamilton-Jacobi equation* is

$$\frac{\partial E_1[\mathbf{x}(\mu), \mu]}{\partial \mu} + H_1^\dagger[\mathbf{x}(\mu), \mu] = 0 \qquad \mathbf{x}(\mu) \in \hat{\mathscr{N}}_1(\mu) \quad (9\text{-}48)$$

where E_1 and $H_1^\dagger$ are defined in (9-30) and (9-43), respectively.

The ordinary differential equations defining the optimum trajectory $\hat{\mathbf{x}}(\mu)$ are derived readily from (9-48). From (9-30), the partial derivative $\partial E_1/\partial \mu$ is found to be

$$\frac{\partial E_1[\mathbf{x}(\mu), \mu]}{\partial \mu} = \left[\hat{p}'_0(\mu) - \sum_{n=1}^{N} \hat{p}_n(\mu)\hat{x}'_n(\mu)\right] + \sum_{n=1}^{N} \hat{p}'_n(\mu)[x_n(\mu) - \hat{x}_n(\mu)] \quad (9\text{-}49)$$

Also, the partial derivative $\partial E_1/\partial x_n$ is found to be

$$\frac{\partial E_1[\mathbf{x}(\mu), \mu]}{\partial x_n(\mu)} = \hat{p}_n(\mu) \quad (9\text{-}50)$$

so that (9-40) becomes

$$\mathscr{H}_1^\dagger[\mathbf{x}(\mu), \mu] = H[\mathbf{x}(\mu), \mathbf{m}^\dagger(\mu), \mu] + \sum_{n=1}^{N} f_n[\mathbf{x}(\mu), \mathbf{m}^\dagger(\mu), \mu]\, \hat{p}_n(\mu) \quad (9\text{-}51)$$

When (9-51) is evaluated at $\hat{\mathbf{x}}(\mu)$, the result is

$$\mathscr{H}_1^\dagger[\hat{\mathbf{x}}(\mu), \mu] = H[\hat{\mathbf{x}}(\mu), \hat{\mathbf{m}}(\mu), \mu] + \sum_{n=1}^{N} \hat{x}'_n(\mu)\hat{p}_n(\mu) \quad (9\text{-}52)$$

due to (9-28) and the property

$$\hat{\mathbf{m}}^\dagger(\mu) = \hat{\mathbf{m}}^*(\mu) = \hat{\mathbf{m}}(\mu) \quad (9\text{-}53)$$

which is obtained from Appendix G. Therefore, (9-49) is rewritten as

$$\frac{\partial E_1[\mathbf{x}(\mu), \mu]}{\partial \mu} = \{\hat{p}'_0(\mu) + H[\hat{\mathbf{x}}(\mu), \hat{\mathbf{m}}(\mu), \mu] - \hat{H}_0(\mu)\} + \sum_{n=1}^{N} \hat{p}'_n(\mu)[x_n(\mu) - \hat{x}_n(\mu)] \quad (9\text{-}54)$$

from (9-44) and (9-52). If (9-43) and (9-54) are substituted into (9-48), then the result is

$$\{\hat{p}'_0(\mu) + H[\hat{\mathbf{x}}(\mu), \hat{\mathbf{m}}(\mu), \mu]\} + \sum_{n=1}^{N} [\hat{p}'_n(\mu) + \hat{H}_n(\mu)][x_n(\mu) - \hat{x}_n(\mu)] = 0 \quad (9\text{-}55)$$

Finally, this equation is satisfied for all $\mathbf{x}(\mu)$ in the neighborhood $\hat{\mathscr{N}}_1(\mu)$ only if all terms appearing in (9-55) vanish independently of $\mathbf{x}(\mu)$ so that

$$\hat{p}'_0(\mu) = -H[\hat{\mathbf{x}}(\mu), \hat{\mathbf{m}}(\mu), \mu] \quad (9\text{-}56)$$

and

$$\hat{p}'_n(\mu) = -\left.\frac{\partial \mathscr{H}_1^\dagger[\mathbf{x}(\mu), \mu]}{\partial x_n(\mu)}\right|_{\mathbf{x}(\mu)=\hat{\mathbf{x}}(\mu)} \qquad n = 1, 2, \ldots, N \quad (9\text{-}57)$$

when the definition given in (9-45) is used. Equation (9-56) is precisely the characteristic equation given in (6-75). Furthermore, (9-28) and (9-57) are the characteristic equations that define the optimum trajectory. The equivalence of (9-29) and (9-57) is shown by differentiating (9-40). When the property which is analogous to (9-26) but for $\mathbf{m}^\dagger(\mu)$ is used, this differentiation yields

$$\frac{\partial \mathscr{H}_1^\dagger[\mathbf{x}(\mu), \mu]}{\partial x_n(\mu)} = \frac{\partial H[\mathbf{x}(\mu), \mathbf{m}^\dagger(\mu), \mu]}{\partial x_n(\mu)} + \sum_{m=1}^{N} \frac{\partial f_m[\mathbf{x}(\mu), \mathbf{m}^\dagger(\mu), \mu]}{\partial x_n(\mu)} \hat{p}_m(\mu) \qquad (9\text{-}58)$$

The evaluation of (9-58) at $\mathbf{x}(\mu) = \hat{\mathbf{x}}(\mu)$, plus the equality given in (9-53), demonstrates this equivalence to (9-27) when evaluated on the optimum trajectory $\hat{\mathbf{x}}(\mu)$.

Therefore, the first-degree expansion of the Hamilton-Jacobi equation yields precisely the characteristic equations presented in Chap. 6. The advantage of developing the characteristic equations in terms of approximations to the minimum-error function is that the method extends directly to higher-degree expansions. On the other hand, extensions of the classical theory presented in Chap. 6 involve considerable mathematical complexity. Furthermore, the approximating properties of the characteristic equations are clarified by this method. In particular, the first-degree expansion of the minimum-error function does not give rise to an extrapolation of the dependence of $\mathbf{m}^\dagger(\mu)$ on the minimum-error function in the neighborhood $\mathscr{N}_1(\mu)$, because $\mathbf{m}^\dagger(\mu)$ is dependent upon $\partial E_1/\partial x_n$. Therefore, higher-degree expansions of the minimum-error function are required for the synthesis of feedback control systems.

9-3 Second-degree expansion of the minimum-error function

The second-degree expansion involves the construction of an elementary quadratic at the point $\hat{\mathbf{x}}(\mu)$. This elementary quadratic is written as

$$\begin{aligned} E_2[\mathbf{x}(\mu), \mu] = \hat{p}_0(\mu) &+ \sum_{n=1}^{N} \hat{p}_n(\mu)[x_n(\mu) - \hat{x}_n(\mu)] \\ &+ \sum_{n=1}^{N} \sum_{m=1}^{N} \hat{p}_{nm}(\mu)[x_n(\mu) - \hat{x}_n(\mu)][x_m(\mu) - \hat{x}_m(\mu)] \qquad (9\text{-}59) \end{aligned}$$

where

$$p_0(\mu) = E[\hat{\mathbf{x}}(\mu), \mu] \qquad (9\text{-}60)$$

$$\hat{p}_n(\mu) = \left.\frac{\partial E[\mathbf{x}(\mu), \mu]}{\partial x_n(\mu)}\right|_{\mathbf{x}(\mu)=\hat{\mathbf{x}}(\mu)} \qquad (9\text{-}61)$$

$$\hat{p}_{nm}(\mu) = \hat{p}_{mn}(\mu) = \left.\frac{1}{2}\frac{\partial^2 E[\mathbf{x}(\mu), \mu]}{\partial x_n(\mu)\,\partial x_m(\mu)}\right|_{\mathbf{x}(\mu)=\hat{\mathbf{x}}(\mu)} \qquad (9\text{-}62)$$

Furthermore, the truncated form of the hamiltonian function used here is written as

$$\mathscr{H}_2^\dagger[\mathbf{x}(\mu), \mu] = \min_{\mathbf{m}(\mu)\in\mathscr{M}(\mu)} \left\{ H[\mathbf{x}(\mu), \mathbf{m}(\mu), \mu] + \sum_{n=1}^{N} f_n[\mathbf{x}(\mu), \mathbf{m}(\mu), \mu] \frac{\partial E_2[\mathbf{x}(\mu), \mu]}{\partial x_n(\mu)} \right\} \qquad (9\text{-}63)$$

The minimization process expressed in (9-63) defines the quasi-optimum control vector $\mathbf{m}^\dagger(\mu)$, for $\mathbf{x}(\mu) \in \mathscr{N}_2(\mu)$, which is a function of $\mathbf{x}(\mu)$, $\hat{\mathbf{x}}(\mu)$, μ, $\hat{p}_n(\mu)$, and

$\hat{p}_{nm}(\mu)$. Finally, the second-degree approximation of $\mathscr{H}_2^\dagger[\mathbf{x}(\mu), \mu]$ is defined as

$$H_2^\dagger[\mathbf{x}(\mu), \mu] = \hat{H}_0(\mu) + \sum_{n=1}^{N} \hat{H}_n(\mu)[x_n(\mu) - \hat{x}_n(\mu)] + \sum_{n=1}^{N} \sum_{m=1}^{N} \hat{H}_{nm}(\mu)[x_n(\mu) - \hat{x}_n(\mu)][x_m(\mu) - \hat{x}_m(\mu)] \tag{9-64}$$

where

$$\hat{H}_0(\mu) = \mathscr{H}_2^\dagger[\hat{\mathbf{x}}(\mu), \mu] \tag{9-65}$$

$$\hat{H}_n(\mu) = \left.\frac{\partial \mathscr{H}_2^\dagger[\mathbf{x}(\mu), \mu]}{\partial x_n(\mu)}\right|_{\mathbf{x}(\mu)=\hat{\mathbf{x}}(\mu)} \tag{9-66}$$

$$\hat{H}_{nm}(\mu) = \hat{H}_{mn}(\mu) = \left.\frac{1}{2}\frac{\partial^2 \mathscr{H}_2^\dagger[\mathbf{x}(\mu), \mu]}{\partial x_n(\mu)\, \partial x_m(\mu)}\right|_{\mathbf{x}(\mu)=\hat{\mathbf{x}}(\mu)} \tag{9-67}$$

The definitions introduced in (9-59), (9-63), and (9-64) are equivalent to those introduced in (9-30), (9-39), and (9-43), respectively, for the first-degree expansion.

If the Taylor-series remainder terms in the expansions of the minimum-error and the hamiltonian functions are introduced as in Sec. 9-2, then the Hamilton-Jacobi equation given in (9-22) reduces to

$$\frac{\partial E_2[\mathbf{x}(\mu), \mu]}{\partial \mu} + \mathscr{H}_2^\dagger[\mathbf{x}(\mu), \mu] + \{\text{third-degree terms in } [x_n(\mu) - \hat{x}_n(\mu)]\} = 0 \tag{9-68}$$

This equation is analogous to (9-42) for the first-degree expansion. Also, a cancellation of second-degree terms occurs when obtaining (9-68), which is analogous to the cancellation of first-degree terms when obtaining (9-42). Now, the expansion of $\mathscr{H}_2^\dagger$ which appears in (9-68) yields

$$\frac{\partial E_2[\mathbf{x}(\mu), \mu]}{\partial \mu} + H_2^\dagger[\mathbf{x}(\mu), \mu] + \{\text{third-degree terms in } [x_n(\mu) - \hat{x}_n(\mu)]\} = 0 \tag{9-69}$$

Finally, the *second-degree expansion of the Hamilton-Jacobi equation*

$$\frac{\partial E_2[\mathbf{x}(\mu), \mu]}{\partial \mu} + H_2^\dagger[\mathbf{x}(\mu), \mu] = 0 \qquad \mathbf{x}(\mu) \in \mathscr{N}_2(\mu) \tag{9-70}$$

is obtained because the first two terms in (9-69) involve only zeroth-, first-, and second-degree terms in $[x_n(\mu) - \hat{x}_n(\mu)]$. This result is analogous to the result obtained in (9-48) for the first-degree expansion of the Hamilton-Jacobi equation.

The explicit ordinary differential equations for the $\hat{p}$ parameters appearing in $E_2[\mathbf{x}(\mu), \mu]$ now are found by evaluating the appropriate partial derivatives. Specifically, these derivatives are

$$\begin{aligned}\frac{\partial E_2[\mathbf{x}(\mu), \mu]}{\partial \mu} &= \left[\hat{p}_0'(\mu) - \sum_{n=1}^{N} \hat{p}_n(\mu)\hat{x}_n'(\mu)\right] \\ &\quad + \sum_{n=1}^{N}\left[\hat{p}_n'(\mu) - 2\sum_{m=1}^{N} \hat{p}_{nm}(\mu)\hat{x}_m'(\mu)\right][x_n(\mu) - \hat{x}_n(\mu)] \\ &\quad + \sum_{n=1}^{N}\sum_{m=1}^{N} \hat{p}_{nm}'(\mu)[x_n(\mu) - \hat{x}_n(\mu)][x_m(\mu) - \hat{x}_m(\mu)]\end{aligned} \tag{9-71}$$

$$\frac{\partial E_2[\mathbf{x}(\mu), \mu]}{\partial x_n(\mu)} = \hat{p}_n(\mu) + 2\sum_{m=1}^{N} \hat{p}_{nm}(\mu)[x_m(\mu) - \hat{x}_m(\mu)] \tag{9-72}$$

When (9-64) and (9-71) are substituted in (9-70) and terms of similar degree in $[x_n(\mu) - \hat{x}_n(\mu)]$ are collected, the result is

$$\{\hat{p}_0'(\mu) + H[\mathbf{x}(\mu), \hat{\mathbf{m}}(\mu), \mu]\}$$
$$+ \sum_{n=1}^{N} \left[\hat{p}_n'(\mu) + \hat{H}_n(\mu) - 2\sum_{m=1}^{N} \hat{p}_{nm}(\mu)\hat{x}_m'(\mu)\right][x_n(\mu) - \hat{x}_n(\mu)]$$
$$+ \sum_{n=1}^{N}\sum_{m=1}^{N} [\hat{p}_{nm}'(\mu) + \hat{H}_{nm}(\mu)][x_n(\mu) - \hat{x}_n(\mu)][x_m(\mu) - \hat{x}_m(\mu)] = 0 \quad (9\text{-}73)$$

In order to satisfy (9-73) for all $\mathbf{x}(\mu) \in \hat{\mathcal{N}}_2(\mu)$,

$$\hat{p}_0'(\mu) = -H[\hat{\mathbf{x}}(\mu), \hat{\mathbf{m}}(\mu), \mu] \quad (9\text{-}74)$$

$$\hat{p}_n'(\mu) = -\left.\frac{\partial \mathscr{H}_2^{\dagger}[\mathbf{x}(\mu), \mu]}{\partial x_n(\mu)}\right|_{\mathbf{x}(\mu)=\hat{\mathbf{x}}(\mu)} + 2\sum_{m=1}^{N} \hat{p}_{nm}(\mu)\hat{x}_m'(\mu) \quad (9\text{-}75)$$

$$\hat{p}_{nm}'(\mu) = -\left.\frac{1}{2}\frac{\partial^2 \mathscr{H}_2^{\dagger}[\mathbf{x}(\mu), \mu]}{\partial x_n(\mu)\,\partial x_m(\mu)}\right|_{\mathbf{x}(\mu)=\hat{\mathbf{x}}(\mu)} \quad (9\text{-}76)$$

are obtained when (9-66) and (9-67) are introduced. Also, these $\hat{p}$ parameters are subject to the boundary conditions

$$\hat{p}_0(T) = \hat{p}_n(T) = \hat{p}_{nm}(T) = 0 \quad (9\text{-}77)$$

due to the free-point terminal-boundary condition $E[\mathbf{x}(T), T] = 0$. Finally, (9-63) and (9-72) give

$$\mathscr{H}_2^{\dagger}[\mathbf{x}(\mu), \mu] = \min_{\mathbf{m}(\mu) \in \mathscr{M}(\mu)} \Bigg(H[\mathbf{x}(\mu), \mathbf{m}(\mu), \mu]$$
$$+ \sum_{n=1}^{N} f_n[\mathbf{x}(\mu), \mathbf{m}(\mu), \mu]\left\{\hat{p}_n(\mu) + 2\sum_{m=1}^{N} \hat{p}_{nm}(\mu)[x_m(\mu) - \hat{x}_m(\mu)]\right\}\Bigg) \quad (9\text{-}78)$$

and the optimum trajectory is given by

$$\hat{x}_n'(\mu) = f_n[\hat{\mathbf{x}}(\mu), \hat{\mathbf{m}}(\mu), \mu] \quad (9\text{-}79)$$

These last five equations completely define the solution of the $\hat{p}$ parameters that appear in the quasi-optimum control equation for $\mathbf{m}^{\dagger}(\mu)$ when $\mathbf{x}(\mu) \in \hat{\mathcal{N}}_2(\mu)$.

An interesting and very important aspect of this method of expanding the minimum-error function is that the expansion can be expressed in a number of forms. This flexibility is a useful attribute in the construction of the quasi-optimum control equation. The alternative form introduced here is the power-series form of the elementary quadratic, which now is written as

$$E_2[\mathbf{x}(\mu), \mu] = \hat{k}(\mu) - 2\sum_{n=1}^{N} \hat{k}_n(\mu)x_n(\mu) + \sum_{n=1}^{N}\sum_{m=1}^{N} \hat{k}_{nm}(\mu)x_n(\mu)x_m(\mu) \quad (9\text{-}80)$$

where

$$\hat{k}(\mu) = \hat{p}_0(\mu) - \sum_{n=1}^{N} \hat{p}_n(\mu)\hat{x}_n(\mu) + \sum_{n=1}^{N}\sum_{m=1}^{N} \hat{p}_{nm}(\mu)\hat{x}_n(\mu)\hat{x}_m(\mu) \quad (9\text{-}81)$$

$$\hat{k}_n(\mu) = -\frac{1}{2}\hat{p}_n(\mu) + \sum_{m=1}^{N} \hat{p}_{nm}(\mu)\hat{x}_m(\mu) \quad (9\text{-}82)$$

$$\hat{k}_{nm}(\mu) = \hat{p}_{nm}(\mu) \quad (9\text{-}83)$$

The quadratic forms given in (9-59) and (9-80) are identical.

If both sides of (9-81) to (9-83) are differentiated with respect to μ and the appropriate substitutions are made, the $\hat{k}$ parameters are given by

$$-\hat{k}'(\mu) = \left\{\mathscr{H}_2^\dagger[\mathbf{x}(\mu), \mu] - \sum_{n=1}^{N} \frac{\partial \mathscr{H}_2^\dagger[\mathbf{x}(\mu), \mu]}{\partial x_n(\mu)} \hat{x}_n(\mu) + \frac{1}{2} \sum_{n=1}^{N} \sum_{m=1}^{N} \frac{\partial^2 \mathscr{H}_2^\dagger[\mathbf{x}(\mu), \mu]}{\partial x_n(\mu)\, \partial x_m(\mu)} \hat{x}_n(\mu)\hat{x}_m(\mu)\right\}_{\mathbf{x}(\mu)=\hat{\mathbf{x}}(\mu)} \tag{9-84}$$

$$-\hat{k}_n'(\mu) = \left\{-\frac{1}{2} \frac{\partial \mathscr{H}_2^\dagger[\mathbf{x}(\mu), \mu]}{\partial x_n(\mu)} + \frac{1}{2} \sum_{m=1}^{N} \frac{\partial^2 \mathscr{H}_2^\dagger[\mathbf{x}(\mu), \mu]}{\partial x_n(\mu)\, \partial x_m(\mu)} \hat{x}_m(\mu)\right\}_{\mathbf{x}(\mu)=\hat{\mathbf{x}}(\mu)} \tag{9-85}$$

$$-\hat{k}_{nm}'(\mu) = \frac{1}{2} \frac{\partial^2 \mathscr{H}_2^\dagger[\mathbf{x}(\mu), \mu]}{\partial x_n(\mu)\, \partial x_m(\mu)}\bigg|_{\mathbf{x}(\mu)=\hat{\mathbf{x}}(\mu)} \tag{9-86}$$

For the free-point terminal-boundary condition, the $\hat{k}$ parameters must satisfy

$$\hat{k}(T) = \hat{k}_n(T) = \hat{k}_{nm}(T) = 0 \tag{9-87}$$

Finally, the form of $\mathscr{H}_2^\dagger[\mathbf{x}(\mu), \mu]$ used in (9-84) to (9-86) is

$$\mathscr{H}_2^\dagger[\mathbf{x}(\mu), \mu] = \min_{\mathbf{m}(\mu) \in \mathscr{M}(\mu)} \left\{H[\mathbf{x}(\mu), \mathbf{m}(\mu), \mu] - 2\sum_{n=1}^{N} f_n[\mathbf{x}(\mu), \mathbf{m}(\mu), \mu]\left[\hat{k}_n(\mu) - \sum_{m=1}^{N} \hat{k}_{nm}(\mu)x_m(\mu)\right]\right\} \tag{9-88}$$

This second-degree power-series expansion of the minimum-error function has appeared previously in the literature. Early work appeared under certain restrictions which are discussed in a subsequent section of this chapter.[22,66] Also, Kipiniak develops these results from the Euler-Lagrange equation; hence control-signal saturation cannot be treated.[67] In fact, the results presented here provide a compact derivation of the equations for linear optimum controls presented in Chap. 7.

For the purposes of illustration, an example defined by

$$H[\mathbf{x}(\mu), \mathbf{m}(\mu), \mu] = \sum_{n=1}^{N} \left\{\left[\sum_{m=1}^{N} a_{nm}x_m(\mu)\right]^2 + m_n^2(\mu)\right\} \tag{9-89}$$

and

$$x_n'(\mu) = \sum_{m=1}^{N} [b_{nm}x_m(\mu)] + m_n(\mu) \tag{9-90}$$

is introduced. For this example, the minimization expressed in (9-88) yields

$$m_n^\dagger(\mu) = \hat{k}_n(\mu) - \sum_{m=1}^{N} \hat{k}_{nm}(\mu)x_m(\mu) \tag{9-91}$$

Therefore, the second-degree form of the truncated hamiltonian becomes

$$\mathscr{H}_2^\dagger[\mathbf{x}(\mu), \mu] = \sum_{n=1}^{N} \left\{\left[\sum_{m=1}^{N} a_{nm}x_m(\mu)\right]^2 - 2\left[\sum_{m=1}^{N} b_{nm}x_m(\mu)\right]\left[\hat{k}_n(\mu) - \sum_{m=1}^{N} \hat{k}_{nm}(\mu)x_m(\mu)\right] - \left[\hat{k}_n(\mu) - \sum_{m=1}^{N} \hat{k}_{nm}(\mu)x_m(\mu)\right]^2\right\} \tag{9-92}$$

Also, the partial derivatives required in (9-85) and (9-86) are

$$\frac{\partial \mathscr{H}_2^\dagger[\mathbf{x}(\mu), \mu]}{\partial x_i(\mu)} = 2\sum_{n=1}^{N} \left(a_{ni} \sum_{m=1}^{N} [a_{nm}x_m(\mu)] - b_{ni}\left[\hat{k}_n(\mu) - \sum_{m=1}^{N} \hat{k}_{nm}(\mu)x_m(\mu)\right] + \hat{k}_{ni}(\mu)\left\{\sum_{m=1}^{N} [b_{nm}x_m(\mu)] + \hat{k}_n(\mu) - \sum_{m=1}^{N} \hat{k}_{nm}(\mu)x_m(\mu)\right\}\right) \tag{9-93}$$

and

$$\frac{\partial^2 \mathscr{H}_2^\dagger[\mathbf{x}(\mu), \mu]}{\partial x_i(\mu)\, \partial x_j(\mu)} = 2\sum_{n=1}^{N} [a_{ni}a_{nj} + b_{ni}\hat{k}_{nj}(\mu) + b_{nj}\hat{k}_{ni}(\mu) - \hat{k}_{nj}(\mu)\hat{k}_{ni}(\mu)] \tag{9-94}$$

The substitution of (9-92) to (9-94) into (9-84) to (9-86) yields

$$-\hat{k}'(\mu) = \sum_{n=1}^{N} [-\hat{k}_n{}^2(\mu)] \tag{9-95}$$

$$-\hat{k}_i'(\mu) = \sum_{n=1}^{N} [b_{ni}\hat{k}_n(\mu) - \hat{k}_n(\mu)\hat{k}_{ni}(\mu)] \tag{9-96}$$

and

$$-\hat{k}_{ij}'(\mu) = \sum_{n=1}^{N} [a_{ni}a_{nj} + b_{ni}\hat{k}_{nj}(\mu) + b_{nj}\hat{k}_{ni}(\mu) - \hat{k}_{ni}(\mu)\hat{k}_{nj}(\mu)] \tag{9-97}$$

These three equations are seen to be independent of the optimum trajectory $\hat{\mathbf{x}}(\mu)$; hence the $\hat{k}$ parameters are independent of the point in state space used for the expansion. As a consequence of this independence, the neighborhood $\mathscr{N}_2(\mu)$ covers all finite points for $\mathbf{x}(\mu)$. On the other hand, if the second-degree Taylor series expansion had been used, the single-subscripted $\hat{p}$ parameters would not be independent of $\hat{\mathbf{x}}(\mu)$. In fact, these parameters are linear functions of the components of $\hat{\mathbf{x}}(\mu)$; hence the neighborhood $\hat{\mathscr{N}}_2(\mu)$ must be kept small for accurate approximations with the second-degree Taylor series expansion. This example illustrates the importance of judiciously selecting an expansion for each design problem.

9-4 *P*th-degree expansion of the minimum-error function

The concepts of the first- and second-degree expansions easily can be generalized to the case of a Pth-degree expansion. In particular, the Pth-degree polynomial form

$$E_P[\mathbf{x}(\mu), \mu] = \hat{k}(\mu) - 2\sum_{n_1=1}^{N} \hat{k}_{n_1}(\mu)x_{n_1}(\mu) + \sum_{n_1=1}^{N}\sum_{n_2=1}^{N} \hat{k}_{n_1n_2}(\mu)x_{n_1}(\mu)x_{n_2}(\mu) + \cdots + \frac{2}{P}\sum_{n_1=1}^{N}\cdots\sum_{n_P=1}^{N} \hat{k}_{n_1\cdots n_P}(\mu)x_{n_1}(\mu)\cdots x_{n_P}(\mu) \tag{9-98}$$

is chosen here. Corresponding to this expansion, the truncated hamiltonian function is taken to be

$$\mathscr{H}_P^\dagger[\mathbf{x}(\mu), \mu] = \min_{\mathbf{m}(\mu)\in\mathscr{M}(\mu)} \left\{ H[\mathbf{x}(\mu), \mathbf{m}(\mu), \mu] + \sum_{n=1}^{N} f_n[\mathbf{x}(\mu), \mathbf{m}(\mu), \mu] \frac{\partial E_P[\mathbf{x}(\mu), \mu]}{\partial x_n(\mu)} \right\} \tag{9-99}$$

Furthermore, the function $\mathscr{H}_P^\dagger[\mathbf{x}(\mu), \mu]$ is expanded in a Pth-degree Taylor series, and the coefficients of this series are grouped according to the various powers of the components of $\mathbf{x}(\mu)$. This power series, which is analogous to (9-64) for the second-degree case, is written as

$$H_P^\dagger[\mathbf{x}(\mu), \mu] = \hat{H}_0(\mu) + \sum_{n_1=1}^{N} \hat{H}_{n_1}(\mu)x_{n_1}(\mu) + \sum_{n_1=1}^{N}\sum_{n_2=1}^{N} \hat{H}_{n_1n_2}(\mu)x_{n_1}(\mu)x_{n_2}(\mu) + \cdots + \sum_{n_1=1}^{N}\cdots\sum_{n_P=1}^{N} \hat{H}_{n_1\cdots n_P}(\mu)x_{n_1}(\mu)\cdots x_{n_P}(\mu) \tag{9-100}$$

The coefficients appearing in (9-100) are given by

$$\hat{H}_{n_1\cdots n_p}(\mu) = \frac{1}{p!}\left\{\frac{\partial^p \mathscr{H}_P^\dagger[\mathbf{x}(\mu), \mu]}{\partial x_{n_1}(\mu)\cdots\partial x_{n_p}(\mu)} - \sum_{n_{p+1}=1}^{N} \frac{\partial^{p+1}\mathscr{H}_P^\dagger[\mathbf{x}(\mu), \mu]}{\partial x_{n_1}(\mu)\cdots\partial x_{n_{p+1}}(\mu)} x_{n_{p+1}}(\mu) + \cdots + \frac{(-1)^{P-p}}{(P-p)!}\sum_{n_{p+1}=1}^{N}\cdots\sum_{n_P=1}^{N}\frac{\partial^P \mathscr{H}_P^\dagger[\mathbf{x}(\mu), \mu]}{\partial x_{n_1}(\mu)\cdots\partial x_{n_P}(\mu)}\hat{x}_{n_{p+1}}(\mu)\cdots\hat{x}_{n_P}(\mu)\right\}_{\mathbf{x}(\mu)=\hat{\mathbf{x}}(\mu)} \tag{9-101}$$

and the coefficient $\hat{H}_0(\mu)$ is found from (9-101) when $p = 0$.

In terms of (9-98) and (9-99), the Hamilton-Jacobi equation becomes

$$\frac{\partial E_P[\mathbf{x}(\mu), \mu]}{\partial \mu} + \mathscr{H}_P^\dagger[\mathbf{x}(\mu), \mu] + \{(P+1)\text{st-degree terms in } [x_n(\mu) - \hat{x}_n(\mu)]\} = 0 \tag{9-102}$$

Also, the collection of the Pth- and lower-degree terms in (9-102), with the use of (9-100), yields

$$\frac{\partial E_P[\mathbf{x}(\mu), \mu]}{\partial \mu} + H_P^\dagger[\mathbf{x}(\mu), \mu] + \{(P+1)\text{st-degree terms in } [x_n(\mu) - \hat{x}_n(\mu)]\} = 0 \tag{9-103}$$

so that, for small regions of state space, the expanded Hamilton-Jacobi equation becomes

$$\frac{\partial E_P[\mathbf{x}(\mu), \mu]}{\partial \mu} + H_P^\dagger[\mathbf{x}(\mu), \mu] = 0 \qquad \mathbf{x}(\mu) \in \hat{\mathscr{N}}_P(\mu) \tag{9-104}$$

When $\partial E_P/\partial\mu$ is found from (9-98) and is substituted into (9-104) along with (9-100), the result is

$$[\hat{k}'(\mu) + \hat{H}_0(\mu)] - 2\sum_{n_1=1}^{N}[\hat{k}'_{n_1}(\mu) - \tfrac{1}{2}\hat{H}_{n_1}(\mu)]x_{n_1}(\mu) + \sum_{n_1=1}^{N}\sum_{n_2=1}^{N}[\hat{k}'_{n_1n_2}(\mu) + \hat{H}_{n_1n_2}(\mu)]x_{n_1}(\mu)x_{n_2}(\mu) + \cdots + \frac{2}{P}\sum_{n_1=1}^{N}\cdots\sum_{n_P=1}^{N}\left[\hat{k}'_{n_1\cdots n_P}(\mu) + \frac{P}{2}\hat{H}_{n_1\cdots n_P}(\mu)\right]x_{n_1}(\mu)\cdots x_{n_P}(\mu) = 0 \tag{9-105}$$

Finally, (9-105) is valid for all $\mathbf{x}(\mu)$ in the neighborhood $\hat{\mathscr{N}}_P(\mu)$ if

$$\hat{k}'(\mu) = -\hat{H}_0(\mu) \tag{9-106}$$

$$\hat{k}'_{n_1}(\mu) = \tfrac{1}{2}\hat{H}_{n_1}(\mu) \tag{9-107}$$

and

$$\hat{k}'_{n_1\cdots n_p}(\mu) = -\frac{p}{2}\hat{H}_{n_1\cdots n_p}(\mu) \qquad p = 2, 3, \ldots, P \tag{9-108}$$

These differential equations are subject to the free-point terminal-boundary conditions

$$\hat{k}(T) = \hat{k}_{n_1\cdots n_p}(T) = 0 \qquad p = 1, 2, \ldots, P$$

The $\hat{k}$ equations for the Pth-degree expansion can be regarded as an extended form of the characteristic equations discussed in Chap. 6. In fact, the equations of Pontryagin, $P = 1$, and the parametric expansion method used for linear optimum control in Chap. 7, $P = 2$, are special cases of the results presented here. However, the class of design problems where the theory applies becomes more restricted with the increasing degree of the expansion. This restriction is due to the requirement for the existence of higher-order partial derivatives of the minimum-error function. Also, the use of these equations requires that the order in which the partial differentiations of the minimum-error function with respect to the various components of $\mathbf{x}(\mu)$ are performed can be interchanged.

Except for these mathematical restrictions, this method of expanding the minimum-error function offers considerable flexibility in the synthesis of the nonlinear optimum control equation. In particular, the degree of the expansion can be selected for each design problem so that the appropriate compromise between the complexity of the control system and the size of $\hat{\mathcal{N}}_P(\mu)$ is established. However, the selection of the expansion with the smallest degree required for an $\hat{\mathcal{N}}_P(\mu)$ of a certain size generally must be determined experimentally by system simulation.

9-5 Example of approximation properties

The previous example presented in this chapter is a special case because the minimum-error function is quadratic. The example presented here is chosen because the minimum-error function is not a polynomial, and approximation errors occur in the neighborhood of the optimum trajectory. This example is typical of a "soft" saturation constraint on the amplitudes of the control signals and is defined by

$$h[\mathbf{q}(\mu), \mathbf{m}(\mu), \mu] = \sum_{n=1}^{N} \{q_n^2(\mu) - 2 \ln [\cos m_n(\mu)]\} \tag{9-109}$$

$$q_n(\mu) = \sum_{m=1}^{N} a_{nm} x_m(\mu) \tag{9-110}$$

$$x_n'(\mu) = \sum_{m=1}^{N} [b_{nm} x_m(\mu)] + m_n(\mu) \tag{9-111}$$

and

$$-\frac{\pi}{2} < m_n(\mu) < \frac{\pi}{2} \tag{9-112}$$

For this example, the truncated hamiltonian function is

$$\mathscr{H}_P^\dagger[\mathbf{x}(\mu), \mu] = \sum_{n=1}^{N} \left(q_n^2(\mu) - 2 \ln [\cos m_n^\dagger(\mu)] + \left\{ \sum_{m=1}^{N} [b_{nm} x_m(\mu)] + m_n^\dagger(\mu) \right\} \frac{\partial E_P[\mathbf{x}(\mu), \mu]}{\partial x_n(\mu)} \right) \tag{9-113}$$

where the quasi-optimum control equation is given by

$$m_n^\dagger(\mu) = \tan^{-1} \left\{ -\frac{1}{2} \frac{\partial E_P[\mathbf{x}(\mu), \mu]}{\partial x_n(\mu)} \right\} \tag{9-114}$$

for all finite values of $\mathbf{x}(\mu)$ and $\hat{\mathbf{x}}(\mu)$. If the quadratic expansion $P = 2$ is used, then the quasi-optimum control equation becomes

$$m_n^\dagger(\mu) = \tan^{-1} \left[\hat{k}_n(\mu) - \sum_{m=1}^{N} \hat{k}_{nm}(\mu) x_m(\mu) \right] \tag{9-115}$$

When the second-degree power series is used, the evaluation of (9-85) and (9-86) yields

$$-\hat{k}_i'(\mu) = \sum_{n=1}^{N} (b_{ni}\hat{k}_n(\mu) - \hat{k}_{ni}(\mu)\{\hat{k}_n(\mu)\cos^2[\hat{m}_n(\mu)] + \hat{m}_n(\mu) - \tfrac{1}{2}\sin[2\hat{m}_n(\mu)]\}) \tag{9-116}$$

and

$$-\hat{k}_{ij}'(\mu) = \sum_{n=1}^{N} [a_{ni}a_{nj} + b_{ni}\hat{k}_{nj}(\mu) + b_{nj}\hat{k}_{ni}(\mu) - \hat{k}_{ni}(\mu)\hat{k}_{nj}(\mu)\cos^2\hat{m}_n(\mu)] \tag{9-117}$$

These equations are strikingly similar to the matrix Riccati equations found for linear optimum controls. Here, however, some of the coefficients appearing in the equations are dependent upon the particular optimum trajectory used in the solution. If the optimum control signal $\hat{m}_n(\mu)$ is very small everywhere along $\hat{\mathbf{x}}(\mu)$, then the first-order effects due to these coefficients enter only the solutions of the single-subscripted $\hat{k}$ parameters. Specifically, (9-116) and (9-117) reduce to

$$-\hat{k}_i'(\mu) \approx \sum_{n=1}^{N} (b_{ni}\hat{k}_n(\mu) - \hat{k}_{ni}(\mu)\{\hat{k}_n(\mu) + \tfrac{2}{3}[\hat{m}_n(\mu)]^3\}) \tag{9-118}$$

and

$$-\hat{k}_{ij}'(\mu) \approx \sum_{n=1}^{N} [a_{ni}a_{nj} + b_{ni}\hat{k}_{nj}(\mu) + b_{nj}\hat{k}_{ni}(\mu) - \hat{k}_{ni}(\mu)\hat{k}_{nj}(\mu)] \tag{9-119}$$

Therefore, the feedback gains of the system are independent of $\hat{\mathbf{x}}(\mu)$, to a first approximation, under these conditions. Under the condition $\hat{m}_n(\mu) = 0$, these $\hat{k}$ parameters are identical to those found for the previous example for a linear optimum control system.

In order to assess the approximations afforded by the second-degree expansion for this example, a simplified case is introduced and is defined by

$$N = a_{11} = 1 \qquad b_{11} = \hat{x}_1(t_0) = 0 \qquad T = \infty \tag{9-120}$$

For these conditions, (9-116) and (9-117) reduce to

$$-\hat{k}_1'(\mu) = -\hat{k}_{11}(\mu)\hat{k}_1(\mu) \tag{9-121}$$

and

$$-\hat{k}_{11}'(\mu) = 1 - \hat{k}_{11}{}^2(\mu) \tag{9-122}$$

because

$$\hat{x}_1(\mu) = \hat{m}_1(\mu) = 0 \tag{9-123}$$

along this particular characteristic curve. Elementary procedures used to solve (9-121) and (9-122) yield

$$\hat{k}_1(\mu) = 0 \qquad \text{and} \qquad \hat{k}_{11}(t) = 1 \tag{9-124}$$

so that (9-115) reduces to

$$m_1^\dagger(t) = -\tan^{-1} x_1(t) \qquad P = 2 \tag{9-125}$$

in the neighborhood of $\hat{x}_1(t) = 0$. Furthermore, the fourth-degree expansion of the minimum-error function, for $\hat{x}_1(t) = 0$, results in the additional $\hat{k}$ equation

$$-\hat{k}_{1111}'(\mu) = \tfrac{1}{3}\hat{k}_{11}{}^4(\mu) - 4\hat{k}_{11}(\mu)\hat{k}_{1111}(\mu) \tag{9-126}$$

The steady-state solution of (9-126) is $\hat{k}_{1111}(t) = \frac{1}{12}$ so that the quasi-optimum control equation for this case becomes

$$m_1^\dagger(t) = -\tan^{-1}[x_1(t) + \tfrac{1}{12}x_1{}^3(t)] \qquad P = 4 \tag{9-127}$$

For the purpose of comparison, the exact solution of the optimum control equation is given by

$$2m_1^*(t)[\tan m_1^*(t)] + 2 \ln [\cos m_1^*(t)] = x_1^2(t) \tag{9-128}$$

for all $x_1(t)$. This exact solution is found by the methods given in Sec. 4-5 for the simplified form of the example considered in this section. Also, the partial derivative $\partial E/\partial x_1$ is given by

$$\frac{\partial E[x_1(t), t]}{\partial x_1(t)} = -2 \tan m_1^*(t) \tag{9-129}$$

Finally, the third-degree Taylor series expansion of the optimum control equation about the point $\hat{x}_1(t) = 0$ is found to be

$$m_1^*(t) \approx -x_1(t) + \tfrac{1}{4}x_1^3(t) \tag{9-130}$$

This approximation, which can be found from either (9-127) or (9-128), is introduced in order to compare various forms of the quasi-optimum control equation.

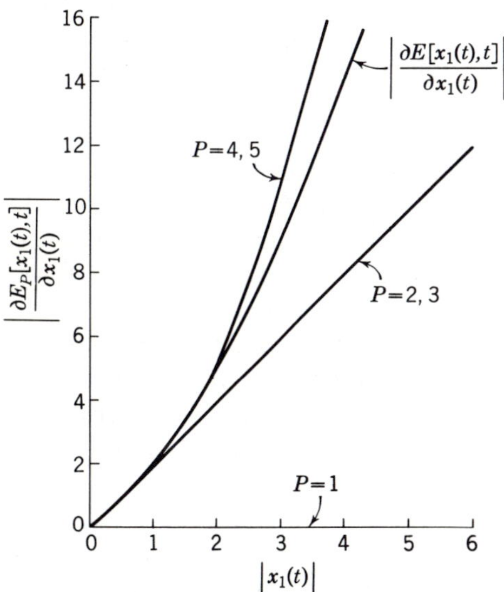

Fig. 9-2 Approximations of minimum-error-function slope for $\hat{x}_1(t_0) = 0$.

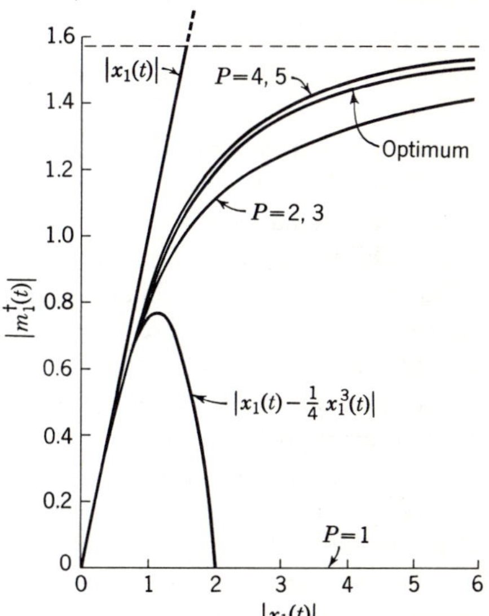

Fig. 9-3 Approximations of optimum control equation for $\hat{x}_1(t_0) = 0$.

The curves shown in Fig. 9-2 illustrate the approximation properties of expanding the minimum-error function in power series of various degrees. Specifically, the approximations to $\partial E/\partial x_1(t)$ are indicated. Because the minimum-error function is an even function about $x_1(t) = 0$, the partial derivative $\partial E/\partial x_1(t)$ is an odd function. Therefore, the expansions for $P = 2$ and $P = 3$ are identical, as are the expansions for $P = 4$ and $P = 5$. The curves in Fig. 9-2 indicate rapid convergence of the expansions.

The curves in Fig. 9-3 illustrate the approximation properties of various forms of the quasi-optimum control equation. The curves corresponding to (9-125) and (9-127) indicate a rapid convergence for the method of expanding the minimum-error function. On the other hand, the Taylor series expansion of the optimum control equation introduces some severe difficulties. For the first-degree expansion, the maximum allowable amplitude of the control signal is exceeded for $|x_1(t)| > 1.572$. Also, the control system becomes unstable for $|x_1(t)| > 2$ in the case of the third-degree expansion resulting in (9-130).

This example serves to illustrate a number of points raised in the preliminary considerations included in Sec. 9-1. Also, (9-114) is a typical illustration of (9-17).

9-6 Stability of quasi-optimum controls

The stability of quasi-optimum controls not only is important from the standpoint of the control system itself but also is important in the validity of the expansions

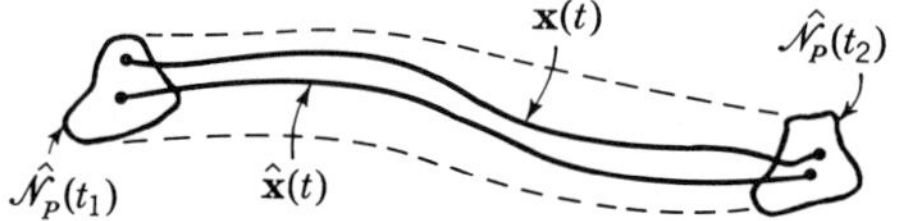

Fig. 9-4 Required behavior of quasi-optimum control system.

presented in this chapter. In particular, given an initial state $\mathbf{x}(t)$ such that $\mathbf{x}(t_1) \in \hat{\mathscr{N}}_P(t_1)$, all subsequent states $\mathbf{x}(t_2)$ for $t_2 > t_1$ must satisfy $\mathbf{x}(t_2) \in \hat{\mathscr{N}}_P(t_2)$. This condition is depicted in Fig. 9-4. If this condition is not satisfied for all t such that $t_1 \leq t \leq t_2$, then the expansions of the minimum-error function may not exist and quasi-optimum system performance may not occur.

The behavior of the system depicted in Fig. 9-4 is called *stable* where the response $\mathbf{x}(t)$ remains in some suitably small neighborhood of $\hat{\mathbf{x}}(t)$. However, asymptotic stability, as defined in Sec. 7-6, of the quasi-optimum control system is discussed initially in this section. That is, behavior which is characterized by $\|\mathbf{x}(t) - \hat{\mathbf{x}}(t)\| \to 0$ as $t \to \infty$ is treated here. In the case of linear optimum systems, asymptotic stability in the large results from the theory. In the case of nonlinear processes and quasi-optimum control, *asymptotic stability in the small* results from the theory.[50] That is, asymptotic stability is guaranteed only for a suitably small neighborhood of $\hat{\mathbf{x}}(t)$. The trajectory $\hat{\mathbf{x}}(t)$ is the optimum trajectory corresponding to the initial state $\hat{\mathbf{x}}(t_1)$. In addition, the trajectory $\hat{\mathbf{x}}(t)$ is the equilibrium trajectory in the context of the stability theory presented here.

For the purposes of investigating the asymptotic stability of quasi-optimum control systems, the second-degree Taylor series expansion of the minimum-error function is investigated. The procedure used is to determine the conditions for which the function $E_2[\mathbf{x}(t), t]$, as given in (9-59), is a Lyapunov function. In order to expedite the derivation, time-function notation is omitted. Also, the class of optimization problems is simplified initially by omitting control-signal saturation.

When the second-degree Taylor series expansion of the minimum-error function is used, the truncated hamiltonian is

$$\mathscr{H}_2^\dagger(\mathbf{x}, \mu) = H(\mathbf{x}, \mathbf{m}^\dagger, \mu) + \sum_{n=1}^{N} f_n(\mathbf{x}, \mathbf{m}^\dagger, \mu)\left[\hat{p}_n + 2\sum_{m=1}^{N} \hat{p}_{nm}(x_m - \hat{x}_m)\right] \qquad (9\text{-}131)$$

For the case where control-signal saturation does not occur, the vector $\mathbf{m}^\dagger$ is defined by

$$\frac{\partial H(\mathbf{x}, \mathbf{m}^\dagger, \mu)}{\partial m_j^\dagger} + \sum_{n=1}^{N} \frac{\partial f_n(\mathbf{x}, \mathbf{m}^\dagger, \mu)}{\partial m_j^\dagger}\left[\hat{p}_n + 2\sum_{m=1}^{N} \hat{p}_{nm}(x_m - \hat{x}_m)\right] = 0 \qquad j = 1, 2, \ldots, M \qquad (9\text{-}132)$$

This condition, which defines $\mathbf{m}^\dagger$, is assumed to correspond to a local minimum; hence the Legendre condition for M dimensions must apply. The Legendre condition for one dimension is discussed in Sec. 5-2. If the variables

$$T_{ij}(\mathbf{x}, \mu) = \frac{\partial^2 H(\mathbf{x}, \mathbf{m}^\dagger, \mu)}{\partial m_i^\dagger\, \partial m_j^\dagger} + \sum_{n=1}^{N} \frac{\partial^2 f_n(\mathbf{x}, \mathbf{m}^\dagger, \mu)}{\partial m_i^\dagger\, \partial m_j^\dagger}\left[\hat{p}_n + 2 \sum_{m=1}^{N} \hat{p}_{nm}(x_m - \hat{x}_m)\right] \quad (9\text{-}133)$$

are defined so that $T_{ij}(\mathbf{x}, \mu) = T_{ji}(\mathbf{x}, \mu)$, then this condition for a local minimum requires that the $M \times M$ matrix $[T_{ij}(\mathbf{x}, \mu)]$ be positive definite. Before the equations which define the quasi-optimum control system are derived, the definition

$$R_{jk}(\mathbf{x}, \mu) = \frac{\partial^2 H(\mathbf{x}, \mathbf{m}^\dagger, \mu)}{\partial x_k\, \partial m_j^\dagger} + \sum_{n=1}^{N} \left\{ \frac{\partial^2 f_n(\mathbf{x}, \mathbf{m}^\dagger, \mu)}{\partial x_k\, \partial m_j^\dagger}\left[\hat{p}_n + 2 \sum_{m=1}^{N} \hat{p}_{nm}(x_m - \hat{x}_m)\right] + 2 \frac{\partial f_n(\mathbf{x}, \mathbf{m}^\dagger, \mu)}{\partial m_j^\dagger} \hat{p}_{nk} \right\} \quad (9\text{-}134)$$

is introduced for a collection of terms which occurs frequently.

The first and second partial derivatives of the truncated hamiltonian with respect to the components of $\mathbf{x}(\mu)$ now are written in terms of these definitions. The first partial derivative becomes

$$\frac{\partial \mathscr{H}_2^\dagger(\mathbf{x}, \mu)}{\partial x_k} = \frac{\partial H(\mathbf{x}, \mathbf{m}^\dagger, \mu)}{\partial x_k} + \sum_{n=1}^{N} \left\{ \frac{\partial f_n(\mathbf{x}, \mathbf{m}^\dagger, \mu)}{\partial x_k}\left[\hat{p}_n + 2 \sum_{m=1}^{N} \hat{p}_{nm}(x_m - \hat{x}_m)\right] + 2 f_n(\mathbf{x}, \mathbf{m}^\dagger, \mu) \hat{p}_{nk} \right\} \quad (9\text{-}135)$$

when the condition given in (9-132) is used. The second partial derivative also becomes

$$\frac{\partial \mathscr{H}_2^\dagger(\mathbf{x}, \mu)}{\partial x_k\, \partial x_l} = \frac{\partial^2 H(\mathbf{x}, \mathbf{m}^\dagger, \mu)}{\partial x_k\, \partial x_l} + \sum_{n=1}^{N} \left\{ \frac{\partial^2 f_n(\mathbf{x}, \mathbf{m}^\dagger, \mu)}{\partial x_k\, \partial x_l}\left[\hat{p}_n + 2 \sum_{m=1}^{N} \hat{p}_{nm}(x_m - \hat{x}_m)\right] + 2 \frac{\partial f_n(\mathbf{x}, \mathbf{m}^\dagger, \mu)}{\partial x_k} \hat{p}_{nl} + 2 \frac{\partial f_n(\mathbf{x}, \mathbf{m}^\dagger, \mu)}{\partial x_l} \hat{p}_{nk} \right\} + \sum_{j=1}^{M} R_{jk}(\mathbf{x}, \mu) \frac{\partial m_j^\dagger}{\partial x_l} \quad (9\text{-}136)$$

when the definition given in (9-134) is used. Furthermore, the partial derivatives $\partial m_j^\dagger / \partial x_l$ are found from differentiating (9-132). The partial derivative of (9-132) with respect to $x_k(\mu)$ yields

$$\sum_{i=1}^{M} T_{ij}(\mathbf{x}, \mu) \frac{\partial m_i^\dagger}{\partial x_k} = -R_{jk}(\mathbf{x}, \mu) \qquad j = 1, 2, \ldots, M \quad (9\text{-}137)$$

when the definitions given in (9-133) and (9-134) are used. For a given value of k, (9-137) is used to find $\partial m_i^\dagger / \partial x_k$ for $i = 1, 2, \ldots, M$, which exist uniquely because the matrix $[T_{ij}(\mathbf{x}, \mu)]$ is positive definite. The set of equations given in (9-137) is solved for each value of k so that $k = 1, 2, \ldots, N$; hence the partial derivatives which appear in (9-136) are specified.

The equations which determine the $\hat{p}$ parameters are found by evaluating (9-135) and (9-136) at $\mathbf{x}(\mu) = \hat{\mathbf{x}}(\mu)$ according to (9-75) and (9-76). These steps yield

$$-\hat{p}_0'(\mu) = H(\hat{\mathbf{x}}, \hat{\mathbf{m}}, \mu) \tag{9-138}$$

$$-\hat{p}_k'(\mu) = \frac{\partial H(\hat{\mathbf{x}}, \hat{\mathbf{m}}, \mu)}{\partial \hat{x}_k} + \sum_{n=1}^{N} \frac{\partial f_n(\hat{\mathbf{x}}, \hat{\mathbf{m}}, \mu)}{\partial \hat{x}_k} \hat{p}_n \tag{9-139}$$

$$\begin{aligned}-\hat{p}_{kl}'(\mu) = \frac{1}{2}\frac{\partial^2 H(\hat{\mathbf{x}}, \hat{\mathbf{m}}, \mu)}{\partial \hat{x}_k\,\partial \hat{x}_l} &+ \sum_{n=1}^{N}\left[\frac{1}{2}\frac{\partial^2 f_n(\hat{\mathbf{x}}, \hat{\mathbf{m}}, \mu)}{\partial \hat{x}_k\,\partial \hat{x}_l}\hat{p}_n\right.\\ &+ \left.\frac{\partial f_n(\hat{\mathbf{x}}, \hat{\mathbf{m}}, \mu)}{\partial \hat{x}_k}\hat{p}_{nl} + \frac{\partial f_n(\hat{\mathbf{x}}, \hat{\mathbf{m}}, \mu)}{\partial \hat{x}_l}\hat{p}_{nk}\right] + \frac{1}{2}\sum_{j=1}^{M} R_{jk}(\hat{\mathbf{x}}, \mu)\frac{\partial \hat{m}_j}{\partial \hat{x}_l}\end{aligned} \tag{9-140}$$

The partial derivatives $\partial \hat{m}_j/\partial \hat{x}_l = \partial \hat{m}_j^\dagger/\partial \hat{x}_l$ are evaluated from (9-137) and are given by

$$\sum_{i=1}^{M} T_{ij}(\hat{\mathbf{x}}, \mu)\frac{\partial \hat{m}_i}{\partial \hat{x}_k} = -R_{jk}(\hat{\mathbf{x}}, \mu) \tag{9-141}$$

where
$$T_{ij}(\hat{\mathbf{x}}, \mu) = \frac{\partial^2 H(\hat{\mathbf{x}}, \hat{\mathbf{m}}, \mu)}{\partial \hat{m}_i\,\partial \hat{m}_j} + \sum_{n=1}^{N}\frac{\partial^2 f_n(\hat{\mathbf{x}}, \hat{\mathbf{m}}, \mu)}{\partial \hat{m}_i\,\partial \hat{m}_j}\hat{p}_n \tag{9-142}$$

and

$$R_{jk}(\hat{\mathbf{x}}, \mu) = \frac{\partial^2 H(\hat{\mathbf{x}}, \hat{\mathbf{m}}, \mu)}{\partial \hat{x}_k\,\partial \hat{m}_j} + \sum_{n=1}^{N}\left[\frac{\partial^2 f_n(\hat{\mathbf{x}}, \hat{\mathbf{m}}, \mu)}{\partial \hat{x}_k\,\partial \hat{m}_j}\hat{p}_n + 2\frac{\partial f_n(\hat{\mathbf{x}}, \hat{\mathbf{m}}, \mu)}{\partial \hat{m}_j}\hat{p}_{nk}\right] \tag{9-143}$$

Finally, the free-point terminal-boundary conditions are

$$\hat{p}_0(T) = \hat{p}_k(T) = \hat{p}_{kl}(T) = 0 \tag{9-144}$$

These equations for the $\hat{p}$ parameters now are used to investigate the stability of the corresponding linearized form of the quasi-optimum control system, which henceforth is referred to as the *optimum linear control system.*

First, the conditions for which the function $E_2(\mathbf{x}, \mu)$ is a Lyapunov function are derived. The conditions for the structural form of a Lyapunov function are given in (7-120) to (7-122). If $E_2(\mathbf{x}, t)$ is to be a Lyapunov function for the equilibrium trajectory $\hat{\mathbf{x}}(t)$, then $E_2(\hat{\mathbf{x}}, t) = 0$ and $E_2(\mathbf{x}, t) > 0$ for $\mathbf{x} \neq \hat{\mathbf{x}}$. These conditions are satisfied only when $\hat{p}_0(t) = 0$ and $\hat{p}_k(t) = 0$ for $k = 1, 2, \ldots, N$. Because of the free-point terminal-boundary conditions, (9-138) and (9-139) provide these required conditions if

$$H(\hat{\mathbf{x}}, \hat{\mathbf{m}}, \mu) = 0 \tag{9-145}$$

and
$$\frac{\partial H(\hat{\mathbf{x}}, \hat{\mathbf{m}}, \mu)}{\partial \hat{x}_k} = 0 \qquad k = 1, 2, \ldots, N \tag{9-146}$$

for all μ on the interval $t \leqq \mu \leqq T$. When (9-146) is valid, the result $\hat{p}_k(\mu) = 0$ is obtained so that (9-140) and (9-141) reduce to

$$\begin{aligned}-\hat{p}_{kl}'(\mu) = \frac{1}{2}\frac{\partial^2 H(\hat{\mathbf{x}}, \hat{\mathbf{m}}, \mu)}{\partial \hat{x}_k\,\partial \hat{x}_l} &+ \sum_{n=1}^{N}\left[\frac{\partial f_n(\hat{\mathbf{x}}, \hat{\mathbf{m}}, \mu)}{\partial \hat{x}_k}\hat{p}_{nl} + \frac{\partial f_n(\hat{\mathbf{x}}, \hat{\mathbf{m}}, \mu)}{\partial \hat{x}_l}\hat{p}_{nk}\right]\\ &- \frac{1}{2}\sum_{i=1}^{M}\sum_{j=1}^{M}\frac{\partial^2 H(\hat{\mathbf{x}}, \hat{\mathbf{m}}, \mu)}{\partial \hat{m}_i\,\partial \hat{m}_j}\frac{\partial \hat{m}_i}{\partial \hat{x}_k}\frac{\partial \hat{m}_j}{\partial \hat{x}_l}\end{aligned} \tag{9-147}$$

and
$$\sum_{i=1}^{M}\frac{\partial^2 H(\hat{\mathbf{x}}, \hat{\mathbf{m}}, \mu)}{\partial \hat{m}_i\,\partial \hat{m}_j}\frac{\partial \hat{m}_i}{\partial \hat{x}_k} = -\frac{\partial^2 H(\hat{\mathbf{x}}, \hat{\mathbf{m}}, \mu)}{\partial \hat{x}_k\,\partial \hat{m}_j} - 2\sum_{n=1}^{N}\frac{\partial f_n(\hat{\mathbf{x}}, \hat{\mathbf{m}}, \mu)}{\partial \hat{m}_j}\hat{p}_{nk} \tag{9-148}$$

Also, the quasi-optimum control vector $\mathbf{m}^\dagger(\mu)$ now is defined by

$$\frac{\partial H(\mathbf{x}, \mathbf{m}^\dagger, \mu)}{\partial m_j^\dagger} + 2\sum_{n=1}^{N}\sum_{m=1}^{N} \frac{\partial f_n(\hat{\mathbf{x}}, \mathbf{m}^\dagger, \mu)}{\partial m_j^\dagger} \hat{p}_{nm}(x_m - \hat{x}_m) = 0 \qquad j = 1, 2, \ldots, M \tag{9-149}$$

which results from (9-132). From (9-149), the optimum control vector $\hat{\mathbf{m}}(\mu)$ along the equilibrium trajectory $\hat{\mathbf{x}}(\mu)$ is found from

$$\frac{\partial H(\hat{\mathbf{x}}, \hat{\mathbf{m}}, \mu)}{\partial \hat{m}_j} = 0 \qquad j = 1, 2, \ldots, M \tag{9-150}$$

Equations (9-147) and (9-148) are written in a form where $T_{ij}(\hat{\mathbf{x}}, \mu)$ and $R_{jk}(\hat{\mathbf{x}}, \mu)$ are eliminated in order to illustrate a most interesting and consequential result. In particular, the quasi-optimum control system is not dependent upon the second derivatives of the dynamic-process state equation when (9-146) holds. This result is seen from (9-147) to (9-149). On the other hand, the quasi-optimum control system is dependent on the second partial derivatives of the error measure. Furthermore, the optimum control vector $\hat{\mathbf{m}}(\mu)$ along the equilibrium trajectory $\hat{\mathbf{x}}(\mu)$ is dependent only upon the partial derivatives $\partial \hat{H}/\partial \hat{m}_j$ and not upon the partial derivatives $\partial \hat{f}_n/\partial \hat{m}_j$. These results allow the direct use of the "second method" of Lyapunov to demonstrate the asymptotic stability of the optimum linear control system under the conditions given in (9-145) and (9-146).

The simplest method for demonstrating asymptotic stability is to introduce an alternative optimization problem with a linear state equation and a quadratic error measure. For suitably small regions of state space, the state equation given in (9-3) is expanded in a first-degree Taylor series about $\hat{\mathbf{x}}(t)$ and $\hat{\mathbf{m}}(t)$ so that $\mathbf{x}'(t)$ is taken to be defined by

$$\frac{d[x_n(t) - \hat{x}_n(t)]}{dt} = \sum_{m=1}^{N} \frac{\partial f_n(\hat{\mathbf{x}}, \hat{\mathbf{m}}, t)}{\partial \hat{x}_m}[x_m(t) - \hat{x}_m(t)] + \sum_{m=1}^{M} \frac{\partial f_n(\hat{\mathbf{x}}, \hat{\mathbf{m}}, t)}{\partial \hat{m}_m}[m_m(t) - \hat{m}_m(t)] \tag{9-151}$$

Equation (9-151) is the linear state equation of this alternative optimization problem. Furthermore, the error measure appearing in (9-5) is expanded in a second-degree Taylor series about $\hat{\mathbf{x}}(t)$ and $\hat{\mathbf{m}}(t)$ so that $H(\mathbf{x}, \mathbf{m}, t)$ is taken to be defined by

$$\begin{aligned} H(\mathbf{x}, \mathbf{m}, t) &= \sum_{n=1}^{N}\sum_{m=1}^{N} \frac{1}{2}\frac{\partial^2 H(\hat{\mathbf{x}}, \hat{\mathbf{m}}, t)}{\partial \hat{x}_n\, \partial \hat{x}_m}[x_n(t) - \hat{x}_n(t)][x_m(t) - \hat{x}_m(t)] \\ &\quad + \sum_{n=1}^{N}\sum_{m=1}^{M} \frac{\partial^2 H(\hat{\mathbf{x}}, \hat{\mathbf{m}}, t)}{\partial \hat{x}_n\, \partial \hat{m}_m}[x_n(t) - \hat{x}_n(t)][m_m(t) - \hat{m}_m(t)] \\ &\quad + \sum_{n=1}^{M}\sum_{m=1}^{M} \frac{1}{2}\frac{\partial^2 H(\hat{\mathbf{x}}, \hat{\mathbf{m}}, t)}{\partial \hat{m}_n\, \partial \hat{m}_m}[m_n(t) - \hat{m}_n(t)][m_m(t) - \hat{m}_m(t)] \end{aligned} \tag{9-152}$$

In this expansion, the conditions given in (9-145), (9-146), and (9-150) are used. Equation (9-152) is the quadratic error measure used for this alternative optimization problem. For the optimization problem defined by (9-151) and (9-152), the application of (9-74) to (9-76) results directly in the conditions given in (9-147) to (9-149). In other words, (9-151) and (9-152) specify a problem in linear optimum controls that is equivalent to nonlinear control problems which satisfy (9-145) and (9-146) if a suitably small region of state space is considered.

A set of sufficient conditions for asymptotic stability now follows almost directly. First, the error measure H is taken to be strictly convex in all components of both $\mathbf{x}(t)$ and $\mathbf{m}(t)$, which is the tacit assumption used in all previous material. A strictly convex function, as defined in Chap. 4, possesses a positive-definite matrix of second derivatives. Therefore, the $(N + M)\times(N + M)$ matrix D, such that

$$D = \left[\begin{array}{c|c} \dfrac{\partial^2 H(\hat{\mathbf{x}}, \hat{\mathbf{m}}, \mu)}{\partial \hat{x}_n\, \partial \hat{x}_m} & \dfrac{\partial^2 H(\hat{\mathbf{x}}, \hat{\mathbf{m}}, \mu)}{\partial \hat{x}_n\, \partial \hat{m}_m} \\ \hline \dfrac{\partial^2 H(\hat{\mathbf{x}}, \hat{\mathbf{m}}, \mu)}{\partial \hat{m}_n\, \partial \hat{x}_m} & \dfrac{\partial^2 H(\hat{\mathbf{x}}, \hat{\mathbf{m}}, \mu)}{\partial \hat{m}_n\, \partial \hat{m}_m} \end{array}\right] \tag{9-153}$$

is positive definite for all μ on the interval $t \leqq \mu \leqq T$. The quadratic form given in (9-152) hence satisfies $H(\mathbf{x}, \mathbf{m}, \mu) > 0$ for all $\mathbf{x} \neq \hat{\mathbf{x}}$ and $\mathbf{m} \neq \hat{\mathbf{m}}$, and $H(\hat{\mathbf{x}}, \hat{\mathbf{m}}, \mu) = 0$.

Now the linear process given in (9-151) and the quadratic error measure given in (9-152) are evaluated for the quasi-optimum control equation where $\mathbf{m} = \mathbf{m}^{\dagger}$. For suitably small regions of state space, the control vector $\mathbf{m}^{\dagger}$ is expressed as

$$m_m^{\dagger}(t) - \hat{m}_m(t) = \sum_{j=1}^{N} \frac{\partial \hat{m}_m}{\partial \hat{x}_j}[x_j(t) - \hat{x}_j(t)] \qquad m = 1, 2, \ldots, M \tag{9-154}$$

which is a first-degree Taylor series expansion of $[m_m^{\dagger}(t) - \hat{m}_m(t)]$. Therefore, the linearized form of the state equation of the quasi-optimum control system becomes

$$\frac{d[x_n(t) - \hat{x}_n(t)]}{dt} = \sum_{m=1}^{N}\left[\frac{\partial f_n(\hat{\mathbf{x}}, \hat{\mathbf{m}}, t)}{\partial \hat{x}_m} + \sum_{j=1}^{M} \frac{\partial f_n(\hat{\mathbf{x}}, \hat{\mathbf{m}}, t)}{\partial \hat{m}_j}\frac{\partial \hat{m}_j}{\partial \hat{x}_m}\right][x_m(t) - \hat{x}_m(t)] \tag{9-155}$$

Also, the quadratic error measure given in (9-152) becomes

$$H(\mathbf{x}, \mathbf{m}^{\dagger}, t) = \sum_{n=1}^{N}\sum_{m=1}^{N} \hat{\lambda}_{nm}(t)[x_n(t) - \hat{x}_n(t)][x_m(t) - \hat{x}_m(t)]$$

where

$$\hat{\lambda}_{nm}(t) = \frac{1}{2}\frac{\partial^2 H(\hat{\mathbf{x}}, \hat{\mathbf{m}}, t)}{\partial \hat{x}_n\, \partial \hat{x}_m} + \sum_{j=1}^{M} \frac{\partial^2 H(\hat{\mathbf{x}}, \hat{\mathbf{m}}, t)}{\partial \hat{x}_n\, \partial \hat{m}_j}\frac{\partial \hat{m}_j}{\partial \hat{x}_m} + \sum_{i=1}^{M}\sum_{j=1}^{M} \frac{1}{2}\frac{\partial^2 H(\hat{\mathbf{x}}, \hat{\mathbf{m}}, t)}{\partial \hat{m}_i\, \partial \hat{m}_j}\frac{\partial \hat{m}_j}{\partial \hat{x}_n}\frac{\partial \hat{m}_i}{\partial \hat{x}_m} \tag{9-156}$$

Because $H(\mathbf{x}, \mathbf{m}, t)$ as given in (9-152) is strictly convex, the matrix $[\hat{\lambda}_{nm}(t)]$ is positive definite. An alternative form of (9-156) is found by the use of (9-147) and (9-148), which give

$$-\hat{\lambda}_{nm}(t) = \hat{p}'_{nm}(t) + 2\sum_{k=1}^{N}\left[\frac{\partial f_k(\hat{\mathbf{x}}, \hat{\mathbf{m}}, t)}{\partial \hat{x}_n} + \sum_{j=1}^{M} \frac{\partial f_k(\hat{\mathbf{x}}, \hat{\mathbf{m}}, t)}{\partial \hat{m}_j}\frac{\partial \hat{m}_j}{\partial \hat{x}_n}\right]\hat{p}_{km} \tag{9-157}$$

The bracketed coefficients in (9-157) readily are identified to be the coefficients of the linearized state equation given in (9-155).

With these relationships available, the asymptotic stability of (9-155) for suitably small regions of state space is found directly. In particular, the function $E_2(\mathbf{x}, t)$ is a Lyapunov function for (9-155). Under the conditions given in (9-145) and (9-146), the function $E_2(\mathbf{x}, t)$ is given by

$$E_2(\mathbf{x}, t) = \sum_{n=1}^{N}\sum_{m=1}^{N} \hat{p}_{nm}(t)[x_n(t) - \hat{x}_n(t)][x_m(t) - \hat{x}_m(t)] \tag{9-158}$$

However, this function is strictly convex because the minimum-error function is strictly convex as a result of the assumption that the matrix D is positive definite, where D is given in (9-153). Therefore, the matrix $[\hat{p}_{nm}(t)]$ is positive definite. This matrix is the matrix of second derivatives of the minimum-error function as given by (9-62). Finally, the total time derivative of $E_2(\mathbf{x}, t)$ is found to be

$$\frac{dE_2(\mathbf{x}, t)}{dt} = -\sum_{n=1}^{N}\sum_{m=1}^{N}\hat{\lambda}_{nm}(t)[x_n(t) - \hat{x}_n(t)][x_m(t) - \hat{x}_m(t)] \tag{9-159}$$

by the differentiation of (9-158). In the derivation of (9-159), (9-155) and (9-157) are used. Because the matrix $[\hat{\lambda}_{nm}(t)]$ is positive definite, the function $E_2(\mathbf{x}, t)$ satisfies the structural requirements given in (7-120) to (7-122) and hence is a Lyapunov function for the linear system defined by (9-155). Therefore, the optimum linear control system is asymptotically stable for suitably small regions of state space given that the matrices $[T_{ij}(\hat{\mathbf{x}}, \mu)]$ and D are positive definite and that (9-145) and (9-146) are valid.

However, asymptotic stability in the "small" through the study of an equivalent linear system requires an additional restriction which has not been mentioned thus far, namely, that the Taylor-series remainder terms associated with the derivation of (9-155) from (9-3) must vanish in the second or higher degree.[50] Usually this condition is stated as

$$\frac{\|G(\mathbf{x} - \hat{\mathbf{x}}, t)[\mathbf{x} - \hat{\mathbf{x}}]\|}{\|\mathbf{x} - \hat{\mathbf{x}}\|} \to 0 \qquad \text{as } \|\mathbf{x} - \hat{\mathbf{x}}\| \to 0 \tag{9-160}$$

where $G(\mathbf{x} - \hat{\mathbf{x}}, t)$ is the matrix of remainder terms. This matrix occurs in the state equation

$$\frac{d[\mathbf{x} - \hat{\mathbf{x}}]}{dt} = F(\hat{\mathbf{x}}, t)[\mathbf{x} - \hat{\mathbf{x}}] + G(\mathbf{x} - \hat{\mathbf{x}}, t)[\mathbf{x} - \hat{\mathbf{x}}]$$

where the elements of the matrix $F(\hat{\mathbf{x}}, t)$ are given by

$$f_{nm}(\hat{\mathbf{x}}, t) = \frac{\partial f_n(\hat{\mathbf{x}}, \hat{\mathbf{m}}, t)}{\partial \hat{x}_m} + \sum_{j=1}^{M}\frac{\partial f_n(\hat{\mathbf{x}}, \hat{\mathbf{m}}, t)}{\partial \hat{m}_j}\frac{\partial \hat{m}_j}{\partial \hat{x}_m} \tag{9-161}$$

in the case of the optimum linear control system.

An interesting point to note is that the available theory for asymptotic stability requires only the linearized version of the dynamic-process state equation and hence only the partial derivatives $\partial \hat{f}_n/\partial \hat{x}_k$ and $\partial \hat{f}_n/\partial \hat{m}_j$. However, the second-degree expansion of the minimum-error function generally involves, in addition, the partial derivatives $\partial^2 \hat{f}_n/\partial \hat{x}_k\, \partial \hat{x}_l$, $\partial^2 \hat{f}_n/\partial \hat{x}_k\, \partial \hat{m}_j$, and $\partial^2 \hat{f}_n/\partial \hat{m}_i\, \partial \hat{m}_j$ except under the condition given in (9-146). The lack of dependence on these second-order partial derivatives in this special case allows the direct use of the available stability theory.

A number of other interesting points can be deduced directly from the results presented thus far. First, quasi-optimum control systems based on the first-degree expansion of the minimum-error function generally do not result in asymptotically stable systems. For $P = 1$, the quasi-optimum control system is constructed with $\hat{p}_{kl}(t)$ set equal to zero, which generally results in $\partial \hat{m}_j^\dagger/\partial \hat{x}_m = 0$. In this case, asymptotic stability of the system requires that the state equation of the dynamic process itself must be asymptotically stable to $\hat{\mathbf{x}}(t)$. Second, expansions of the minimum-error function for $P > 2$ are of little or no use in demonstrating the asymptotic stability of a quasi-optimum control system. In the situation where $E_P[\mathbf{x}(t), t] = E[\mathbf{x}(t), t]$ and $E_P[\mathbf{x}(t), t]$ is a Lyapunov function, the control system based on $E_P[\mathbf{x}(t), t]$ is not quasi-optimum but is exactly optimum and also is asymptotically stable. On the other hand, if $E_P[\mathbf{x}(t), t] \neq E[\mathbf{x}(t), t]$ for all $\mathbf{x}(t)$ and hence $E_P[\mathbf{x}(t), t]$ does not equal the value of the error index for all $\mathbf{x}(t)$, then $E_P[\mathbf{x}(t), t]$ can be used only to demonstrate asymptotic stability in the small of the corresponding quasi-optimum control system. Asymptotic stability in the small using available theories is based on the equivalent linear quasi-optimum control system, and the procedures revert to those of this section.

As pointed out in Sec. 9-1, the nonlinear design problems where $E_P[\mathbf{x}(t), t] = E[\mathbf{x}(t), t]$ for all $\mathbf{x}(t)$ are specialized and relatively unimportant so that asymptotic stability in the large of quasi-optimum control systems generally cannot be demonstrated. However, if the dynamic process is linear and if the linearized form of the quasi-optimum control equation, as given by (9-154), is used for control purposes, then the control system is asymptotically stable in the large when $E_2[\mathbf{x}(t), t]$ is a Lyapunov function.

The discussion of asymptotic stability thus far excludes the possibility of control-signal saturation. Control-signal saturation poses a number of difficulties for which adequate theory is not available. In general, asymptotic stability cannot be demonstrated and also does not occur when control signals saturate. In order to substantiate these points, suppose that all components of both $\hat{\mathbf{m}}(t)$ and $\mathbf{m}^\dagger(t)$ are on a particular boundary of $\mathscr{M}(t)$. During the period of time that this situation exists, the components of the optimum control vector are independent functions of time; hence $\partial \hat{m}_j(t)/\partial \hat{x}_k(t) = 0$ for $j = 1, 2, \ldots, M$ and $k = 1, 2, \ldots, N$. Therefore, the dynamic process itself must be asymptotically stable with respect to $\hat{\mathbf{x}}(t)$ if the quasi-optimum system is to be asymptotically stable. Note that, during this period of time, the control system is open-loop and the dynamic process is subjected to only a driving function. Of course, asymptotic stability during total saturation is meaningful only if this situation continues indefinitely. However, if at some subsequent time total saturation ceases to exist so that $\hat{\mathbf{m}}(t) \in \mathscr{M}(t)$, $\mathbf{m}^\dagger(t) \in \mathscr{M}(t)$, and $\mathbf{x}(t) \in \hat{\mathscr{N}}_2(t)$, then the quasi-optimum control system could be asymptotically stable in the same sense as problems not involving saturation, provided that saturation does not occur again. One difficulty that arises in such cases is the verification that $\mathbf{x}(t) \in \hat{\mathscr{N}}_2(t)$ holds at the instant of time when total saturation ends. Furthermore, problems where saturation of one or more components of $\hat{\mathbf{m}}(t)$ reoccurs an indefinite number of times during system operation do not lend themselves to available methods for demonstrating asymptotic stability. For instance, the minimum-error function may be infinite even for finite $\hat{\mathbf{x}}(t)$; hence the matrix of second derivatives of the minimum-error function would not exist. This possibility is typical of design problems where insufficient control effort is available for controlling the dynamic process.

Perhaps the more significant aspects of asymptotic stability of quasi-optimum control systems involve engineering design considerations. In particular, neighboring optimum trajectories normally possess properties associated with stable systems. That is, the optimum trajectories satisfy $\|{}_1\hat{\mathbf{x}}(t) - {}_2\hat{\mathbf{x}}(t)\| < \delta$, where two adjacent trajectories are defined as ${}_1\hat{\mathbf{x}}(t)$ and ${}_2\hat{\mathbf{x}}(t)$. This property is deduced from the geometrical concept introduced in Chap. 6 for characteristic curves, namely, that the integral surface can be constructed as a family of characteristic curves. If the integral surface possesses elementary continuity properties, then neighboring characteristic curves must possess the properties associated with stable systems.§ In this situation, disturbances occurring during system operation eventually cause a system state vector that does not correspond to the original optimum trajectory, say ${}_1\hat{\mathbf{x}}(t)$. Then the system must be altered so as to correspond to the new optimum trajectory, say ${}_2\hat{\mathbf{x}}(t)$. Unless the optimum control equation can be constructed for a large region of state space, the control-equation construction for a small region of state space must be altered in order to correspond to ${}_2\hat{\mathbf{x}}(t)$. Generally, this alteration requires on-line computations of the new characteristic curve and the parameters of the expansion about this curve. Such on-line computations impose severe restrictions on the construction of the control system because of the complexity of the computational problem.

On the other hand, if neighboring optimum trajectories possess the property $\|{}_1\hat{\mathbf{x}}(t) - {}_2\hat{\mathbf{x}}(t)\| \to 0$ as $t \to \infty$, which is associated with asymptotically stable systems, then the problem of system construction may be simplified considerably. In this situation, if disturbances have occurred so that the new trajectory is ${}_2\hat{\mathbf{x}}(t)$, then the optimum system eventually reverts to the original trajectory ${}_1\hat{\mathbf{x}}(t)$. Furthermore, if the new trajectory is such that ${}_2\hat{\mathbf{x}}(t) \in {}_1\hat{\mathscr{N}}_P(t)$ where ${}_1\hat{\mathscr{N}}_P(t)$ is the permissible neighborhood associated with ${}_1\hat{\mathbf{x}}(t)$, then the quasi-optimum control system for $P > 1$ also is asymptotically stable to ${}_1\hat{\mathbf{x}}(t)$. Under these conditions, on-line computations of the new characteristic curve and the parameters of the expansion about this curve are not necessary, thereby greatly simplifying the system. The interest in the asymptotic stability of quasi-optimum systems is mainly due to this potential simplification.

The class of disturbances that give rise to incremental departures from ${}_1\hat{\mathbf{x}}(t)$ in the case of asymptotically stable systems is restrictive indeed. In fact, adequate theory in this area is not available. In order to indicate some of these difficulties and restrictions on the permissible class of disturbances, the following situation is considered. Suppose that the linearized version of the quasi-optimum system is constructed for a linear dynamic process and $P = 2$. Also, the assumption is made that (9-146) holds so that this system is asymptotically stable in the large. If the definition

$$\mathbf{z}(t) = \mathbf{x}(t) - \hat{\mathbf{x}}(t) \tag{9-162}$$

is introduced, then the state equation for this control system is given by

$$\mathbf{z}'(t) = F(t)\mathbf{z}(t) + \mathbf{u}(t) \tag{9-163}$$

where $\mathbf{u}(t)$ is defined as a load-disturbance vector.¶ The elements of the matrix $F(t)$ are given by (9-161). Now the conditions for which $|z_n(t)| < \delta_n(t)$ is valid,

§ These continuity properties, namely, $|\hat{p}_{kl}(\mu)| < \infty$, may not occur even in design problems where the error measure, state equations, and their required derivatives are continuous and bounded. The lack of these properties invalidates not only the second- and higher-degree expansions presented here but also the derivation of the characteristic equations presented in Chap. 6. This situation is unusual, but an example of it is given in Chap. 10, where the variables $\hat{p}_{kl}(\mu)$ are found from actual computation.

¶ For the purposes of this discussion, the assumption is made that $\mathbf{u}(t)$ is not known a priori and is not included in the state equation leading to $\hat{\mathbf{x}}(t)$.

given that $\mathbf{z}(t_0) = \mathbf{0}$, are sought. For this purpose, a special class of load disturbances are defined. That is, these signals are assumed to be bounded so that

$$|u_m(\sigma)| = \begin{cases} \leqq U_m & t_1 \leqq \sigma \leqq t_2 \\ = 0 & t_0 \leqq \sigma \leqq t_1,\ t_2 < \sigma \leqq t \end{cases} \tag{9-164}$$

where U_m is a positive but finite constant. A typical signal within this class is sketched in Fig. 9-5.

For the linear state equation given in (9-163), the components of the incremental state vector $\mathbf{z}(t)$ are written in terms of a sum of superposition integrals as

$$z_n(t) = \sum_{m=1}^{N} \int_{t_0}^{t} w_{nm}(t, \sigma) u_m(\sigma)\, d\sigma \tag{9-165}$$

If the optimum linear control system is asymptotically stable, then the impulse responses are bounded exponentially so that

$$|w_{nm}(t, \sigma)| \leqq w_{nm}(t_0) e^{-\sigma_{nm}(t_0)(t-\sigma)} \qquad t \geqq \sigma \tag{9-166}$$

where $w_{nm}(t_0)$ is a positive but finite constant and $\sigma_{nm}(t_0)$ is a positive but nonzero constant.[51] These constants are shown to be functions of the starting time because

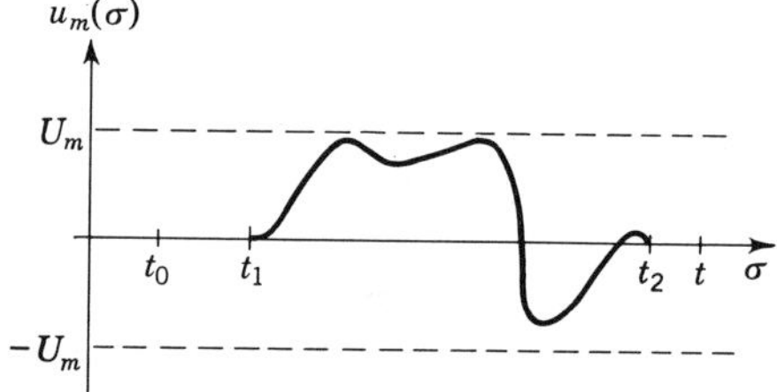

Fig. 9-5 Sketch of permissible load-disturbance signal.

the state equation of the system is, in general, time-varying. The use of the properties

$$|y_1(\sigma) + y_2(\sigma)| \leqq |y_1(\sigma)| + |y_2(\sigma)| \qquad \left| \int_a^b y(\sigma)\, d\sigma \right| \leqq \int_a^b |y(\sigma)|\, d\sigma \tag{9-167}$$

leads to the relationship

$$|z_n(t)| \leqq \sum_{m=1}^{N} \frac{U_m w_{nm}(t_0)}{\sigma_{nm}(t_0)} (1 - e^{-\sigma_{nm}(t_0)(t_2 - t_1)}) \tag{9-168}$$

This relationship is found from (9-165) when (9-164) and (9-166) hold. If the state vector satisfies $\mathbf{x}(t) \in \hat{\mathcal{N}}_P(t)$, when $\hat{\mathcal{N}}_P(t)$ is defined by $|x_n(t) - \hat{x}_n(t)| < \delta_n(t)$, then (9-168) yields

$$\sum_{m=1}^{N} \frac{U_m w_{nm}(t_0)}{\sigma_{nm}(t_0)} (1 - e^{-\sigma_{nm}(t_0)(t_2 - t_1)}) < \delta_n(t) \tag{9-169}$$

This inequality, then, can be considered as a condition that defines a class of permissible disturbances for quasi-optimum control of linear dynamic processes if the undisturbed system is asymptotically stable.

The condition given in (9-169) is pessimistic and can be weakened considerably. Also, this condition is of doubtful value for practical purposes because of the restricted class of disturbances treated here and because of the relative inaccessibility of the constants $w_{nm}(t_0)$ and $\sigma_{nm}(t_0)$. However, (9-169) can be used to demonstrate certain restrictions and difficulties. In particular, this equation clearly indicates that the maximum amplitude U_m and the duration $(t_2 - t_1)$ of the disturbance combine in restricting the class of permissible disturbance signals. Roughly speaking, these signals must be of bounded energy as opposed to bounded power. Perhaps this relationship between amplitude and duration is seen more readily from

$$\sum_{m=1}^{N} (t_2 - t_1)U_m w_{nm}(t_0) < \delta_n(t) \tag{9-170}$$

which is found from (9-169) when $(t_2 - t_1)$ becomes arbitrarily small.

Thus far, only the problem of deterministic disturbances has been discussed. In the case of statistical disturbances, a strict inequality of the type given in (9-169) occurs only when a restriction of the type $|u_m(\sigma)| \leqq U_m$ with a probability of 1 is imposed. A square wave of amplitude $\pm U_m$ with random switching times satisfies this restriction. On the other hand, the common case of a gaussian signal does not satisfy this restriction. If this type of amplitude restriction is not satisfied, then recourse to criteria such as convergence or stability in the mean must be made. Unfortunately, these criteria are not altogether satisfactory from an engineering point of view because stable operation may not be achieved for a given member of the disturbance ensemble.

The problem of disturbances has been introduced in the case where the quasi-optimum system is linear and asymptotically stable in the large. In this situation, incremental departures from the optimum trajectory involve only the accuracy of the system performance in approximating the optimum performance according to the nonquadratic error measure. On the other hand, when the dynamic process is nonlinear or when the quasi-optimum control equation is not linearized, the question of disturbances mainly involves the stability on the nonlinear system. The limits $\delta_n(t)$ bounding the incremental departures from the optimum trajectory in this latter case must be chosen so that stability is maintained in some small region of state space. Analytical methods for selecting $\delta_n(t)$ generally are not available. Hence, the engineer must resort to system simulation in these cases in order to determine whether a satisfactory solution to the design problem has been obtained.

9-7 Design of quasi-optimum control systems for small regions of state space with deterministic signals

In spite of the many impending difficulties pointed out in Sec. 9-6, quasi-optimum control systems have important applications. Primarily, these applications are design problems where the cost and complexities of digital components in the control system must be avoided if at all possible. Of course, this approach must be abandoned altogether unless the region of state space where system operation occurs is suitably small so that system stability is maintained and the degradation from optimum performance remains within suitable bounds. Alternatives to system synthesis for small regions of state space involve some form of table-lookup system, and hence a digital memory, or some form of on-line computing system discussed in Chap. 11 and hence digital memory and computations.

The first step in the design for a small region of state space is the selection of a nominal set of initial conditions $\hat{\mathbf{x}}(t_0)$ and other design parameters. This nominal set of initial conditions determines the optimum trajectory used in the expansion. The second step is the selection of the degree of the expansion used in the approximation of the optimum control equation. This selection is based on a compromise between system complexity and the size of the region in state space where the system must operate. Ideally, the minimum acceptable degree for the expansion should be used for a specified region of operation. However, there is no direct means for finding this minimum.

Generally speaking, the second-degree expansion is used for an initial feasibility study unless this expansion is known to be unacceptable. This selection is based on the knowledge that the number of distinct $\hat{k}$ parameters increases rapidly with the degree of the expansion. The number of first-order differential equations that must be solved given $\hat{\mathbf{x}}(t)$ is equal to the number of distinct $\hat{k}$ parameters.

The number of distinct parameters in the pth-degree terms of the power series is $(N-1+p)!/(N-1)!p!$. Therefore, the total number of distinct $\hat{k}$ parameters for the Pth-degree expansion of the minimum-error function that determine the quasi-optimum control equation is

$$K = \sum_{p=1}^{P} \frac{(N-1+p)!}{(N-1)!p!} \tag{9-171}$$

This number does not include $\hat{k}$. Table 9-1 is included in order to indicate the

Table 9-1 *Total number of distinct $\hat{k}$ parameters: K*

P	N						
	1	2	3	4	5	6	7
1	1	2	3	4	5	6	7
2	2	5	9	14	20	27	
3	3	9	19	34	55		
4	4	14	34	69			
5	5	20	55				
6	6	27					
7	7						

behavior of K for expansions of different degree and for systems of different order. Clearly, the practicality of high-degree expansions is limited to systems of low order. Because of these considerations, the remainder of the discussion in this section is limited to second-degree expansions.

The remainder of the design procedure mainly involves the evaluation of system performance and the validity of the expansion chosen. Principally, this phase of the problem concerns the stability of the system. However, the second-degree expansion must be valid, which requires that the matrix $[T_{ij}(\hat{\mathbf{x}}, \mu)]$, with elements defined in (9-142), be positive definite. This matrix can be checked for positive definiteness along $\hat{\mathbf{x}}(\mu)$ during the trajectory computation. If this condition is not satisfied, then the error measure must be modified for use in the design of the quasi-optimum control system.

Presumably, this modification should not change the optimum trajectory $\hat{\mathbf{x}}(\mu)$ and should alter the characteristics of the system as little as possible. These considerations suggest a modified error measure

$$H_m(\mathbf{x}, \mathbf{m}, \mu) = H(\mathbf{x}, \mathbf{m}, \mu) + \sum_{n=1}^{M} \sum_{m=1}^{M} \hat{h}_{nm}(\mu)[m_n(\mu) - \hat{m}_n(\mu)][m_m(\mu) - \hat{m}_m(\mu)] \quad (9\text{-}172)$$

for use in the design of the quasi-optimum control system. The $\hat{h}_{nm}(\mu)$ coefficients are selected so that the matrix $[T_{ij}(\hat{\mathbf{x}}, \mu) + \hat{h}_{ij}(\mu)]$ is positive definite for all μ on the interval $t \leqq \mu \leqq T$. Clearly, the selection of these coefficients is not unique. However, the coefficients should be chosen to be as close to zero as possible in order to alter the system characteristics from the exact optimum as little as possible. Furthermore, unnecessarily large values for these coefficients tend to decrease the incremental feedback gains of the quasi-optimum control system, thereby accentuating the stability problem and generally decreasing the allowable size of the region of state space where the system can operate. The principal effect of introducing the $\hat{h}_{nm}(\mu)$ coefficients is the reduction of infinite incremental feedback gains to finite gains, thereby establishing a local continuity of the control equation. In the remainder of this section, the validity of the second-degree expansion is assumed tacitly.

In design problems where the dynamic process is linear, system stability ceases to be a problem under certain conditions; that is, if the state equation is linear, saturation constraints are avoided, and the linear form of the quasi-optimum control equation is used, then the system is stable. On the other hand, if any one of these three conditions is not true, then system stability must be evaluated very carefully. Generally this problem of stability for the region of system operation in state space must be resolved by system simulation, a time-consuming and costly process at best.

Because of these difficulties, a number of possible approximations may be employed. The material in Chap. 8 and Appendix E describes the consequences of approximations by selecting a quadratic error measure in conjunction with a linear dynamic process.

Another form of approximation can be used for a linear state equation but where the control and/or state signals tend to saturate. In this type of design problem, the "hard" constraints can be approximated by "soft" constraints. Soft constraints generally are introduced by adding appropriate terms to the error measure. For instance, suppose that the saturation constraint $-X_n \leqq x_n(t) \leqq X_n$ is imposed. Then, the error measure used in the minimization is written as

$$H[\mathbf{x}(t), \mathbf{m}(t), t] + H_n\left[\frac{x_n(t)}{X_n}\right]$$

where $H_n(y)$ is the penalty function used for approximating the hard constraint. Functions which have been found useful for this purpose are

$$H_n(y) = y^{2m} \text{ (m very large)} \qquad \text{and} \qquad -\ln\left[\cos\left(\frac{\pi}{2}y\right)\right] \quad (9\text{-}173)$$

This indirect means of imposing constraints can be very useful, and generally close approximations to the exact constraints can be obtained.[67] When this method is used in conjunction with a linear state equation and the linear form of the quasi-optimum control equation, the appropriate increases and decreases in the control-equation gains required for the constraints are introduced automatically along the optimum trajectory. Furthermore, there is a suitably small region of state space about this trajectory, where the system is linear and the saturation constraints are not violated.

The reader should note that this method of approximating hard saturation constraints is not limited to dynamic processes described by a linear state equation. As a matter of fact, the soft constraints are of special significance for nonlinear state equations.

In design problems where the dynamic process is described by a nonlinear state equation, the problem of system stability is most severe. The severity of this case is due to the combination of two difficulties. First, the boundaries of the region of state space where the quasi-optimum system is stable generally are not known. Second, perturbations due to load disturbances, measurement noise, and model inaccuracies cannot be avoided completely in the actual system. Depending on the design problem, satisfactory system operation may not be achieved, or satisfactory system operation with a very small margin of safety may be achieved. In either case, an approximation can be introduced in order to help ensure system stability. Specifically, the error measure can be modified so that the quasi-optimum control system is asymptotically stable to the precomputed optimum trajectory.§ This modification would be introduced in order to minimize the possibility of the development of an unstable situation during system operation.

The construction of the asymptotically stable system generally involves an additional step because the sufficient conditions for asymptotic stability, which were derived in Sec. 9-6, seldom occur for a particular optimum trajectory. That is, $\partial\hat{H}/\partial\hat{x}_k \neq 0$ and hence $\partial\hat{H}/\partial\hat{m}_j \neq 0$ except in very specialized design problems.

The first step in the construction of the asymptotically stable system is to find the optimum trajectory $\hat{\mathbf{x}}(t)$ and the corresponding $\hat{\mathbf{m}}(t)$ for a given $\hat{\mathbf{x}}(t_0)$, $H(\mathbf{x}, \mathbf{m}, \mu)$, and dynamic process. This trajectory is always found with soft-constraint approximations for saturating signals. Saturation must be avoided in general if asymptotic stability is to be obtained. The second step is the computation of a set of control-equation parameters that give asymptotic stability to this optimum trajectory. The sufficient conditions for asymptotic stability found in Sec. 9-6 require only the use of a positive-definite measure of error about this optimum trajectory. Therefore, the selection of the error measure for designing the quasi-optimum control system subject to the requirement of asymptotic stability is not unique.

Presumably, the error measure selected for this purpose should resemble the original error measure as much as possible, thereby preserving many of the properties of the exact optimum control system. According to this reasoning, the original error measure is modified only to the extent required to satisfy the sufficient conditions for asymptotic stability. The modified error measure then is chosen to be

$$\begin{aligned} H_m(\mathbf{x}, \mathbf{m}, \mu) = H(\mathbf{x}, \mathbf{m}, \mu) - H(\hat{\mathbf{x}}, \hat{\mathbf{m}}, \mu) &- \sum_{n=1}^{N} \frac{\partial H(\hat{\mathbf{x}}, \hat{\mathbf{m}}, \mu)}{\partial \hat{x}_n} [x_n(\mu) - \hat{x}_n(\mu)] \\ &- \sum_{m=1}^{M} \frac{\partial H(\hat{\mathbf{x}}, \hat{\mathbf{m}}, \mu)}{\partial \hat{m}_n} [m_n(\mu) - \hat{m}_n(\mu)] \qquad (9\text{-}174) \end{aligned}$$

§ Even though the mathematical concept applies only to the case $T = \infty$, the conditions associated with the *constraint of asymptotic stability* also should be imposed for finite T when the design considerations described here are pertinent.

This error measure satisfies the conditions

$$H_m(\hat{\mathbf{x}}, \hat{\mathbf{m}}, \mu) = 0 \tag{9-175}$$

$$\frac{\partial H_m(\hat{\mathbf{x}}, \hat{\mathbf{m}}, \mu)}{\partial \hat{x}_k} = 0 \qquad \frac{\partial H_m(\hat{\mathbf{x}}, \hat{\mathbf{m}}, \mu)}{\partial \hat{m}_j} = 0 \tag{9-176}$$

and

$$\frac{\partial^2 H_m(\hat{\mathbf{x}}, \hat{\mathbf{m}}, \mu)}{\partial \hat{x}_k\, \partial \hat{x}_l} = \frac{\partial^2 H(\hat{\mathbf{x}}, \hat{\mathbf{m}}, \mu)}{\partial \hat{x}_k\, \partial \hat{x}_l} \qquad \frac{\partial^2 H_m(\hat{\mathbf{x}}, \hat{\mathbf{m}}, \mu)}{\partial \hat{x}_k\, \partial \hat{m}_j} = \frac{\partial^2 H(\hat{\mathbf{x}}, \hat{\mathbf{m}}, \mu)}{\partial \hat{x}_k\, \partial \hat{m}_j}$$
$$\frac{\partial^2 H_m(\hat{\mathbf{x}}, \hat{\mathbf{m}}, \mu)}{\partial \hat{m}_j\, \partial \hat{m}_i} = \frac{\partial^2 H(\hat{\mathbf{x}}, \hat{\mathbf{m}}, \mu)}{\partial \hat{m}_j\, \partial \hat{m}_i} \tag{9-177}$$

Because of (9-176) and

$$\hat{\mathbf{x}}'(\mu) = \mathbf{f}(\hat{\mathbf{x}}, \hat{\mathbf{m}}, \mu) \tag{9-178}$$

the quasi-optimum control system based on the error measure given in (9-174) is asymptotically stable in the small to the trajectory $\hat{\mathbf{x}}(\mu)$. The basis for this result is established in Sec. 9-6.

The quasi-optimum control equation under the constraint of asymptotic stability now is found readily. In order to distinguish between the control equations with and without the constraint of asymptotic stability, the second-degree expansion of the minimum-error function associated with the error measure H_m is defined as

$$E_{m2}[\mathbf{x}(\mu), \mu] = \sum_{n=1}^{N} \sum_{m=1}^{N} \hat{P}_{nm}(\mu)[x_n(\mu) - \hat{x}_n(\mu)][x_m(\mu) - \hat{x}_m(\mu)] \tag{9-179}$$

In addition, the quasi-optimum control vector is defined as $\mathbf{m}^{\ddagger}(\mu)$ and satisfies the property $\hat{\mathbf{m}}^{\ddagger}(\mu) = \hat{\mathbf{m}}(\mu)$. Furthermore, the $\hat{P}$ parameters appearing in (9-179) are given by

$$-\hat{P}'_{kl}(\mu) = \frac{1}{2}\frac{\partial^2 H(\hat{\mathbf{x}}, \hat{\mathbf{m}}, \mu)}{\partial \hat{x}_k\, \partial \hat{x}_l} + \sum_{n=1}^{N}\left[\frac{\partial f_n(\hat{\mathbf{x}}, \hat{\mathbf{m}}, \mu)}{\partial \hat{x}_k}\hat{P}_{nl} + \frac{\partial f_n(\hat{\mathbf{x}}, \hat{\mathbf{m}}, \mu)}{\partial \hat{x}_l}\hat{P}_{nk}\right]$$
$$- \frac{1}{2}\sum_{i=1}^{M}\sum_{j=1}^{M}\left[\frac{\partial^2 H(\hat{\mathbf{x}}, \hat{\mathbf{m}}, \mu)}{\partial \hat{m}_i\, \partial \hat{m}_j}\frac{\partial \hat{m}_i^{\ddagger}}{\partial \hat{x}_k}\frac{\partial \hat{m}_j^{\ddagger}}{\partial \hat{x}_l}\right] \tag{9-180}$$

where the partial derivatives $\partial \hat{m}_i^{\ddagger}/\partial \hat{x}_k$ are given by

$$\sum_{i=1}^{M}\frac{\partial^2 H(\hat{\mathbf{x}}, \hat{\mathbf{m}}, \mu)}{\partial \hat{m}_i\, \partial \hat{m}_j}\frac{\partial \hat{m}_i^{\ddagger}}{\partial \hat{x}_k} = -\frac{\partial^2 H(\hat{\mathbf{x}}, \hat{\mathbf{m}}, \mu)}{\partial \hat{x}_k\, \partial \hat{m}_j} - 2\sum_{n=1}^{N}\frac{\partial f_n(\hat{\mathbf{x}}, \hat{\mathbf{m}}, \mu)}{\partial \hat{m}_j}\hat{P}_{nk} \tag{9-181}$$

Finally, the quasi-optimum control vector $\mathbf{m}^{\ddagger}(t)$ is defined by

$$\frac{\partial H(\mathbf{x}, \mathbf{m}^{\ddagger}, t)}{\partial m_j^{\ddagger}} - \frac{\partial H(\hat{\mathbf{x}}, \hat{\mathbf{m}}, t)}{\partial \hat{m}_j} + 2\sum_{n=1}^{N}\sum_{m=1}^{N}\frac{\partial f_n(\mathbf{x}, \mathbf{m}^{\ddagger}, t)}{\partial m_j^{\ddagger}}\hat{P}_{nm}(t)[x_m(t) - \hat{x}_m(t)] = 0$$
$$j = 1, 2, \ldots, M \tag{9-182}$$

or by the linearized form of the equation

$$m_j^{\ddagger}(t) = \hat{m}_j(t) + \sum_{m=1}^{N}\frac{\partial \hat{m}_j^{\ddagger}}{\partial \hat{x}_m}[x_m(t) - \hat{x}_m(t)] \qquad j = 1, 2, \ldots, M \tag{9-183}$$

In general, the quasi-optimum control systems with and without the design constraint of asymptotic stability exhibit somewhat different characteristics. First, the system responses are different because $\mathbf{m}^{\ddagger}(t) \neq \mathbf{m}^{\dagger}(t)$ except when $\mathbf{x}(t) = \hat{\mathbf{x}}(t)$.

Furthermore, the incremental feedback gains of the linearized equations are different; that is, $\partial \hat{m}_i^\ddagger/\partial \hat{x}_k \neq \partial \hat{m}_i/\partial \hat{x}_k$ because $\hat{P}_{kl}(t) \neq \hat{p}_{kl}(t)$. This statement is verified readily by comparing (9-140) and (9-141) with (9-180) and (9-181). The constraint of asymptotic stability generally requires a redistribution of the values of the $\hat{p}_l$'s and $\hat{p}_{kl}$'s so that increased feedback (and hence a corresponding improvement in the stability properties of the system) is obtained.

The equations presented here for the design of asymptotically stable systems possess a direct analogy to the k equations used in Chap. 7 for linear optimum control systems. In order to demonstrate this analogy, the results of this section first are specialized to the case treated in Chap. 7. Specifically, the assumption is made that

$$\frac{\partial^2 H(\hat{\mathbf{x}}, \hat{\mathbf{m}}, \mu)}{\partial \hat{m}_i \, \partial \hat{m}_j} = 0 \qquad \text{for } i \neq j$$

$$\frac{\partial^2 H(\hat{\mathbf{x}}, \hat{\mathbf{m}}, \mu)}{\partial \hat{x}_k \, \partial \hat{m}_j} = 0 \qquad \text{for all } k, j \tag{9-184}$$

$$\frac{\partial f_n(\hat{\mathbf{x}}, \hat{\mathbf{m}}, \mu)}{\partial \hat{m}_k} = 0 \qquad \text{for } n \neq k$$

$$M = N$$

Furthermore, sets of time functions are defined as

$$\hat{M}_n(\mu) = \hat{m}_n(\mu) \qquad \hat{k}_{kl}(\mu) = \hat{P}_{kl}(\mu) \qquad \hat{k}_k(\mu) = \sum_{l=1}^{N} \hat{P}_{kl}(\mu)\hat{x}_l(\mu)$$

$$\hat{c}_{nn}(\mu) = \frac{\partial f_n(\hat{\mathbf{x}}, \hat{\mathbf{m}}, \mu)}{\partial \hat{m}_n} \qquad \hat{b}_{nk}(\mu) = \frac{\partial f_n(\hat{\mathbf{x}}, \hat{\mathbf{m}}, \mu)}{\partial \hat{x}_k} \tag{9-185}$$

$$\sum_{n=1}^{N} \hat{\phi}_{nn}(\mu)\hat{a}_{nk}(\mu)\hat{a}_{nl}(\mu) = \frac{1}{2}\frac{\partial^2 H(\hat{\mathbf{x}}, \hat{\mathbf{m}}, \mu)}{\partial \hat{x}_k \, \partial \hat{x}_l} \qquad \hat{\psi}_{jj}(\mu) = \frac{1}{2}\frac{\partial^2 H(\hat{\mathbf{x}}, \hat{\mathbf{m}}, \mu)}{\partial \hat{m}_j^2}$$

The choice of notation used in (9-185) becomes clear upon the substitution of these time functions into (9-180) and (9-181). When these two equations are combined, the result is

$$-\hat{k}'_{kl}(\mu) = \sum_{n=1}^{N} \left[\hat{\phi}_{nn}(\mu)\hat{a}_{nk}(\mu)\hat{a}_{nl}(\mu) + \hat{b}_{nk}(\mu)\hat{k}_{nl}(\mu) + \hat{b}_{nl}(\mu)\hat{k}_{nk}(\mu) - \frac{\hat{c}_{nn}{}^2(\mu)}{\hat{\psi}_{nn}(\mu)} \hat{k}_{nk}(\mu)\hat{k}_{nl}(\mu) \right] \tag{9-186}$$

Furthermore, the linearized form of the quasi-optimum control equation given in (9-183) becomes

$$m_j^\ddagger(t) = \hat{M}_j(t) + \frac{\hat{c}_{jj}(t)}{\hat{\psi}_{jj}(t)} \left[\hat{k}_j(t) - \sum_{m=1}^{N} \hat{k}_{jm}(t)x_m(t) \right] \tag{9-187}$$

If (9-186) and (9-187) are compared with (7-34) and (7-26), then the design of asymptotically stable quasi-optimum systems is seen to be almost identical to the design of linear optimum control systems.[66] The differences are due to the fact that the time functions defined in (9-185) are dependent upon the optimum trajectory, which is also the equilibrium trajectory, whereas the time functions associated with linear optimum control systems are completely independent of a trajectory.

The equilibrium trajectory referred to thus far in this section is a trajectory found from system optimization. Even though the design of quasi-optimum control systems for small regions of state space involves second- or higher-degree expansions, the use of Pontryagin's equations is not precluded for finding the optimum trajectory $\hat{\mathbf{x}}(\mu)$ and the corresponding $\hat{\mathbf{m}}(\mu)$ and is necessary when the matrix $[T_{ij}(\hat{\mathbf{x}}, \mu)]$ is not positive definite. As a matter of fact, any equilibrium trajectory can be used for the design of quasi-optimum control systems. An acceptable equilibrium trajectory must be a solution to the state equation, by definition, which corresponds to control signals generated by a stable control system; but it does not have to be the solution to the optimization problem.

9-8 Design of quasi-optimum control systems for small regions of state space with statistical signals

Thus far, the discussion of the design of quasi-optimum control systems has been limited to design problems with deterministic signals. When statistical signals enter the design problem, additional restrictions must be introduced if on-line digital equipment is to be avoided. Furthermore, approximations must be made with respect to the statistical estimators which are available for design purposes. In this section, the assumption is made that only the mean and the conditional mean of the statistical signals are available because of equipment limitations. Also, the variances of these signals are assumed to be small so that approximations of the type

$$\overline{F[\mathbf{s}(\mu), \mu]}^{\mathbf{s}|\mathbf{v}} \approx F[\overline{\mathbf{s}(\mu)}, \mu] + \sum_{i=1}^{S} \frac{\partial F[\overline{\mathbf{s}(\mu)}, \mu]}{\partial \overline{s_i(\mu)}} \overline{[s_i(\mu) - \overline{s_i(\mu)}]}^{s_i|\mathbf{v}}$$

are valid.

If the approach of quasi-optimum control in a small region of state space about a precomputed trajectory is to be used, then this trajectory must be computed on the basis of a priori statistical estimators. For instance, the trajectory could be computed by using the mean values of the statistical signals. On the other hand, conditional means cannot be used because they are functionals of measured signals which occur during system operation. The procedure used here is to find the precomputed trajectory by minimizing the error index, with the statistical signals being replaced by the mean values of these signals. This trajectory is denoted as $\bar{\mathbf{x}}(\mu)$, and the corresponding control vector is denoted as $\bar{\mathbf{m}}(\mu)$. Then conditions are derived for which the optimum trajectory $\hat{\mathbf{x}}(\mu)$ and the corresponding $\hat{\mathbf{m}}(\mu)$ are equal to $\bar{\mathbf{x}}(\mu)$ and $\bar{\mathbf{m}}(\mu)$ to a first approximation. Under these conditions, the optimum trajectory would not be recomputed during system operation because only small perturbations in the statistical signals, and hence small perturbations in the required statistical estimators, occur. These conditions are found to be very restrictive and hence establish clearly the need for a more general approach to many design problems.

The conditions sought here are derived for a slightly specialized case defined by

$$\begin{aligned} \frac{\partial^2 H}{\partial m_i\, \partial m_j} &= \frac{\partial^2 f_n}{\partial m_i\, \partial m_j} = 0 \qquad \text{for } i \neq j \\ \frac{\partial^2 H}{\partial x_k\, \partial m_j} &= 0 \end{aligned} \tag{9-188}$$

The specialization given in the first equation of (9-188) is introduced in order to avoid a matrix inversion such as is required in the solution of (9-141). The specialization in the second equation of (9-188) is given merely for convenience and so that the results correspond to those of Chap. 7.

As stated previously, the precomputed trajectory is found subject to the constraint that the error measure used in the trajectory computation is $H[\overline{\mathbf{s}(\mu)}, \mathbf{x}(\mu), \mathbf{m}(\mu), \mu]$. Corresponding to this constraint, the minimum-error function is defined as

$$\bar{E}[\mathbf{x}(t), t] = \min_{\mathbf{m}(\sigma)} \int_t^T H(\bar{\mathbf{s}}, \mathbf{x}, \mathbf{m}, \sigma)\, d\sigma \tag{9-189}$$

Furthermore, the second-degree expansion of $\bar{E}(\mathbf{x}, \mu)$ is introduced notationally as

$$\bar{E}_2[\mathbf{x}(\mu), \mu] = \bar{k}(\mu) - 2 \sum_{n=1}^{N} \bar{k}_n(\mu)x_n(\mu) + \sum_{n=1}^{N} \sum_{m=1}^{N} \bar{k}_{nm}(\mu)x_n(\mu)x_m(\mu) \tag{9-190}$$

In a similar fashion, the instantaneous minimum-error function is defined as

$$\mathscr{E}[\mathbf{x}(t), t] = \min_{\mathbf{m}(\sigma)} \int_t^T \overline{H(\mathbf{s}, \mathbf{x}, \mathbf{m}, \sigma)}^{\mathbf{s}|\mathbf{v}}\, d\sigma \tag{9-191}$$

and the corresponding second-degree expansion is defined as

$$\mathscr{E}_2[\mathbf{x}(\mu), \mu] = \hat{k}(\mu) - 2\sum_{n=1}^{N} \hat{k}_n(\mu)x_n(\mu) + \sum_{n=1}^{N} \sum_{m=1}^{N} \hat{k}_{nm}(\mu)x_n(\mu)x_m(\mu) \tag{9-192}$$

If $\bar{\mathbf{x}}(\mu)$ is to serve as a precomputed trajectory, then $\hat{\mathbf{x}}(\mu) = \bar{\mathbf{x}}(\mu)$ to a first approximation must be established.

The conditions which validate this approximation can be established directly from suitable structural requirements for $\bar{E}[\mathbf{x}(\mu), \mu]$. Specifically, this function must be insensitive to, or rather approximately independent of, $\mathbf{x}(\mu)$ in the neighborhood of $\bar{\mathbf{x}}(\mu)$. Because second-degree expansions are being used in this development, the function $\bar{E}[\mathbf{x}(\mu), \mu]$ must be an even function in the components of $\mathbf{x}(\mu)$ about $\bar{\mathbf{x}}(\mu)$ up to fifth-degree terms. Therefore, the restrictions

$$\left.\frac{\partial \bar{E}[\mathbf{x}(\mu), \mu]}{\partial x_n(\mu)}\right|_{\mathbf{x}(\mu)=\bar{\mathbf{x}}(\mu)} = 0 \tag{9-193}$$

and

$$\left.\frac{\partial^3 \bar{E}[\mathbf{x}(\mu), \mu]}{\partial x_n(\mu)\, \partial x_m(\mu)\, \partial x_k(\mu)}\right|_{\mathbf{x}(\mu)=\bar{\mathbf{x}}(\mu)} = 0 \tag{9-194}$$

are imposed so that the $\bar{k}_{nm}$ parameters are independent of small perturbations in the components of $\bar{\mathbf{x}}(\mu)$. These restrictions are satisfied when

$$\frac{\partial \bar{H}}{\partial \bar{x}_k} = \frac{\partial \bar{H}}{\partial \bar{m}_j} = 0 \tag{9-195}$$

and

$$\begin{gathered} \frac{\partial^3 \bar{H}}{\partial \bar{x}_i\, \partial \bar{x}_j\, \partial \bar{x}_k} = \frac{\partial^3 \bar{H}}{\partial \bar{x}_i\, \partial \bar{x}_j\, \partial \bar{m}_k} = \frac{\partial^3 \bar{H}}{\partial \bar{x}_i\, \partial \bar{m}_j\, \partial \bar{m}_k} = \frac{\partial^3 \bar{H}}{\partial \bar{m}_i\, \partial \bar{m}_j\, \partial \bar{m}_k} = 0 \\ \frac{\partial^2 \bar{f}_n}{\partial \bar{x}_i\, \partial \bar{x}_j} = \frac{\partial^2 \bar{f}_n}{\partial \bar{x}_i\, \partial \bar{m}_j} = \frac{\partial^2 \bar{f}_n}{\partial \bar{m}_i\, \partial \bar{m}_j} = 0 \end{gathered} \tag{9-196}$$

where the notations $\bar{H} = H(\bar{\mathbf{s}}, \bar{\mathbf{x}}, \bar{\mathbf{m}}, \mu)$ and $\bar{f}_n = f_n(\bar{\mathbf{x}}, \bar{\mathbf{m}}, \mu)$ are used. The conditions given in (9-195) and (9-196) are sufficient for avoiding on-line trajectory computations and hence for avoiding solutions of a two-point boundary-value problem if the perturbations

$$\overline{s_i(\mu)}^{s_i|\mathbf{v}} - \overline{s_i(\mu)}$$

are small. For the limited class of design problems that satisfy these conditions, the statistical variations of

$$\overline{s_i(\mu)}^{s_i|\mathbf{v}}$$

give rise to the need for the on-line solution of just the $\hat{k}_k$ and $\hat{k}_{kl}$ parameters.

In order to indicate the approximations that are necessary to substantiate these conclusions, a slightly more restricted case is pursued. That is, the condition

$$\frac{\partial^3 \bar{H}}{\partial \bar{x}_i \, \partial \bar{x}_j \, \partial \bar{s}_k} = \frac{\partial^3 \bar{H}}{\partial \bar{m}_j^{\,2} \, \partial \bar{s}_k} = 0 \tag{9-197}$$

is imposed, which results in the $\hat{k}_{kl}$ parameters being independent of $\overline{s_i(\mu)}^{s_i|\mathbf{v}}$ to a first approximation. For this condition, only the $\hat{k}_k$ parameters require on-line computation. When the notations

$$\hat{H} = \overline{H(\mathbf{s}, \hat{\mathbf{x}}, \hat{\mathbf{m}}, \mu)}^{\mathbf{s}|\mathbf{v}} \qquad \text{and} \qquad \hat{f}_n = f_n(\hat{\mathbf{x}}, \hat{\mathbf{m}}, \mu)$$

are used, the $\hat{k}$ parameters are defined by

$$-\hat{k}_k' = -\frac{1}{2}\frac{\partial \hat{H}}{\partial \hat{x}_k} + \sum_{n=1}^{N} \frac{\partial \hat{f}_n}{\partial \hat{x}_k}\left(\hat{k}_n - \sum_{m=1}^{N} \hat{k}_{nm}\hat{x}_m\right) - \sum_{n=1}^{N} \hat{f}_n \hat{k}_{nk} - \sum_{l=1}^{N} \hat{k}_{kl}'\hat{x}_l \tag{9-198}$$

$$\begin{aligned} -\hat{k}_{kl}' = {} & \frac{1}{2}\frac{\partial^2 \hat{H}}{\partial \hat{x}_k \, \partial \hat{x}_l} - \sum_{n=1}^{N} \frac{\partial^2 \hat{f}_n}{\partial \hat{x}_k \, \partial \hat{x}_l}\left(\hat{k}_n - \sum_{m=1}^{N} \hat{k}_{nm}\hat{x}_m\right) \\ & + \sum_{n=1}^{N}\left(\frac{\partial \hat{f}_n}{\partial \hat{x}_k}\hat{k}_{nl} + \frac{\partial \hat{f}_n}{\partial \hat{x}_l}\hat{k}_{nk}\right) - \sum_{j=1}^{M} \frac{\hat{R}_{jk}\hat{R}_{jl}}{2\hat{T}_{jj}} \end{aligned} \tag{9-199}$$

where
$$\hat{R}_{jk} = 2\sum_{n=1}^{N}\left[\frac{\partial \hat{f}_n}{\partial \hat{m}_j}\hat{k}_{nk} - \frac{\partial^2 \hat{f}_n}{\partial \hat{x}_k \, \partial \hat{m}_j}\left(\hat{k}_n - \sum_{m=1}^{N} \hat{k}_{nm}\hat{x}_m\right)\right]$$

and
$$\hat{T}_{jj} = \frac{\partial^2 \hat{H}}{\partial \hat{m}_j^{\,2}} - 2\sum_{n=1}^{N} \frac{\partial^2 \hat{f}_n}{\partial \hat{m}_j^{\,2}}\left(\hat{k}_n - \sum_{m=1}^{N} \hat{k}_{nm}\hat{x}_m\right)$$

These equations are derived under the conditions given in (9-188). Furthermore, the equations that define the $\bar{k}$ parameters are found to be

$$\begin{aligned} -\bar{k}_k' = {} & -\sum_{n=1}^{N} \bar{f}_n \bar{k}_{nk} + \sum_{l=1}^{N} \bar{\phi}_{kl}\bar{x}_l + \sum_{l=1}^{N}\left[\sum_{n=1}^{N}(\bar{b}_{nl}\bar{k}_{nk} + \bar{b}_{nk}\bar{k}_{nl})\right]\bar{x}_l \\ & - \sum_{l=1}^{N}\sum_{j=1}^{M}\sum_{n=1}^{N}\sum_{m=1}^{N} \frac{\bar{c}_{nj}\bar{c}_{mj}}{\bar{\psi}_{jj}} \bar{k}_{nk}\bar{k}_{ml}\bar{x}_l \end{aligned} \tag{9-200}$$

and
$$-\bar{k}_{kl}' = \bar{\phi}_{kl} + \sum_{n=1}^{N} (\bar{b}_{nk}\bar{k}_{nl} + \bar{b}_{nl}\bar{k}_{nk}) - \sum_{j=1}^{M}\sum_{n=1}^{N}\sum_{m=1}^{N} \frac{\bar{c}_{nj}\bar{c}_{mj}}{\bar{\psi}_{jj}} \bar{k}_{nk}\bar{k}_{ml} \tag{9-201}$$

Equations (9-200) and (9-201) are derived for the conditions given in (9-188) and (9-195), where the property

$$\bar{k}_n = \sum_{m=1}^{M} \bar{k}_{nm}\bar{x}_m \tag{9-202}$$

has been used. Also, the definitions

$$\bar{\phi}_{kl} = \frac{1}{2}\frac{\partial^2 \bar{H}}{\partial \bar{x}_k\, \partial \bar{x}_l} \qquad \bar{\psi}_{jj} = \frac{1}{2}\frac{\partial^2 \bar{H}}{\partial \bar{m}_j^{\,2}} \tag{9-203}$$

and

$$\bar{b}_{nm} = \frac{\partial \bar{f}_n}{\partial \bar{x}_m} \qquad \bar{c}_{nm} = \frac{\partial \bar{f}_n}{\partial \bar{m}_m} \tag{9-204}$$

are introduced into the derivation of (9-200) and (9-201).

The remainder of the derivation involves finding approximate equations that define $\bar{K}_k(\mu)$ and $\bar{K}_{kl}(\mu)$. These perturbations in $\hat{k}_k(\mu)$ and $\hat{k}_{kl}(\mu)$, respectively, are written as

$$\bar{K}_k(\mu) = \hat{k}_k(\mu) - \bar{k}_k(\mu) \qquad \bar{K}_{kl}(\mu) = \hat{k}_{kl}(\mu) - \bar{k}_{kl}(\mu) \tag{9-205}$$

The additional perturbations

$$\bar{X}_n(\mu) = \hat{x}_n(\mu) - \bar{x}_n(\mu) \qquad \bar{M}_j(\mu) = \hat{m}_j(\mu) - \bar{m}_j(\mu) \qquad \bar{S}_i(\mu) = \overline{s_i(\mu)}^{\,s_i(\mu)|\mathbf{v}} - \overline{s_i(\mu)} \tag{9-206}$$

are introduced for the sake of convenience. Finally, the functions

$$\bar{\Phi}_{ki} = \frac{1}{2}\frac{\partial^2 \bar{H}}{\partial \bar{x}_k\, \partial \bar{s}_i} \qquad \bar{\Psi}_{ji} = \frac{1}{2}\frac{\partial^2 \bar{H}}{\partial \bar{m}_j\, \partial \bar{s}_i} \tag{9-207}$$

are introduced. Now the functions $\hat{H}$, $\hat{\mathbf{f}}$, and the various partial derivatives of these functions which appear in (9-198) and (9-199) are expanded in first-degree Taylor series about $\bar{H}$, $\bar{\mathbf{f}}$, etc. According to the conditions given in (9-195) through (9-197) and the definitions given in (9-203), (9-204), (9-206), and (9-207), the first-degree approximations of the functions are

$$\begin{gathered}
\frac{1}{2}\frac{\partial \hat{H}}{\partial \hat{x}_k} \approx \sum_{i=1}^{N} \bar{\phi}_{ki}\bar{X}_i + \sum_{i=1}^{S} \bar{\Phi}_{ki}\bar{S}_i \qquad \frac{1}{2}\frac{\partial^2 \hat{H}}{\partial \hat{x}_k\, \partial \hat{x}_l} \approx \bar{\phi}_{kl} \\
\frac{1}{2}\frac{\partial \hat{H}}{\partial \hat{m}_j} \approx \bar{\psi}_{jj}\bar{M}_j + \sum_{i=1}^{S} \bar{\Psi}_{ji}\bar{S}_i \qquad \frac{1}{2}\frac{\partial^2 \hat{H}}{\partial \hat{m}_j^{\,2}} \approx \bar{\psi}_{jj} \\
\hat{f}_n \approx \bar{f}_n + \sum_{m=1}^{N} \bar{b}_{nm}\bar{X}_m + \sum_{m=1}^{M} \bar{c}_{nm}\bar{M}_m \qquad \frac{\partial \hat{f}_n}{\partial \hat{x}_m} \approx \bar{b}_{nm} \qquad \frac{\partial \hat{f}_n}{\partial \hat{m}_m} \approx \bar{c}_{nm} \\
\frac{\partial^2 \hat{f}_n}{\partial \hat{x}_i\, \partial \hat{x}_j} \approx \sum_{k=1}^{N} \frac{\partial^3 \bar{f}_n}{\partial \bar{x}_i\, \partial \bar{x}_j\, \partial \bar{x}_k}\bar{X}_k + \sum_{k=1}^{M} \frac{\partial^3 \bar{f}_n}{\partial \bar{x}_i\, \partial \bar{x}_j\, \partial \bar{m}_k}\bar{M}_k \\
\frac{\partial^2 \hat{f}_n}{\partial \hat{x}_i\, \partial \hat{m}_j} \approx \sum_{k=1}^{N} \frac{\partial^3 \bar{f}_n}{\partial \bar{x}_i\, \partial \bar{m}_j\, \partial \bar{x}_k}\bar{X}_k + \sum_{k=1}^{M} \frac{\partial^3 \bar{f}_n}{\partial \bar{x}_i\, \partial \bar{m}_j\, \partial \bar{m}_k}\bar{M}_k \\
\frac{\partial^2 \hat{f}_n}{\partial \hat{m}_i\, \partial \hat{m}_j} \approx \sum_{k=1}^{N} \frac{\partial^3 \bar{f}_n}{\partial \bar{m}_i\, \partial \bar{m}_j\, \partial \bar{x}_k}\bar{X}_k + \sum_{k=1}^{M} \frac{\partial^3 \bar{f}_n}{\partial \bar{m}_i\, \partial \bar{m}_j\, \partial \bar{m}_k}\bar{M}_k
\end{gathered} \tag{9-208}$$

These approximations now are substituted into (9-199) along with the definitions given in (9-205). Furthermore, if the assumption is made that the components of $\bar{\mathbf{S}}(\mu)$ are small so that the increments $\bar{K}_k$, $\bar{K}_{kl}$, $\bar{X}_k$, and $\bar{M}_k$ are also small, then the second- and higher-degree terms in these increments can be neglected. When these manipulations are performed, the first-degree approximation of (9-199) becomes

$$-\bar{K}'_{kl} \approx \sum_{n=1}^{N} (\bar{b}_{nk}\bar{K}_{nl} + \bar{b}_{nl}\bar{K}_{nk}) - \sum_{j=1}^{M} \sum_{n=1}^{N} \sum_{m=1}^{N} \frac{\bar{c}_{nj}\bar{c}_{mj}}{\bar{\psi}_{jj}} (\bar{k}_{nk}\bar{K}_{nl} + \bar{k}_{nl}\bar{K}_{mk}) \quad \text{(9-209)}$$

Equation (9-201) is used in these manipulations. Because of the free-point terminal-boundary conditions, $\bar{K}_{kl}(T) = 0$ so that

$$\bar{K}_{kl}(\mu) \approx 0 \quad \text{(9-210)}$$

is the result of (9-209). Therefore, $\hat{k}_{kl}(\mu) = \bar{k}_{kl}(\mu)$ to a first approximation, and the on-line computation of the double-subscripted $\hat{k}$ parameters is not required.

This approximation procedure now is applied to (9-198) in order to derive the differential equations for the $\bar{K}_k$ parameters. This procedure yields

$$-\bar{K}'_k \approx -\sum_{i=1}^{S} \bar{\Phi}_{ki}\bar{S}_i + \sum_{n=1}^{N} \bar{b}_{nk}\bar{K}_n - \sum_{n=1}^{N} \sum_{m=1}^{M} \bar{c}_{nm}\bar{k}_{nk}\bar{M}_m - \sum_{l=1}^{N} \sum_{j=1}^{M} \sum_{n=1}^{N} \sum_{m=1}^{N} \frac{\bar{c}_{nj}\bar{c}_{mj}}{\bar{\psi}_{jj}} \bar{k}_{nk}\bar{k}_{ml}\bar{X}_l \quad \text{(9-211)}$$

However, the condition that defines the optimum control vector $\hat{\mathbf{m}}(\mu)$ is

$$\frac{\partial \hat{H}}{\partial \hat{m}_j} - 2\sum_{n=1}^{N} \frac{\partial \hat{f}_n}{\partial \hat{m}_j}\left(\hat{k}_n - \sum_{m=1}^{N} \hat{k}_{nm}\hat{x}_m\right) = 0 \quad \text{(9-212)}$$

and the first-degree approximation of this condition yields

$$\bar{M}_j \approx -\sum_{i=1}^{S} \frac{\bar{\Psi}_{ji}}{\bar{\psi}_{jj}} \bar{S}_i + \sum_{n=1}^{N} \frac{\bar{c}_{nj}}{\bar{\psi}_{jj}} K_n - \sum_{n=1}^{N} \sum_{m=1}^{N} \frac{\bar{c}_{nj}}{\bar{\psi}_{jj}} \bar{k}_{nm}\bar{X}_m \quad \text{(9-213)}$$

When (9-211) and (9-213) are combined, the result is

$$-\bar{K}'_k \approx \sum_{i=1}^{S}\left(-\bar{\Phi}_{ki} + \sum_{j=1}^{M} \sum_{n=1}^{N} \frac{\bar{\Psi}_{ji}}{\bar{\psi}_{jj}} \bar{c}_{nj}\bar{k}_{nk}\right)\bar{S}_i + \sum_{n=1}^{N}\left(\bar{b}_{nk} - \sum_{j=1}^{M} \sum_{m=1}^{N} \frac{\bar{c}_{nj}\bar{c}_{mj}}{\bar{\psi}_{jj}} \bar{k}_{mk}\bar{K}_n\right) \quad \text{(9-214)}$$

This equation is seen to be independent of the perturbation vector $\bar{\mathbf{X}}(\mu)$, and therefore the trajectory $\hat{\mathbf{x}}(\mu)$ is independent to a first approximation of small perturbations in the components of $\bar{\mathbf{S}}(\mu)$.

In this section, sufficient conditions are obtained for the design of quasi-optimum control systems with statistical signals so that the on-line recomputation of optimum trajectories can be avoided. Although these conditions are indeed restrictive, significant system simplifications are obtained when the appropriate approximations can be made without an objectionable decrease in system performance. Furthermore, the approximations that result in (9-210) avoid the necessity of the recomputation of the $\hat{k}_{kl}$ parameters. The corresponding equation for the $\hat{k}_k$ parameters given in (9-214) is linear. Hence, if the components of the vector $\overline{\mathbf{s}(\mu)}^{\mathbf{s}|\mathbf{v}}$ are separable, then the procedure discussed in Chap. 7 can be used here. In this case, no on-line computations are required, thereby eliminating the need for digital equipment in the construction of the system.

9-9 Summary

The class of design problems where the dynamic process is nonlinear, where the error measure is nonquadratic, but where system operation is confined to a suitably small region of state space is treated in this chapter. For this class of problems, a reasonably versatile synthesis procedure is developed whereby a compromise between system complexity and system performance can be selected.

System complexity involves the number of terms appearing in the control equation and a corresponding number of ordinary differential equations that must be solved numerically. These ordinary differential equations comprise a one-point boundary-value problem and hence are amenable to standard numerical techniques.

System performance involves both the stability of the quasi-optimum control system and the accuracy with which the exact optimum system is approximated by the quasi-optimum system. The accuracy of this approximation increases with increased system complexity. Also, the size of the region of state space where system stability is maintained generally increases with increased system complexity. Unfortunately, the complexity required for stable operation in a region of specific dimensions cannot be obtained, in general, from theoretical considerations and hence must be determined from system simulation.

Because of many similarities between the numerical and theoretical aspects of linear optimum and quasi-optimum control systems, the design considerations discussed in Chap. 8 apply to the class of systems discussed in this chapter. For instance, the weighting-factor selection procedure given for linear optimum controls can be used for selecting weighting factors in the error measure used for quasi-optimum controls. The squares of maximum allowable errors, etc., which appear in (8-105), (8-107), and (8-110), would be replaced by the mathematical functions of the errors appearing in the error measure.

However, a number of additional design considerations must be introduced for quasi-optimum control with saturation constraints and nonlinear state equations. Principally, these considerations concern system stability, and the quasi-optimum system often is designed subject to the constraint of asymptotic stability. In order to impose this constraint, hard saturation constraints are replaced by soft saturation constraints. Finally, a treatment of statistical signals in quasi-optimum control systems is presented for a class of design problems where on-line optimum trajectory computations can be avoided. If the statistical signals are separable, then all on-line computations are avoided for this class of design problems, thereby eliminating the need for digital system components.

An illustration of a number of these design considerations is given in Sec. 10-9 via a simple example. This example is included in the next chapter because the computation of $\hat{\mathbf{x}}(\mu)$ is required.

10

Computation Methods for Solving Two-point Boundary-value Problems

The computational problems associated with the solution of two-point boundary-value problems have thus far been avoided. In the case of linear optimum controls, the two-point boundary-value problem completely decouples into a sequence of two one-point boundary-value problems. In the case of quasi-optimum controls for small regions of state space, the material is presented in Chap. 9 on the basis that an optimum trajectory has already been found as the solution of a two-point boundary-value problem. Therefore, if the design of quasi-optimum controls is to be carried to completion, then suitable computational procedures must be developed.

10-1 The two-point boundary-value problem

The two-point boundary-value problem of interest here concerns the numerical solution of the two sets of differential equations

$$\mathbf{x}'(\sigma) = \mathbf{f}[\mathbf{x}(\sigma), \mathbf{m}(\sigma), \sigma] \tag{10-1}$$

and

$$\mathbf{y}'(\sigma) = \mathbf{F}[\mathbf{x}(\sigma), \mathbf{y}(\sigma), \mathbf{m}(\sigma), \sigma] \tag{10-2}$$

The vector $\mathbf{y}(\sigma)$ is used to denote either the set of $\hat{p}$ parameters or the set of $\hat{k}$ parameters obtained from the Pth-degree expansion of the minimum-error function for $P \geqq 1$. The solutions of these differential equations must satisfy the boundary conditions

$$\mathbf{x}(t) = \hat{\mathbf{x}}(t) \qquad \mathbf{y}(T) = \hat{\mathbf{y}}(T) \tag{10-3}$$

where the components of $\hat{\mathbf{x}}(t)$ and $\hat{\mathbf{y}}(T)$ are assumed to be given sets of numbers. Also, these differential equations are coupled by

$$\mathbf{m}(\sigma) = \mathbf{P}[\mathbf{x}(\sigma), \mathbf{y}(\sigma), \sigma] \qquad t \leqq \sigma \leqq T \tag{10-4}$$

Clearly, any solutions of (10-1) and (10-2) must be performed simultaneously with (10-4). However, these solutions cannot be initiated at either $\sigma = t$ or $\sigma = T$ because neither $\mathbf{y}(t)$ nor $\mathbf{x}(T)$ is known.

An obvious approach to resolving this computational dilemma is to select a set of boundary conditions on a trial-and-error basis. For instance, suppose that the boundary conditions $\mathbf{x}(T) = \mathbf{x}^{(0)}(T)$ are chosen arbitrarily and (10-1), (10-2), and (10-4) are solved simultaneously in the backward-time direction. If the corresponding solution is denoted as $\mathbf{x}^{(0)}(\sigma)$, then the vector $\mathbf{x}^{(0)}(t)$ can be compared with the required vector $\hat{\mathbf{x}}(t)$. Presumably, the components of these two vectors are not identical, and additional trials must be made. Suppose that $N + 1$ trials with the choices labeled as $\mathbf{x}^{(i)}(T)$ for $i = 0, 1, \ldots, N$ are performed and the corresponding $\mathbf{x}^{(i)}(t)$ are recorded. Then further selections can be made from a linear approximation to the actual relationship between $\mathbf{x}(T)$ and $\mathbf{x}(t)$. More explicitly, this approximation is defined as

$$x_n(T) = l_n + \sum_{m=1}^{N} l_{nm} x_m(t) \qquad n = 1, 2, \ldots, N \tag{10-5}$$

where the coefficients l_n and l_{nm} are determined from the data obtained from the first $N + 1$ trials. If the optimum system is in fact linear, then (10-5) is exact and the $(N + 2)$nd trial gives precisely the same required result as that discussed in Sec. 4-4. However, the cases of interest are nonlinear, and (10-5) is only an approximation to the set of nonlinear equations

$$\mathbf{x}(T) = \mathbf{L}[\mathbf{x}(t)] \tag{10-6}$$

where the approximation is determined from $N + 1$ points on the set of functions $\mathbf{L}$.

This approach of *boundary-condition iteration* seems to offer some hope at the outset. For instance, a more detailed investigation of the problem leads to the conclusion that problems can be restated as "hill-climbing" problems; hence a large body of material on trial-and-error minimization methods can be brought to bear.[68, 69] Such a restatement is accomplished by defining an appropriate distance measure or norm $\|\hat{\mathbf{x}}(t) - \mathbf{x}^{(i)}(t)\|$. This norm then is used to define a surface S as

$$S = \|\hat{\mathbf{x}}(t) - \mathbf{L}^{-1}[\mathbf{x}^{(i)}(T)]\| \tag{10-7}$$

where the functions $\mathbf{L}^{-1}$ are used to denote the inverse of the relationship given in (10-6). The selection problem now is stated as the selection of $\mathbf{x}^{(i)}(T)$ in order to find the minimum value, namely, $S = 0$, of the surface.

However, a number of difficulties occur in the application of trial-and-error minimization methods to this problem. First, there is a "high cost per trial" because each point on the surface is found from a numerical solution of (10-1), (10-2), and (10-4). Second, the convexity of the surface S is dependent upon both the norm chosen and the functions $\mathbf{L}^{-1}$ associated with the particular design problem at hand. Of course, these functions are not known a priori. If the surface S is not convex—or, more strongly, not strictly convex—then the convergence of practical trial-and-error minimization methods cannot be guaranteed. Moreover, because of the high cost per trial, a rapidly converging method is very desirable. Finally, numerical accuracy in the computation of the points on the surface S plays an important role in the efficiency of trial-and-error methods. As discussed in Chap. 4, the equations that result from the calculus of variations—or, more generally, the characteristic equations—are always unstable. This instability makes the incidence of significant numerical errors very difficult to avoid. When significant errors are present, the points on the surface S are not obtained precisely, thereby further jeopardizing the convergence of the trial-and-error method being used.

The proponents of dynamic programming point to many of these difficulties in obtaining the numerical solution of variational problems from the characteristic equations. Therefore, the alternative approach of discrete dynamic programming must be considered as a numerical procedure. Discrete dynamic programming involves both the quantizing of each component of the state vector and the truncation of the time interval into an appropriate number of segments. These discrete variables lead to the use of first differences for derivatives and sums for integrals as approximations. The discrete version of the dynamic-programming condition for minimum error then is solved for all discrete points in state space formed by the quantization of $\mathbf{x}(t)$ at each point in time. The problem is solved in the backward-time direction because the only known boundary condition is $E[\mathbf{x}(T), T] = 0$ for the free-point case. Also, the step from one point in time to the next requires the computer storage of the discrete version of the minimum-error function from the previous step.

This *flooding* procedure, where the solutions are found for all discrete points in state space, leads to significant practical difficulties. The primary difficulty is the computer memory which is required. In particular, the temporary storage of the minimum-error function requires Q^N computer memory locations, where Q is the number of quantizing levels assumed for each state signal and N is the order of the dynamic process. This memory requirement becomes excessive even for reasonably simple systems. For instance, suppose that 1 per cent sensors are to be used in the system so that Q is taken to be greater than or equal to 100. Then the computer must have a memory size something in excess of 1 million registers for a third-order system, a truly discouraging prospect to base a general computing procedure on. Other difficulties arise; for example, maximum accuracy generally requires a grid size for state space that varies as a function of how far each point in time is from the terminal time.

A number of artificial means have been suggested for eliminating the problems of dimensionality associated with discrete dynamic programming. For instance, Lagrange multipliers can be introduced in order to reduce the effective order of the dynamic process or to restrict the magnitude of the state signals that occur during the solution and hence reduce the number of points required for a given grid size.[70] However, if the Lagrange multipliers are fixed a priori, constraints are introduced and only suboptimum solutions are obtained. If the Lagrange multipliers are to be adjusted so that no effective constraints are introduced in the case of dimensionality reduction, then recourse to a trial-and-error selection procedure for the Lagrange multiplier must be made. This selection procedure requires an iterative solution of the problem, and a compromise between computer memory size and the computer time required in obtaining the solution has been introduced.

Another approach to the memory-requirement problem is the approximation of the discrete points obtained for the minimum-error function by an orthogonal series at each point in time.[71] Instead of the points themselves being stored, the coefficients of the series for each point in time are stored temporarily for use in the next step. Presumably, the number of coefficients required in the expansion is significantly less than the original number of grid points for an acceptable truncation of the orthogonal series. Because of considerations of the rate of convergence and the algebraic simplicity of computing the coefficients, Chebyshev polynomials have been suggested for use in this approximation. The number of coefficients required in a Pth degree expansion of an N-dimensional function is $(P + 1)^N$. This number may also be very large if P must be large in order to obtain sufficient accuracy. Furthermore, additional computing time is required to compute the coefficients of the series. Another point of difficulty arises because of the approximating properties of orthogonal functions. Specifically, expansions in terms of orthogonal functions minimize the mean-square-error between the approximation and the points found on the minimum-error function. However, the optimum control signals are dependent only upon the partial derivatives $\partial E/\partial x_n$. Even though the points may be matched exactly by taking a high enough degree of expansion, the partial derivatives of the series expansion on which the control equation is based may be in large error at every point in state space. These errors may give rise to limit cycles and other instability problems as well as to a general degradation in performance of the system constructed from these approximations.

In addition to the computing difficulties associated with flooding and methods which require the "over solution" of the problem, especially in the case of design for small regions of state space, discrete dynamic programming generally requires a gross departure from a fundamentally continuous problem. This departure introduces many questions concerning the propagation of truncation and quantizing errors during solution. These questions are unresolved for the most part, and significant errors may be encountered. Therefore, only methods involving the continuous form of the two-point boundary-value problem are examined from here on in this chapter.

The continuous computational procedure discussed in detail here is developed from properties and simplifications observed in the case of linear optimum controls. In the case of linear optimum controls, the two-point boundary-value problem decouples into a sequence of two one-point boundary-value problems. Also, when this decoupling occurs, the set of unstable differential equations decouples into two sets of stable differential equations when one set is solved in the forward-time direction and the other set is solved in the backward-time direction. In the case of nonlinear optimum controls, (10-1) and (10-2) are decoupled artificially by neglecting the minimization process that specifies (10-4). Instead, the control vector is treated as a set of independent time functions.

Suppose that the control vector used in the solution of (10-1) is the set of known independent time functions $\mathbf{m}^{(i)}(\sigma)$. Then the corresponding state vector, denoted as $\mathbf{x}^{(i)}(\sigma)$, is found by integrating

$$\mathbf{x}^{(i)\prime}(\sigma) = \mathbf{f}[\mathbf{x}^{(i)}(\sigma), \mathbf{m}^{(i)}(\sigma), \sigma] \tag{10-8}$$

in the forward-time direction from the known boundary condition

$$\mathbf{x}^{(i)}(t) = \hat{\mathbf{x}}(t) \tag{10-9}$$

After this forward-time computation has been performed, the data required for a solution of (10-2), namely, $\mathbf{m}(\sigma)$ and $\mathbf{x}(\sigma)$ now entering as sets of independent time functions, are available. Therefore, the vector $\mathbf{y}(\sigma)$, which corresponds to $\mathbf{x}^{(i)}(\sigma)$ and $\mathbf{m}^{(i)}(\sigma)$, is denoted as $\mathbf{y}^{(i)}(\sigma)$ and is found by integrating

$$\mathbf{y}^{(i)\prime}(\sigma) = \mathbf{F}[\mathbf{x}^{(i)}(\sigma), \mathbf{y}^{(i)}(\sigma), \mathbf{m}^{(i)}(\sigma), \sigma] \tag{10-10}$$

in the backward-time direction from the known boundary conditions

$$\mathbf{y}^{(i)}(T) = \hat{\mathbf{y}}(T) \tag{10-11}$$

Therefore, (10-1) and (10-2) can be solved as a sequence of two one-point boundary-value problems if the control vector is treated as a set of independent time functions.

In general, the assumed vector $\mathbf{m}^{(i)}(\sigma)$ and the solutions $\mathbf{x}^{(i)}(\sigma)$ and $\mathbf{y}^{(i)}(\sigma)$ do not satisfy (10-4); hence the optimum set of time functions has not been found. If another set of time functions $\mathbf{m}^{(i+1)}(\sigma)$ can always be found such that system performance is improved, then the two-point boundary-value problem can be solved to any degree of accuracy desired by solving a sequence of one-point boundary-value problems. This method of neglecting the minimization process which gives (10-4) is a *relaxation* procedure whereby a sequence of control vectors $\mathbf{m}^{(i)}(\sigma)$ is constructed. This sequence is constructed so that each successive vector decreases the error index without violating the boundary conditions and the design constraints and hence improves system performance.

10-2 Introduction to relaxation methods based on first variations

The method of relaxation, which is introduced informally in Sec. 10-1, is developed here for the first-order dynamic process

$$x'(\sigma) = f[x(\sigma), m(\sigma), \sigma] \tag{10-12}$$

and the error index

$$e(t) = \int_t^T H[x(\sigma), m(\sigma), \sigma]\, d\sigma \tag{10-13}$$

The results obtained here for methods based on first variations are essentially the same as those of Kelley[72] and Bryson.[73] However, the derivation presented here is considerably different from those of Kelley and Bryson. Specifically, this derivation does not require the use of adjoint operators and hence readily extends to a method based on second variations which possesses important implications.

As discussed in Sec. 10-1, the heart of the proposed relaxation procedure is the selection algorithm for obtaining a set of time functions $\mathbf{m}^{(i+1)}(\sigma)$ from the data of the previous iteration—namely, $\mathbf{m}^{(i)}(\sigma)$, $\mathbf{x}^{(i)}(\sigma)$, and $\mathbf{y}^{(i)}(\sigma)$—such that a sequence convergent to the optimum is obtained. The convergence of this sequence is examined with the integral

$$e^{(i+1)} = \int_t^T [H^{(i+1)} + p^{(i+1)}(f^{(i+1)} - x^{(i+1)\prime})]\, d\sigma \tag{10-14}$$

The shorthand notation

$$H^{(i+1)} = H[x^{(i+1)}(\sigma), m^{(i+1)}(\sigma), \sigma] \qquad f^{(i+1)} = f[x^{(i+1)}(\sigma), m^{(i+1)}(\sigma), \sigma] \quad (10\text{-}15)$$

is adopted for convenience in (10-14), and time-function notation is dropped. The time function $p^{(i+1)}$ is a Lagrange multiplier, with the notation being selected in conformity with Chaps. 6 and 9. The Lagrange multiplier, of course, is adjusted so that eventually the error index given in (10-13) is minimized subject to the constraint given in (10-12).

The integral given in (10-14) now is expanded in a Taylor series about the previous iteration, namely, the ith iteration. This expansion is facilitated by the introduction of the incremental variables

$$M^{(i)}(\sigma) = m^{(i+1)}(\sigma) - m^{(i)}(\sigma) \qquad X^{(i)}(\sigma) = x^{(i+1)} - x^{(i)}(\sigma) \quad (10\text{-}16)$$

The Taylor series expansions of $H^{(i+1)}$ and $f^{(i+1)}$ are taken to be first-degree and are written as

$$H^{(i+1)} \approx H^{(i)} + \frac{\partial H^{(i)}}{\partial x^{(i)}} X^{(i)} + \frac{\partial H^{(i)}}{\partial m^{(i)}} M^{(i)} \quad (10\text{-}17)$$

and

$$f^{(i+1)} \approx f^{(i)} + \frac{\partial f^{(i)}}{\partial x^{(i)}} X^{(i)} + \frac{\partial f^{(i)}}{\partial m^{(i)}} M^{(i)} \quad (10\text{-}18)$$

Similarly, the Lagrange multiplier is expanded as

$$p^{(i+1)} \approx p^{(i)} + \frac{\partial p^{(i)}}{\partial x^{(i)}} X^{(i)} \quad (10\text{-}19)$$

because the multiplier is dependent only upon the process state variable. Finally, the relationships

$$X^{(i)\prime} = x^{(i+1)\prime} - x^{(i)\prime} \quad (10\text{-}20)$$

and

$$x^{(i)\prime} = f^{(i)} \quad (10\text{-}21)$$

result from previous definitions. Equation (10-21) is equivalent to (10-8) and is the equation used for the forward-time computation.

These expansions now are substituted into (10-14), and all second-degree terms in $M^{(i)}$ and $X^{(i)}$ are neglected on the assumption that these incremental qualities are small. These manipulations yield

$$e^{(i+1)} \approx e^{(i)} + \int_t^T \left[-p^{(i)} X^{(i)\prime} + \left(\frac{\partial H^{(i)}}{\partial x^{(i)}} + p^{(i)} \frac{\partial f^{(i)}}{\partial x^{(i)}} \right) X^{(i)} + \left(\frac{\partial H^{(i)}}{\partial m^{(i)}} + p^{(i)} \frac{\partial f^{(i)}}{\partial m^{(i)}} \right) M^{(i)} \right] d\sigma \quad (10\text{-}22)$$

The backward-time equation, which corresponds to (10-10), is taken to be

$$-p^{(i)\prime} = \frac{\partial H^{(i)}}{\partial x^{(i)}} + p^{(i)} \frac{\partial f^{(i)}}{\partial x^{(i)}} \quad (10\text{-}23)$$

This relationship is chosen so that the forward- and backward-time equations are also the characteristic equations. When (10-23) is substituted into (10-22), the result is

$$e^{(i+1)} \approx e^{(i)} + \int_t^T (-p^{(i)} X^{(i)\prime} - p^{(i)\prime} X^{(i)} + S^{(i)} M^{(i)})\, d\sigma \quad (10\text{-}24)$$

where the definition

$$S^{(i)} = \frac{\partial H^{(i)}}{\partial m^{(i)}} + p^{(i)} \frac{\partial f^{(i)}}{\partial m^{(i)}} \tag{10-25}$$

also has been introduced. Also, the first term appearing in (10-24) is integrated by parts, thereby obtaining

$$e^{(i+1)} \approx e^{(i)} - [p^{(i)}X^{(i)}]_{\sigma=t}^{\sigma=T} + \int_t^T S^{(i)}M^{(i)}\,d\sigma \tag{10-26}$$

However, the fixed-point initial-boundary condition requires that $x^{(i)}(t) = \hat{x}(t)$, and no constraint due to the dynamic process is imposed at the terminal time for a free-point terminal-boundary condition. Therefore, the conditions

$$X^{(i)}(t) = 0 \qquad p^{(i)}(T) = 0 \tag{10-27}$$

which correspond to (10-9) and (10-11), are imposed, and (10-26) reduces to

$$e^{(i+1)} \approx e^{(i)} + \int_t^T S^{(i)}M^{(i)}\,d\sigma \tag{10-28}$$

This is the equation sought for the purpose of selecting $m^{(i+1)}$.

The procedure used to arrive at (10-28) is the expansion of $e^{(i+1)}$ in terms of first variations along the lines used for the calculus of variations in Chap. 4. The $X^{(i)}$ and $p^{(i)}$ components of the first variations vanish because of the use of (10-21) and (10-23). One might suppose that the $M^{(i)}$ component of the first variations can be made to vanish by setting $S^{(i)} = 0$ and thereby obtaining conditions for an exact minimum. Unfortunately, this is not the case if the original difficulties associated with two-point boundary-value problems are to be avoided. The primary concept of the proposed relaxation procedure is that $m^{(i)}$ and hence $x^{(i)}$ and $p^{(i)}$ are specified previous to the time when $m^{(i+1)}$ is selected by solving first (10-21) and then (10-23). Therefore, $S^{(i)}$ is treated as a completely predetermined time function in (10-28).

The success of the proposed relaxation procedure is dependent upon finding a selection algorithm for $M^{(i)}$ that guarantees convergence to the optimum. In order to decrease the value of the error index, that is, to make $e^{(i+1)} \leqq e^{(i)}$, the approximation given in (10-28) is used to impose the inequality condition

$$e^{(i+1)} - e^{(i)} \approx \int_t^T S^{(i)}M^{(i)}\,d\sigma \leqq 0 \tag{10-29}$$

Equation (10-29) is written with the understanding that the equality holds only when $S^{(i)} = 0$ everywhere on the interval $t \leqq \sigma \leqq T$. A sufficient condition for satisfying (10-29) is selecting $M^{(i)}$ so that the integrand is negative or zero everywhere. This condition leads to

$$M^{(i)} = -\delta^{(i)} \operatorname{sgn} S^{(i)} \tag{10-30}$$

where $\delta^{(i)}$ is defined as $\delta^{(i)} = \delta(x^{(i)}, m^{(i)}, p^{(i)}, \sigma)$ subject to the condition

$$\delta(x^{(i)}, m^{(i)}, p^{(i)}, \sigma) \geqq 0 \tag{10-31}$$

where the equality can occur only for $S^{(i)} = 0$. Also, the sgn function appearing in (10-30) is defined as

$$\operatorname{sgn} x = \begin{cases} 1 & x > 0 \\ 0 & x = 0 \\ -1 & x < 0 \end{cases} \tag{10-32}$$

The reader should note that this equation for selecting $M^{(i)}$ possesses the correct properties at the exact optimum where the sequence terminates; that is, $M^{(i)} = 0$ everywhere on the interval $t \leqq \sigma \leqq T$ when the first variations vanish according to $S^{(i)} = 0$ everywhere.

10-3 Considerations in the selection of step size

The function $\delta^{(i)}$ is shown to be an undetermined function of the data $x^{(i)}$, $m^{(i)}$, and $p^{(i)}$ obtained from the previous iteration. Furthermore, there is no unique determination of this function. However, the selection of $\delta^{(i)}$ is successful only when a converging sequence of control signals is established. In order to test the acceptability of $\delta^{(i)}$ for each iteration, the values of the error index between successive iterations are compared. Specifically, the function $\delta^{(i)}$ is considered acceptable only when

$$e^{(i+1)} \leqq e^{(i)} \tag{10-33}$$

where the equality holds only when $S^{(i)} = 0$ everywhere on $t \leqq \sigma \leqq T$. If $\delta^{(i)}$ has been chosen so that $e^{(i+1)} < e^{(i)}$, then $m^{(i+1)}$ is a better solution to the variational problem than $m^{(i)}$. Presumably, the sequence of control signals obtained from this relaxation procedure converges to the optimum control signal if $\delta^{(i)}$ can always be chosen so that (10-33) is satisfied. If $\delta^{(i)}$ is chosen so that (10-33) is satisfied, then this relaxation procedure is said to "guarantee convergence." However, the question of convergence must be examined in somewhat more detail.§ Furthermore, the rate of convergence, that is, the number of iterations required, is extremely important in evaluating the efficiency of relaxation methods for solving two-point boundary-value problems.

The problem of obtaining convergence relates to the approximations used in the steps leading to the conclusion that $S^{(i)}M^{(i)} \leqq 0$ must be established. Specifically, if the approximately equal relationship appearing in (10-28) is instead an exact equality, then convergence is guaranteed. However, the first-degree Taylor series expansions used to obtain an expression for $e^{(i+1)}$ involve errors which generally increase with increases in the magnitudes of $M^{(i)}$ and $X^{(i)}$. Therefore, the errors in the expression for $e^{(i+1)}$ should be kept tolerably small by keeping the magnitude of $M^{(i)}$, and hence $X^{(i)}$, small. If these errors are small enough, then presumably a converging selection of $M^{(i)}$ is obtained. The problem of selecting a converging step then is the problem of finding a sufficiently small $\delta^{(i)}$. The acceptability of a particular $\delta^{(i)}$ can be determined by computing $e^{(i)}$ and $e^{(i+1)}$ and then performing the test given in (10-33).

On the other hand, the function $\delta^{(i)}$ should be kept as large as possible in order to maintain large step size and hence the largest possible rate of convergence. The selection of $\delta^{(i)}$ for a large step size but maintaining convergence again is something of an art and is dependent upon the particular problem at hand. However, the

§ The convergence problem treated in this chapter concerns only the condition

$$e^{(i+1)} < e^{(i)}$$

such that $\{e^{(i)}\}$ forms a monotonically decreasing sequence of positive numbers. Such a sequence possesses a limit point. The conditions for which this limit point is also e^* are not treated.

actual variation of the error index and the first variation, as given in (10-29), can be compared for the purpose of increasing the rate of convergence after an acceptable $\delta^{(i)}$ has been found. If the minimization problem possesses positive-definite second variations, then the approximation appearing in (10-29) is rewritten as

$$e^{(i+1)} - e^{(i)} \geqq \int_t^T S^{(i)} M^{(i)} \, d\sigma \tag{10-34}$$

where the greater-than property is due to the assumption concerning second variations. Because the rate of convergence should be kept as large as possible with large step sizes, the linear approximations which lead to the expression for first variations should be kept as gross as possible. Therefore, the function $\delta^{(i)}$ is considered to be suitably large if

$$\frac{e^{(i+1)} - e^{(i)}}{\int_t^T S^{(i)} M^{(i)} \, d\sigma} - 1 \geqq \epsilon_s \tag{10-35}$$

where ϵ_s is a positive number. However, the test suggested by (10-35) cannot be checked until after $e^{(i+1)}$ has been computed and hence is not of interest until a successful iteration has been performed, that is, when $\delta^{(i)}$ is suitably small. On the other hand, (10-35) can be used as a basis for selecting a larger step size for the $(i + 2)$nd iteration and hence increasing $\delta^{(i+2)}$. The number ϵ_s is arbitrary although small and must be selected from experience. The only difficulty with selecting ϵ_s too large is that $\delta^{(i+2)}$ may have to be decreased again in order to make $e^{(i+2)} \leqq e^{(i+1)}$, thereby wasting computer time.

The most elementary function for selecting step size is a constant. That is, a simple form for $\delta^{(i)}$ is

$$\delta^{(i)} = \varepsilon \tag{10-36}$$

For this form, a simple procedure for satisfying (10-33) is repeated halving of the constant ε after each trial $M^{(i)}$ that does not give convergence. Also, if (10-35) is not satisfied, then the constant can be doubled for use in the next iteration. The simplified flow diagram shown in Fig. 10-1 summarizes the previous discussions concerning the iterative solution of two-point boundary-value problems using first variations. The diagram indicates the sequence of computations, the required temporary data storage, and the tests for finding an acceptable step size. The algorithm is based on (10-36), and the selection of ε is based on (10-33) and (10-35). The reader should note that the notation used in this diagram (such as the equal sign denoting replacement) is similar to that used in Appendix F. Also, the initialization of the sequence of control signals should be noted. Specifically, the initial signal $m^{(0)}$ is arbitrary and is chosen by the engineer, using any means at his disposal. The details of numerical integration, data storage, and data interpolation are not shown in Fig. 10-1. A flow diagram which includes these details is presented in Appendix F, but a number of alterations should be made if this program is to be used efficiently in conjunction with Fig. 10-1. These alterations are primarily due to the large number of tabulated data points required in this relaxation procedure.

As mentioned previously, the selection of a mathematical form for $\delta^{(i)}$ is somewhat arbitrary. In an effort to increase the rate of convergence, a number of alternatives to (10-36) can be postulated. For instance, the step size, for a constant

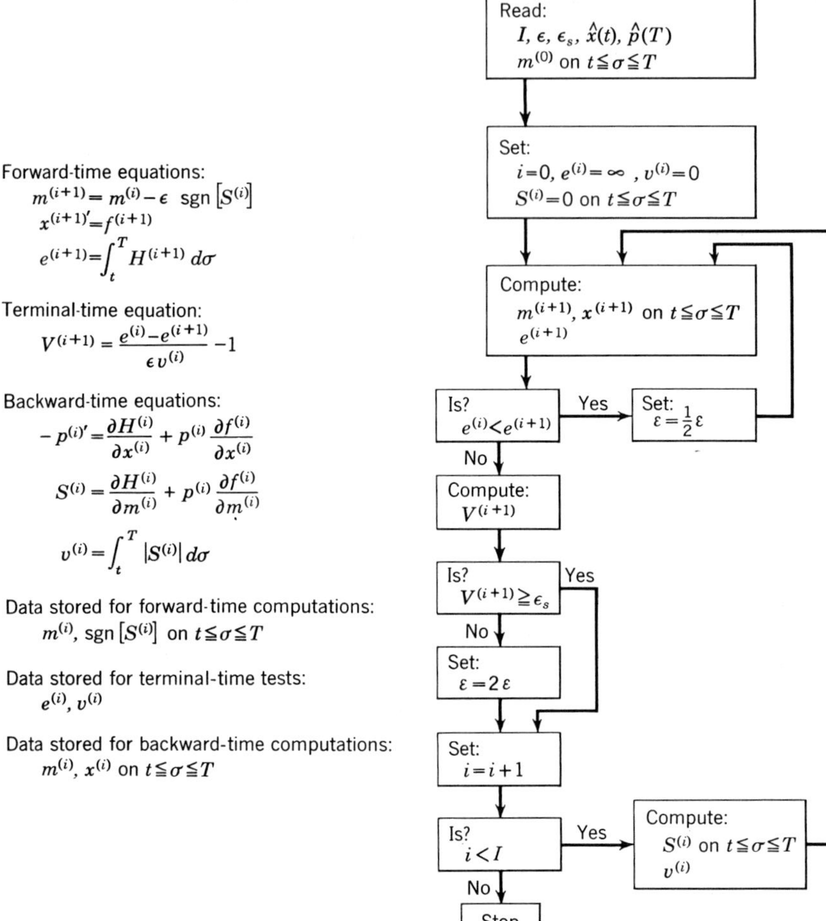

Fig. 10-1 Simplified flow diagram for the solution of two-point boundary-value problems using a relaxation procedure based on first variations.

ε, is a constant over the interval $t \leqq \sigma \leqq T$. However, one might be led to believe that a step size which varies over this interval might be more desirable and that the size at each point in time should reflect the distance between the exact optimum, $S^{(i)} = 0$, and the present solution. These considerations suggest

$$\delta^{(i)} = \varepsilon\, |S^{(i)}| \tag{10-37}$$

and hence

$$M^{(i)} = -\,\varepsilon S^{(i)} \tag{10-38}$$

for the determination of step size.§ Unfortunately, the possible improvements afforded by (10-37) over (10-36) cannot be determined a priori because they are dependent upon the problem at hand.

§ This iteration algorithm is equivalent to the steepest-descent method associated with trial-and-error minimization procedures.

Another possible form for $\delta^{(i)}$ is derived from considering the function $\mathscr{H}$ defined as

$$\mathscr{H} = H(x^{(i)}, m, \sigma) + p^{(i)}f(x^{(i)}, m, \sigma) \tag{10-39}$$

This function is similar to the hamiltonian function used in previous chapters; however, $\mathscr{H}$ is treated here as a function of m and σ only because $x^{(i)}$ is assumed fixed from the previous iteration. Also, the notation

$$\mathscr{H}^{(i)} = H(x^{(i)}, m^{(i)}, \sigma) + p^{(i)}f(x^{(i)}, m^{(i)}, \sigma) \tag{10-40}$$

is used here in conjunction with $\mathscr{H}$. The function $\mathscr{H}$ is sketched in Fig. 10-2 and is shown as a strictly convex function for reasons that will become apparent. Also,

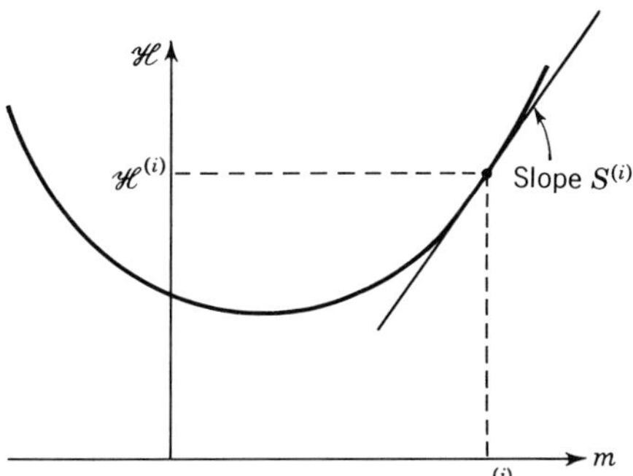

Fig. 10-2 Sketch of $\mathscr{H}$.

the geometrical interpretation of $S^{(i)}$ is shown. This sketch immediately suggests selecting $m^{(i+1)}$ in order to correspond to the minimum point of $\mathscr{H}$. However, this value of $m^{(i+1)}$ generally is too big a step size because of the gross approximating properties of first variations.

As an alternative to attempting to locate $m^{(i+1)}$ at the minimum point, the partial derivative $\partial S^{(i)}/\partial m^{(i)}$ can be used as an added refinement for determining the rate at which the condition $S^{(i)} = 0$ is being achieved. This refinement is introduced with the second-degree Taylor series about $m^{(i)}$:

$$\mathscr{H} \approx \mathscr{H}^{(i)} + S^{(i)}(m - m^{(i)}) + \frac{1}{2}\frac{\partial S^{(i)}}{\partial m^{(i)}}(m - m^{(i)})^2 \tag{10-41}$$

When m is adjusted to minimize the right-hand side of (10-41), the result is

$$m - m^{(i)} = \frac{-S^{(i)}}{\partial S^{(i)}/\partial m^{(i)}}$$

Finally, the step size is restricted by taking some fraction of this difference, thereby obtaining the iteration algorithm §

$$M^{(i)} = \frac{-\epsilon S^{(i)}}{\partial S^{(i)}/\partial m^{(i)}} \tag{10-42}$$

or, rather,

$$\delta^{(i)} = \frac{\varepsilon\,|S^{(i)}|}{\partial S^{(i)}/\partial m^{(i)}} \tag{10-43}$$

§ This iteration algorithm is equivalent to Newton's method associated with trial-and-error minimization procedures.

The division appearing in this iteration formula causes no difficulties under the condition of positive-definite second variations, $\partial S^{(i)}/\partial m^{(i)} > 0$, because of the Legendre condition when $x^{(i)}$ is in a suitably small neighborhood of the optimum trajectory.

As before, the formula given in (10-43) may or may not be superior to the simplest case given in (10-36) for a particular design problem. However, the relaxation procedure based on (10-43) has worked as well as or better than the previously discussed procedures on test problems. Primarily, this performance has been due to the fact that the problem of specifying an initial value for ε is simplified, and the derivative $\partial S^{(i)}/\partial m^{(i)}$ tends to inhibit a tendency to select too large a step size, thereby causing unnecessary repeated halving of ε. The initial value of ε generally is constrained by $0 < \varepsilon < 1$ because $\varepsilon = 1$ is the full step to the minimum value of the approximated $\mathscr{H}$ function or, rather, $S^{(i+1)} \approx 0$.

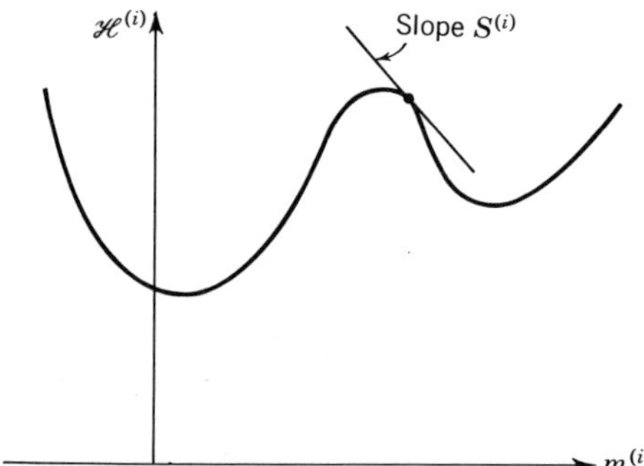

Fig. 10-3 Sketch of a multimodel $\mathscr{H}^{(i)}$.

In addition to providing a somewhat improved iteration algorithm, consideration of the function $\mathscr{H}$ and the geometrical interpretation of $S^{(i)}$ points out a difficulty with stepwise minimization procedures in general. Specifically, if $\mathscr{H}$ is a multimodal function, such as sketched in Fig. 10-3, the slope $S^{(i)}$ is zero for more than one value of $m^{(i)}$. Because $M^{(i)} = 0$ for $S^{(i)} = 0$, the relaxation procedure may terminate at a local minimum or at a local maximum and hence not converge to an absolute minimum. A sufficient condition for convergence to an absolute minimum of $\mathscr{H}$ is that the second variations be positive semidefinite, that is, $\partial S^{(i)}/\partial m^{(i)} \geqq 0$. A sufficient condition for a unique absolute minimum is that the second variation be positive definite and hence $\partial S^{(i)}/\partial m^{(i)} > 0$. As noted previously, this condition is analogous to the Legendre condition. Because of these considerations, further discussion of relaxation procedures is restricted to minimization problems where the Legendre condition is satisfied. Furthermore, only problems where $S^{(i)}$ is continuous are treated.

The final important aspect of selecting step size occurs in conjunction with control-signal saturation. Suppose now that the original minimization problem defined by (10-12) and (10-13) is modified by the addition of the constraint

$$M^-(\sigma) \leqq m(\sigma) \leqq M^+(\sigma) \tag{10-44}$$

The addition of this constraint does not alter the derivation leading to (10-29) but does alter the selection of $M^{(i)}$. The selection procedure for saturation constraints is is established easily from the geometrical relationships depicted in Fig. 10-4. Specifically, the selection of $m^{(i+1)}$ should always be in the direction indicated by the sign

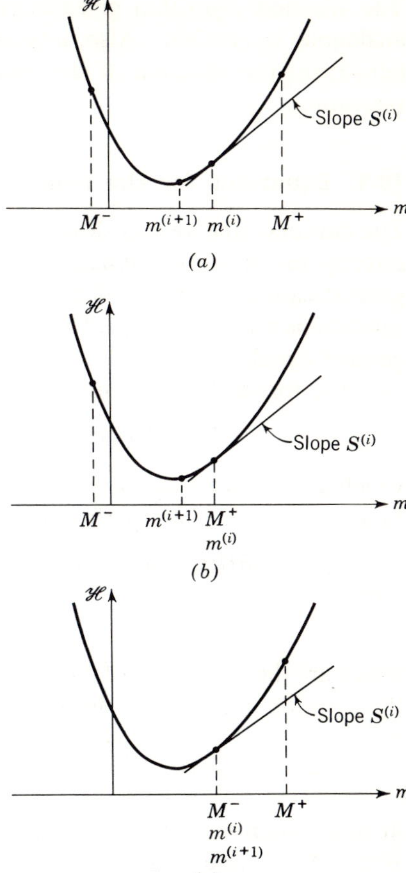

Fig. 10-4 Geometrical relationship between saturation boundary and slope $S^{(i)}$. (*a*) Not on boundary; (*b*) leaving boundary; (*c*) remaining on boundary.

of $S^{(i)}$, provided that a boundary is not encountered. This procedure is expressed as

$$m^{(i+1)} = \begin{cases} M^+ & M^+ \leqq s^{(i)} \\ s^{(i)} & M^- < s^{(i)} < M^+ \\ M^- & s^{(i)} \leqq M^- \end{cases} \tag{10-45}$$

where the switching signal $s^{(i)}$ is obtained from (10-30) and is written as

$$s^{(i)} = m^{(i)} - \delta^{(i)} \operatorname{sgn} S^{(i)} \tag{10-46}$$

Furthermore, this procedure is reexpressed in terms of $M^{(i)}$ by subtracting $m^{(i)}$ from both sides of (10-45), thereby obtaining

$$M^{(i)} = \begin{cases} M^+ - m^{(i)} & M^+ - m^{(i)} \leqq -\delta^{(i)} \operatorname{sgn} S^{(i)} \\ -\delta^{(i)} \operatorname{sgn} S^{(i)} & M^- - m^{(i)} < -\delta^{(i)} \operatorname{sgn} S^{(i)} < M^+ - m^{(i)} \\ M^- - m^{(i)} & -\delta^{(i)} \operatorname{sgn} S^{(i)} \leqq M^- - m^{(i)} \end{cases} \tag{10-47}$$

The selection algorithm given in (10-47) for control-signal saturation constraints is analogous to (10-30). Also, any of the forms of $\delta^{(i)}$ given in (10-36), (10-38), and (10-43) for the selection of step size can be used in conjunction with control-signal saturation.

10-4 Equations for the relaxation procedure based on first variations

The concepts and results developed for a first-order dynamic process extend almost directly to Nth-order dynamic processes. Therefore, the equations for the more general case are developed here as briefly as possible except where additional considerations are necessary. These additional considerations arise in regard to multiple control signals.

The forward-time computations are performed with

$$x_n^{(i)\prime}(\sigma) = f_n(\mathbf{x}^{(i)}, \mathbf{m}^{(i)}, \sigma) \qquad n = 1, 2, \ldots, N \tag{10-48}$$

which are the state equations of the dynamic process. Furthermore, the backward-time computations are performed with

$$-p_n^{(i)\prime}(\sigma) = \frac{\partial H(\mathbf{x}^{(i)}, \mathbf{m}^{(i)}, \sigma)}{\partial x_n^{(i)}} + \sum_{m=1}^{N} p_m^{(i)}(\sigma) \frac{\partial f_m(\mathbf{x}^{(i)}, \mathbf{m}^{(i)}, \sigma)}{\partial x_n^{(i)}} \qquad n = 1, 2, \ldots, N \tag{10-49}$$

which are the second set of equations comprising the characteristic equations. Again the iteration algorithm is derived from the expansion of

$$e^{(i+1)} = \int_t^T \left[H^{(i+1)} + \sum_{n=1}^{N} p_n^{(i+1)}(f_n^{(i+1)} - x_n^{(i+1)\prime}) \right] d\sigma \tag{10-50}$$

in first variations. In this expansion, the first-degree Taylor series expansions of $H^{(i+1)}$, $f_n^{(i+1)}$, and $p_n^{(i+1)}$ are introduced; (10-48) and (10-49) are used; and the fixed-point initial-boundary condition and the free-point terminal-boundary condition are imposed. Then, the first-degree expansion of (10-50) becomes

$$e^{(i+1)} \approx e^{(i)} + \int_t^T \sum_{n=1}^{M} S_n^{(i)} M_n^{(i)}\, d\sigma \tag{10-51}$$

when the definitions

$$S_n^{(i)} = \frac{\partial H(\mathbf{x}^{(i)}, \mathbf{m}^{(i)}, \sigma)}{\partial m_n^{(i)}} + \sum_{m=1}^{N} p_m^{(i)}(\sigma) \frac{\partial f_m(\mathbf{x}^{(i)}, \mathbf{m}^{(i)}, \sigma)}{\partial m_n^{(i)}} \qquad n = 1, 2, \ldots, M \tag{10-52}$$

and

$$M_n^{(i)} = m_n^{(i+1)} - m_n^{(i)} \qquad X_n^{(i)} = x_n^{(i+1)} - x_n^{(i)} \tag{10-53}$$

are introduced.

A sufficient condition for convergence, such that $e^{(i+1)} \leqq e^{(i)}$ as given by (10-51), is

$$\sum_{n=1}^{M} S_n^{(i)} M_n^{(i)} \leqq 0 \tag{10-54}$$

However, an explicit form for $M_n^{(i)}$ must be obtained such as found for the first-order case. A simple method for satisfying (10-54) is to make each term negative thereby obtaining the sufficient condition

$$S_n^{(i)} M_n^{(i)} \leqq 0 \qquad n = 1, 2, \ldots, M \tag{10-55}$$

This condition is written alternatively as

$$M_n^{(i)} = -\delta_n^{(i)} \operatorname{sgn} S_n^{(i)} \tag{10-56}$$

where

$$\delta_n(\mathbf{x}^{(i)}, \mathbf{m}^{(i)}, \mathbf{p}^{(i)}, \sigma) \geqq 0 \tag{10-57}$$

Suitable selections for $\delta_n^{(i)}$ are

$$\delta_n^{(i)} = \varepsilon_n \qquad \delta_n^{(i)} = \varepsilon_n\,|S_n^{(i)}| \qquad \delta_n^{(i)} = \frac{\varepsilon_n\,|S_n^{(i)}|}{\partial S_n^{(i)}/\partial m_n^{(i)}} \tag{10-58}$$

which are analogous to (10-36), (10-37), and (10-43). Finally, the control-signal saturation constraints

$$M_n^-(\sigma) \leqq m_n(\sigma) \leqq M_n^+(\sigma) \tag{10-59}$$

gives rise to the selection formula

$$M_n^{(i)} = \begin{cases} M_n^+ - m_n^{(i)} & M_n^+ - m_n^{(i)} \leqq -\,\delta_n^{(i)} \operatorname{sgn} S_n^{(i)} \\ -\delta_n^{(i)} \operatorname{sgn} S_n^{(i)} & M_n^- - m_n^{(i)} < -\,\delta_n^{(i)} \operatorname{sgn} S_n^{(i)} < M_n^+ - m_n^{(i)} \\ M_n^- - m_n^{(i)} & -\delta_n^{(i)} \operatorname{sgn} S_n^{(i)} \leqq M_n^- - m_n^{(i)} \end{cases}$$

$$n = 1, 2, \ldots, M \tag{10-60}$$

which is analogous to (10-47).

When multiple control signals occur, the problem of step-size selection becomes somewhat more difficult. In particular, there is no direct method for selecting different values of the various ε_n's appearing in (10-58). Therefore, the common practice is to equate these constants so that $\varepsilon_n = \varepsilon$ for $n = 1, 2, \ldots, M$. The step size now is determined by the single constant ε and is adjusted for guaranteed convergence. This simplification in step-size selection has the disadvantage that the rate of convergence generally is decreased.

In an effort to avoid the selection of multiple arbitrary constants and yet not sacrifice rate of convergence, (10-54) is used directly to determine a step-size selection formula. The function $\mathscr{H}$, which is defined in (10-39) for a first-order dynamic process, is expanded in the second-degree series

$$\mathscr{H}^{(i+1)} \approx \mathscr{H}^{(i)} + \sum_{n=1}^{M} S_n^{(i)} M_n^{(i)} + \sum_{n=1}^{M} \sum_{m=1}^{M} \frac{1}{2} T_{nm}^{(i)} M_n^{(i)} M_m^{(i)} \tag{10-61}$$

when the definition

$$T_{nm}^{(i)} = T_{mn}^{(i)} = \frac{\partial^2 H^{(i)}}{\partial m_n^{(i)}\,\partial m_m^{(i)}} + \sum_{k=1}^{N} p_k^{(i)} \frac{\partial^2 f_k^{(i)}}{\partial m_n^{(i)}\,\partial m_m^{(i)}} \tag{10-62}$$

is introduced. The approximate value of $\mathscr{H}^{(i+1)}$ now is minimized with respect to the components of $\mathbf{M}^{(i)}$, assuming that control-signal saturation does not enter the design problem. This minimization yields

$$\sum_{m=1}^{M} T_{nm}^{(i)} M_m^{(i)} = -\varepsilon S_n^{(i)} \qquad n = 1, 2, \ldots, M \tag{10-63}$$

when $\varepsilon = 1$. However, this constant is chosen to be on the interval $0 < \varepsilon < 1$ in order to restrict step size according to the reasoning presented in Sec. 10-3. Now an explicit expression for $M_n^{(i)}$ is obtained by using the notation for the matrix $T^{(i)} = [T_{nm}^{(i)}]$. Specifically, the properties of matrices are used to rewrite (10-63) as

$$M_n^{(i)} = -\varepsilon \sum_{m=1}^{M} \frac{|T^{(i)}|_{nm}}{|T^{(i)}|} S_m^{(i)} \qquad n = 1, 2, \ldots, M \tag{10-64}$$

where $|T^{(i)}|$ is defined as the determinant of $T^{(i)}$ and $|T^{(i)}|_{nm}$ is defined as the cofactor of the nm element of $T^{(i)}$. Because of the Legendre condition, the matrix $T^{(i)}$ is positive definite, and hence $|T^{(i)}| > 0$, when $\mathbf{x}^{(i)}$ is in a suitably small neighborhood of the optimum trajectory.

The selection formula given in (10-64) for no saturation of control signals is a generalization of (10-42) to the case of M control signals. This formula does not in general satisfy the sufficient condition for guaranteed convergence given in (10-55). On the other hand, the less restrictive condition given in (10-54) is satisfied. This is demonstrated from

$$\sum_{n=1}^{M} S_n^{(i)} M_n^{(i)} = -\varepsilon \sum_{n=1}^{M} \sum_{m=1}^{M} \frac{|T^{(i)}|_{nm}}{|T^{(i)}|} S_n^{(i)} S_m^{(i)} \leqq 0 \tag{10-65}$$

because the coefficients of quadratic function appearing in (10-65) form a positive-definite matrix when the matrix $T^{(i)}$ is positive definite. In the special case where the elements $T_{nm}^{(i)} = 0$ for $n \neq m$, (10-64) reduces to the selection formula previously postulated in conjunction with the third form of $\delta_n^{(i)}$ given in (10-58). Therefore, any improvement in the rate of convergence afforded by (10-64) is due to the allowances made for the interaction between the various control signals.

The use of (10-64) as an iteration formula may present some serious difficulties. First, the direct use of (10-64) for design problems with control-signal saturation does not satisfy either (10-54) or (10-55). For instance, one might propose that (10-60) be used, where $\delta_n^{(i)}$ is defined from (10-64) as

$$\delta_n^{(i)} = \varepsilon(\operatorname{sgn} S_n^{(i)}) \sum_{m=1}^{M} \frac{|T^{(i)}|_{nm}}{|T^{(i)}|} S_m^{(i)} \tag{10-66}$$

The function $\delta_n^{(i)}$ does not satisfy $\delta_n^{(i)} \geqq 0$ because (10-64) does not satisfy (10-55). Also, when either M_n^- or M_n^+ is encountered, (10-54) in general is not satisfied either, because the boundary requires $0 \leqq |M_n^{(i)}| < |\delta_n^{(i)}|$, thereby invalidating the quadratic form of (10-65). A simple expedient is to use $|\delta_n^{(i)}|$ instead of $\delta_n^{(i)}$, as given by (10-66), in (10-60). This ensures guaranteed convergence according to (10-55). Unfortunately, this iteration algorithm neglects part of the interaction between the various control signals, and the improvement in the rate of convergence may not warrant the added computations required for $T_{nm}^{(i)}$ and the required matrix inversions. Furthermore, these matrices may prove to be ill-conditioned because of numerical errors in the computations even though the inversions are valid from a theoretical standpoint. The use of (10-64) should be preceded by a very careful consideration of the problem at hand.

10-5 Introduction to relaxation methods based on second variations

The primary factor in the effectiveness of relaxation procedures for solving two-point boundary-value problems is the guarantee of convergence to the optimum control vector. In addition, discrete approximations without an adequate theory of error, as in the case of dynamic programming, are not required. Even though considerable computer memory is required for the temporary storage of the time functions $\mathbf{m}^{(i)}$, $\mathbf{x}^{(i)}$, and $\boldsymbol{\delta}^{(i)}$, the overwhelming memory requirements of dynamic programming also are avoided. Only the problem of selecting an efficient step size and hence obtaining a high rate of convergence remains as an area of further investigation. However, the problem of selecting an "optimum" step size cannot be resolved completely, a dilemma that also occurs in trial-and-error minimization methods.

In most trial-and-error computational methods, an attempt is made to find a suitable compromise between required computer memory and the computational time required for obtaining a solution. Computation time roughly is equal to the time required per iteration times the number of iterations required to find an acceptable solution. The number of iterations, or rather what has been referred to here as the *rate of convergence*, primarily is determined by the efficient selection of step size. Therefore, an attempt is made here to decrease the number of iterations at the expense of computer memory but without an undue increase in computation time required per iteration. Provided that the memory requirements do not increase exponentially with the order of the dynamic process, as in the case of dynamic programming, more complicated relaxation methods may be feasible and more efficient. These are the hypotheses under which the following results must be viewed.

The further development of relaxation procedures is based on a number of observations stemming from the results presented in Chap. 9. When the Legendre condition holds, the optimum control equation is locally continuous and this equation is approximately linear to any degree of accuracy required for a suitably small region of state space. Relaxation methods based on first variations require the validity of the Legendre condition for guaranteed convergence, but the linearity property for small regions of state space is not used. In other words, there is no reason to conclude that the rate of convergence increases when the solution $\mathbf{x}^{(i)}(\sigma)$ approaches the optimum solution $\hat{\mathbf{x}}(\sigma)$. With a properly constructed relaxation procedure, the hope is that regions where the control equation is quasi-linear can be traversed very quickly and hence decrease the number of iterations required. Moreover, if the solution $\mathbf{x}^{(i)}(\sigma)$ and the optimum solution $\hat{\mathbf{x}}(\sigma)$ are both in a region where the optimum control equation is exactly linear on the entire interval $t \leqq \sigma \leqq T$, one-step convergence should be obtained. From these arguments, a relaxation procedure that directly depends on the degree of linearity of the optimum control equation is sought.

The results presented in Chap. 9 demonstrate that linear optimum controls, and the corresponding complete decoupling of the two-point boundary-value problem into a sequence of two one-point boundary-value problems, occur with the second degree expansion of the minimum-error function. Therefore, the backward-time computation in a relaxation procedure which takes into account the degree of linearity of the optimum control equation is based on the second-degree expansion. This expansion occurs in conjunction with the first and second variations of the error index.

The relaxation procedure based on second variations is introduced in terms of a first-order dynamic process; hence the forward-time computation is performed with

$$x^{(i+1)\prime} = f(x^{(i+1)}, m^{(i+1)}, \sigma) \tag{10-67}$$

The backward-time computation is based on both the single-subscripted and the double-subscripted p parameters. The equations for these parameters are given in (9-139) and (9-140). If the assumption is made that control-signal saturation does not occur, then the backward-time computations are performed with

$$-p_1^{(i)\prime} = \frac{\partial H^{(i)}}{\partial x^{(i)}} + p_1^{(i)} \frac{\partial f^{(i)}}{\partial x^{(i)}} \tag{10-68}$$

and

$$-p_{11}^{(i)\prime} = \frac{1}{2}\frac{\partial^2 H^{(i)}}{\partial x^{(i)2}} + \frac{1}{2} p_1^{(i)} \frac{\partial^2 f^{(i)}}{\partial x^{(i)2}} + 2p_{11}^{(i)} \frac{\partial f^{(i)}}{\partial x^{(i)}} - \frac{1}{2}\frac{R^{(i)2}}{T^{(i)}} \tag{10-69}$$

The notation used in (10-68) and (10-69) is defined in (10-15), but the $p^{(i)}$ parameters are subscripted for the purpose of identification. The time function $T^{(i)}$ is defined to be

$$T^{(i)} = \frac{\partial^2 H^{(i)}}{\partial m^{(i)2}} + p_1^{(i)} \frac{\partial^2 f^{(i)}}{\partial m^{(i)2}} \tag{10-70}$$

which corresponds to the $T_{11}^{(i)}$ element defined in (10-62). Also, the time function $R^{(i)}$ is defined as

$$R^{(i)} = \frac{\partial^2 H^{(i)}}{\partial x^{(i)}\, \partial m^{(i)}} + p_1^{(i)} \frac{\partial^2 f^{(i)}}{\partial x^{(i)}\, \partial m^{(i)}} + 2p_{11}^{(i)} \frac{\partial f^{(i)}}{\partial m^{(i)}} \tag{10-71}$$

Finally, the definition

$$S^{(i)} = \frac{\partial H^{(i)}}{\partial m^{(i)}} + p_1^{(i)} \frac{\partial f^{(i)}}{\partial m^{(i)}} \tag{10-72}$$

is used in the subsequent derivation.

The conditions for selecting the increment $M^{(i)}$ are found by expanding

$$e^{(i+1)} = \int_t^T [H^{(i+1)} + p_1^{(i+1)}(f^{(i+1)} - x^{(i+1)\prime})]\, d\sigma \tag{10-73}$$

in first and second variations. This expansion is based on the second-degree Taylor series.

$$H^{(i+1)} \approx H^{(i)} + \frac{\partial H^{(i)}}{\partial x^{(i)}} X^{(i)} + \frac{\partial H^{(i)}}{\partial m^{(i)}} M^{(i)} + \frac{1}{2}\frac{\partial^2 H^{(i)}}{\partial x^{(i)2}} X^{(i)2} + \frac{\partial^2 H^{(i)}}{\partial x^{(i)}\, \partial m^{(i)}} X^{(i)} M^{(i)} + \frac{1}{2}\frac{\partial^2 H^{(i)}}{\partial m^{(i)2}} M^{(i)2} \tag{10-74}$$

and

$$f^{(i+1)} \approx f^{(i)} + \frac{\partial f^{(i)}}{\partial x^{(i)}} X^{(i)} + \frac{\partial f^{(i)}}{\partial m^{(i)}} M^{(i)} + \frac{1}{2}\frac{\partial^2 f^{(i)}}{\partial x^{(i)2}} X^{(i)2} + \frac{\partial^2 f^{(i)}}{\partial x^{(i)}\, \partial m^{(i)}} X^{(i)} M^{(i)} + \frac{1}{2}\frac{\partial^2 f^{(i)}}{\partial m^{(i)2}} M^{(i)2} \tag{10-75}$$

and on the first-degree Taylor series

$$p_1^{(i+1)} \approx p_1^{(i)} + 2p_{11}^{(i)} X^{(i)} \tag{10-76}$$

The property

$$p_{11}^{(i)} = \frac{1}{2}\frac{\partial p_1^{(i)}}{\partial x^{(i)}} \tag{10-77}$$

is used in (10-76) and is derived from (9-61) and (9-62). When these expansions are used in (10-73) and third-degree terms in the increments $M^{(i)}$ and $X^{(i)}$ are neglected, the result is

$$e^{(i+1)} \approx e^{(i)} - \int_t^T (p_1^{(i)\prime} X^{(i)} + p_1^{(i)} X^{(i)\prime})\, d\sigma - \int_t^T (p_{11}^{(i)\prime} X^{(i)2} + 2p_{11}^{(i)} X^{(i)} X^{(i)\prime})\, d\sigma$$
$$+ \int_t^T \left[\frac{1}{2} T^{(i)} M^{(i)2} + (S^{(i)} + R^{(i)} X^{(i)}) M^{(i)} + \frac{1}{2} \frac{R^{(i)2}}{T^{(i)}} X^{(i)2}\right] d\sigma \quad (10\text{-}78)$$

Furthermore, the integration by parts of one term in each of the first two integrals appearing in (10-78) yields

$$e^{(i+1)} \approx e^{(i)} - (p_1^{(i)} X^{(i)} + p_{11}^{(i)} X^{(i)2})_{\sigma=t}^{\sigma=T}$$
$$+ \int_t^T \left[\frac{1}{2} T^{(i)} M^{(i)2} + (S^{(i)} + R^{(i)} X^{(i)}) M^{(i)} + \frac{1}{2} \frac{R^{(i)2}}{T^{(i)}} X^{(i)2}\right] d\sigma \quad (10\text{-}79)$$

Finally, (10-79) reduces to

$$e^{(i+1)} \approx e^{(i)} + \int_t^T \left[\frac{1}{2} T^{(i)} M^{(i)2} + (S^{(i)} + R^{(i)} X^{(i)}) M^{(i)} + \frac{1}{2} \frac{R^{(i)2}}{T^{(i)}} X^{(i)2}\right] d\sigma \quad (10\text{-}80)$$

when the fixed-point initial-boundary conditions and the free-point terminal-boundary conditions are imposed. The integral appearing in this result is the sum of the nonvanishing components associated with first and second variations.

The selection of an iteration algorithm based on first and second variations unfortunately does not proceed in the straightforward manner associated with first variations only. The property of guaranteed convergence is achieved when $M^{(i)}$ can be selected small enough so that the approximations leading to (10-80) are valid and when $e^{(i+1)} < e^{(i)}$. Therefore, algorithms giving rise to this property are selected according to the condition

$$\int_t^T \left[\frac{1}{2} T^{(i)} M^{(i)2} + (S^{(i)} + R^{(i)} X^{(i)}) M^{(i)} + \frac{1}{2} \frac{R^{(i)2}}{T^{(i)}} X^{(i)2}\right] d\sigma < 0 \quad (10\text{-}81)$$

Furthermore, a more restricted class of algorithms can be defined by requiring that the integrand of (10-81) be negative so that

$$\frac{1}{2} T^{(i)} M^{(i)2} + (S^{(i)} + R^{(i)} X^{(i)}) M^{(i)} + \frac{1}{2} \frac{R^{(i)2}}{T^{(i)}} X^{(i)2} < 0 \qquad t \leqq \sigma \leqq T \quad (10\text{-}82)$$

At first glance, the condition given in (10-82) is analogous to (10-30) for first variations. In fact, if the increments $M^{(i)}$ and $X^{(i)}$ are sufficiently small so that the second-degree terms are negligible, then (10-82) reduces to (10-30).

The selection of an appropriate algorithm for $M^{(i)}$ unfortunately is somewhat more difficult than in the case of first variations. The primary difficulties arise because of the dependence of $X^{(i)}$ at time $\sigma = \mu$ on the choice of $M^{(i)}$ on the previous interval $t \leqq \sigma \leqq \mu$. This dependence is established from the process state equation, which is written here in linearized form as

$$X^{(i)\prime} = \frac{\partial f^{(i)}}{\partial x^{(i)}} X^{(i)} + \frac{\partial f^{(i)}}{\partial m^{(i)}} M^{(i)} \quad (10\text{-}83)$$

and the initial condition $X^{(i)}(t) = 0$. For instance, (10-82) is satisfied for real values of $M^{(i)}$ only when

$$S^{(i)}(S^{(i)} + 2R^{(i)} X^{(i)}) > 0 \quad (10\text{-}84)$$

Because of the arbitrary nature of $S^{(i)}$ and $R^{(i)}$, the algorithm for $M^{(i)}$ which gives guaranteed convergence according to (10-82) cannot be chosen indiscriminately. An acceptable algorithm is found by minimizing the left-hand side of (10-82) given a value for $X^{(i)}$. This minimization leads to

$$M^{(i)} = -\varepsilon\left(\frac{S^{(i)}}{T^{(i)}} + \frac{R^{(i)}}{T^{(i)}} X^{(i)}\right) \tag{10-85}$$

for $\varepsilon = 1$. However, the constant ε is introduced in order to restrict step size. The acceptability of (10-85) is due to the fact that ε can be chosen small enough that (10-82) is satisfied if $S^{(i)} \neq 0$ everywhere on $t < \sigma \leqq T$. This property is demonstrated with the use of the linear state equation given in (10-83) to show that $X^{(i)}$ is proportional to ε as $\varepsilon \to 0$. Therefore, the selection of step size for (10-85) is restricted not only by the validity of the approximations leading to (10-80) but also by the inequalities in either (10-81) or (10-82). This latter consideration generally requires a value for ε that is unnecessarily small; hence (10-85) is somewhat undesirable.

If relaxation procedures based on second variations are to form the basis for efficient iterations, then step size should be restricted only by the validity of (10-80). This goal is achieved by the selection of an algorithm from considering the integral of the sum of the first and second variations. Specifically, this integral is minimized subject to a restricted step size; hence (10-81) is the condition used for guaranteed convergence.

The integral to be minimized is defined notationally as

$$v^{(i)}(\mu) = \int_{\mu}^{T} \left[\frac{1}{2} T^{(i)} M^{(i)2} + (S^{(i)} + R^{(i)} X^{(i)}) M^{(i)} + \frac{1}{2} \frac{R^{(i)2}}{T^{(i)}} X^{(i)2}\right] d\sigma \tag{10-86}$$

and the corresponding minimum-error function is defined as

$$V^{(i)}[X^{(i)}(\mu), \mu] = \min_{\substack{M^{(i)}(\sigma) \\ [\mu, T]}} v^{(i)}(\mu) \tag{10-87}$$

subject to the boundary condition

$$V^{(i)}[X^{(i)}(T), T] = 0 \tag{10-88}$$

Because the error measure appearing in (10-86) is quadratic, the linearized state equation given in (10-83) is used in this minimization problem. Therefore, the minimum-error function $V^{(i)}$ is quadratic and is defined as

$$V^{(i)}[X^{(i)}(\mu), \mu] = g^{(i)}(\mu) + g_1{}^{(i)}(\mu) X^{(i)}(\mu) + g_{11}{}^{(i)}(\mu)[X^{(i)}(\mu)]^2 \tag{10-89}$$

where

$$g^{(i)}(T) = g_1{}^{(i)}(T) = g_{11}{}^{(i)}(T) = 0 \tag{10-90}$$

The methods presented in either Chap. 7 or Chap. 9 can be used to show that

$$-g^{(i)\prime}(\mu) = -\frac{1}{2T^{(i)}}\left(S^{(i)} + \frac{\partial f^{(i)}}{\partial m^{(i)}} g_1{}^{(i)}\right)^2 \tag{10-91}$$

$$-g^{(i)\prime}(\mu) = -\frac{R^{(i)}S^{(i)}}{T^{(i)}} - \left(\frac{\partial f^{(i)}}{\partial m^{(i)}} \frac{R^{(i)}}{T^{(i)}} - \frac{\partial f^{(i)}}{\partial x^{(i)}}\right) g_1{}^{(i)} \tag{10-92}$$

and

$$g_{11}{}^{(i)}(\mu) = 0 \tag{10-93}$$

Hence, the minimum value of $v^{(i)}(t)$ becomes

$$V^{(i)} = -\int_t^T \frac{1}{2T^{(i)}}\left(S^{(i)} + \frac{\partial f^{(i)}}{\partial m^{(i)}} g_1{}^{(i)}\right)^2 d\sigma \tag{10-94}$$

as given by (10-87) for the boundary condition $X^{(i)}(t) = 0$. The occurrence of $g_{11}{}^{(i)}(\mu) = 0$ is somewhat unusual and is due to the special form of minimization problem involved here. Specifically, the error measure appearing in (10-86) is convex but not strictly convex. This structural property is demonstrated by showing that the matrix of second derivatives of the integrand is only positive semidefinite. The minimization procedure resulting in (10-91) to (10-93) therefore is only a necessary condition for a minimum value of (10-86) subject to (10-83) and may not be a sufficient condition according to (4-21).

As discussed in Chap. 4 on the calculus of variations, sufficiency conditions involve the investigation of the second variations. When the second variations are positive definite for all possible perturbations, a sufficient condition for a minimum has been obtained. In establishing a selection algorithm for $M^{(i)}$, all possible variations are not of interest. On the other hand, only the perturbations caused by restricting step size are of interest. A selection algorithm for $M^{(i)}$ can be chosen in conjunction with the control equation used to derive (10-91) to (10-93). The control equation which gives rise to a stationary value of (10-86) is

$$M^{(i)} = -\varepsilon\left(\frac{\partial f^{(i)}}{\partial m^{(i)}}\frac{g_1{}^{(i)}}{T^{(i)}} + \frac{S^{(i)}}{T^{(i)}}\right) - \left(\frac{R^{(i)}}{T^{(i)}}\right)X^{(i)} \tag{10-95}$$

when $\varepsilon = 1$. As before, the constant ε is introduced in order to restrict step size and thereby ensure the validity of the second variations.

The variations due to restricting step size are investigated by substituting (10-95) into (10-86), thereby obtaining

$$v^{(i)} = -\varepsilon\int_t^T \frac{1}{2T^{(i)}}\left[S^{(i)2} + 2S^{(i)}R^{(i)}\chi^{(i)} - \left(\frac{\partial f^{(i)}}{\partial m^{(i)}}\right)^2 g_1{}^{(i)2}\right] d\sigma + \varepsilon(1-\varepsilon)V^{(i)} \tag{10-96}$$

In the derivation of (10-96), (10-94) and the definition

$$\chi^{(i)}(\sigma) = \frac{X^{(i)}(\sigma)}{\varepsilon} \tag{10-97}$$

have been introduced. If the state and control equations given in (10-83) and (10-95) are combined and if the definition for $\chi^{(i)}$ is introduced, then the result is

$$\chi^{(i)\prime} = -\left(\frac{\partial f^{(i)}}{\partial m^{(i)}}\frac{R^{(i)}}{T^{(i)}} - \frac{\partial f^{(i)}}{\partial x^{(i)}}\right)\chi^{(i)} - \frac{\partial f^{(i)}}{\partial m^{(i)}}\left(\frac{\partial f^{(i)}}{\partial m^{(i)}}\frac{g_1{}^{(i)}}{T^{(i)}} + \frac{S^{(i)}}{T^{(i)}}\right) \tag{10-98}$$

This equation shows that $\chi^{(i)}$, and hence the integral appearing in (10-96), is independent of ε. Furthermore, the value of the error index given in (10-96) becomes

$$v^{(i)}\Big|_{\varepsilon=1} = -\int_t^T \frac{1}{2T^{(i)}}\left[S^{(i)2} + 2S^{(i)}R^{(i)}\chi^{(i)} - \left(\frac{\partial f^{(i)}}{\partial m^{(i)}}\right)^2 g_1{}^{(i)2}\right] d\sigma \tag{10-99}$$

for $\varepsilon = 1$. The integrals appearing in (10-94) and (10-99), however, must be identical because the case $\varepsilon = 1$ corresponds to the stationary value of (10-86). The dependence of the first and second variations on ε for the selection algorithm given in

(10-95) finally reduces to

$$v^{(i)} = (\varepsilon^2 - 2\varepsilon)(-V^{(i)}) \tag{10-100}$$

where $V^{(i)}$ is a negative number. Finally, the value $\varepsilon = 1$ uniquely minimizes (10-100) as seen from the conditions

$$\left.\frac{\partial v^{(i)}}{\partial \varepsilon}\right|_{\varepsilon=1} = 0 \qquad \left.\frac{\partial^2 v^{(i)}}{\partial \varepsilon^2}\right|_{\varepsilon=1} > 0 \tag{10-101}$$

The algorithm given in (10-95) eliminates the difficulties found in conjunction with (10-85). In particular, the improvement in performance $(e^{(i)} - e^{(i+1)})$ increases monotonically with increasing ε on the interval $0 < \varepsilon \leqq 1$ provided that the expansions used to derive (10-80) are valid. Therefore, step size and hence the selection of ε are restricted only by the validity of the first and second variations. In design problems where the error measure and process state equation are in fact quadratic and linear, respectively, the relationship given in (10-80) is exact for all values of the increments $X^{(i)}$ and $M^{(i)}$. Under these conditions, (10-95) with $\varepsilon = 1$ gives one-step convergence, a primary goal in developing a relaxation procedure based on second variations.

10-6 Numerical considerations in the selection of a relaxation method

Another primary goal in developing a relaxation procedure based on second variations concerns the stability of the differential equations used in the computations. Although not mentioned previously, both the forward-time and the backward-time computations in the case of first variations are performed with unstable differential equations when the state equations of the dynamic process are unstable. On the other hand, all computations are performed with stable equations when the state equations of the dynamic process are stable. This property is due to the fact that the homogeneous forms of (10-68) and (10-83) are identical when the sign of the derivative in (10-68) is changed for the backward-time computations. Of course, the driving functions in (10-68) and (10-83) do not affect the stability of these equations provided that the incremental variables remain small. This similarity between the differential equations used in both the forward-time and the backward-time computations is recognized sometimes by the term *co-state* variables, used in conjunction with $p_1^{(i)}$.

In design problems where the state equations of the dynamic process are unstable, numerically accurate solutions to these differential equations may be difficult or impossible to obtain. This difficulty is compounded in design problems where the interval of time $(T - t)$ is very large. In the case of second variations, the use of the algorithm given in (10-95) tends to stabilize the differential equations solved during the forward-time computations. The stabilized incremental state equation becomes

$$X^{(i)\prime} = -\left(\frac{\partial f^{(i)}}{\partial m^{(i)}}\frac{R^{(i)}}{T^{(i)}} - \frac{\partial f^{(i)}}{\partial x^{(i)}}\right)X^{(i)} - \frac{\varepsilon}{T^{(i)}}\frac{\partial f^{(i)}}{\partial m^{(i)}}\left(S^{(i)} + \frac{\partial f^{(i)}}{\partial m^{(i)}}g_1^{(i)}\right) \tag{10-102}$$

when (10-83) and (10-95) are combined. The coefficient of $X^{(i)}$ in (10-102) generally is positive except for short intervals of time and for time near the terminal time. Therefore, the forward-time computations of $x^{(i+1)}$ are based on stabilized differential equations for small regions of state space about $x^{(i)}$.

The stabilizing effects of using the algorithm given in (10-95) also are experienced in the backward-time computations. First, the variable $g_1^{(i)}$ is seen from (10-92) to be the co-state variable of (10-102). Furthermore, the $p_{11}^{(i)}$ equation given in (10-69) is a stable Riccati equation for $p_{11}^{(i)} \geqq 0$, which is ensured by a strictly convex minimum-error function. The $p_1^{(i)}$ equation, on the other hand is the co-state equation of the dynamic-process state equation. However, an alternative equation can be used for the backward-time computation and is found by manipulating (10-68). Specifically, (10-68) and (10-72) are combined to give

$$-p_1^{(i)\prime} = \frac{S^{(i)}R^{(i)}}{T^{(i)}} + \left(\frac{\partial H^{(i)}}{\partial x^{(i)}} - \frac{R^{(i)}}{T^{(i)}}\frac{\partial H^{(i)}}{\partial m^{(i)}}\right) - \left(\frac{\partial f^{(i)}}{\partial m^{(i)}}\frac{R^{(i)}}{T^{(i)}} - \frac{\partial f^{(i)}}{\partial x^{(i)}}\right)p_1^{(i)} \qquad (10\text{-}103)$$

If $S^{(i)}$, $R^{(i)}$, and $T^{(i)}$ were completely independent of $p_1^{(i)}$, then this equation would be stabilized in the backward-time direction and $p_1^{(i)}$ would be the co-state variable of (10-102). However, the time functions $R^{(i)}$ and $T^{(i)}$ are dependent upon $p_1^{(i)}$ except along the optimum trajectory which corresponds to the condition $S^{(i)} = 0$. This dependence is observed from

$$\frac{R^{(i)}}{T^{(i)}} = \frac{\left[\dfrac{\partial f^{(i)}}{\partial m^{(i)}}\dfrac{\partial^2 H^{(i)}}{\partial m^{(i)}\,\partial x^{(i)}} - \dfrac{\partial^2 f^{(i)}}{\partial m^{(i)}\,\partial x^{(i)}}\dfrac{\partial H^{(i)}}{\partial m^{(i)}} + 2\left(\dfrac{\partial f^{(i)}}{\partial m^{(i)}}\right)^2 p_{11}^{(i)}\right] + \left(\dfrac{\partial^2 f^{(i)}}{\partial x^{(i)}\,\partial m^{(i)}}\right)S^{(i)}}{\left(\dfrac{\partial f^{(i)}}{\partial m^{(i)}}\dfrac{\partial^2 H^{(i)}}{\partial m^{(i)2}} - \dfrac{\partial^2 f^{(i)}}{\partial m^{(i)2}}\dfrac{\partial H^{(i)}}{\partial m^{(i)}}\right) + \left(\dfrac{\partial^2 f^{(i)}}{\partial m^{(i)2}}\right)S^{(i)}} \qquad (10\text{-}104)$$

which is found by combining (10-70) and (10-71) with (10-72) so that $p_1^{(i)}$ is eliminated. The residual dependence of (10-103) on $p_1^{(i)}$ now is completely due to the dependence of $S^{(i)}$ on $p_1^{(i)}$, as seen from (10-72). Therefore, (10-103) is exactly the co-state equation which corresponds to the stabilized incremental state equation of the dynamic process along the optimum trajectory.

The form of the $p_1^{(i)}$ equation given in (10-103) is preferred to that given in (10-68) because of numerical considerations. In particular, truncation and round-off errors arise from the numerical integration of differential equations on a digital computer, making the condition $S^{(i)} = 0$ on the interval $t \leqq \sigma \leqq T$ impossible to achieve. Furthermore, the condition $S^{(i)} \approx 0$ on this interval corresponds to a trajectory that is nearly optimum and is particularly sensitive to these numerical errors. This sensitivity is due to the small numerical difference between two generally large numbers in the computation of $S^{(i)}$. Therefore, the practice of setting $S^{(i)} = 0$ in the solution of (10-103) when $|S^{(i)}| \leqq \epsilon_r$ is adopted. This procedure stabilizes the solution for $p_1^{(i)}$ in the region where the computation is the most sensitive to numerical errors. The value of ϵ_r adopted for the computations is based on both the accuracy desired in the solution of the two-point boundary-value problem and the expected size of unavoidable numerical errors. However, the size of unavoidable numerical errors generally cannot be determined a priori. Therefore, an approximate selection procedure for ϵ_r is based on the accuracy desired in the solution of the two-point boundary-value problem. In turn, this accuracy is more or less meaningless unless it can be achieved with components available for the construction of the control system. If ϵ_m is defined as the magnitude of unavoidable errors in the control signal due to component inaccuracies, then the value of ϵ_r may

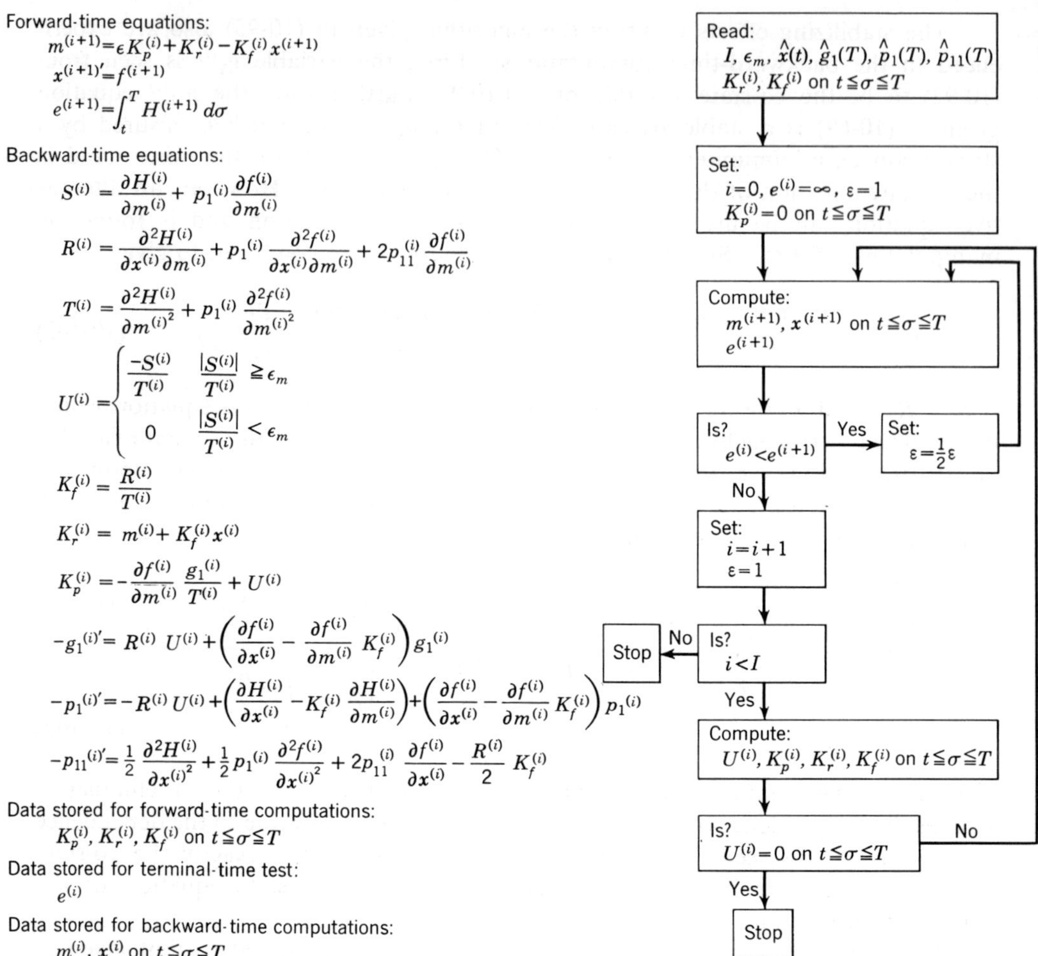

Fig. 10-5 Simplified flow diagram for the solution of two-point boundary-value problems using a relaxation procedure based on second variations.

be chosen from

$$\epsilon_r = \epsilon_m T^{(i)} \tag{10-105}$$

This expression is found from (10-95) by considering only the term $S^{(i)}/T^{(i)}$ and by setting $\varepsilon = 1$ for the full step size. The number ϵ_m generally is known, and (10-105) provides a time-varying threshold for inhibiting the propagation of numerical errors in the solution of (10-103) and for obtaining solutions that correspond to engineering design considerations.

The simplified flow diagram shown in Fig. 10-5 summarizes the previous discussions concerning the iterative solution of two-point boundary-value problems using second variations. The diagram indicates the sequence of computations, the required temporary data storage, and the tests for finding an acceptable step size. The step size is decreased in the same manner as used for first variations. On the other hand, the full step size $\varepsilon = 1$ is used at the beginning of each forward-time computation as a method of obtaining the maximum possible rate of convergence and eventually one-step convergence. However, this procedure can be modified if there is some a priori reason for suspecting that an unnecessarily large number of unsuccessful forward-time computations would occur. The step-size selection procedure indicated in Fig. 10-5 is based on the premise that the backward-time computations consume more computer time than the forward-time computations. Therefore, unsuccessful forward-time computations are tolerated in an effort to obtain a maximum rate of convergence. The backward-time computations indicated in Fig. 10-5 are based on the threshold condition given in (10-105) for a nearly optimum trajectory. This condition provides a test for stopping the program when the desired accuracy has been achieved.

Figures 10-1 and 10-5 provide a comparison of the required computations and memory for relaxation methods based on first and second variations, respectively, with a first-order dynamic process. The difference between the computations required mainly involves the number of differential equations that must be solved. The backward-time computations with second variations are performed with three equations, whereas one equation occurs with first variations. However, the computer time required for the added equations in the case of second variations must be balanced against the possible decrease in computer time due to the improved stability of the differential equations and the improved rate of convergence, and hence fewer steps in the numerical integrations and fewer iterations, respectively. These compromises in computer time can be ascertained only in relationship to the particular problem at hand. The difference between the computer memory required also involves only the data supplied from the backward-time computations. Specifically, three time functions must be stored for second variations as opposed to two time functions for first variations with a first-order dynamic process.

Unlike the case of first variations, the relaxation procedures based on second variations, which are discussed here, cannot be extended to control-signal saturation. The difficulties arise in the minimization of (10-86), which leads to (10-95). In particular, the minimization of (10-86) gives rise to a second two-point boundary-value problem unless the points in time where the control signal leaves and enters the boundaries are identical for the ith and $(i + 1)$st iterations. This difficulty also arises in the conventional methods of the calculus of variations and is discussed in Chap. 4. If guaranteed convergence is to be obtained, then these points in time cannot be held fixed. Furthermore, the attractiveness of second variations is diminished considerably when control-signal saturation occurs, because the stabilizing properties of (10-95) do not occur during saturation. If the state equation of the dynamic process is unstable, then the saturation constraints probably must be approximated with soft constraints along the lines outlined in Chap. 9. This type of approximation allows the use of second variations and potential improvements in numerical accuracy. The reader must remember that the guarantee of convergence afforded by the relaxation procedures discussed in this chapter is valid only when the solutions of the state and co-state equations are numerically accurate.

10-7 Equations for the relaxation procedure based on second variations

The concepts and pertinent equations developed in Sec. 10-6 extend readily to the case of an Nth-order dynamic process without control-signal saturation. The forward-time computations are performed with (10-48). Also, the expansion of (10-50) in second variations yields

$$e^{(i+1)} \approx e^{(i)} + \int_t^T \Bigg[\sum_{n=1}^{N} \sum_{m=1}^{M} \tfrac{1}{2} T_{nm}{}^{(i)} M_n{}^{(i)} M_m{}^{(i)} + \sum_{n=1}^{N} \left(S_n{}^{(i)} + \sum_{m=1}^{N} R_{nm}{}^{(i)} X_m{}^{(i)} \right) M_n{}^{(i)} + \sum_{n=1}^{M} \sum_{m=1}^{N} \sum_{k=1}^{N} \tfrac{1}{2} R_{nm}{}^{(i)} K_{nk}{}^{(i)} X_m{}^{(i)} X_k{}^{(i)} \Bigg] d\sigma \quad (10\text{-}106)$$

where $S_n{}^{(i)}$ and $T_{nm}{}^{(i)}$ are defined in (10-52) and (10-62), respectively. The additional time functions appearing in this expansion are defined as

$$R_{nm}{}^{(i)} = \frac{\partial^2 H^{(i)}}{\partial m_n{}^{(i)}\, \partial x_m{}^{(i)}} + \sum_{k=1}^{N} p_k{}^{(i)} \frac{\partial^2 f_k{}^{(i)}}{\partial m_n{}^{(i)}\, \partial x_m{}^{(i)}} + 2 \sum_{k=1}^{N} p_{km}{}^{(i)} \frac{\partial f_k{}^{(i)}}{\partial m_n{}^{(i)}}$$

$$n = 1, 2, \ldots, M;\; m = 1, 2, \ldots, N \quad (10\text{-}107)$$

and

$$\sum_{m=1}^{M} T_{nm}{}^{(i)} U_m{}^{(i)} = -S_n{}^{(i)} \qquad \sum_{m=1}^{M} T_{nm}{}^{(i)} K_{ml}{}^{(i)} = R_{nl}{}^{(i)} \qquad n = 1, 2, \ldots, M \quad (10\text{-}108)$$

One set of backward-time equations is found by the substitution of these definitions into (9-139) and (9-140) and by performing manipulations similar to those used to reduce (10-68) to (10-103). These equations are

$$-p_k{}^{(i)\prime} = -\sum_{n=1}^{M} R_{nk}{}^{(i)} U_n{}^{(i)} + \left(\frac{\partial H^{(i)}}{\partial x_k{}^{(i)}} - \sum_{m=1}^{M} \frac{\partial H^{(i)}}{\partial m_m{}^{(i)}} K_{mk}{}^{(i)} \right) + \sum_{n=1}^{N} \left(\frac{\partial f_n{}^{(i)}}{\partial x_k{}^{(i)}} - \sum_{m=1}^{M} \frac{\partial f_n{}^{(i)}}{\partial m_m{}^{(i)}} K_{mk}{}^{(i)} \right) p_n{}^{(i)} \qquad k = 1, 2, \ldots, N \quad (10\text{-}109)$$

and

$$-p_{kl}{}^{(i)\prime} = \frac{1}{2} \frac{\partial^2 H^{(i)}}{\partial x_k{}^{(i)}\, \partial x_l{}^{(i)}} + \frac{1}{2} \sum_{n=1}^{N} \frac{\partial^2 f_n{}^{(i)}}{\partial x_k{}^{(i)}\, \partial x_l{}^{(i)}} p_n{}^{(i)} + \sum_{n=1}^{N} \left(\frac{\partial f_n{}^{(i)}}{\partial x_k{}^{(i)}} p_{nl}{}^{(i)} + \frac{\partial f_n{}^{(i)}}{\partial x_l{}^{(i)}} p_{nk}{}^{(i)} \right) - \frac{1}{2} \sum_{n=1}^{M} R_{nk}{}^{(i)} K_{nl}{}^{(i)}$$

$$k = 1, 2, \ldots, l;\; l = 1, 2, \ldots, N \quad (10\text{-}110)$$

The other set of backward-time equations is derived from the minimization of the integral appearing in (10-106) subject to the linearized state equation. These equations are

$$-g_k{}^{(i)\prime} = \sum_{n=1}^{M} R_{nk}{}^{(i)} U_n{}^{(i)} + \sum_{n=1}^{N} \left(\frac{\partial f_n{}^{(i)}}{\partial x_k{}^{(i)}} - \sum_{m=1}^{M} \frac{\partial f_n{}^{(i)}}{\partial m_m{}^{(i)}} K_{mk}{}^{(i)} \right) g_n{}^{(i)} \qquad k = 1, 2, \ldots, N \quad (10\text{-}111)$$

Finally, the iteration algorithm for the vector $\mathbf{m}^{(i+1)}$ is found from the minimization problem which results in (10-111). This equation is written as

$$m_n{}^{(i+1)} = \varepsilon G_n{}^{(i)} + K_n{}^{(i)} - \sum_{k=1}^{N} K_{nk}{}^{(i)} x_k{}^{(i+1)} \qquad n = 1, 2, \ldots, M \quad (10\text{-}112)$$

where the additional time functions appearing in this equation are defined as

$$\sum_{m=1}^{M} T_{nm}{}^{(i)} G_m{}^{(i)} = -S_n{}^{(i)} - \sum_{k=1}^{N} \frac{\partial f_k{}^{(i)}}{\partial m_n{}^{(i)}} g_k{}^{(i)} \qquad n = 1, 2, \ldots, M \quad (10\text{-}113)$$

and $$K_n^{(i)} = m_n^{(i)} + \sum_{k=1}^{N} K_{nk}^{(i)} x_k^{(i)} \qquad n = 1, 2, \ldots, M \tag{10-114}$$

The parameter ε is introduced in order to restrict step size and is adjusted on the interval $0 < \varepsilon \leqq 1$. When step size is restricted in this manner, the expansion given in (10-106) becomes

$$e^{(i+1)} \approx e^{(i)} + \frac{\varepsilon^2 - 2\varepsilon}{2} \int_t^T \sum_{n=1}^{M} \sum_{m=1}^{M} T_{nm}^{(i)} G_n^{(i)} G_m^{(i)} \, d\sigma \tag{10-115}$$

Hence, convergence is ensured when the matrix $[T_{nm}^{(i)}]$ is positive definite.

This completes the equations for the relaxation procedure based on second variations except for introducing the threshold on the variables $S_n^{(i)}$ when a nearly optimum trajectory has been found. The introduction of this threshold is accomplished in the manner described in Sec. 10-6.

The complete nature of the compromise between computer memory and number of equations vs. step size and numerical accuracy in the case of unstable dynamic processes becomes apparent only in the Nth-order case. The number of differential equations solved in the backward-time direction is $2N + N(N + 1)/2$ for second variations and N for first variations. The number of time functions stored in order to perform the forward-time computations is, in the most general case, $M(N + 2)$ for second variations and $2M$ for first variations. Furthermore, the matrix inversions indicated in (10-108) and (10-113) can add considerable computation time, depending on the problem at hand.

10-8 Example comparison of relaxation methods

In order to indicate the relative efficiencies of relaxation methods based on first variations and on second variations, the numerical solution of a simple design problem is presented here. The design problem is somewhat artificial but is chosen so that both the stability of the dynamic process and the linearity of the optimum control equation can be altered readily. This design problem could be considered a regulator problem where the trajectory $x_1(t) = x_2(t) = 0$ for all real time is the desired trajectory and also the steady-state solution. Also, four cases for the design problem are treated in order to illustrate a number of properties of the two relaxation methods.

Case I. Quadratic error measure and linear dynamic process

This case is introduced in order to illustrate the one-step convergence property of the method based on second variations. Specifically, the error measure is taken to be

$$H = 2x_1^2 + x_2^2 + m_1^2 \tag{10-116}$$

and the linear second-order dynamic process with poles on the j axis is defined by

$$x_1' = -x_2 + m_1 \qquad x_2' = x_1 \tag{10-117}$$

Case II. Quadratic error measure and nonlinear dynamic process

This case is defined by the error measure

$$H = x_1^2 + x_2^2 + m_1^2 \tag{10-118}$$

and the dynamic process

$$x_1' = (1 - x_2^2)x_1 - x_2 + m_1 \qquad x_2' = x_1 \tag{10-119}$$

The homogeneous form of these state equations forms the nonlinear van der Pol equation, which possesses a bounded limit cycle enclosing the desired trajectory $x_1 = x_2 = 0$.

Case III. Nonquadratic error measure and linear dynamic process

This case is defined by the error measure

$$H = 2x_1^2 + \left(\frac{x_1}{0.4}\right)^{32} + x_2^2 + m_1^2 \tag{10-120}$$

and the linear dynamic process given in (10-117). The error measure given in (10-120) is used to approximate the state-signal saturation constraint $|x_1| \leqq 0.4$.

Case IV. Nonquadratic error measure and linear dynamic process

This case is defined by the error measure

$$H = 2x_1^2 + x_2^2 - \frac{2}{\pi^2} \ln [\cos (\pi m_1)] \qquad |m_1| < 0.5 \tag{10-121}$$

and the linear dynamic process given in (10-117). The error measure given in (10-121) is used to impose the control-signal saturation constraint $|m_1| < 0.5$ with a penalty function that is quasi-quadratic and strictly convex about any admissible value of the control signal.

The numerics appearing in the error measures for these four cases have been selected so that the optimum linear control systems possess similar properties in the neighborhood of the desired trajectory $x_1 = x_2 = 0$. Specifically, the damping ratios and resonant frequencies of the closed-loop systems are identical in the four cases. In order to illustrate these considerations, the general equations given in Sec. 10-7 for the method based on second variations are specialized here for case I. The backward-time equations given in (10-109) and (10-110) become

$$-p_1^{(i)\prime} = -2p_{11}U_1^{(i)} + 4x_1^{(i)} - p_{11}(2m_1^{(i)} + p_1^{(i)}) + p_2^{(i)} \tag{10-122}$$

$$-p_2^{(i)\prime} = -2p_{12}U_1^{(i)} + 2x_2^{(i)} - p_{12}(2m_1^{(i)} + p_1^{(i)}) - p_1^{(i)} \tag{10-123}$$

$$-p_{11}' = 2 + 2p_{12} - p_{11}^2 \tag{10-124}$$

$$-p_{12}' = -p_{11} + p_{22} - p_{11}p_{12} \tag{10-125}$$

and

$$-p_{22}' = 1 - 2p_{12} - p_{12}^2 \tag{10-126}$$

Also for case I, the backward-time equations given in (10-111) reduce to

$$-g_1^{(i)\prime} = 2p_{11}U_1^{(i)} - p_{11}g_1^{(i)} + g_2^{(i)} \tag{10-127}$$

and

$$-g_2^{(i)\prime} = 2p_{12}U_1^{(i)} - (1 + p_{12})g_1^{(i)} \tag{10-128}$$

The algorithm given in (10-112) also becomes

$$m_1^{(i+1)} = \varepsilon(U_1^{(i)} - \tfrac{1}{2}g_1^{(i)}) + K_1^{(i)} - K_{11}x_1^{(i+1)} - K_{12}x_2^{(i+1)} \tag{10-129}$$

where

$$K_1^{(i)} = m_1^{(i)} + K_{11}x_1^{(i)} + K_{12}x_2^{(i)} \qquad K_{11} = p_{11} \qquad K_{12} = p_{12} \tag{10-130}$$

Finally, the variable $U_1^{(i)}$ is given by (10-108), but a threshold ϵ_m is introduced so that

$$U_1^{(i)} = \begin{cases} -m_1^{(i)} - \frac{1}{2}p_1^{(i)} & |m_1^{(i)} + \frac{1}{2}p_1^{(i)}| > \epsilon_m \\ 0 & |m_1^{(i)} + \frac{1}{2}p_1^{(i)}| \leqq \epsilon_m \end{cases} \tag{10-131}$$

The superscript notation, denoting the ith iteration, on the double-subscripted p and K variables has been dropped here because (10-124) to (10-126) do not depend upon the trajectory. The equations corresponding to case I for the relaxation method based on first variations are found from (10-122), (10-123), and (10-129) by simply setting p_{11}, p_{12}, and $g_1^{(i)}$ equal to zero.

The backward-time equations for case I are used to perform a preliminary analysis of the closed-loop control system. In particular, the infinite-interval solution, namely, $T = \infty$, of (10-124) to (10-126) yields

$$K_{11} = 2^{\frac{3}{4}} \qquad K_{12} = 2^{\frac{1}{2}} - 1 \tag{10-132}$$

These values of the iteration-equation feedback gains yield the damping ratio and resonant frequency

$$\zeta = 2^{-\frac{1}{2}} \qquad \omega_s = 2^{\frac{1}{4}} \tag{10-133}$$

of the closed-loop system when the method based on second variations is used.

Sets of equations similar to (10-122) through (10-128) could be derived for cases II to IV but are omitted here. In addition, the closed-loop systems for these cases can be analyzed for the special conditions $T = \infty$, $x_1(0) = x_2(0) = 0$. Under these conditions, the damping ratio and resonant frequency of the closed-loop systems for all cases are given by (10-133) for a suitably small region of state space about $x_1(0) = x_2(0) = 0$. Also, in this region, the feedback gains for cases III and IV are given by (10-132). However, the feedback gains for case II are

$$K_{11} = 2^{\frac{3}{4}} + 1 \qquad K_{12} = 2^{\frac{1}{2}} - 1 \tag{10-134}$$

for this small region of state space. The numerics used in the error measures for these four cases are purposely chosen in order to normalize the control systems to the conditions given in (10-133). This normalization tends to isolate the effects due to state and control-signal saturation and to make possible approximately equal numerical accuracy in the four cases with a fixed increment size for numerical integration.

The iteration algorithm used here for the method based on first variations is given by (10-63). Therefore, a comparison of the two methods should demonstrate the effects of adding first- and second-degree extrapolation terms in the state signals with the method based on second variations. The method used for adjusting step size ε in this comparison is taken to be identical for both methods. The method used for decreasing ε is the same as that depicted in Figs. 10-1 and 10-5. The method used for increasing ε, however, is somewhat different from those discussed previously. The initial trial ε used for the $(i + 1)$st iteration is set equal to twice the value found to be successful for the ith iteration so that $0 < \varepsilon \leqq 1$ always is satisfied. Also, the initial trial is taken to be $\varepsilon = 1$ for the second iteration. This procedure for selecting step size represents a compromise for both relaxation methods, and the convergence rates of both methods could be improved with alternative but different selection procedures of step size for the two relaxation methods.

The computer program for comparing these two relaxation methods is written so that unnecessary computations, in any of the four cases or using either of the two methods, are not performed. The numerical integration is performed by using the Gill method, which is summarized in Appendix F, but the increment size for integration is held constant. Finally, an accurate record of total elapsed computer time, which does not include program and input-data reading and output-data conversion and printing, is kept.

The design-problem parameters used in this comparison for all four cases are

$$x_1(0) = 0 \qquad x_2(0) = 1 \qquad T = 5 \tag{10-135}$$

The numerical solutions are obtained with an increment size of 0.01 used for numerical integration. Therefore, the total number of tabulated data points for the required time functions is 2,505 for first variations and 3,507 for second variations. Also, the numerical solutions are obtained with $\epsilon_m = 0$ for first variations and with $\epsilon_m = 0.0001$ for second variations. This selection of ϵ_m in the case of first variations is chosen because the iterations are terminated in all four cases before a comparable accuracy in obtaining the optimum trajectory is achieved.

The format chosen here for data presentation is selected so as to compare the rates between the convergence of $e^{(i)}$ to e^* and the convergence of $m_1{}^{(i)}$ to m_1^*. The variables

$$\Delta e^{(i)} = e^{(i)} - e_a^* \qquad \Delta m_1{}^{(i)} = m_1{}^{(i)} - m_a^* \tag{10-136}$$

are introduced for this purpose. The variables e_a^* and m_a^* are approximate values of e^* and m_1^* and are set equal to $e^{(i)}$ and $m_1{}^{(i)}$ obtained from a suitable number of iterations using the method based on second variations. Because analytical solutions are not possible and numerical errors cannot be avoided completely, the precise optimum cannot be obtained and hence e^* and m_1^* cannot be computed exactly. Therefore, a lower bound for e^* is computed in each of the four cases for which data are presented. This lower bound is obtained from (10-115) when specialized to this example so that

$$e^{(i+1)} = e^{(i)} + \frac{\varepsilon^2 - 2\varepsilon}{2} \int_0^5 T_{11}{}^{(i)}(G_1{}^{(i)})^2 \, d\sigma \tag{10-137}$$

where the equality is taken because of the assumption that the ith trajectory is in a suitably small neighborhood of the optimum trajectory. In addition, the inequality

$$e^{(i+1)} \geqq e^{(i)} - \tfrac{5}{2}[T_{11}{}^{(i)}(G_1{}^{(i)})^2]_{\max} \tag{10-138}$$

is obtained, where the maximum value occurring on the interval $0 \leqq \sigma \leqq 5$ of the integrand which appears in (10-137) is used in (10-138). Then a lower bound for e^*, denoted as e_{lb}^*, is defined in terms of (10-138) as

$$e_{lb}^* = e^{(i)} - \tfrac{5}{2}[T_{11}{}^{(i)}(G_1{}^{(i)})^2]_{\max} \tag{10-139}$$

Finally, the data are presented in terms of the following definitions:

τ = total accumulated elapsed computer time (in minutes for IBM 704)
j = total accumulated unsuccessful selections of step size ε

Case I

The quasi-optimum trajectory for this case is shown in Fig. 10-6, and the control signal shown on this figure is defined to be m_a^* for this case. Also, this trajectory is

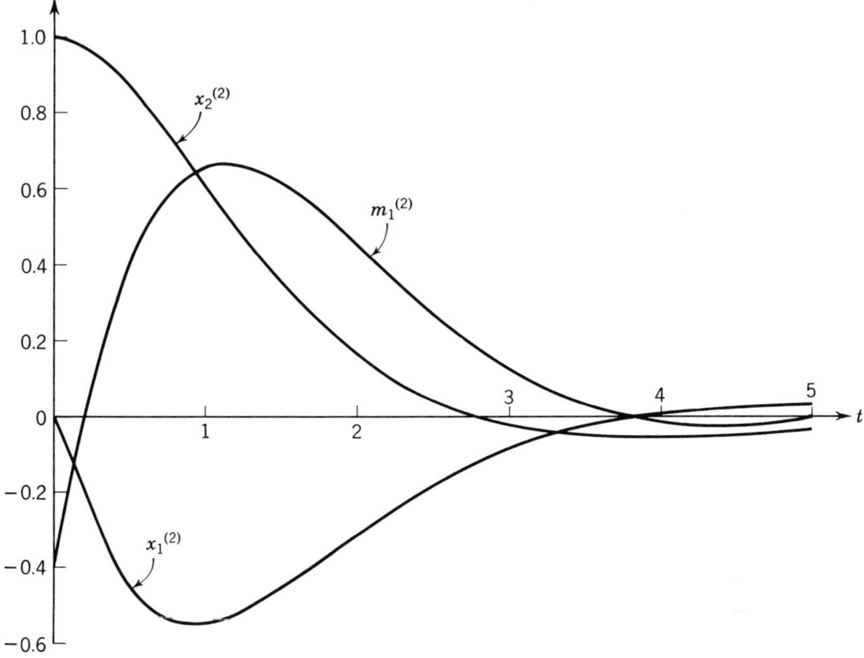

Fig. 10-6 Quasi-optimum trajectory for case I.

the second iteration found with the relaxation method based on second variations which should correspond to one-step convergence. The computed control signal $m_1^{(2)}$ approximates m_1^* to six significant decimal digits in this case. The time-varying parameters of the linear optimum control equation, given by (10-129) with $\varepsilon = 0$, are shown in Fig. 10-7 for this case. These data are included in order to illustrate the numerical accuracy obtained from the digital computations and to provide a basis for comparing the results obtained for the other three cases.

The data shown in Fig. 10-8 illustrate the convergence properties of the relaxation method based on first variations. Also, comparative data for second variations,

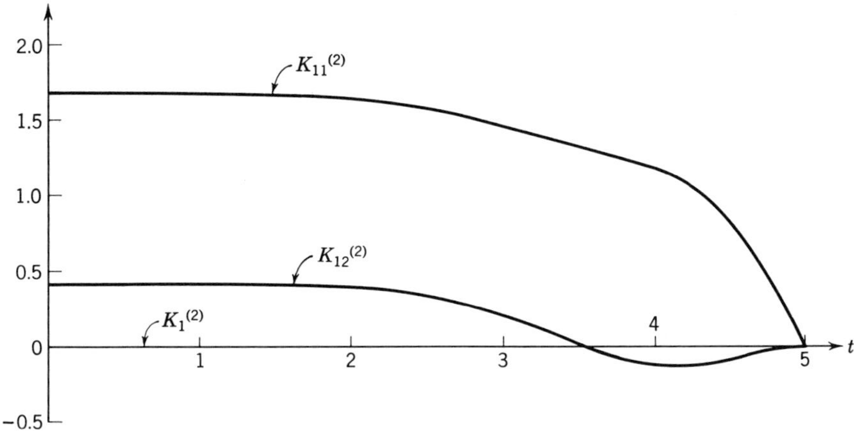

Fig. 10-7 Linear-control-equation parameters for case I.

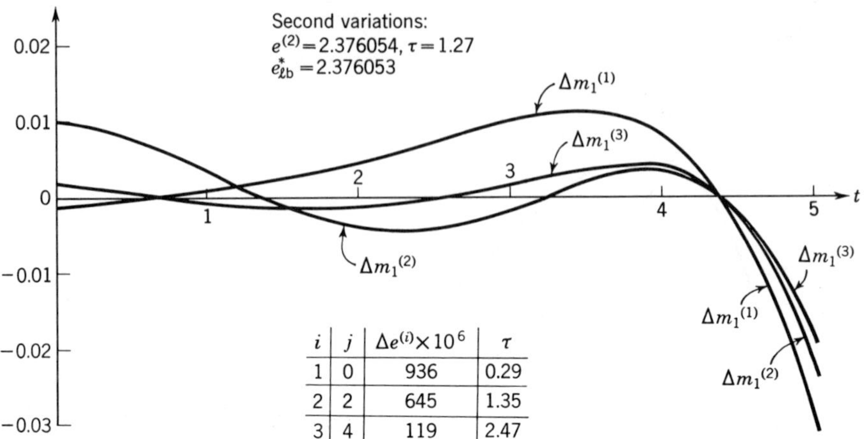

i	j	$\Delta e^{(i)} \times 10^6$	τ
1	0	936	0.29
2	2	645	1.35
3	4	119	2.47

Fig. 10-8 Control-signal error using first variations for case I: $K_1^{(0)} = 0$, $K_{11}^{(0)} = 2^{\frac{3}{4}}$, $K_{12}^{(0)} = 2^{\frac{1}{2}} - 1$.

namely, $e^{(i)}$, τ, and e^*_{lb}, are shown in Fig. 10-8. The priming trajectory, namely, $m_1{}^{(1)}$, is based on the control equation given in (10-129) with $\varepsilon = 0$ and the control-equation parameters given in (10-132), which would be optimum for $T = \infty$. Two important difficulties of the method based on first variations are illustrated in Fig. 10-8. First, the convergence of $e^{(i)}$ to e^* is considerably faster than that of $m_1{}^{(i)}$ to m_1^*. This convergence-rate problem with respect to $m_1{}^{(i)}$ is particularly serious with respect to peak errors. Second, the inability to select an acceptable value for ε may make the elapsed computer time per iteration with first variations comparable to

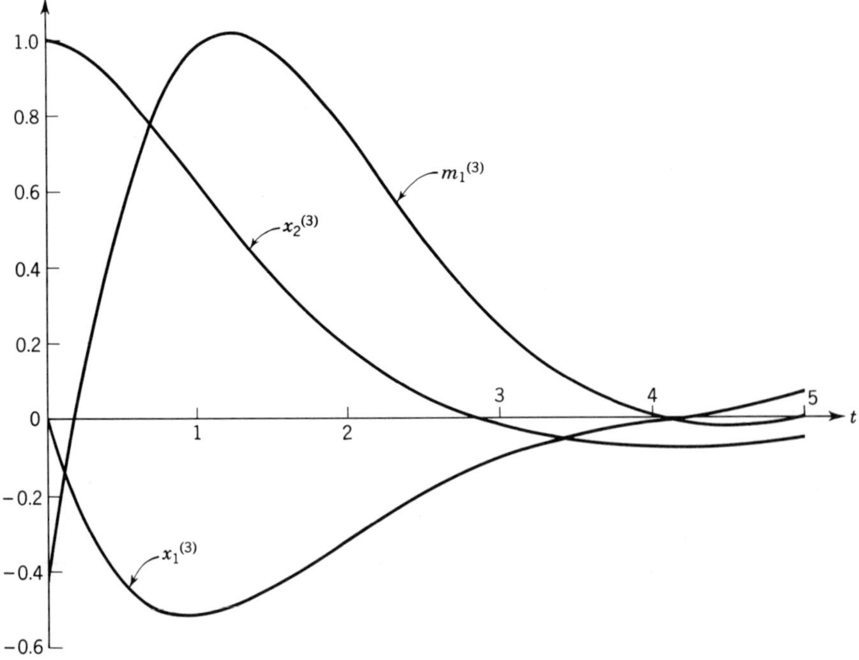

Fig. 10-9 Quasi-optimum trajectory for case II.

that with second variations where additional backward-time computations per iteration are performed. These two difficulties make the method based on first variations unattractive, particularly in the neighborhood of the optimum trajectory.

Case II

The quasi-optimum trajectory for this nonlinear dynamic process is shown in Fig. 10-9, and the control signal shown in this figure is defined to be m_a^* for this case. The primary difference between this trajectory and the trajectory for case I is that a considerably larger control signal is required here. Also, the time-varying parameters of the optimum linear control equation, given by (10-129) with $\varepsilon = 0$, are shown in Fig. 10-10 for the nonlinear dynamic process. Of course, these parameters are optimum only for the initial state $x_1(0) = 0$ and $x_2(0) = 1$. Primarily,

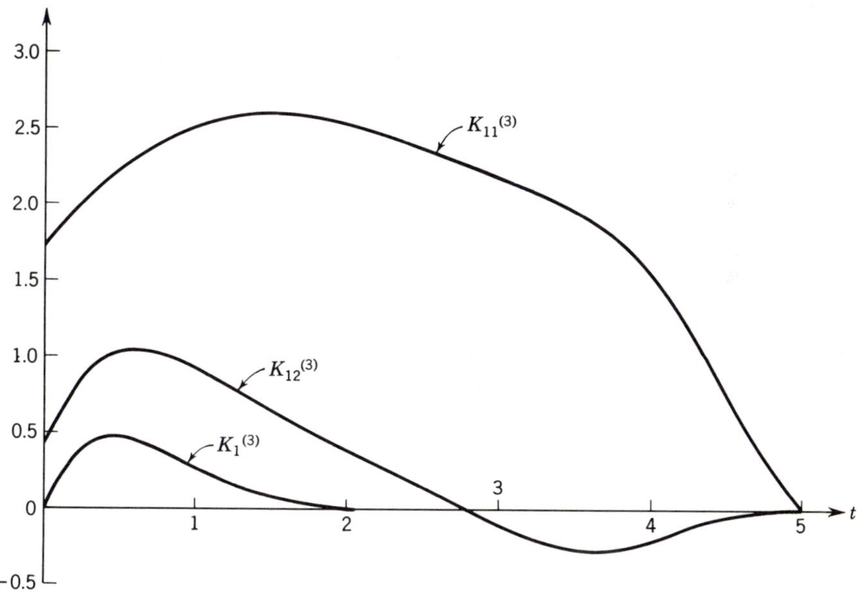

Fig. 10-10 Linear-control-equation parameters for case II.

a large rate feedback gain K_{11} is required here in order to compensate for the nonlinear damping term associated with the van der Pol equation. An interesting point of note is the behavior of these parameters at $t = 0$ where $\partial f_1/\partial x_1 = 0$ and $\partial f_1/\partial x_2 = -1$ because of the initial state $x_1(0) = 0$ and $x_2(0) = 1$. These values for the partial derivatives are identical to those for case I; hence, the time-varying parameters shown in Fig. 10-10 at $t = 0$ approach the steady-state values of the corresponding parameters shown in Fig. 10-7.

The data shown in Figs. 10-11 and 10-12 compare the convergence properties of the two relaxation methods for the nonlinear dynamic process. The priming trajectory for both methods is found with the numerics given in (10-134), which are optimum for the conditions $x_1(0) = x_2(0) = 0$ and $T = \infty$. The method based on second variations gives rise not only to a higher rate of convergence of $e^{(i)}$ to e^* but also to a much higher rate of convergence of $m_1^{(i)}$ to m_1^*. Again peak errors in the control signal persist after many iterations using the method based on first variations.

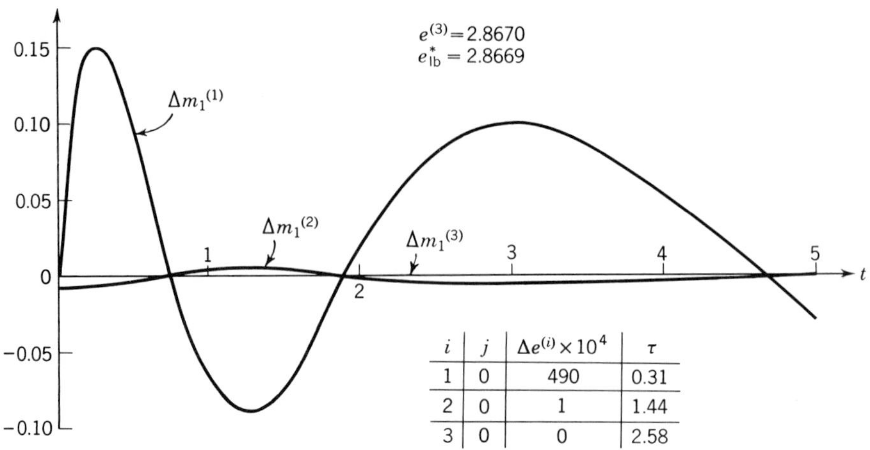

i	j	$\Delta e^{(i)} \times 10^4$	τ
1	0	490	0.31
2	0	1	1.44
3	0	0	2.58

Fig. 10-11 Control-signal error using second variations for case II: $K_1^{(0)} = 0$, $K_{11}^{(0)} = 1 + 2^{3/4}$, $K_{12}^{(0)} = 2^{1/2} - 1$.

On the other hand, this example is useful for demonstrating the possibility of extreme difficulties that may arise with the method based on second variations. This difficulty is numerical in nature. Also, this example is useful in demonstrating the importance of suitable priming trajectories. Specifically, the matrix Riccati equations are stable when the matrix $[p_{kl}{}^{(i)}]$ is positive definite, a condition that may not occur when the dynamic process is nonlinear and unstable or even stable with a bounded limit cycle. The curves shown in Fig. 10-13 for the variable $p_{22}{}^{(i)}$ are obtained for the slight modification of case II defined by

$$H = x_1{}^2 + x_2{}^2 + m_1{}^2 \qquad T = \frac{\pi}{2} \tag{10-140}$$

These data are obtained with the initial parameters $K_1{}^{(0)} = K_{11}{}^{(0)} = K_{12}{}^{(0)} = 0$ for the control equation so that $m_1{}^{(1)} = 0$. Clearly, the matrix $[p_{kl}{}^{(1)}]$ is not positive definite for all t because $p_{22}{}^{(1)}$ is not positive everywhere. Even in this example,

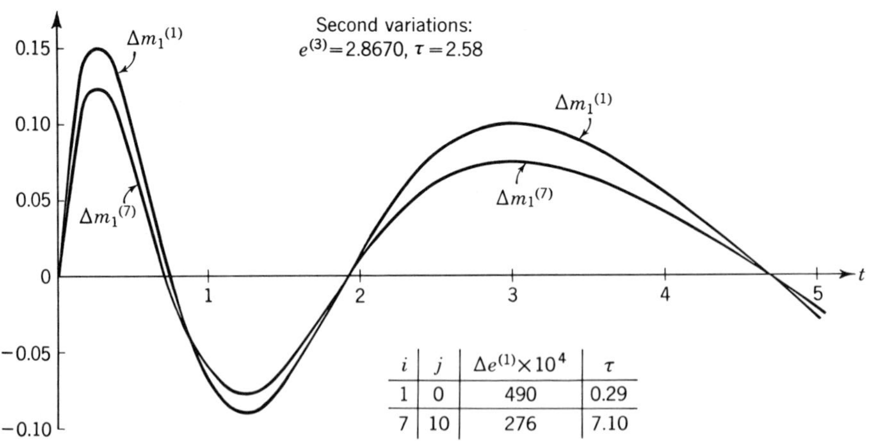

i	j	$\Delta e^{(1)} \times 10^4$	τ
1	0	490	0.29
7	10	276	7.10

Fig. 10-12 Control-signal error using first variations for case II: $K_1^{(0)} = 0$, $K_{11}^{(0)} = 1 + 2^{3/4}$, $K_{12}^{(0)} = 2^{1/2} - 1$.

however, rapid convergence is obtained, as indicated in Fig. 10-13. On the other hand, the matrix Riccati equations become unstable on the first iteration, and attempts to solve this example but for $T = 5$ with $m_1^{(1)} = 0$ and the method based on second variations result in complete failure. For all practical purposes, the numerical solutions for $p_{kl}^{(1)}(t)$ become infinite in this case at $(T - t) \approx 3.84$, thereby

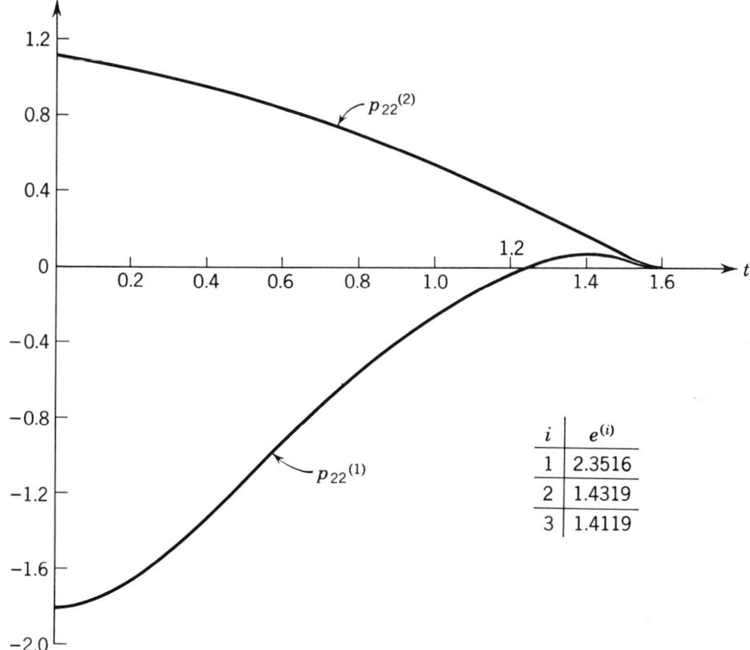

Fig. 10-13 Example of $[p_{kl}^{(i)}]$ not positive definite for case II: $T = \pi/2$, $K_1^{(0)} = K_{11}^{(0)} = K_{12}^{(0)} = 0$.

making the solution over the whole interval impossible. The propagation of numerical errors prior to reaching the point 3.84 makes the identification of the precise nature of the difficulty somewhat nebulous. However, the matrix Riccati equations are nonlinear and are characterized by a finite escape time under some conditions.§

§ At the point in time where the escape occurs, denoted as t_c, the $\hat{p}_{kl}$ variables simultaneously become $|\hat{p}_{kl}(t_c)| = \infty$. The point t_c is identical to the *conjugate point* associated with the classical theory of the calculus of variations. In addition to the Euler-Lagrange and Legendre conditions, a trajectory with continuous time derivatives gives rise to a local minimum only if the second variations about $\hat{\mathbf{x}}(\mu)$ are finite. This condition implies that no conjugate points are allowed on the interval $t \leqq \mu \leqq T$ and is called the *Jacobi condition*.[37] The occurrence of conjugate points along $\hat{\mathbf{x}}(\mu)$ also invalidates the second-degree expansions given in Chap. 9. The necessary and sufficient conditions for no conjugate points involve the examination of a matrix of second derivatives of the hamiltonian function along the trajectory $\hat{\mathbf{x}}(\mu)$. This matrix requires the solution of the characteristic equations; hence the direct examination of this matrix is extremely difficult or impossible in practical problems. Therefore, the validity of the second-degree expansions is ensured by the simple expedient of obtaining bounded numerical solutions to the matrix Riccati equations with sufficient accuracy.

This numerical difficulty is eliminated if a priming control equation can be found such that the differential equations used in the forward-time computations are stable without limit cycles. This conclusion is supported by an examination of $[p_{kl}^{(1)}]$ associated with the data presented for $K_1^{(0)} = 0$, $K_{11}^{(0)} = 1 + 2^{\frac{3}{4}}$, and $K_{12}^{(0)} = 2^{\frac{1}{2}} - 1$ in Fig. 10-11. Of course, this difficulty never occurs with a linear dynamic process, because then $\partial^2 f/\partial x_n\, \partial x_m = \partial^2 f/\partial x_n\, \partial m_m = \partial^2 f/\partial m_n\, \partial m_m = 0$.

Case III

The quasi-optimum trajectory for a soft saturation constraint on the state signal x_1 is shown in Fig. 10-14, and the control signal shown in this figure is defined to be m_a^* for this case. Of course, the primary difference between this trajectory and the

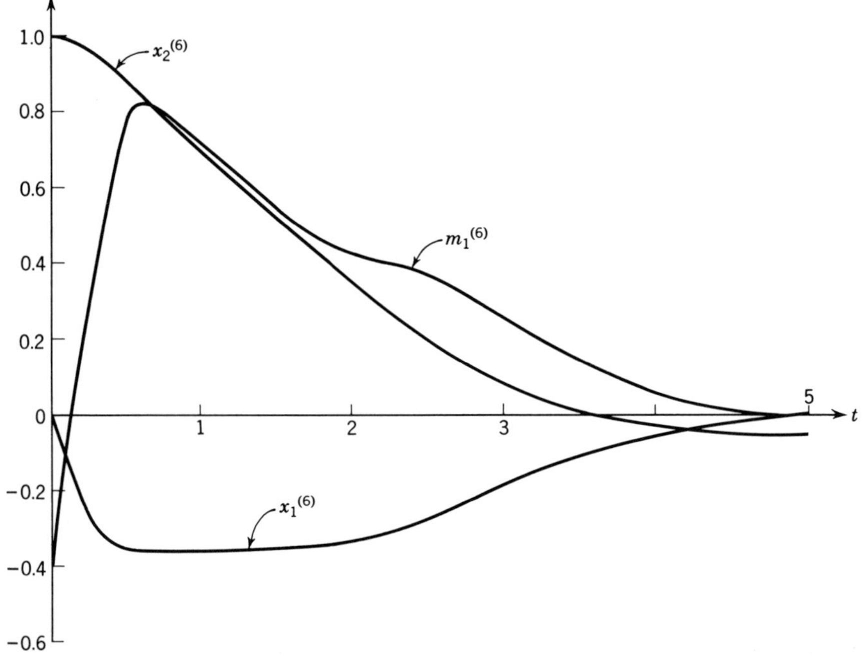

Fig. 10-14 Quasi-optimum trajectory for case III.

trajectory for case I is that here $|x_1^{(i)}| < 0.4$. Also, the time-varying parameters of the optimum linear control equation are shown in Fig. 10-15 for rate saturation. As expected, the rate feedback gain, namely, K_{11}, becomes very large during the period of time when x_1 is near its allowable boundary. This behavior of K_{11} tends to maintain small perturbations in x_1 when load disturbances or initial-condition errors are present. The reader should note again the asymptotic behavior of these parameters to those of case I during periods of time where x_1 is not near its allowable boundary.

The data appearing in Figs. 10-16 and 10-17 compare the convergence properties of the two relaxation methods. Roughly speaking, second variations possess the same advantages here that occurred in case II. However, the reader should note the use of $K_1^{(0)} = 0$, $K_{11}^{(0)} = 10$, and $K_{12}^{(0)} = 0$ for the priming control equation. The large rate gain $K_{11}^{(0)} = 10$ is used in order to ensure $|x_1^{(1)}| < 0.4$ so that

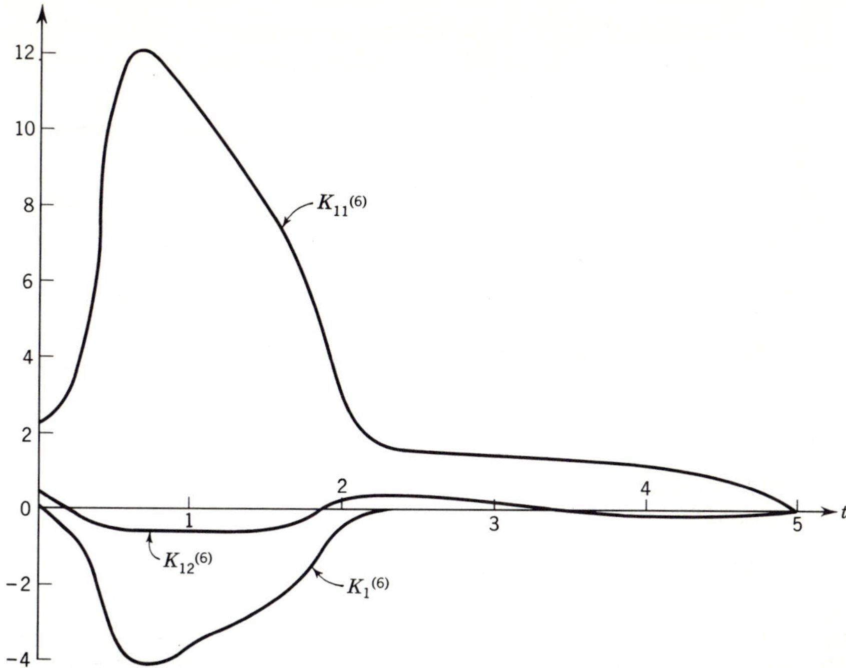

Fig. 10-15 Linear-control-equation parameters for case III.

$(x_1{}^{(1)}/0.4)^{32}$ appearing in the error measure does not result in numbers larger than 10^{36}, which would result in computer overflow. When penalty functions are used to approximate saturation constraints, the priming trajectory—and hence the priming-control-equation parameters—must be selected carefully.

Case IV

The quasi-optimum trajectory for a saturation constraint on the control signal m_1 is shown in Fig. 10-18, and the control signal shown in this figure is defined to be

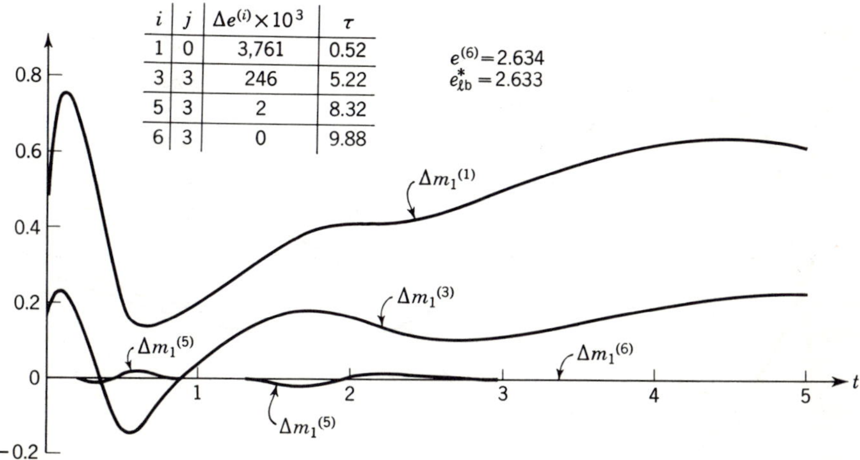

Fig. 10-16 Control-signal error using second variations for case III: $K_1^{(0)} = 0$, $K_{11}^{(0)} = 10$, $K_{12}^{(0)} = 0$.

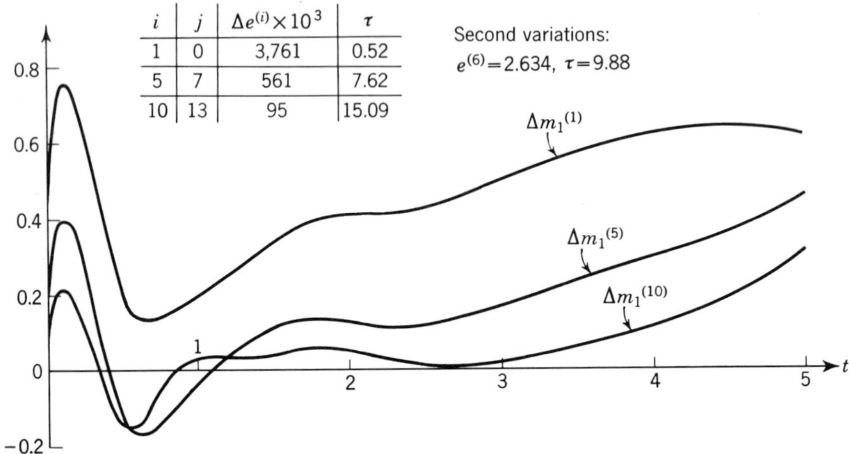

Fig. 10-17 Control-signal error using first variations for case III: $K_1^{(0)} = 0$, $K_{11}^{(0)} = 10$, $K_{12}^{(0)} = 0$.

m_a^* for this case. Here, the control signal is restricted so that $|m_1{}^{(i)}| < 0.5$, thereby significantly altering the results of case I. Also, the time-varying parameters of the optimum linear control equation are shown in Fig. 10-19 for control-signal saturation. During the period of time where the saturation boundary is important, the feedback gains are reduced to small values in order to maintain possible perturbations in m_1 small. The large peak in K_{11} at about $t = 0.15$ results from anticipating the entry into the control boundary. For time greater than about 3, these parameters are asymptotic to those of case I.

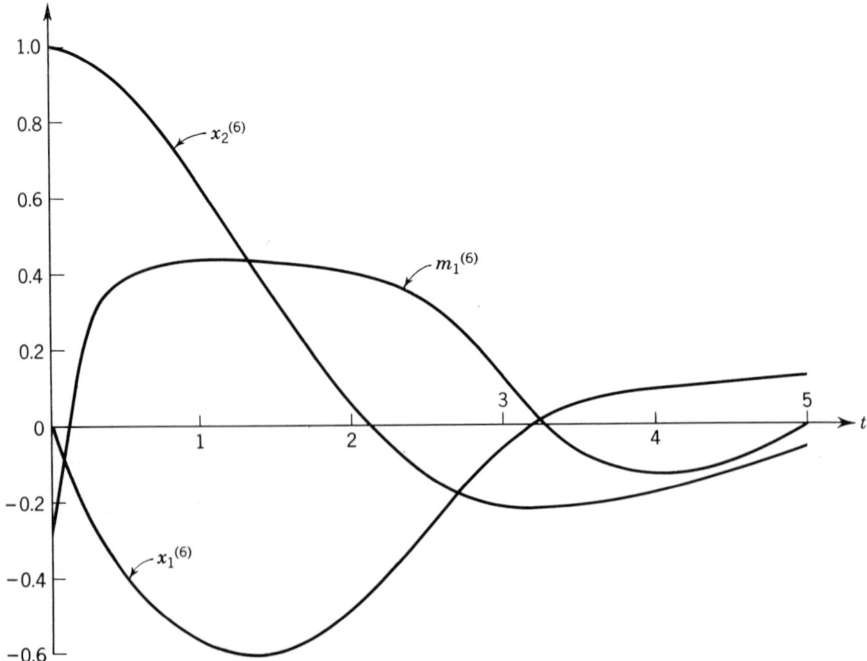

Fig. 10-18 Quasi-optimum trajectory for case IV.

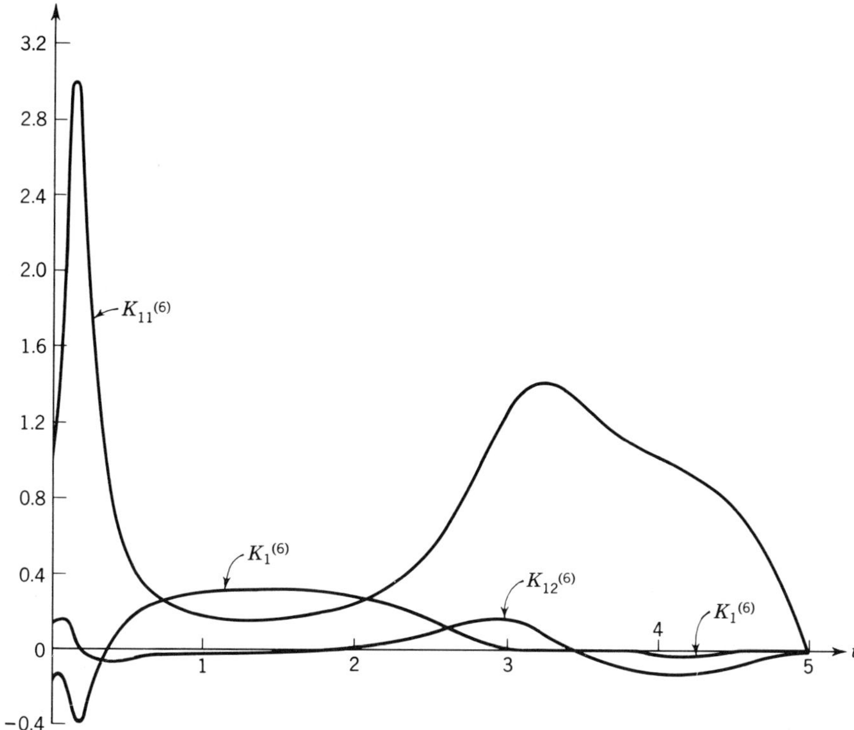

Fig. 10-19 Linear-control-equation parameters for case IV.

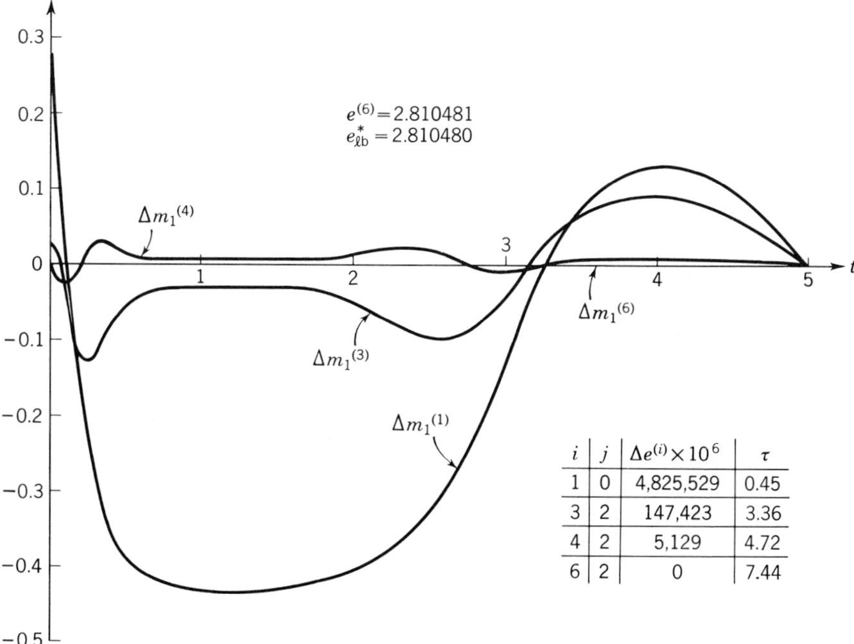

i	j	$\Delta e^{(i)} \times 10^6$	τ
1	0	4,825,529	0.45
3	2	147,423	3.36
4	2	5,129	4.72
6	2	0	7.44

Fig. 10-20 Control-signal error using second variations for case IV: $K_1^{(0)} = K_{11}^{(0)} = K_{12}^{(0)} = 0$.

The data appearing in Figs. 10-20 and 10-21 compare the convergence properties of the two relaxation methods. Both methods are initiated with $K_1^{(0)} = K_{11}^{(0)} = K_{12}^{(0)} = 0$ in order to maintain $m_1^{(1)}$ inside its allowable boundaries. Also, the step size ε must be decreased whenever $|m_1^{(i)}|$ exceeds 0.5 because of the particular penalty function appearing in (10-121). This case is particularly interesting because the method based on first variations yields a higher convergence rate for the first two iterations than the method based on second variations. This result is primarily due to the fact that the first two iterations with second variations require a step size $\varepsilon = \frac{1}{2}$ so that $|m_1^{(i)}| < 0.5$ is maintained. Also, the two methods possess identical extrapolation properties on the penalty function associated with the control signal.

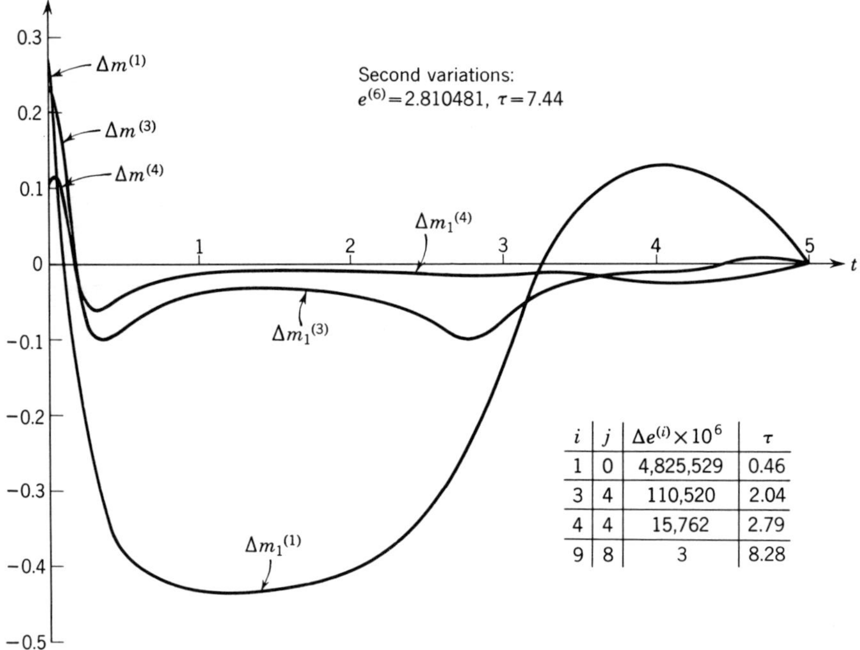

i	j	$\Delta e^{(i)} \times 10^6$	τ
1	0	4,825,529	0.46
3	4	110,520	2.04
4	4	15,762	2.79
9	8	3	8.28

Fig. 10-21 Control-signal error using first variations for case IV: $K_1^{(0)} = K_{11}^{(0)} = K_{12}^{(0)} = 0$.

Once the ith trajectory is in a suitably small neighborhood of the optimum trajectory, however, the method based on second variations again proves to be superior to the method based on first variations. These observations suggest that the combination of both methods, with second variations being used when the ith trajectory is in a suitably small neighborhood of the optimum trajectory, would yield a convergence rate which is larger than those obtained with either method used exclusively.

10-9 Computation of linear-feedback-control-equation parameters

The iterative methods discussed thus far primarily are directed at the problem of trajectory optimization, or rather to finding an acceptable approximation to $\hat{\mathbf{x}}(\sigma)$ by computational means. The vector $\hat{\mathbf{x}}(\sigma)$ is defined in Chap. 9 as the optimum state vector corresponding to the initial state $\hat{\mathbf{x}}(t)$. The question now arises as to the computation of the parameters appearing in the feedback control equations discussed in the previous chapter. However, only the computations required for the optimum linear control equation are discussed here.

The desire to construct a feedback control equation after an acceptable solution to the trajectory optimization problem has been obtained does affect the choice of iteration methods used in the trajectory optimization problem. Specifically, the computer program used for trajectory iterations should be compatible, and also usable with minor modifications, for the computation of the linear-control-equation parameters and for the evaluation of the feedback-system trajectories resulting from small initial-condition errors. Such compatibility would result in considerable savings of money and time associated with computer programming. Moreover, the desire to construct a feedback control equation determines in part the restrictions placed on what is called an acceptable solution to the trajectory optimization problem. Even though the ith trajectory is an *efficient trajectory*, in the sense that $e^{(i)}$ approximates e^* to the degree of accuracy required, an acceptable solution to the two-point boundary-value problem may not have been obtained when the dynamic process is nonlinear. In other words, the desire to construct a linear control equation determines, in part, when the iterations can be terminated and may require many more iterations than would otherwise be necessary.

The optimum linear control equation is precisely the iteration equation based on second variations when the ith trajectory also is the optimum trajectory $\hat{\mathbf{x}}(\sigma)$. However, numerical errors make this condition impossible to achieve; hence the threshold ϵ_m must be introduced. All trajectories that satisfy the condition $\mathbf{U}^{(i)} = \mathbf{0}$ everywhere on the interval $t \leqq \sigma \leqq T$ subject to the value of ϵ_m chosen are called *quasi-optimum* when the iteration equation based on second variations is being used for trajectory optimization. Of course, the condition $\mathbf{U}^{(i)} = \mathbf{0}$ results in the termination of trajectory iterations. Now, suppose that a particular quasi-optimum trajectory, called the *nominal trajectory*, is used in the second-degree expansions that lead to the optimum linear control equation discussed in Chap. 9. Specifically, the trajectory $\hat{\mathbf{x}}(\sigma)$ is replaced by $\mathbf{x}^{(i)}(\sigma)$; hence the nominal trajectory is treated as the optimum trajectory of a modified minimization problem. This *modified minimization problem* corresponds to the modified error measure

$$H_m(\mathbf{x}, \mathbf{m}, \sigma) = H(\mathbf{x}, \mathbf{m}, \sigma) - \sum_{n=1}^{M} S_n^{(i)}(m_n - m_n^{(i)}) \tag{10-141}$$

In addition, the optimum linear control equation for this modified minimization problem is precisely the iteration equation based on second variations when evaluated on the nominal trajectory. Hence, the parameters of this control equation are computed with the backward-time equations normally solved for the iteration equation based on second variations. The backward-time equations, in fact, simplify somewhat for these computations. Specifically, $\mathbf{U}^{(i)} = \mathbf{0}$ and hence $g_k^{(i)} = G_k^{(i)} = 0$ because the nominal trajectory is defined as a particular quasi-optimum trajectory. Therefore, the $g_k^{(i)}$ equations can be omitted entirely in the computation of these control-equation parameters. These parameters, therefore, can be computed without writing an additional computer program and are used for the linear control equation in a suitably small neighborhood about the nominal trajectory under certain restrictions.

The restrictions placed on the use of these control-equation parameters concern the stability of the feedback system and the accuracy with which the nominal trajectory approximates the optimum trajectory. The linear control equation must give rise to system stability in a suitably small but nonvanishing neighborhood about the nominal trajectory in order to accommodate small initial-condition errors, etc. Therefore, the nominal trajectory must be within a suitably small neighborhood of the optimum trajectory where the nominal and neighboring quasi-optimum trajectories exhibit the properties of stable systems, as discussed in Chap. 9. In this neighborhood, the modified minimization problem is strictly convex so that the matrices $[T_{nm}{}^{(i)}]$ and $[p_{kl}{}^{(i)}]$ are positive definite and positive semidefinite, respectively, everywhere on the interval $t \leqq \sigma \leqq T$. The nominal trajectory is made to satisfy these conditions by choosing the value of ϵ_m used in the trajectory iterations to be suitably small when the dynamic-process state equation is nonlinear. Even though the convexity of the error measure is not modified by (10-141), the second partials of the state equations with respect to the elements of $\mathbf{m}^{(i)}$ and $\mathbf{x}^{(i)}$ modify the matrices $[T_{nm}{}^{(i)}]$ and $[p_{kl}{}^{(i)}]$ when the dynamic process is nonlinear.

The neighborhood about the optimum trajectory where the original minimization problem, and hence the modified minimization problem, is strictly convex unfortunately is extremely small for many design problems. In other words, the value of ϵ_m that would be used to establish a nominal trajectory acceptable for the computation of the parameters appearing in the linear control equation may be very small. Such a situation could cause numerous trajectory iterations after an efficient trajectory had been found. In addition, severe situations may require decreasing the increment size used for the numerical integration of the forward- and backward-time equations in order to obtain a degree of numerical accuracy which is compatible with ϵ_m. Decreasing the increment size adds a proportionate increase in elapsed computer time per iteration. In fact, round-off errors due to the finite word length of the digital computer may make it impossible to obtain a sufficiently accurate solution to the two-point boundary-value problem in especially difficult situations.

In order to alleviate these difficulties, an alternative set of parameters for the linear control equation can be computed. The terms involving the second partials of the dynamic-process state equation with respect to the elements of $\mathbf{m}^{(i)}$ and $\mathbf{x}^{(i)}$, which also cause the lack of convexity in the modified minimization problem, vanish under the constraint of asymptotic stability. This property is introduced in Chap. 9. Suppose that the value of ϵ_m is increased to the point where most of the efficient trajectories are included in the neighborhood of the optimum trajectory so that $\mathbf{U}^{(i)} = \mathbf{0}$. Then, a second modified minimization problem is introduced, which corresponds to the modified error measure

$$H_m(\mathbf{x}, \mathbf{m}, \sigma) = H(\mathbf{x}, \mathbf{m}, \sigma) - \sum_{n=1}^{M} \frac{\partial H^{(i)}}{\partial m_n{}^{(i)}} (m_n - m_n{}^{(i)}) - \sum_{n=1}^{N} \frac{\partial H^{(i)}}{\partial x_n{}^{(i)}} (x_n - x_n{}^{(i)}) \tag{10-142}$$

The ith trajectory again is taken to be the nominal trajectory, and the condition $p_k{}^{(i)} = 0$ is obtained for this second modified minimization problem. Furthermore, the parameters of the linear control equation for the error measure given in (10-142) and the iteration equation based on second variations are made identical by simple modifications in the backward-time equations.

After the ith trajectory has been obtained and is deemed to be efficient for the purposes of a nominal trajectory, the conditions $\epsilon_m = \infty$ and $p_k^{(i)} = 0$ are imposed. These conditions allow the neglecting of the $g_k^{(i)}$ and $p_k^{(i)}$ equations normally solved in the backward-time direction when the iteration equation based on second variations is used for trajectory optimization. Furthermore, the parameters of the linear iteration equation give rise to asymptotic stability of the feedback control system in a suitably small neighborhood of the nominal trajectory. Because no restrictions are placed on the nominal trajectory for the case of asymptotic stability except that it be efficient, the computational difficulties in obtaining a set of suitable linear-control-equation parameters are eased considerably. These difficulties may be an important factor governing the synthesis of feedback control systems for many design problems where the state equation of the dynamic process is nonlinear.

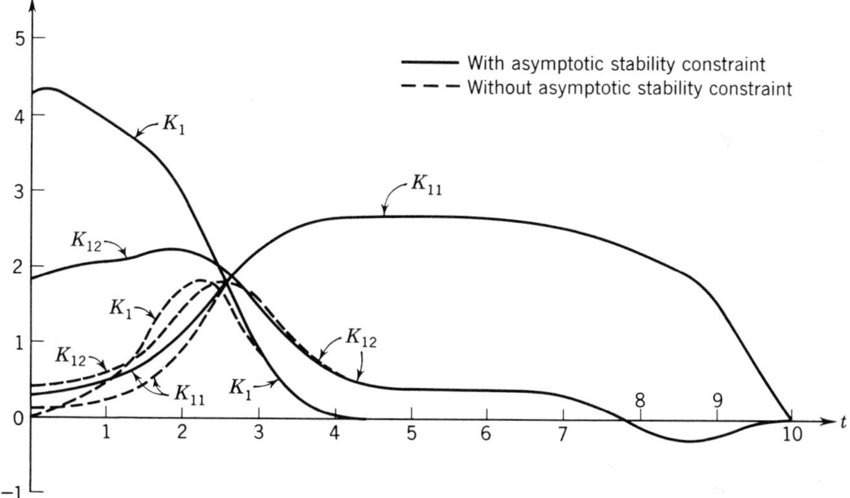

Fig. 10-22 Linear-control-equation parameters for case II: $x_1(0) = 0$, $x_2(0) = 3$, $T = 10$.

The example given in Sec. 10-8 under the conditions of case II provides a simple illustration of the results of computing these two sets of linear-control-equation parameters. This example also is used to illustrate some of the design considerations presented in Chap. 9. Specifically, the size and shape of the neighborhood about the nominal trajectory where stability is maintained are very similar to those of the neighborhood about the optimum trajectory where the minimization problem is convex.

These two sets of parameters are shown in Fig. 10-22 for case II under the conditions $x_1(0) = 0$, $x_2(0) = 3$, and $T = 10$. The nominal trajectory used in the computations is quasi-optimum for $\epsilon_m = 0.001$. The primary differences in these two sets of parameters occur where the nominal trajectory is near or outside the limit cycle associated with the homogeneous form of the van der Pol equation. In this region of state space, the second partials of $f_1^{(i)}$ with respect to $x_1^{(i)}$ and $x_2^{(i)}$ are significant, and the effects of setting $p_1^{(i)} = p_2^{(i)} = 0$ are appreciable. Also, the feedback gains are generally larger in this region for the case of the asymptotic stability constraint and hence give rise to improved system stability. The reference signal K_1, in the case of the asymptotic stability constraint, is a rough approximation to the shape of $x_2(t)$ associated with the nominal trajectory. This combination of control-equation parameters gives rise to the return of the system from neighboring points in state space to the nominal trajectory.

The position variable x_2 for the nominal and two perturbed trajectories is shown in Fig. 10-23 for the case where the asymptotic stability constraint is not imposed. Similarly, the corresponding responses are shown in Fig. 10-24 for the case with the asymptotic stability constraint. These two cases are denoted where necessary with the subscripts w/o and w, respectively. The trajectory $A_{w/o}$, gives rise to a value of

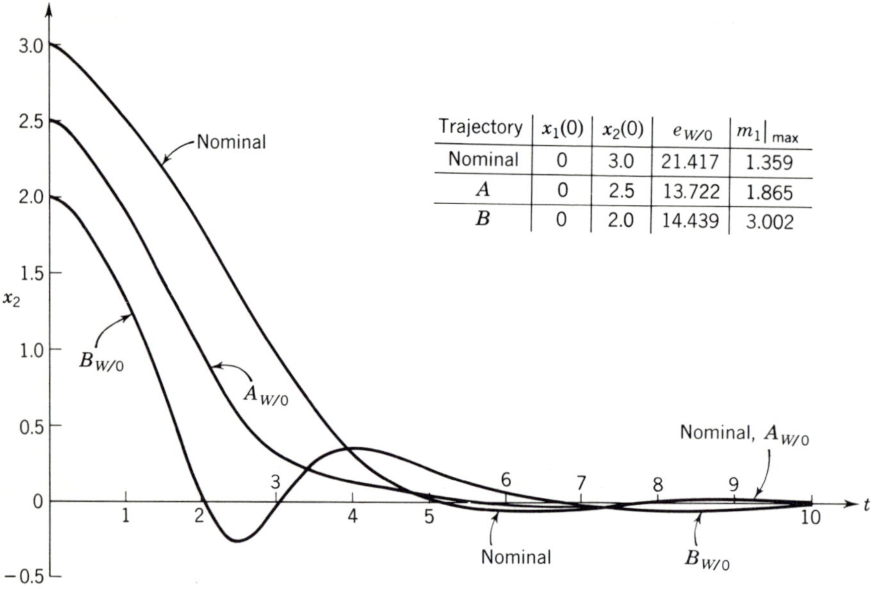

Trajectory	$x_1(0)$	$x_2(0)$	$e_{W/O}$	$m_1\|_{max}$
Nominal	0	3.0	21.417	1.359
A	0	2.5	13.722	1.865
B	0	2.0	14.439	3.002

Fig. 10-23 Response of system without asymptotic stability constraint for case II.

the error index $e_{w/o}$, corresponding to the error measure given in (10-118), that is smaller than the value e_w for A_w. This result is expected, because $A_{w/o}$ approximates the neighboring quasi-optimum trajectory that corresponds to the initial conditions $x_1(0) = 0$, $x_2(0) = 2.5$, whereas A_w is constrained to return to the nominal trajectory. On the other hand, the trajectory $B_{w/o}$ causes a value $e_{w/o}$ that is larger than the value e_w for B_w. This phenomenon is due to the fact that the trajectory $B_{w/o}$ is outside the neighborhood about the nominal trajectory where accurate approximations to neighboring quasi-optimum trajectories are obtained. For the initial point in state space which corresponds to $B_{w/o}$, the linear control equation without the asymptotic stability constraint causes considerable overshoot and a nearly unstable situation to develop. However, the improved stability properties of the linear control equation with the asymptotic stability constraint give rise to the same damped behavior of the nominal trajectory for this initial point in state space.

Also of interest are the peak values of the control variable, denoted as $m_1|_{max}$, occurring for these trajectories. This information is given in Figs. 10-23 and 10-24 and shows that the neighboring quasi-optimum trajectory $A_{w/o}$ requires a larger value of $m_1|_{max}$ than either the nominal or the A_w trajectory. This is contrary to what might be expected. However, the nonlinear damping term associated with the van der Pol equation tends to drive the system toward the limit cycle when the point in state space lies within the limit cycle. For the perturbed trajectories shown, the nominal trajectory is in the general direction of the limit cycle; hence less peak control effort is required in the case of the asymptotic stability constraint. For initial points in state space in the region $x_1(0) = 0$ and $3 < x_2(0) \leqq 4$, the desired trajectory $x_1 = x_2 = 0$, the nominal trajectory, and the limit cycle are all roughly in the same direction from the perturbed trajectories. Therefore, little difference exists

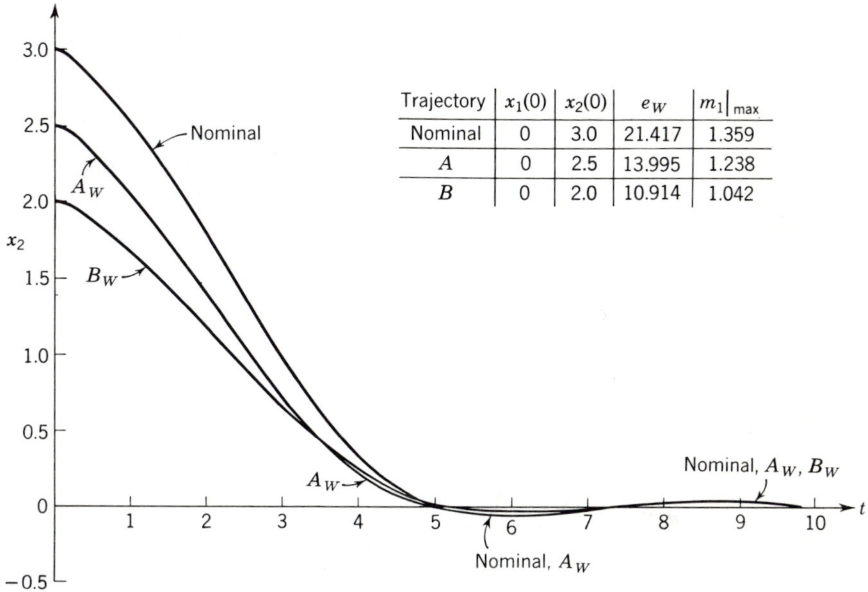

Trajectory	$x_1(0)$	$x_2(0)$	e_W	$m_1\|_{max}$
Nominal	0	3.0	21.417	1.359
A	0	2.5	13.995	1.238
B	0	2.0	10.914	1.042

Fig. 10-24 Response of system with asymptotic stability constraint for case II.

between the linear control equations with and without the asymptotic stability constraint in this region.

As mentioned in Chap. 9, the boundaries of the neighborhood where neighboring quasi-optimum trajectories can be approximated with a linear control equation unfortunately cannot be established analytically. A limited number of trajectory evaluations for different initial points in state space, however, can be performed computationally. Actually, such trajectory evaluations can be performed conveniently because they merely involve the forward-time equations solved for the iteration algorithm based on second variations. However, only a limited number of trajectory evaluations are feasible, and not every point on the boundary of this neighborhood can be identified or even investigated. A convenient method of characterizing a suitable interior of this neighborhood involves the computation of a distance measure with respect to the nominal trajectory and establishing a threshold. Then the linear control equation without the asymptotic stability constraint can be used whenever the distance measure is less than or equal to this threshold.

A distance measure $d^2(t)$ which is suitable for this purpose can be defined as

$$d^2(t) = \sum_{k=1}^{N} \sum_{l=1}^{N} d_{kl}^{(i)}(t)[x_k(t) - x_k^{(i)}(t)][x_l(t) - x_l^{(i)}(t)] \tag{10-143}$$

where $[d_{kl}^{(i)}(t)]$ is any positive-definite symmetric matrix and $\mathbf{x}^{(i)}(t)$ is treated now as the state vector of the nominal trajectory. The selection of a quadratic form for the distance measure, although somewhat arbitrary, is made because the minimization

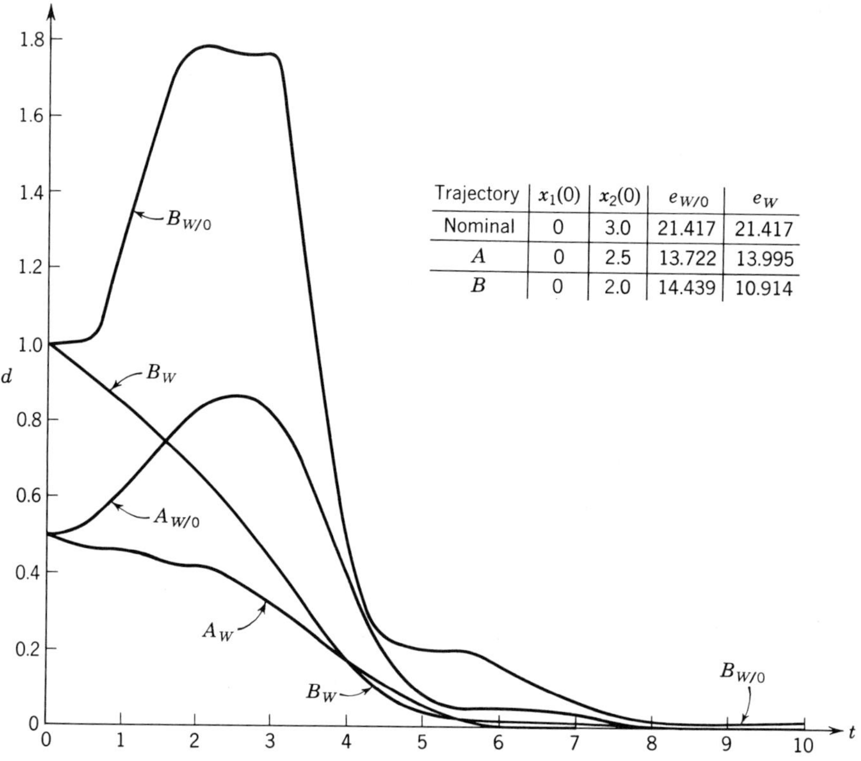

Trajectory	$x_1(0)$	$x_2(0)$	$e_{W/O}$	e_W
Nominal	0	3.0	21.417	21.417
A	0	2.5	13.722	13.995
B	0	2.0	14.439	10.914

Fig. 10-25 Vector distance between nominal and perturbed trajectories with and without the asymptotic stability constraint for case II.

problem is strictly convex and hence quasi-quadratic in the neighborhood of the nominal trajectory. Because the error measure is strictly convex, the matrix elements

$$d_{kl}^{(i)}(t) = \frac{1}{2} \frac{\partial^2 H^{(i)}}{\partial x_k^{(i)}\, \partial x_l^{(i)}} \tag{10-144}$$

satisfy the positive-definite requirement. Also, the matrix elements

$$d_{kl}^{(i)}(t) = \frac{1}{2} \sum_{n=1}^{M} R_{nk}^{(i)} K_{nl}^{(i)} \tag{10-145}$$

satisfy this requirement because the nominal trajectory is in a strictly convex region. However, the matrix elements given in (10-144) may be preferable because they are independent of whether or not the asymptotic stability constraint is being used.

The distance measure $d(t)$, when evaluated from (10-144), is shown in Fig. 10-25 for the trajectories which are plotted in Figs. 10-23 and 10-24. This particular distance measure decreases monotonically with time for the trajectories with the asymptotic stability constraint, whereas monotonic behavior is not obtained for trajectories without the asymptotic stability constraint. According to the limited data presented thus far, the threshold associated with $d(t)$ should be set somewhere between the curves in Fig. 10-25, which are labeled $A_{w/o}$ and $B_{w/o}$. The proper setting of this threshold results in satisfactory system performance of the linear control equation without the asymptotic stability constraint when $d(t)$ is less than or equal to the time-varying threshold chosen.

Unfortunately, the use of such a distance measure and threshold completely neglects the lack of symmetry normally associated with the boundaries of the neighborhood where neighboring quasi-optimum trajectories can be approximated with a

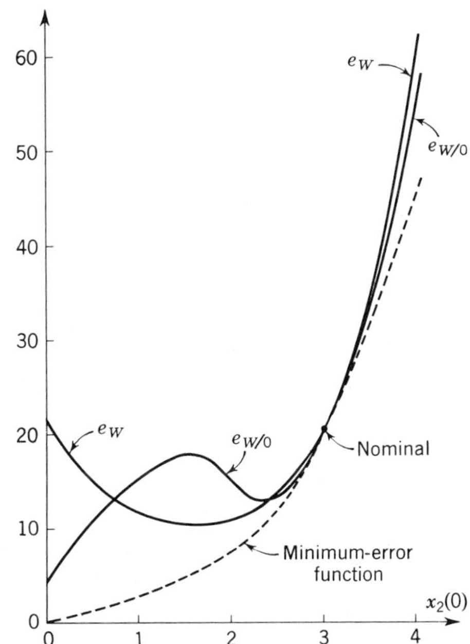

Fig. 10-26 Error indices for perturbed systems with and without the constraint of asymptotic stability: $x_1(0) = 0$, $T = 10$.

linear control equation. This lack of symmetry is illustrated in Fig. 10-26, where the error index is plotted for the linear control equations with and without the constraint of asymptotic stability. These data indicate that for $x_1(0) = 0$ and the range $2.35 \leqq x_2(0) \leqq 4$ the linear control equation without the constraint of asymptotic stability is adequate and, in fact, superior to the linear control equation with the constraint of asymptotic stability. In a practical situation where the added difficulties of computing and storing both sets of time-varying parameters are not excessive, the range of initial conditions for which acceptable system response is obtained can be extended. For instance, the linear control equation can be used satisfactorily for $x_1(0) = 0$ and the range $1.5 \leqq x_2(0) \leqq 2.35$ by switching to the parameters computed with the constraint of asymptotic stability. In the range $0 \leqq x_2(0) \leqq 1.5$ for $x_1(0) = 0$, however, both sets of parameters result in appreciable overshoot of the type indicated in Fig. 10-23. Therefore, the linear control equation based on the nominal trajectory with the initial point in state space $x_1(0) = 0$, $x_2(0) = 3$ is inadequate for this range.

The desire to obtain only an efficient trajectory and the desire to use the linear-control-equation parameters which approximate neighboring quasi-optimum trajectories may be incompatible. As indicated in this example, the differences between the error indices obtained with the two sets of control-equation parameters are relatively small in the neighborhood of the nominal trajectory. These differences also may be of the order of magnitude which corresponds to the differences obtained between efficient trajectories and a nominal trajectory that lies within a strictly convex region about the optimum. Furthermore, when this strictly convex region is small, the region of allowable initial-condition errors, where the parameters computed without the asymptotic stability constraint are sufficient to maintain system stability, also is small in general. These computational and design considerations strongly suggest the use of the asymptotic stability constraint in many design problems where the state equation of the dynamic process is nonlinear.

10-10 Summary

The ability to obtain numerical solutions is a primary consideration in the usefulness of optimization theory for designing feedback control systems. Certainly, this method of design requires access to a large-scale and high-speed digital computer for significant engineering problems. However, relaxation methods provide a reasonably flexible computational means for approximating the solution of two-point boundary-value problems with any degree of accuracy desired. As opposed to boundary-condition iteration and discrete dynamic-programming methods, relaxation methods guarantee convergence to the optimum without requiring a nearly unbounded computer memory. Furthermore, the computations are performed with ordinary differential equations and simple iteration algorithms. Therefore, these methods can be undertaken with known numerical methods and without a large amount of preliminary study for each design problem.

One initial numerical consideration, however, is the stability of the differential equations to be solved when relaxation methods are used. The stability of these equations is determined by both the stability and the nonlinearities of the dynamic process and by the method used for finding a priming trajectory. When the ith trajectory is in a suitably small neighborhood of the optimum trajectory, the differential equations used with the method based on second variations are stable when the threshold ϵ_m also is used. On the other hand, the differential equations used with the method based on first variations are unstable when the dynamic-process state equations are unstable even in this suitably small region. Therefore, recourse to the relaxation method based on second variations is advisable in this region when the state equations are unstable. The stabilization of these equations afforded by second variations eliminates many impending difficulties with numerical accuracy. In addition, when the ith trajectory is in a suitably small neighborhood of the optimum trajectory, the optimum control equation is quasi-linear and the method based on second variations yields essentially one-step convergence. This property gives rise to a very large rate of convergence with the method based on second variations once an efficient priming trajectory has been obtained. The superiority of the method based on second variations is accentuated when the error measure is highly sensitive to perturbations in the state signals. Finally, the computations associated with the method based on second variations directly provide the time-varying parameters appearing in the optimum linear control equation and would have to be performed for a design based on a small region of state space.

On the other hand, the method based on first variations probably is more efficient for the first few iterations when only a very gross priming trajectory is available. The computational results presented for case IV in Sec. 10-8 illustrate such a situation. In addition, recourse to first variations for the first few iterations may be necessary when a suitable priming control equation cannot be found for a nonlinear and unstable, or stable with a bounded limit cycle, dynamic process. Under these conditions, the matrix Riccati equations may be unstable, which would cause insurmountable numerical problems. The computational results presented for case II in Sec. 10-8 illustrate such a situation. As a result of these considerations, a hybrid method based on first variations for the first few iterations and then on second variations once an efficient trajectory has been found is recommended as a general computational means for solving two-point boundary-value problems. Any of the possible algorithms can be used for the iterations based on first variations. However, the algorithm given in (10-63) is a specialization of the algorithm based on second variations and is a logical choice for the hybrid method. The use of (10-63) requires only a simple transfer in the computer program to a branch where (10-110) and (10-111) are not computed and where the terms involving $g_k^{(i)}$, $R_{nk}^{(i)}$, and $K_{nk}^{(i)}$ in (10-109) and in (10-112) to (10-114) are neglected.

In order to use this hybrid method automatically on the computer, a means for determining whether an iteration can be completed successfully with second variations is required. Supposedly, second variations should be used when the matrix Riccati equations are stable. The stability of these equations is ensured by iterating with second variations only if the matrix $[p_{kl}^{(i)}]$ is positive semidefinite everywhere on the interval $t \leqq \sigma \leqq T$. However, the requirement that $p_{kk}^{(i)} \geqq 0$, for $k = 1, 2, \ldots, N$, everywhere on the interval $t \leqq \sigma \leqq T$ for the use of second variations has been found to be sufficient for avoiding the numerical problems associated with unstable Riccati equations. This latter test requires a much smaller number of computations and hence requires less computer time than testing the matrix $[p_{kl}^{(i)}]$ for positive definiteness.

Also, the reader should be warned that both the methods based on second variations and first variations using (10-63) require that the matrix $[T_{nm}^{(i)}]$ be positive definite for guaranteed convergence. This matrix may not be positive definite everywhere on the interval $t \leqq \sigma \leqq T$ when the state equations are nonlinear, so that $\partial^2 f_k^{(i)}/\partial m_n^{(i)}\,\partial m_m^{(i)} \neq 0$, unless the ith trajectory is in a suitably small neighborhood of the optimum trajectory. When this matrix is not positive definite, then an alternative algorithm in conjunction with first variations is used in the hybrid method. For instance, any of the algorithms given in (10-58) are acceptable alternatives. On the other hand, the form of (10-63) can be preserved by replacing $T_{nm}^{(i)}$ with $\tau_{nm}^{(i)}$, where $[\tau_{nm}^{(i)}]$ is chosen as any positive-definite symmetric matrix. Preserving the form of (10-63) is desirable when $[\tau_{nm}^{(i)}]$ can be selected so that two conditions are satisfied. First, a threshold can be introduced so that $[T_{nm}^{(i)}]$ is replaced by $[\tau_{nm}^{(i)}]$ before $[T_{nm}^{(i)}]$ ceases to be positive definite. Then the portion of backward-time integration near the point where the singularity occurs would not have to be repeated. Second, the transition to $[\tau_{nm}^{(i)}]$ should not cause a step discontinuity in the components of $\mathbf{M}^{(i)}$. Both these conditions are imposed in order to preserve numerical accuracy.

An important property of relaxation procedures is that the fundamentally continuous character of the design problem does not have to be disregarded. Also, all intermediate solutions in the iteration are found for the desired fixed-point initial-boundary condition without violating the design constraints. These properties imply that the intermediate solutions are meaningful, although suboptimum, and can be used for design purposes. In many industrial applications of advanced automatic control systems, the engineer must demonstrate the potential economic return afforded by these systems in competition with existing manual control and operating practices. If pertinent records of past system operation are available, then the system response due to manual control or elementary automatic control can be used as the priming or initial trajectory in the relaxation solution of the optimization problem. Therefore, all expenditures for computer time and engineering effort are devoted to making improvements on existing system performance. Also, the point of diminishing returns of these expenditures readily is identified. Because of these considerations, relaxation methods for solving two-point boundary-value problems are suited eminently for efficient feasibility studies.

11

Synthesis of Nonlinear Optimum Control Systems for Large Regions of State Space

As mentioned in Chap. 9, design problems are divided conveniently into two classes for the purposes of system synthesis. This division is employed because the synthesis problem must be treated as an approximation problem, and approximation methods roughly fall into two classes. On the one hand, Taylor-type methods are suitable for localized approximations. On the other hand, discrete dynamic programming and refinements such as Chebyshev polynomial expansions appear to be suitable for global approximations.

The synthesis procedures presented in Chap. 9 for small regions of state space require severe restrictions to be placed on the design problems. First, many important design problems involve systems where operation is not confined to a localized region. In addition, the boundaries of the region of state space, where convergence and system stability are maintained, are not known precisely and in general must be identified from system simulation. Finally, the occurrence of statistical signals and other disturbances cannot be treated adequately. The first two of these difficulties illustrate the need for a synthesis procedure that results in a predesigned or non-computing type of system suitable for large regions of state space. The third of these difficulties illustrates the need for a synthesis procedure that results in a computing type of system suitable for large regions of state space and changing design conditions such as statistical signals.

Even though the synthesis problem for large regions of state space is of extreme importance, the prospects for pertinent results in the approximation of multidimensional functions are somewhat discouraging. In addition to the difficulties discussed in Chap. 9, pertinent methods of approximation must satisfy two properties. First, the approximation must converge uniformly to the optimum control system with increasing complexity of the approximation. Second, when the approximation is truncated at any degree of complexity, the resulting control system must be stable without unwanted limit cycles. At this time, no series or other known form of approximation possesses both these required properties. In particular, the approaches of discrete dynamic programming and Chebyshev polynomial expansions for a noncomputing type of control system do not possess the second property in general; hence these approaches must be investigated carefully for each design problem undertaken. Moreover, discrete dynamic programming and Chebyshev polynomial expansions do not appear to be feasible in design problems where a computing type of control system is required, because of the difficulties discussed in Chaps. 9 and 10.

The material presented in this chapter describes a new approach to the synthesis problem for large regions of state space. This approach utilizes a computer as a component in the control system and is more suited to design problems where a computing type of system is required in any event because of changing design conditions. However, this approach may have some value in design problems where a predesigned system would have been suitable because of possible compromises between the size of the computer memory and arithmetic facilities when compared with the requirements associated with discrete dynamic programming and related methods. The synthesis method introduced here follows almost directly from the material in Chaps. 9 and 10.

11-1 Introduction to synthesis methods utilizing a computer

The primary difficulty in the synthesis problem for large regions of state space is the overwhelming amount of data that must be computed and encoded into a suitable control equation. Of course, the immensity of this task is due to the almost infinite number of points in state space that may be pertinent. However, at any given instant of time during the operation of the system, only one point in state space is of interest. A logical approach, then, would be to construct a control system that performs computations so that the operation of the system eventually becomes optimum given the existing state of the system. Of course, the price paid in system performance for this approach is the suboptimum performance that must be tolerated during the time required for convergence to the optimum.

The basis for this approach is established from three observations on the results in Chap. 10. First, every intermediate trajectory generated from the relaxation procedure is an acceptable trajectory. That is, each trajectory satisfies the required initial state of the dynamic process and the system constraints, and each successive trajectory improves system performance. Second, in design problems where the method based on second variations can be used, a sequence of control equations is generated. The general form for this sequence of linear equations is given in

(10-112). Third, these control equations stabilize the dynamic process for a suitably small region of state space. In order to maintain feedback control at all times, only design problems that are amenable to the method using second variations are treated in this chapter. That is, amplitude constraints on control and state signals must be approximated by soft constraints.

If the relaxation method based on second variations is to be used to establish a control system that both stabilizes the process and eventually optimizes the system performance during system operation, then the relaxation method must be reformulated so that the optimization takes place sequentially in time. The sketch in Fig. 11-1 is useful for informally establishing a suitable *sequential optimization* procedure. For the present, disturbances are neglected completely.

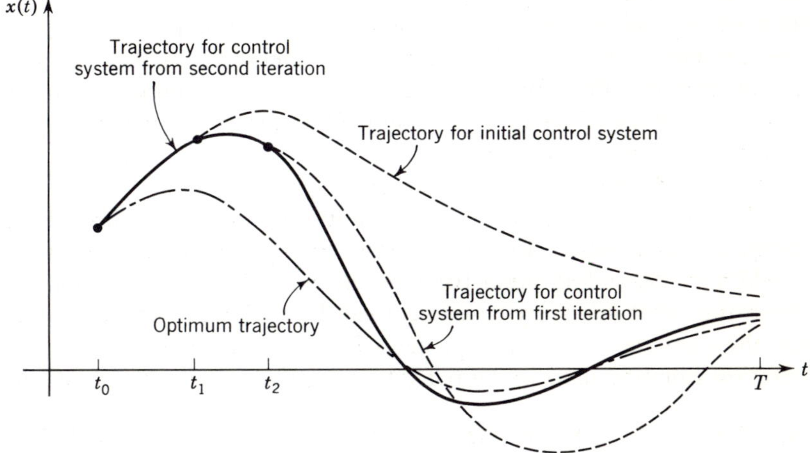

Fig. 11-1 Sketch of trajectory converging to the optimum equilibrium trajectory.

Suppose that the process is initiated at time t_0 and that one successful iteration of the relaxation method requires $(t_1 - t_0)$ sec of real time because of a finite computation rate. Then an improved control system is established at time t_1 if, at time t_0, the process is being controlled with an initial control equation which is acceptable for the interval $t_0 \leqq t \leqq T$. Now the required computations that are performed during the elapse of $(t_1 - t_0)$ sec of real time must be established.

First, an initial trajectory is needed in order to compute the parameters of the new control equation. Therefore, suppose that $\mathbf{x}(t)$ is computed on a compressed time scale for the interval $t_0 \leqq t \leqq T$, with $\mathbf{m}(t)$ being computed from the initial control equation. This computation requires some fraction of $(t_1 - t_0)$ sec of real time. Also, suppose that the error index is computed as part of this computation. However, the error measure need be integrated only over the interval $t_1 \leqq t \leqq T$, because the trajectory is fixed on the interval $t_0 \leqq t \leqq t_1$. This value of the error index is needed for establishing an improved control system which is to be used at time t_1 and is denoted as $e^{(0)}(t_1)$.

Second, the parameters which are required in the iteration algorithm based on second variations must be computed on the interval $t_1 \leqq t \leqq T$. Of course, these computations are performed in the backward-time direction, and a second fraction of $(t_1 - t_0)$ sec of real time elapses during this computation.

Finally, a second trajectory must be computed on the interval $t_1 \leq t \leq T$, where the control vector $\mathbf{m}(t)$ is computed from the iteration algorithm based on second variations. Also, the error index is computed for this new control equation and is denoted by $e^{(1)}(t_1)$. Presumably, the step size has been selected properly so that $e^{(1)}(t_1) < e^{(0)}(t_1)$ with only one trial ε being required.

At the instant these three blocks of computations have been completed, the process is assumed to be at time t_1. Also, the algorithm based on second variations is an improved control equation by virtue of $e^{(1)}(t_1) < e^{(0)}(t_1)$, and the initial control system can be replaced at time t_1. This new control system could be used until the terminal time is reached, thereby improving system performance. On the other hand, the new control system could be used just over the next $(t_2 - t_1)$ sec of real time, during which a second iteration of the relaxation procedure is being performed. The computational procedure is basically the same as that for the first iteration except that the first forward-time trajectory computation is not required. The second trajectory of the first iteration is identical to the first trajectory of the second iteration and hence need not be recomputed. Of course, these iterations can be continued throughout the operation of the process.

The physical construction of a computing system of this type cannot be presented in detail here. In general, these computers would be special-purpose and would vary depending upon the application. However, the functions of the computer and other necessary equipment can be outlined. The computer supplies data for the construction of a sequence of linear control equations. Specifically, the iteration equation used in the relaxation method based on second variations is

$$m_n^{(i+1)}(t) = \varepsilon G_n^{(i)}(t) + K_n^{(i)}(t) - \sum_{m=1}^{N} K_{nm}^{(i)}(t) x_m^{(i+1)}(t) \qquad n = 1, 2, \ldots, M \tag{11-1}$$

where
$$K_n^{(i)}(t) = m_n^{(i)}(t) + \sum_{m=1}^{N} K_{nm}^{(i)}(t) x_m^{(i)}(t) \qquad n = 1, 2, \ldots, M \tag{11-2}$$

The parameters $G_n^{(i)}$, $K_n^{(i)}$, and $K_{nm}^{(i)}$ are computed in what is called the *optimization computer*. The actual control equation which generates the control signals to the fixed member of the system is also linear and is expressed notationally as

$$m_n(t) = R_n(t) - \sum_{m=1}^{N} K_{nm}(t) x_m(t) \qquad n = 1, 2, \ldots, M \tag{11-3}$$

where $x_m(t)$ are the measured state signals. The time functions $R_n(t)$ and $K_{nm}(t)$ are constructed from appropriate segments of the time functions appearing in (11-1). According to the arguments associated with Fig. 11-1, the time functions appearing in (11-3) are given by

$$R_n(t) = \varepsilon G_n^{(i)}(t) + K_n^{(i)}(t) \qquad \text{on } t_i \leq t \leq t_{i+1} \tag{11-4}$$

and
$$K_{nm}(t) = K_{nm}^{(i)}(t) \qquad \text{on } t_i \leq t \leq t_{i+1} \tag{11-5}$$

where t_i and t_{i+1} are the points in real time when the ith and $(i + 1)$st iterations begin.

The construction process denoted in (11-4) and (11-5) for the control-system parameters requires a temporary memory device which accepts data from the optimization computer and retains these data during the period of real time when the next iteration is being performed in the optimization computer. Also, the discrete form of the data being generated from numerical integration must be interpolated for use in the continuous time operation of the process. These two functions are denoted by the *memory buffer and interpolator* and are shown in Fig. 11-2. Finally, the matrix multiplication and summation, also shown in Fig. 11-2, are considered to be continuous analog operations for purposes of discussion here. With these

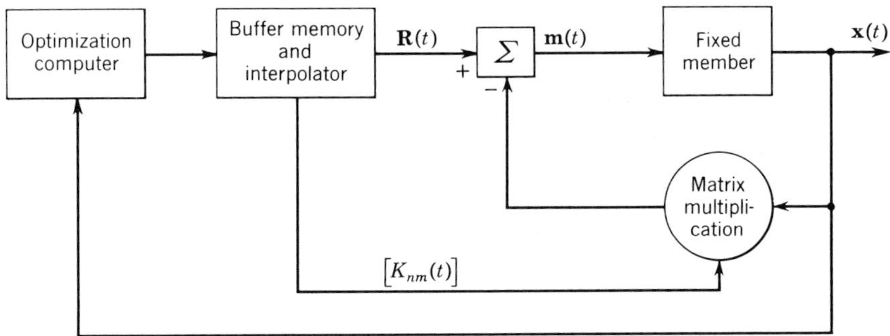

Fig. 11-2 Simplified block diagram of control system using a computer.

assumptions concerning the construction of the control system, the control signals are piecewise continuous, as indicated in Fig. 11-3. The discontinuity occurs at the *transition point* t_{i+1} from the ith to the $(i + 1)$st iterations of the relaxation procedure. The size of the discontinuity is due only to improvement in system performance resulting from the iteration if the system is disturbance-free so that $\mathbf{x}(t_{i+1}) = \mathbf{x}^{(i)}(t_{i+1})$. Also, the slope discontinuities indicated in Fig. 11-3 result from assumed straight-line interpolation of the discrete form of $G_n{}^{(i)}(t)$, $K_n{}^{(i)}(t)$, and $K_{nm}{}^{(i)}(t)$.

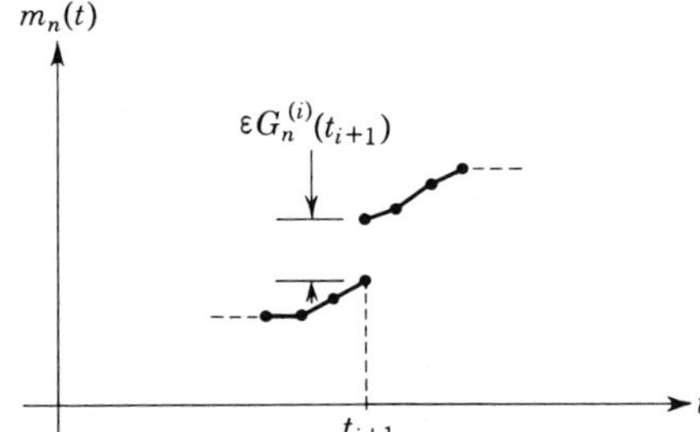

Fig. 11-3 Behavior of control signal at transition point for no disturbance.

Before the detailed operation of the optimization computer is discussed, a number of questions should be raised. The chief question concerns the conditions for which this computing type of system results in the convergence of system performance to the optimum. There are two factors that govern convergence, but these factors are not completely independent of each other.

First, the computation rate of the optimization computer determines both whether the system converges and the rate of convergence. If the computation rate is infinite, then no real time is required to perform the number of iterations of the relaxation procedure required to obtain an optimum trajectory. However, more and more real time is required for convergence as the computation rate decreases. Clearly, convergence or even partial convergence cannot be obtained when the computation rate is decreased to the point where the remaining time interval of system operation $(T - t_i)$ is required in order to obtain a successful iteration of the relaxation procedure. Therefore, the applicability of the synthesis method outlined here is limited technologically by the computation rates that can be achieved with available computers. In the main, the only problem treated in detail here is the requirement on computation rates.

The second factor in the convergence of this type of control system is the accuracy of the trajectory computations performed in the optimization computer. The considerations presented in Chap. 9 demonstrate that quasi-optimum control based on the second-degree expansion of the minimum-error function is restricted to a small region of state space. The control equation given in (11-3) is based on the second-degree expansion of the error function about $\mathbf{x}^{(i)}(t)$ and hence also is applicable only to a small region of state space. Therefore, the actual system trajectory $\mathbf{x}(t)$ and the computed trajectory $\mathbf{x}^{(i)}(t)$ must be neighboring in state space. This behavior required of the system is depicted in Fig. 11-4, along with the requirements

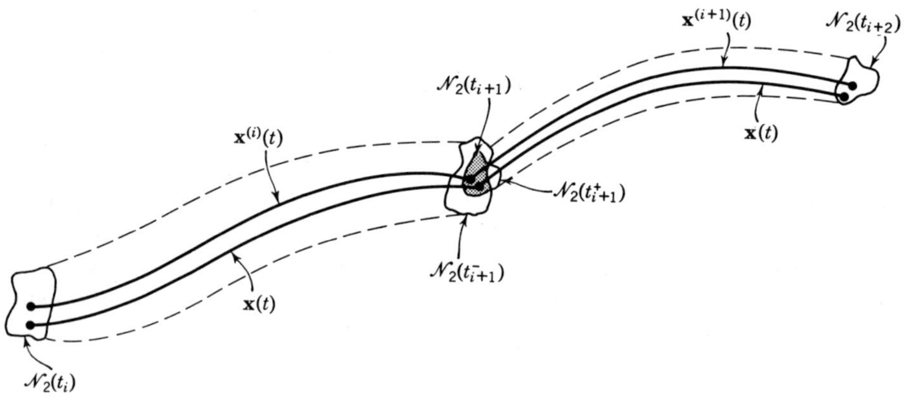

Fig. 11-4 Required behavior of control system.

at the transition point t_{i+1}. Specifically, the vector $\mathbf{x}(t_{i+1})$ must lie in the region formed by the intersection of the regions that are suitable for the $\mathbf{x}^{(i)}(t_{i+1})$ and $\mathbf{x}^{(i+1)}(t_{i+1})$ vectors. The incidence of inaccuracies in the computation of $\mathbf{x}^{(i)}(t)$ is due to a number of sources. Of course, numerical errors due to numerical integration and interpolation may be significant. Perhaps more serious, however, are the errors due to inaccuracies in the model of the system fixed member (namely, the assumed state equation), measurement of state [that is, the knowledge of $\mathbf{x}(t_i)$ at the beginning of an iteration], and disturbances occurring during the time interval $t_i \leq t \leq t_{i+1}$ which is required to perform the ith iteration. The difficulties in convergence due to disturbances are related intimately to the computation rate of the optimization computer. An introduction to the pertinent considerations for disturbances is given in Sec. 9-6.

The problems of convergence due to inaccuracies in the trajectories raise many unsolved questions to which research effort presently is being directed. Therefore, the emphasis of the remaining material is on the rate of convergence associated with computation rates. Before this aspect is discussed in detail, however, the functions and the organization of the optimization computer must be established.

The organization of the optimization computer is visualized easily when the required memory is divided into three parts. These parts are identified according to their functional operation so that the principal steps in the computational process and the flow of data are clarified. These functions of the optimization computer are shown in Fig. 11-5.

The data memory, denoted as DM, is used to store the current computed trajectory $\mathbf{m}^{(i)}$, $\mathbf{x}^{(i)}$ and the parameters $\mathbf{G}^{(i)}$, $\mathbf{K}^{(i)}$, $[K_{nm}{}^{(i)}]$ needed for the control equation associated with this trajectory. Also, design-problem parameters such as initial time and terminal time, denoted as t_0 and T, respectively, are stored in the DM.

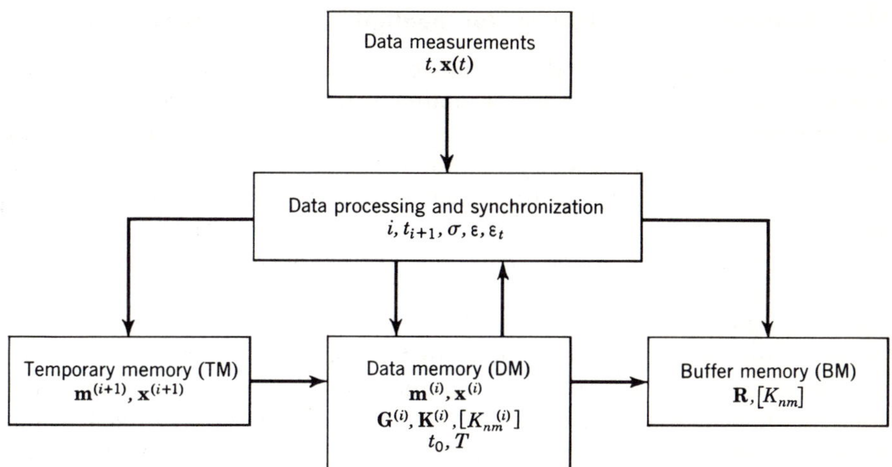

Fig. 11-5 Data flow and storage in the optimization computer.

The temporary memory, denoted as TM, is used to store the trajectory $\mathbf{m}^{(i+1)}$, $\mathbf{x}^{(i+1)}$, which is computed for a trial step size ε. This trajectory is stored until the value of ε is determined to be satisfactory. If the trial ε is satisfactory, then $\mathbf{m}^{(i+1)}$, $\mathbf{x}^{(i+1)}$ become the computed trajectory and replace $\mathbf{m}^{(i)}$, $\mathbf{x}^{(i)}$ in the DM. The buffer memory, denoted as BM, is used to store the current values of the control-equation parameters and also is the device depicted in Fig. 11-2. The feedback gains $[K_{nm}{}^{(i)}]$ are transferred from the DM to the BM, and the reference signals $\mathbf{R}$ are first computed and then transferred to the BM when a satisfactory step size has been determined.

In addition to memory functions, provisions for data processing and synchronization must be made in the optimization computer. The data processing includes all the arithmetic and logical operations involved with numerical integration (on the dummy time variable σ), step-size selection (ε and ε_t, which is a temporarily stored value of ε), etc. The data synchronization mainly involves maintaining the required relationships between real time t in the dynamic process and the interval of real time $(t_{i+1} - t)$ consumed while the computations are being performed for each iteration. This function of synchronization is required for transferring data to the memory buffer and the initiation of each iteration. Finally, the computer must have provisions for accepting measured data from the dynamic process. The measured data are real time t, which is needed for computer synchronization, and the process state vector $\mathbf{x}(t)$, which is needed for the initiation of the trajectory computations.

In the actual construction and programming of the optimization computer, many artifices may be introduced, depending upon the application, in order to conserve computer memory. Specifically, actual memory locations may be common to all three of the blocks of memory depicted in Fig. 11-5. These artifices include relocating and overwriting data as well as time-sharing the arithmetic unit between numerical integration in the optimization computer and interpolation in the device shown in Fig. 11-2. Therefore, the previous description of the optimization computer is strictly tutorial, and the subsequent simplified flow diagrams for sequential optimization do not constitute efficient computer programs.

11-2 Sequential optimization for negligible disturbances

In order to investigate the relationships between computation rate of the computer and the rate of convergence to the optimum performance, the case where process disturbances are negligible in the programming of the optimization computer is considered here. The trajectory computations performed by the optimization computer are coupled loosely to the actual trajectory of the dynamic process in this situation. In effect, the assumption is made that the state vector of the dynamic process does not have to be monitored and compared with the computed state vector except at the initiation of the control process, because this assumption implies that $\mathbf{x}^{(i)}(t)$ and $\mathbf{x}(t)$ remain within the neighborhood $\mathcal{N}_2(t)$ at all times. This assumption somewhat negates the need for feedback control and hence the use of the relaxation method based on second variations. However, this case results in simple computer logic and also gives rise to the maximum rate of convergence because the need for monitoring, comparison, and possible recomputation of the trajectories is eliminated. In cases where disturbances cannot be neglected, these additional computational functions are needed in order to maintain system stability and ensure convergence to the optimum performance.

For the assumption of negligible disturbances, the diagram shown in Fig. 11-6 outlines many details of the optimization-computer operation. The notation used in this diagram conforms to the notation introduced in Sec. 11-1, with one exception. The computed error index is denoted as $x_0(t)$ and is defined by

$$x_0(t) = \int_{t_0}^{t} H(\mathbf{x}(\sigma), \mathbf{m}(\sigma), \sigma]\, d\sigma \tag{11-6}$$

so that the vectors $\mathbf{x}$ and $\mathbf{x}^{(i)}$ used in Fig. 11-6 have $N + 1$ components. In addition, the method used here for adjusting step size ε is the same as that used in the example of Sec. 10-8.

This simplified flow diagram is useful in demonstrating the need for synchronizing the flow of data in the computer. Specifically, the READ operation indicated in Fig. 11-6 is used to establish the coincidence of real time t with the instant of real time t_{i+1} when the ith iteration, or the appropriate part of the computation, is completed. In general, the computer must be halted at this step until $t = t'_{i+1}$. The variables $\tau_f(t)$, $\tau_b(t)$, and $\tau_{fb}(t)$ are defined as

$\tau_f(t)$ = interval of real time required for a forward-time computation of the trajectory and associated logic

$\tau_b(t)$ = interval of real time required for a backward-time computation of the control-equation parameters and associated logic

$\tau_{fb}(t)$ = interval of real time required for the computations associated with both $\tau_f(t)$ and $\tau_b(t)$

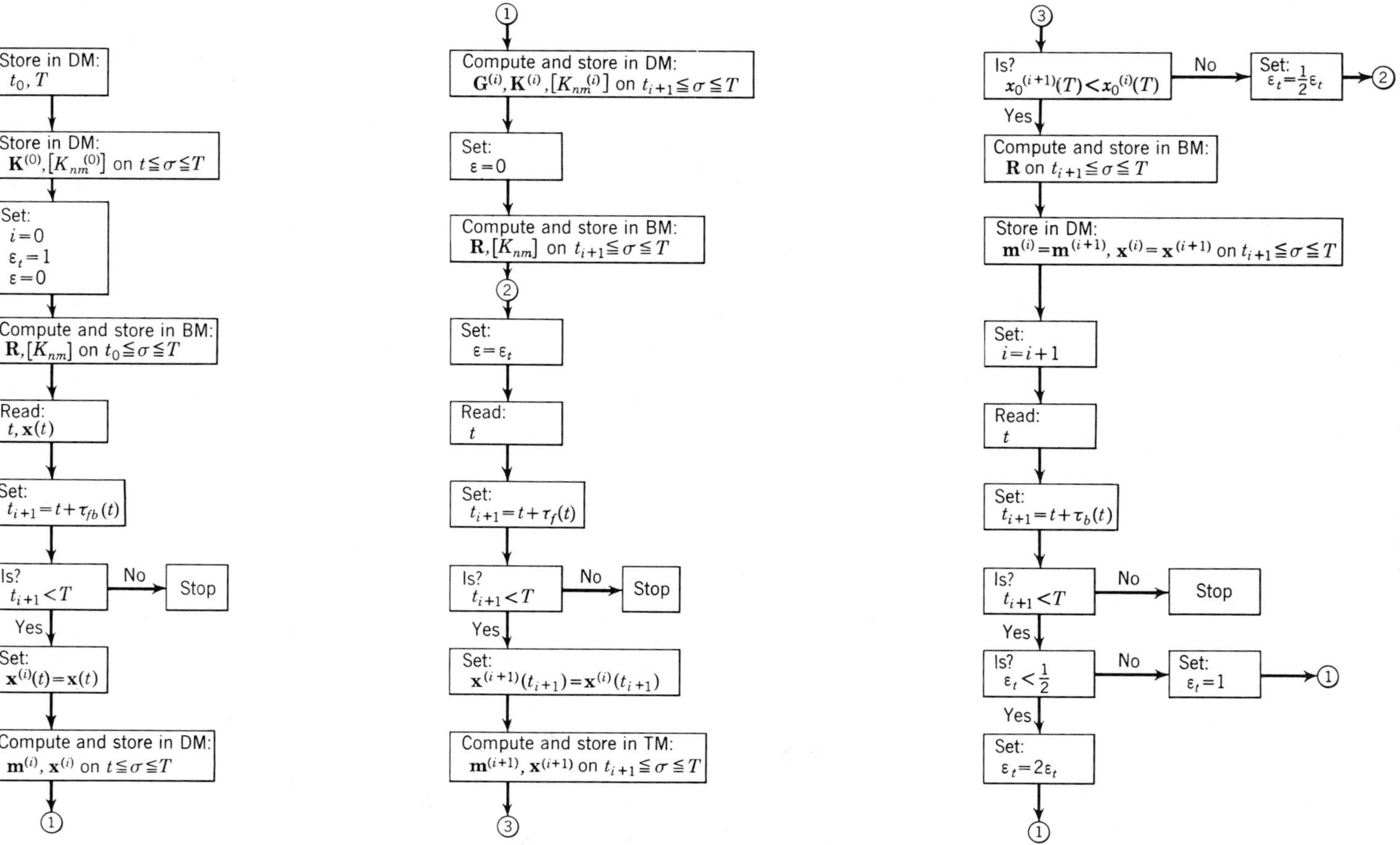

Fig. 11-6 Simplified flow diagram of the optimization computer for negligible disturbances.

Presumably, these variables can be specified for the particular computer and computer program used in the system and should be chosen as upper bounds for the required time intervals so that the computations can be synchronized. Also, these variables are shown to be functions of real time t because the length of real time required for numerical integration is proportional to time-to-go $(T - t)$.

Finally, the computation of $\mathbf{R}$ and $[K_{nm}]$ with $\varepsilon = 0$ after the backward-time computation, as shown in Fig. 11-6, is performed as the result of a compromise between rate of convergence and maintaining control-system stability. After an acceptable value of ε has been found, the control system is perturbed by the introduction of $\varepsilon\mathbf{G}^{(i)}$ which in turn perturbs the trajectory of the dynamic process so that $\mathbf{x}(t) = \mathbf{x}^{(i+1)}(t)$. However, at this point in the control process, the feedback gains remain $K_{nm}(t) = K_{nm}{}^{(i)}(t)$ because the backward-time computations corresponding to $\mathbf{x}^{(i+1)}(t)$ have not yet been performed. Because of this situation, these perturbations tend to drive $\mathbf{x}(t)$ out of the neighborhood $\mathcal{N}_2(t)$, which is associated with $\mathbf{x}^{(i)}(t)$, where stability is maintained. In order to decrease the length of time during which this undesirable situation exists, the feedback gains should be updated as soon as possible without appreciably decreasing the rate of convergence. Therefore, these gains are updated to the current trajectory before trial values of ε are computed along with the time-consuming forward-time computations which would be required. This method of updating requires the computation and storage of $\mathbf{R}$ in the buffer memory with $\varepsilon = 0$ so that no further perturbations occur in the system trajectory until an acceptable value for ε is determined. On the other hand, if the computer had been programmed so that the perturbations are not introduced until the feedback gains associated with the perturbed trajectory also are computed, then the perturbation would be introduced at a latter point in real time, thereby decreasing the rate of convergence. Of course, if the disturbances were completely negligible in the dynamic process as well as in the programming of the optimization computer, then stability would not be a consideration.

11-3 Simulation of a simple system

The optimization computer defined by the flow diagram given in Fig. 11-6 is useful for a brief investigation of the effects of computation rates. For this purpose, both the optimization computer and the fixed member can be simulated on a general-purpose digital computer. The control problem treated here is the simple problem also treated in Sec. 10-8.

The simulation consists mainly in selecting the variables $\tau_f(t)$, $\tau_b(t)$, and $\tau_{fb}(t)$. These variables are used to simulate the effects of finite computation rate, but the momentary halting of the optimization computer required for synchronization is neglected. Suppose that the computation of $\mathbf{m}^{(i+1)}$, $\mathbf{x}^{(i+1)}$ over r units of the dummy time variable σ consumes 1 unit of real time. From this supposition, a forward-time computation over $(T - t_{i+1})$ units of σ requires $(T - t_{i+1})/r$ units of real time. Also, suppose that the logical operations associated with a forward-time computation consume an additional fixed c_f/r units of real time. Then the total consumption of real time for forward-time computations, when the dynamic process is at time t, is

$$t_{i+1} - t = \frac{c_f}{r} + \frac{T - t_{i+1}}{r} \tag{11-7}$$

This relationship directly results in

$$\tau_f(t) = \frac{c_f + (T - t)}{r + 1} \tag{11-8}$$

Similar reasoning is used to establish

$$t_{i+1} - t = \frac{c_b}{r} + \frac{r_b(T - t_{i+1})}{r} \tag{11-9}$$

for backward-time computations where r_b is the ratio of real time required for backward-time integration to real time required for forward-time integration per unit of σ. Therefore, the variable $\tau_b(t)$ is given by

$$\tau_b(t) = \frac{c_b + r_b(T - t)}{r + r_b} \tag{11-10}$$

Finally, the interval of real time required for first a forward-time computation over the interval $(T - t)$ and then a backward-time computation over the interval $(T - t_{i+1})$ is found to be

$$\tau_{fb}(t) = \frac{(c_{fo} + c_{bo}) + (1 + r_b)(T - t)}{r + r_b} \tag{11-11}$$

from simple manipulations. The *time compression ratio* r is the variable parameter studied in this simulation. The parameters c_{fo}, c_f, c_{bo}, c_b, and r_b are estimated and fixed from the actual computer program used in the simulation. The time compression ratio then is used to simulate different computation rates and is adjusted so that

$$r > \frac{c_{fo} + c_{bo}}{T - t_0} + 1 \tag{11-12}$$

This lower limit for the time compression ratio is found from (11-11) and is required if the optimization computer is to reach the point of updating the buffer memory.

As mentioned previously, the design problem treated here is stated in Sec. 10-8. However, data are presented only for cases I and II. The numerics given in Sec. 10-8 also are used. In addition, the parameters of the optimization computer being simulated are taken to be

$$c_f = c_{fo} = 0.03 \qquad c_b = c_{bo} = 0.02 \qquad r_b = 2.32 \tag{11-13}$$

which are estimates based on the computer program used for the simulation. The time compression ratio must satisfy $r > 1.01$ according to (11-12) for these numerics if partial convergence is to be obtained.

Typical trajectories for cases I and II are shown in Figs. 11-7 and 11-8, respectively. Also, the parameters of the control equation that correspond to these trajectories are shown in Figs. 11-9 and 11-10. In case I for a linear dynamic process and a quadratic error measure, only one iteration is required in order to obtain the optimum parameters for the control equation. However, the error index e is somewhat greater than e^* because of the period of real time required to obtain the optimum parameters. In case II for a nonlinear dynamic process and a quadratic error measure, only two iterations are required to obtain nearly optimum parameters for the control equation. Also, two iterations are required when $r = \infty$, as can be seen in Fig. 10-11. However, finite time compression ratios may require added

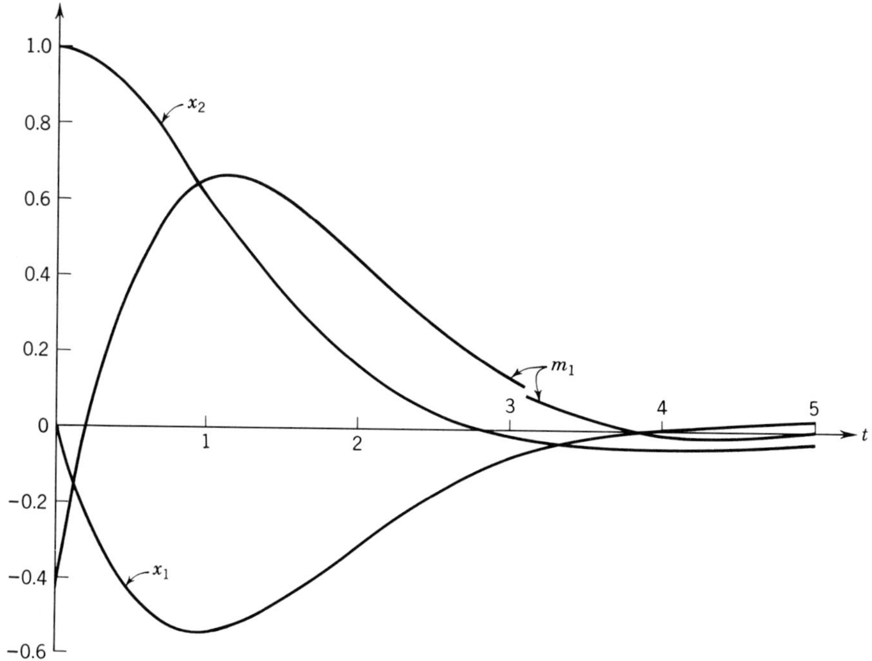

Fig. 11-7 Trajectory for case I: $r = 2$.

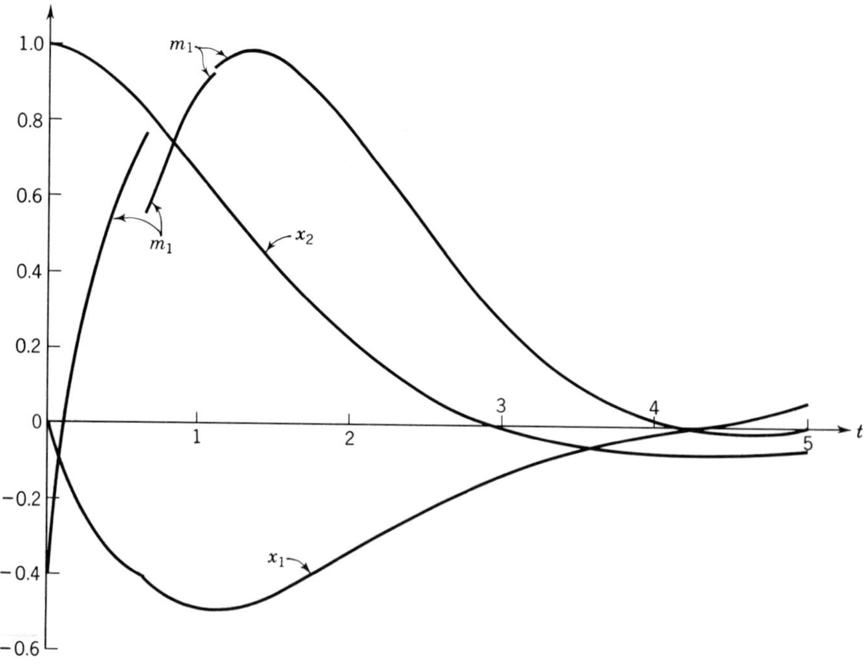

Fig. 11-8 Trajectory for case II: $r = 30$.

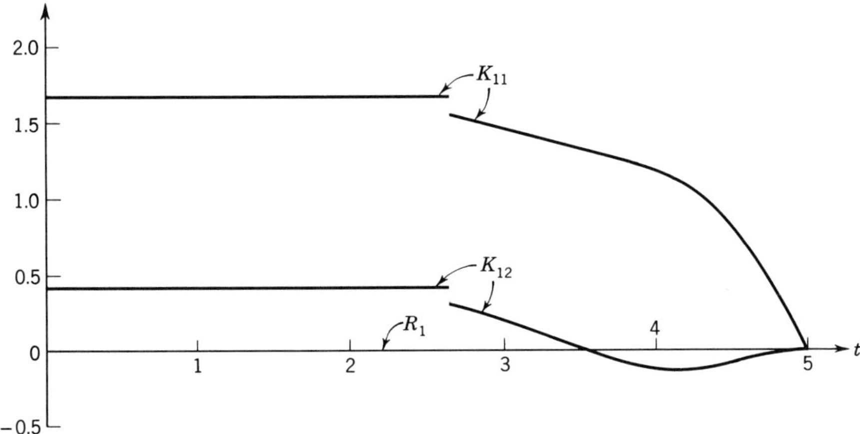

Fig. 11-9 Control-equation feedback gains and reference signal for case I: $r = 2$.

iterations, in general, because of the shifting initial point and the perturbations from the ith to the $(i + 1)$st trajectory. In particular, each iteration is effectively the initial iteration on a new and slightly altered variational problem. Finally, a comparison of either Figs. 11-7 and 11-9 or Figs. 11-8 and 11-10 illustrates the construction of $\mathbf{R}(t)$ and $[K_{nm}(t)]$, as defined in Fig. 11-6, and the step discontinuities in $\mathbf{m}(t)$ which are introduced in Fig. 11-3.

The effects of the time compression ratio for cases I and II are shown in Figs. 11-11 and 11-12, respectively. These two cases serve to illustrate the fact that the

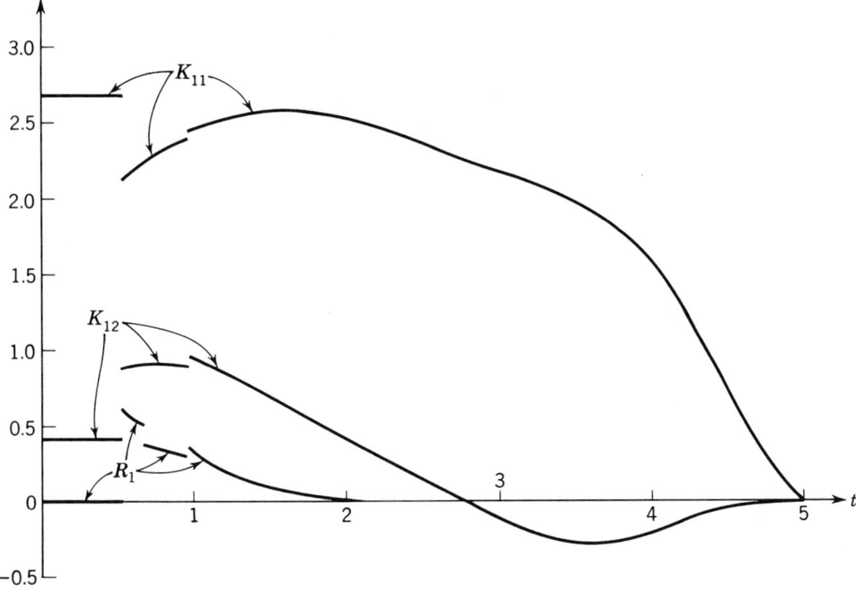

Fig. 11-10 Control-equation feedback gains and reference signal for case II: $r = 30$.

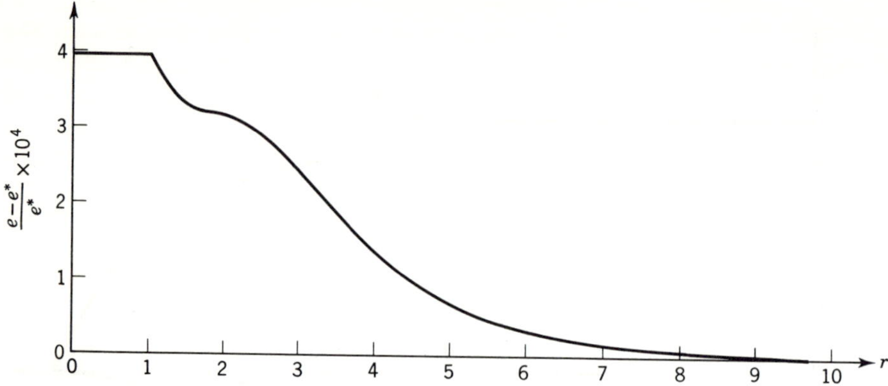

Fig. 11-11 Error index for case I.

time compression ratios required to approach the performance associated with $r = \infty$ vary significantly from problem to problem. Also, the initial values of the parameters appearing in the control equation strongly influence the value of the time compression ratio required for asymptotic behavior to $r = \infty$. In case I, the optimum parameters of the control equation differ from their steady-state values only near the terminal time, as can be seen in Fig. 10-7. Hence, relatively small time compression ratios are acceptable for case I. On the other hand, significant variations in the optimum parameters of the linear control equation occur near the initial time for case II, as can be seen in Fig. 10-10. Hence, relatively large time compression ratios are required for the nonlinear dynamic process with the initial state $x_1(0) = 0$ and $x_2(0) = 1$.

11-4 Sequential optimization for small disturbances

The assumptions used in Sec. 11-3 are somewhat unrealistic in engineering situations, because disturbances cannot be neglected completely in the operation of the optimization computer. In particular, the use of sequential optimization in practical design problems must include facilities for observing and accounting for at least

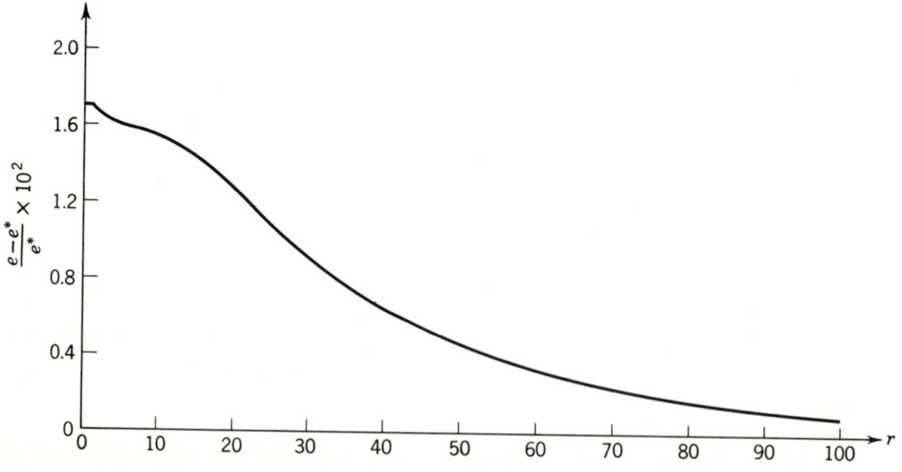

Fig. 11-12 Error index for case II.

small errors in the measured and computed state signals. In this section, the assumption is made that small but not negligible errors between the measured and computed state signals occur because of disturbances. These errors are assumed to be due to unmeasurable load disturbances and errors in previous measurements of the state signals. The assumption of small errors is stated more explicitly by imposing the restriction

$$\delta_l < \|\mathbf{x}(t) - \mathbf{x}^{(i)}(t)\| < \delta_u \tag{11-14}$$

where $\delta_l > 0$. The number δ_u is suitably small so that the approximations for the small region of state space that are afforded by the second-degree expansion are valid. Also, the number δ_l is suitably large so that the errors are not negligible, thereby reverting to the case treated in Sec. 11-3.

The chief difficulty in establishing a sequential optimization procedure when errors due to disturbances occur is obtaining a converging sequence of values of the error index to an *irreducible minimum.* In the case of unmeasurable load disturbances, the irreducible minimum, of course, is larger than the minimum value of the error index obtained in the disturbance-free case. Also, the value of the irreducible minimum is dependent upon the practical constraints imposed on the information and computational procedures which can be used in the design of the system. In this regard, a priori statistics and other specifications for the disturbances are neglected here, so that the modifications required in the computational procedure are emphasized.

The required modifications in the computational procedure are based on two considerations. First, the test $e^{(i+1)} < e^{(i)}$, as presently constituted, is required for establishing convergence and must be performed with $\mathbf{x}^{(i+1)}(t) = \mathbf{x}^{(i)}(t)$. However, the computed state signals must be modified if the optimization computer is to be updated in order to take account of accumulated errors in the state signals. For instance, suppose that the optimization computer is updated by introducing $\mathbf{x}^{(i+1)}(t) = \mathbf{x}(t)$ so that now $\mathbf{x}^{(i+1)}(t) \neq \mathbf{x}^{(i)}(t)$. Then the test $e^{(i+1)} < e^{(i)}$ is no longer valid unless modifications can be introduced. In order to maintain a reasonable rate of convergence, the recomputation of the trajectory $\mathbf{m}^{(i)}$, $\mathbf{x}^{(i)}$ should be avoided if at all possible. The significance of $\mathbf{x}^{(i+1)}(t) \neq \mathbf{x}^{(i)}(t)$ is that the relaxation procedure is no longer based on fixed-point initial-boundary conditions.

The derivation of the relaxation procedure based on second variations is altered easily for the conditions arising here. Specifically, the second-degree expansion of $e^{(i+1)}(t)$ becomes

$$e^{(i+1)} \approx e^{(i)}(t) + \sum_{n=1}^{N} p_n^{(i)}(t) X_n^{(i)}(t) + \sum_{n=1}^{N} \sum_{m=1}^{N} p_{nm}^{(i)}(t) X_n^{(i)}(t) X_m^{(i)}(t) + v^{(i)}(t) \tag{11-15}$$

where $v^{(i)}(t)$ is the integral appearing in (10-106). In addition, when the iteration equation given in (10-112) is used, this integral becomes

$$v^{(i)}(t) = \sum_{n=1}^{N} g_n^{(i)}(t) X_n^{(i)}(t) + \frac{\varepsilon^2 - 2\varepsilon}{2}(-V^{(i)}) \tag{11-16}$$

where $V^{(i)}$ is the minimum value of $v^{(i)}(t)$ under the condition $\mathbf{X}^{(i)}(t) = \mathbf{0}$. Finally, the error index $e(t)$ is expressed in terms of $x_0(T)$, which is computed conveniently in sequential optimization, in accordance with the definition given in (11-6) such that

$$e(t) = x_0(T) - x_0(t) \tag{11-17}$$

Now the expansion given in (11-15) is rewritten in terms of (11-16) and (11-17), thereby giving

$$x_0^{(i+1)}(T) \approx x_0^{(i)}(T) + X_0^{(i)}(t) + \frac{\varepsilon^2 - 2\varepsilon}{2}(-V^{(i)}) \tag{11-18}$$

where

$$X_0^{(i)}(t) = [x_0^{(i+1)}(t) - x_0^{(i)}(t)] + \sum_{n=1}^{N} [p_n^{(i)}(t) + g_n^{(i)}(t)]X_n^{(i)}(t) + \sum_{n=1}^{N} \sum_{m=1}^{N} p_{nm}^{(i)}(t)X_n^{(i)}(t)X_m^{(i)}(t) \tag{11-19}$$

The expansion given in (11-18) now can be used to establish a condition for convergence. Specifically, the term $X_0^{(i)}(t)$ is fixed at time t because the state of the dynamic process $\mathbf{x}(t)$, where $\mathbf{x}^{(i+1)}(t) = \mathbf{x}(t)$, and the computed state $\mathbf{x}^{(i)}(t)$ are fixed. Also, the last term in (11-18) is negative and is varied according to step size ε. Therefore, the condition for convergence is taken to be

$$x_0^{(i+1)}(T) < x_0^{(i)}(T) + X_0^{(i)}(t) \tag{11-20}$$

when errors due to disturbances are suitably small so that the expansions used to derive $X_0^{(i)}(t)$ are valid. The sequence formed by $x_0^{(i)}(T)$ does not decrease monotonically if disturbance errors persist. However, if these errors vanish for $i > I$, then convergence is obtained in the same sense that arises when disturbances are neglected altogether.

The simplified flow diagram given in Fig. 11-13 depicts the programming of the optimization computer used for small disturbances and errors between the measured and computed state signals. The use of the test given in (11-20) requires no recomputation of the trajectory $\mathbf{m}^{(i)}$, $\mathbf{x}^{(i)}$, thereby maintaining a reasonable rate of convergence. However, the trajectory $\mathbf{m}^{(i+1)}$, $\mathbf{x}^{(i+1)}$ must be computed on the interval $t \leqq \sigma \leqq t_{i+2}$, which would be unnecessary for negligible errors. This computation is performed in order to establish $\mathbf{x}^{(i+1)}(t_{i+2})$, which is needed for computing $X_0^{(i)}(t_{i+2})$ in the test for convergence.

The simplified flow diagram given in Fig. 11-13 also illustrates another consideration in sequential optimization where errors due to disturbances are not negligible. In particular, the feedback gains transferred to the buffer memory are computed for the perturbed trajectory, namely, $\mathbf{m}^{(i+1)}$, $\mathbf{x}^{(i+1)}$, before the perturbation is introduced into the control system. This procedure is adopted when the disturbances are not negligible so that the neighborhood $\mathcal{N}_2(t)$ is associated at all times with the trajectory of the control system and not the trajectory $\mathbf{m}^{(i)}$, $\mathbf{x}^{(i)}$. Of course, this modification is introduced so that as many of the stability problems caused by errors as possible are eliminated without abandoning the relaxation procedure based on second variations.

The reader should note that the additional integration needed for $X_0^{(i)}(t_{i+2})$ and the method of delaying the transfer of the data to the buffer memory require additional computer time and hence decrease the rate of convergence. In other words, a decrease in the rate of convergence is the price paid for computational accuracy and system stability. Also, the reader should note that the two methods depicted in Figs. 11-6 and 11-13 can be integrated easily into the programming of the optimization computer. The combining of these two diagrams is accomplished by introducing δ_l, which appears in (11-14), as a threshold for neglecting the errors.

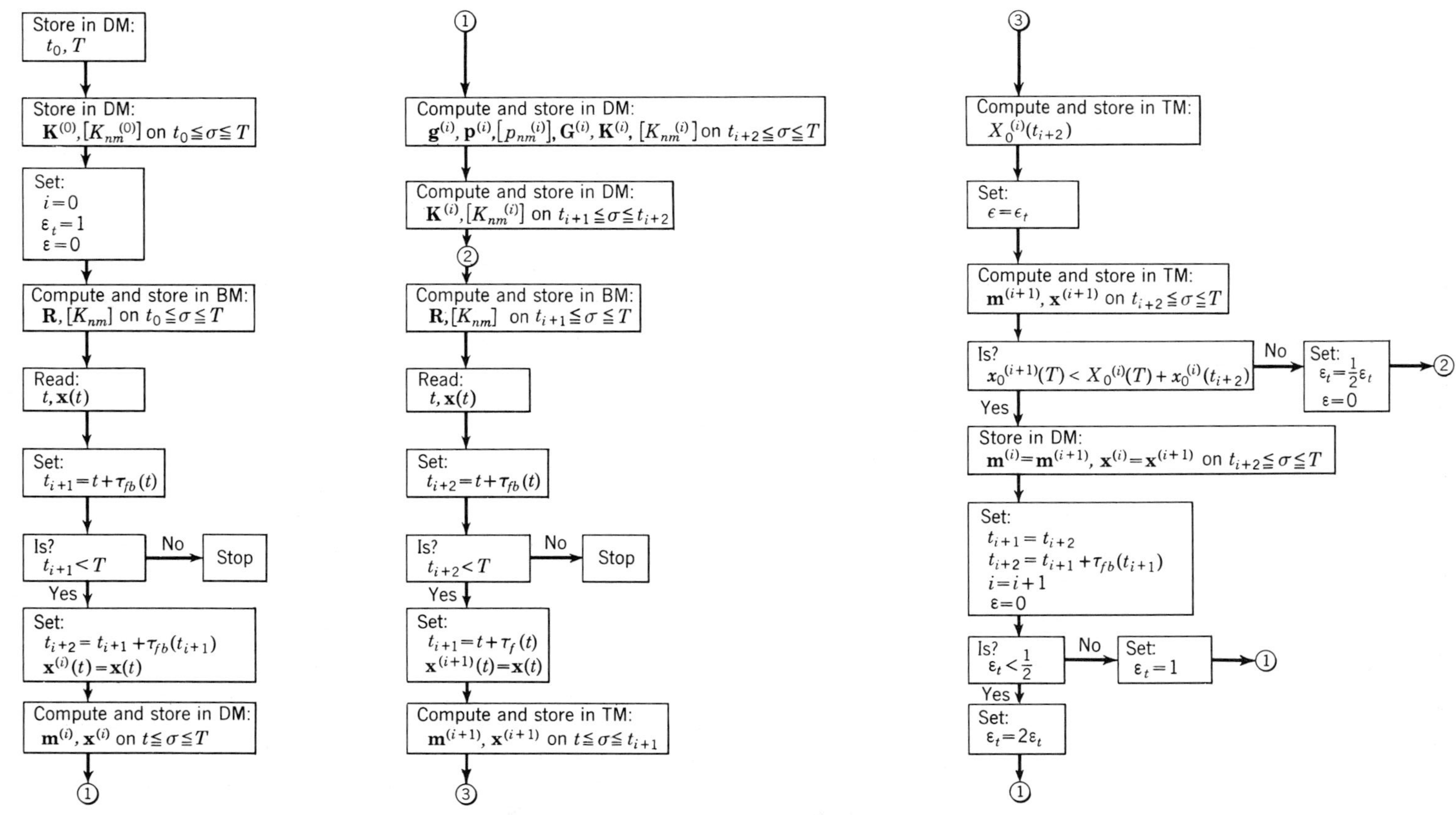

Fig. 11-13 Simplified flow diagram of the optimization computer for small disturbances.

The use of this threshold is advantageous in most practical problems, because the errors may be negligible during part of the operation of the system, and the rate of convergence can be increased while this condition exists.

In design problems where the errors are appreciable and the problem of maintaining system stability is critical, the condition postulated in (11-14) would not be satisfied. An approach to this class of design problems is to abandon second variations and resort to first variations as a relaxation procedure. In addition, the control equation is formed with the $P_{kl}{}^{(i)}$ parameters computed for the constraint of asymptotic stability, which is discussed in Sec. 9-7. The constraint of asymptotic stability is imposed on the control system in order to combat the effects of disturbances during the trajectory computation so that $\|\mathbf{x}(t) - \mathbf{x}^{(i)}(t)\|$ is kept small. Of course, this procedure of using first variations in sequential optimization further decreases the rate of convergence. The programming of the optimization computer for this procedure is almost identical to that depicted in Fig. 11-13. Also, a threshold δ_u, which appears in (11-14), can be introduced so that this method can be combined with the other sequential optimization procedures discussed previously.

A prime difficulty faced by the engineer is the assignment of numerical values to the thresholds δ_l and δ_u for the design problem at hand. No adequate theory for the selection of these thresholds exists, so the engineer must resort to experimentation and simulation.

11-5 Summary

The approach of sequential optimization to control-equation synthesis for large regions of state space raises many unresolved questions. In fact, the approach is still very much the subject of active research. The primary purpose of including this material is that the approach is a natural interpretation and combination of the material presented in Chaps. 9 and 10. Also, this approach differs in concept from the synthesis method that is an outgrowth of discrete dynamic programming and which does not require a computer except for table lookup.

Of course, the primary practical limitation of sequential optimization is due to the technological limitations of available digital computers. Specifically, numerical integration is a time-consuming process, and limited computation rates greatly restrict the complexity of the problems and problem time scales that can be undertaken with present equipment. Also, the computer-memory requirements can become a severe restriction when core memory must be used for airborne applications. However, important applications of sequential optimization have been made in both commercial and military areas.

A number of additional topics in sequential optimization are under consideration but are not treated in detail here. One topic of particular interest is the floating-interval problem where $T = t + \tau$. This type of problem not only arises in certain applications but also is introduced artificially for approximating problems where time-to-go $(T - t)$ is very large with respect to the response time of the closed-loop control system. The pertinent considerations here are introduced in terms of linear optimum controls in Sec. 8-5. The motivation for artificially introducing the floating-interval problem in conjunction with sequential optimization is that the interval $t \leqq \sigma \leqq T$ is reduced in length, thereby reducing the time required for numerical integration. When floating-interval problems are of interest, the need for a computing type of control system and hence sequential optimization is clear. In particular, the control system must be altered to correspond to the segment of the desired-response and desired-control signals that occurs on the interval $t \leqq \sigma \leqq t + \tau$ during the operation of the system. Another topic of interest in sequential optimization involves statistical signals. Here the statistical estimators change during the operation of the system, because the segments of the measured input signals on the interval $-\infty < \sigma \leqq t$ change. In design problems involving statistical signals, approximations such as those introduced in Sec. 9-8 may be useful in conjunction with sequential optimization.

The synthesis problem for large regions of state space continues to be the most difficult and least resolved aspect of feedback control optimization. In fact, two important classes of control optimization problems, namely, minimum time[74] and state signal saturation,[41,75,76] have not been pursued in detail in this book because of these difficulties in synthesis and additional computational difficulties.[77,78] However, these two classes of problems are of intense interest from the research and mathematical points of view, and the reader is urged to expect new results on these classes of problems.

appendix a

Basic Nomenclature

$a_{nm}(t)$	Element in A matrix
$a_{nm}{}^0$	Binary equivalent of $a_{nm}(\mu)$
A	Matrix in linear response equation
$b(t)$	Bandwidth signal
$b_{nm}(t)$	Element in B matrix
$b_{nm}{}^0$	Binary equivalent of $b_{nm}(\mu)$
B	Matrix in linear state equation
$B_{11}(\eta)$	Normalized value of $b_{11}(\mu)$
$c_{nm}(t)$	Element in C matrix
$c_{nn}{}^0$	Binary equivalent of $c_{nn}(\mu)$
C	Matrix in linear state equation
$e(t)$	Error index
$e_c(t)$	Constrained error index
$e_m(t)$	Error measure
$e^*(t)$, $e_*(t)$	Minimum value of error index
$E[\mathbf{x}(t), t]$	Minimum-error function
$E_c[\mathbf{x}(t), t]$	Constrained minimum-error function
$E_P(\mathbf{x}, \mu)$	Pth-degree expansion of the minimum-error function
$\mathscr{E}[\mathbf{x}(t), t]$	Instantaneous minimum-error function

$\mathbf{f}(\mathbf{x}, \mathbf{m}, \sigma)$	Vector value of derivatives of the state signals
F	Function value of control signal in terms of state signals
$F_n(t)$	Linearly independent solutions of homogeneous linear differential equations
$F_m(t, T)$	Value of $k_m(t)$ as a function of terminal time
$F_{mn}(t, T)$	Value of $k_{mn}(t)$ as a function of terminal time
$\mathscr{F}$	State-vector transition functional
$\mathbf{g}(\mathbf{x}, \sigma)$	Vector value of the response signals
$g_n^{(i)}(\sigma)$	Variable in the iteration algorithm based on second variations
G	Function value of response signal in terms of state signals
$G_n(\sigma)$	Parameters in the impulse response of linear differential equations
$G_n^{(i)}(\sigma)$	Perturbation signals in the iteration algorithm based on second variations
$\mathscr{G}$	Response-vector transition functional
$h(\mathbf{q}, \mathbf{m}, \sigma)$	Function value of error measure
$\bar{h}(\mathbf{q}, \mathbf{m}, \sigma, t)$	Instantaneous error measure for statistical signals
$H(\mathbf{x}, \mathbf{m}, \sigma)$	Function value of error measure in terms of state signals
$H_c(\mathbf{x}, \mathbf{m}, \boldsymbol{\lambda}, \sigma)$	Function value of constrained error measure in terms of state signals
$H_P^\dagger(\mathbf{x}, \mu)$	Pth-degree expansion of the hamiltonian function
$\mathscr{H}(\mathbf{x}, \mathbf{m}, \mathbf{p}, \sigma)$	Hamiltonian function
$\mathscr{H}^*(\mathbf{x}, \mu)$	Hamiltonian function evaluated along an optimum trajectory
$\mathscr{H}_P^\dagger(\mathbf{x}, \mu)$	Truncated hamiltonian function based on $E_P(\mathbf{x}, \mu)$
i	Summation index
i_μ	Amplitude of $i(\sigma)$ at $\sigma = \mu$
$i(\sigma)$	Arbitrary signal
$I(x, x', \sigma)$	Integrand of integral constraint
j	Summation index
k	Summation index
$k(\mu)$, $k_n(\mu)$, $k_{nm}(\mu)$	Parameters in quadratic form of the minimum-error function
$\hat{k}(\mu)$, $\hat{k}_n(\mu)$, $\hat{k}_{nm}(\mu)$	Parameters in power-series form of $E_2(\mathbf{x}, \mu)$
$k_m^{\ r}$	Binary equivalent of $k_m(\mu)$ at rth transition
$k_{mk}^{\ p}$	Binary equivalent of $k_{mk}(\mu)$ at pth transition
$K_1(\eta)$	Normalized value of $k_1(\mu)$
$K_{11}(\eta)$	Normalized value of $k_{11}(\mu)$
$K_{nk}(t)$	Feedback gain in linear optimum control equation
$K_n^{(i)}(t)$	Reference signal in iteration algorithm based on second variations
$K_{nk}^{(i)}(\sigma)$	Feedback gain in iteration algorithm based on second variations
$l_m(t)$	Precomputed unseparated component of $k_m(t)$
$L_i(t)$	Precomputed reference signal in linear optimum control equation
m	Summation index
$\mathbf{m}(t)$	Control-signal vector
$\mathbf{m}^*(\mu)$	Optimum control vector
$\mathbf{m}^\dagger(\mu)$	Quasi-optimum control vector

$\mathbf{m}^{\ddagger}(\mu)$	Quasi-optimum control vector subject to the constraint of asymptotic stability
M	Number of control signals in dynamic process
$\mathbf{M}(t)$	Desired-control-signal vector
$\bar{\mathbf{M}}(\mu)$	Control-vector perturbation due to conditional mean
$\mathbf{M}^{(i)}(\sigma)$	Incremental control vector between the ith and $(i+1)$st iterations
$\overline{M_{ij}(t)}$	Separated component signal of $\overline{M_i(\mu)}^t$
$\mathscr{M}(t)$	Allowable region of control-signal vector
$\mathscr{M}_i(\mu)$	Unseparated driving component of $\overline{M_i(\mu)}^t$
$\mathscr{M}_{ij}(\mu, t)$	Separated driving component of $\overline{M_i(\mu)}^t$
$\mathscr{M}_j^*(\mathbf{x}, \mu)$	Partial derivative of the hamiltonian function with respect to $m_j^*(\mu)$ in the neighborhood of $\hat{\mathbf{x}}(\mu)$
$\mathscr{M}_j^{\dagger}(\mathbf{x}, \mu)$	Partial derivative of the truncated hamiltonian function with respect to $m_j^{\dagger}(\mu)$ in the neighborhood of $\hat{\mathbf{x}}(\mu)$
n	Summation index
$n(t)$	Noise signal
N	Number of state and load-disturbance signals in dynamic process
$\mathscr{N}$	Normalizing factor for time variable
$\mathscr{N}_m$	Normalizing factor for $k_m(t)$
$\mathscr{N}_{mk}$	Normalizing factor for $k_{mk}(t)$
$\hat{\mathscr{N}}_P(\mu)$	Neighborhood in state space about $\hat{\mathbf{x}}(\mu)$ associated with $E_P(\mathbf{x}, \mu)$
$\mathbf{p}$	Adjustable parameter vector
$\mathbf{p}(\mu)$	Momentum vector in hamiltonian function
$\hat{p}_0(\mu)$, $\hat{p}_m(\mu)$, $\hat{p}_{mk}(\mu)$	Parameters in Taylor-series form of $E_2(\mathbf{x}, \mu)$
$p_{mij}(t)$	Precomputed component of $k_m(\mu)$ associated with $\overline{M_{ij}(t)}^t$
$p(v_t; t)$	Amplitude-probability density
$p(v_t \mid i_\mu; t, \mu)$	Conditional-probability density
$p(v_t, i_\mu; t, \mu)$	Joint-probability density
P	Degree expansion used for the minimum-error function
P_i	Number of separated components of $\overline{M_i(\mu)}^t$
$P_{mij}(t)$	Precomputed feedforward gain multiplying $\overline{M_{ij}(t)}^t$
$P(x, y, z)$	Coefficient of quasi-linear equation
$\mathbf{q}(t)$	Response-signal vector
Q	Number of response signals
$\overline{Q_{ij}(t)}^t$	Separated-component signal of $\overline{Q_i(\mu)}^t$
$Q(x, y, z)$	Coefficient of quasi-linear equation
$\mathscr{Q}_i(\mu)$	Unseparated driving component of $\overline{Q_i(\mu)}^t$
$\mathscr{Q}_{ij}(\mu, t)$	Separated driving component of $\overline{Q_i(\mu)}^t$
r	Time compression ratio
$r_{mij}(\mu)$	Precomputed separated component of $k_m(\mu)$ associated with $\overline{Q_{ij}(t)}^t$

R_i	Number of separated components of $\overline{Q_i(\mu)}^t$
$R_i(t)$	Reference signal in linear optimum and optimum linear control equations
$R_{jk}(\mathbf{x}, \mu)$	Second partial derivative of the truncated hamiltonian function with respect to $m_j^\dagger(\mu)$ and $x_k(\mu)$ in the neighborhood of $\hat{\mathbf{x}}(\mu)$
$R_{jk}{}^{(i)}(\sigma)$	Second partial derivative of the hamiltonian function with respect to $m_j(\sigma)$ and $x_k(\sigma)$ when evaluated on the ith iteration
$R_{mij}(t)$	Precomputed feedforward gain multiplying $\overline{Q_{ij}(t)}^t$
$R(x, y, z)$	Coefficient of quasi-linear equation
$\hat{\mathscr{R}}(\mu)$	Region of state space where the series required in the expansion of the hamiltonian function converges uniformly given $\hat{\mathbf{x}}(\mu)$
$\hat{\mathscr{R}}^*(\mu)$	Region of state space where $\mathscr{M}^*(\mathbf{x}, \mu) = 0$ given $\hat{\mathbf{x}}(\mu)$
$\hat{\mathscr{R}}^\dagger(\mu)$	Region of state space where $\mathscr{M}^\dagger(\mathbf{x}, \mu) = 0$ given $\hat{\mathbf{x}}(\mu)$
s	Complex-frequency variable
$s_{mij}(\mu)$	Precomputed separated component of $k_m(\mu)$ associated with $U_{ij}(t)$
$S(t)$	Allowable region of state-signal derivative
$\bar{\mathbf{S}}(\mu)$	Vector difference between conditional mean and mean value of $\mathbf{s}(\mu)$
S_i	Number of separated components of $u_i(\mu)$
$S_n{}^{(i)}(\sigma)$	Partial derivative of the hamiltonian function with respect to $m_n(\sigma)$ when evaluated on the ith iteration
$S_y{}^x(z)$	Sensitivity function
$S_{mij}(t)$	Precomputed feedforward gain multiplying $U_{ij}(t)$
t	Real time
$t(t)$	Test signal
T	Terminal time
T_n	Normalized time-to-go
$T_{nm}(\mathbf{x}, \mu)$	Second partial derivative of the truncated hamiltonian function with respect to $m_n^\dagger(\mu)$ and $m_m^\dagger(\mu)$ in the neighborhood of $\hat{\mathbf{x}}(\mu)$
$T_{nm}{}^{(i)}(\sigma)$	Second partial derivative of the hamiltonian function with respect to $m_n(\sigma)$ and $m_m(\sigma)$ when evaluated on the ith iteration
$\mathbf{u}(t)$	Load-disturbance-signal vector
$u_0(\tau)$	Unit-impulse function at $\tau = 0$
$u(x, y, z)$	Solution surface
$U_n{}^{(i)}(\sigma)$	Threshold function used in the iteration algorithm based on second variations
$U_{ij}(t)$	Separated-component signal of $u_i(\mu)$
$\mathscr{U}_i(\mu)$	Unseparated driving component of $u_i(\mu)$
$\mathscr{U}_{ij}(\mu, t)$	Separated component of $u_i(\mu)$
$\mathbf{v}(\sigma)$	Statistical-input-signal vector
v_t	Amplitude of $v(\sigma)$ at $\sigma = t$
$v^{(i)}(t)$	Integral of sum of first and second variations between the ith and $(i + 1)$st iterations
$V^{(i)}(\mathbf{X}, \mu)$	Minimum value of $v^{(i)}(\mu)$
$w(t, \sigma)$	Impulse response of linear differential equations

$w_c(t, \sigma)$	Impulse response of linear compensation member
$w_f(t, \sigma)$	Impulse response of linear fixed member
$w^*(t, \sigma)$	Impulse response of linear Wiener filter
$w_{ci}(t - \sigma)$	Impulse response of optimum linear time-invariant compensation filter
$W(s)$	Linear-system transfer function
$W^*(s)$	Transfer function of linear Wiener filter
$W_{ci}(s)$	Transfer function of optimum linear compensation filter
$W_{ni}(s)$	Transfer function of linear dynamic process from m_i to x_n
$\mathscr{W}$	Energy-storage functional
$\mathbf{x}(t)$	State-signal vector
$\hat{\mathbf{x}}(\mu)$	Optimum trajectory corresponding to the initial state $\hat{\mathbf{x}}(t)$
$\overline{x_n(t_0)}$	Mean value of unmeasurable state signal
$\mathbf{X}(t)$	Equilibrium state vector
$\bar{\mathbf{X}}(\mu)$	State-vector perturbation due to conditional mean
$\mathbf{X}^{(i)}(\sigma)$	Incremental state vector between the ith and $(i + 1)$st iterations
$\mathscr{X}(t)$	Allowable region of state-signal vector
$\mathbf{y}(t)$	Incremental control vector from equilibrium trajectory
$\mathbf{z}(t)$	Incremental state vector from equilibrium trajectory
α	Dummy variable of integration
a	Real part of complex-frequency variable
β	Dummy variable of integration
γ	Dummy variable of integration
δ_{mi}	Kronecker delta function
$\delta^{(i)}$	Function which determines the incremental step between the ith and $(i + 1)$st iterations
ϵ	Arbitrary parameter used in variational mathematics
ε	Parameter determining step size
ζ	Damping ratio
η	Dummy variable of integration
$\eta(\sigma)$	Unrestricted perturbation in the calculus of variations
$\boldsymbol{\lambda}(\sigma)$	Lagrange-multiplier vector
μ	Dummy real-time variable
$\mu(x, y, z)$	Primitive of partial differential equation
ξ	Dummy variable of integration
σ	Dummy time variable
τ	Time interval where response errors are weighted
$\phi_n(\sigma)$, $\phi_{nn}(\sigma)$	Response-error weighting factor
$\phi_{vi}(t, t + \beta)$	Cross-correlation function
$\phi_{vv}(t, t + \beta)$	Autocorrelation function
$\phi_{nn}{}^0$	Binary equivalent of $\phi_{nn}(\mu)$
$\Phi_{vi}(s)$	Spectral density
$\psi_n(\sigma)$, $\psi_{nn}(\sigma)$	Control-error weighting factor
$\psi_{nn}{}^0$	Binary equivalent of $\psi_{nn}(\mu)$
ω	Imaginary part of complex-frequency variable
$\in$	Notation for *contained within or on the boundary of*
$\notin$	Notation for *not contained within and not on the boundary of*

appendix b

References

1. James, H. M., N. B. Nichols, and R. S. Phillips: "Theory of Servomechanisms," McGraw-Hill Book Company, Inc., New York, 1947.
2. Brown, G. S., and D. P. Campbell: "Principles of Servomechanisms," John Wiley & Sons, Inc., New York, 1948.
3. Chestnut, H., and R. W. Mayer: "Servomechanisms and Regulating System Design," vol. I, John Wiley & Sons, Inc., New York, 1951.
4. Truxal, J. G.: "Automatic Feedback Control System Synthesis," McGraw-Hill Book Company, Inc., New York, 1955.
5. Flugge-Lotz, I.: "Discontinuous Automatic Control," Princeton University Press, Princeton, N.J., 1953.
6. Ku, Y. H.: "Analysis and Control of Nonlinear Systems," The Ronald Press Company, New York, 1958.
7. Korn, G. A., and T. M. Korn: "Electronic Analog Computers," 2d ed., McGraw-Hill Book Company, Inc., New York, 1956.
8. Soroka, W. W.: "Analog Methods in Computation and Simulation," McGraw-Hill Book Company, Inc., New York, 1954.
9. Ragazzini, J. R., and G. F. Franklin: "Sampled-data Control Systems," McGraw-Hill Book Company, Inc., New York, 1958.
10. Kalman, R. E., and J. E. Bertram: General Synthesis Procedures for Computer Control of Single and Multiloop Linear Systems, *Trans. AIEE*, pt. II, vol. 77, pp. 602–609, 1958.

11. Laning, J. H., and R. H. Battin: "Random Processes in Automatic Control," McGraw-Hill Book Company, Inc., New York, 1956.
12. Newton, G. C., Jr., L. A. Gould, and J. F. Kaiser: "Analytical Design of Linear Feedback Controls," John Wiley & Sons, Inc., New York, 1957.
13. Malkin, I. G.: "Theory of Stability of Motion," State Publishing House of Technical-Theoretical Literature, Moscow, Leningrad, 1952; translation by United States Atomic Energy Commission, no. AEC-tr 3352.
14. Kalman, R. E.: On the General Theory of Control Systems, *Proc. IFAC*, Butterworth & Co. (Publishers), Ltd., London 1960.
15. Bellman, R.: "Dynamic Programming," Princeton University Press, Princeton, N.J., 1957.
16. Rozonoer, L. I.: The Maximum Principle of L. S. Pontriagin in Optimal System Theory—Parts I, II, III, *Autom. Telemekhan.*, vol. 20, 1959; translation in *J. Automation Remote Control.*
17. Booton, R. C., Jr.: Optimum Design of Final Value Control Systems, *Proc. Polytech. Inst. Brooklyn Symp. Nonlinear Control Systems*, 1956.
18. Booton, R. C., Jr.: Final Value Systems with Gaussian Inputs, *IRE Trans. Inform. Theory, Symp. Inform. Theory*, 1956.
19. Mathews, M. V., and C. W. Steeg, Jr.: Final Value Controller Synthesis, *IRE Trans. Autom. Control*, vol. PGAC-2, pp. 6–16, 1957.
20. Merriam, C. W., III: Mathematics for Linear Systems, chap. 4 in W. W. Seifert and C. W. Steeg, Jr. (eds.), "Control Systems Engineering," McGraw-Hill Book Company, Inc., New York, 1960.
21. Hildebrand, F. B.: "Advanced Calculus for Engineers," Prentice-Hall, Inc., Englewood Cliffs, N.J., 1950.
22. Merriam, C. W., III: A Class of Optimum Control Systems, *J. Franklin Inst.*, vol. 267, no. 4, April, 1959.
23. Hildebrand, F. B.: "Methods of Applied Mathematics," Prentice-Hall, Inc., Englewood Cliffs, N.J., 1952.
24. Wiener, N.: "The Fourier Integral and Certain of Its Applications," Cambridge University Press, New York, 1933.
25. Wiener, N.: "Extrapolation, Interpolation, and Smoothing of Stationary Time Series," The Technology Press of the Massachusetts Institute of Technology, Cambridge, Mass., 1949.
26. Booton, R. C., Jr.: An Optimization Theory for Time Varying Linear Systems and Non-stationary Statistical Inputs, *Proc. IRE*, vol. 40, no. 8, pp. 977–981, August, 1952.
27. Lee, Y. W.: "Statistical Theory of Communication," John Wiley & Sons, Inc., New York, 1960.
28. Booton, R. C., Jr.: Nonlinear and Time Varying Systems, unpublished class notes, Department of Electrical Engineering, Massachusetts Institute of Technology, Cambridge, 1956.
29. Bendat, J. S.: "Principles and Applications of Random Noise Theory," John Wiley & Sons, Inc., New York, 1958.
30. Amara, R.: The Linear Least Squares Synthesis of Continuous and Sampled Data Multivariable Systems, *Stanford Electron. Lab. Tech. Rept.* 40, July, 1958.
31. Kaiser, J. F.: Constraints and Performance Indices in the Analytical Design of Linear Controls, *MIT Servomechanisms Lab. Rept.* 7849-R-6, 7967-R-1, February, 1959.
32. Cramer, H.: "Mathematical Methods of Statistics," Princeton University Press, Princeton, N.J., 1946.
33. Wax, N.: "Selected Papers on Noise and Stochastic Processes," Dover Publications, Inc., New York, 1954.
34. Anderson, T. W.: "Introduction to Multivariate Statistical Analysis," John Wiley & Sons, Inc., New York, 1958.

35. Wiener, N.: "Nonlinear Problems in Random Theory," John Wiley & Sons, Inc., New York, 1958.
36. Bose, A. G.: A Theory of Nonlinear Systems, *MIT Res. Lab. Electron. Tech. Rept.* 309, 1956.
37. Morse, M.: "The Calculus of Variations in the Large," vol. 18, American Mathematical Society Colloquium Publications, New York, 1934.
38. Kipiniak, W.: Optimum Nonlinear Controllers, *MIT Servomechanisms Lab. Rept.* 7793-R-2, 1958.
39. Weinstock, R.: "Calculus of Variations," McGraw-Hill Book Company, Inc., New York, 1952.
40. Lanzos, C.: "The Variational Principles of Mechanics," University of Toronto Press, Toronto, Canada, 1957.
41. Boltyanskii, V. G., R. V. Gamkrelidze, E. F. Mishchenko, and L. S. Pontryagin: The Maximum Principle in the Theory of Optimal Processes of Control, *Proc. Intern. Federation Autom. Control Congr.*, Moscow, 1960.
42. Sneddon, I. N.: "Elements of Partial Differential Equations," McGraw-Hill Book Company, Inc., New York, 1957.
43. John, F.: "Partial Differential Equations," New York University Institute of Mathematical Sciences, New York, 1953.
44. Katz, S.: "Best Temperature Profiles in Plug Flow Reactors: Methods of the Calculus of Variations," Annals of New York Academy of Sciences, 1960.
45. Breakwell, J.: Optimization of Trajectories, *J. Soc. Ind. Appl. Math.*, vol. 7, no. 2, 1959.
46. Volterra, V.: "Theory of Functionals and of Integral and Integrodifferential Equations," Dover Publications, Inc., New York, 1959.
47. Hohn, F. E.: "Elementary Matrix Algebra," The Macmillan Company, New York, 1958.
48. Merriam, C. W., III: Use of a Mathematical Error Criterion in the Design of Adaptive Control Systems, *Trans. AIEE, Appl. Ind., Paper* 59-1158, January, 1960.
49. Merriam, C. W., III: Computation Considerations for a Class of Optimum Control Systems, *Autom. Remote Control*, Moscow, 1960.
50. LaSalle, J., and S. Lefschetz: "Stability by Liapunov's Direct Method," Academic Press Inc., New York, 1961.
51. Kalman, R. E., and J. E. Bertram: Control System Analysis and Design via the Second Method of Lyapunov, *J. Basic Eng.*, *ASME Paper* 59-NAC-2, August, 1959.
52. Merriam, C. W., III: Optimum Control with Statistical Signals and Parametric Uncertainties, *GE Res. Lab. Rept.* 60-RL-2407 E, Schenectady, N.Y., June, 1960.
53. Levin, J. J.: On the Matrix Riccati Equation, *Trans. Am. Math. Soc.*, vol. 10, 1959.
54. Bode, H. W.: "Network Analysis and Feedback Amplifier Design," D. Van Nostrand Company, Inc., Princeton, N.J., 1945.
55. Kalman, R. E.: Contributions to the Theory of Optimal Control, *Bol. Soc. Math. Mex.*, 1960.
56. Merriam, C. W., III: Synthesis of Adaptive Controls, *MIT Servomechanisms Lab. Rept.* 7793-R-3, February, 1959.
57. Ellert, F. J.: Design of a Linear Time Varying Feedback Control System Using Optimization Theory Methods, *GE Gen. Eng. Lab. Rept.* 61GL67, Schenectady, N.Y., April, 1961.
58. Perkins, C. D., and R. E. Hage: "Airplane Performance Stability and Control," John Wiley & Sons, Inc., New York, 1949.
59. O'Hearn, E. A., and R. K. Smyth: Terminal Control Applications, *IRE Trans. Autom. Control*, vol. AC-6, no. 2, May, 1961.
60. Ralston, A., and H. S. Wilf: "Mathematical Methods for Digital Computers," John Wiley & Sons, Inc., New York, 1960.

61. Gill, S.: A Process for the Step-by-Step Integration of Differential Equations in an Automatic Digital Computing Machine, *Proc. Cambridge Phil. Soc.*, vol. 47, pt. 1, June, 1950.
62. "Reference Manual: FORTRAN II Automatic Coding System for the IBM 704 Data Processing System," International Business Machines Corporation, New York.
63. Collatz, L.: "The Numerical Treatment of Differential Equations," 3d ed., Springer-Verlag OHG, Berlin, 1960.
64. Tannas, L. E., and F. J. Ellert: Design of a Linear Time-varying Feedback Control System Using Optimization Theory, Part II—System Evaluation, *GE Gen. Eng. Lab. Rept.* 61GL261, Schenectady, N.Y., March, 1962.
65. Wilson, E. B.: "Advanced Calculus," Ginn and Co., Boston, 1912.
66. Merriam, C. W., III: An Optimization Theory for Feedback Control System Design, *Inform. Control*, vol. 3, no. 1, pp. 32–59, March, 1960.
67. Kipiniak, W.: "Dynamic Optimization and Control," John Wiley & Sons, Inc., New York, 1961.
68. Spang, H. A., III: A Review of Minimization Techniques for Nonlinear Functions, *SIAM Rev.*, vol. 4, no. 4, October, 1962.
69. Wolfe, P.: Recent Developments in Nonlinear Programming, *Rand Corp. Rept.* R-401-PR, May, 1962.
70. Bellman, R., and R. Kalaba: Reduction of Dimensionality, Dynamic Programming, and Control Processes, *Rand Corp. Rept.* R-1964, June, 1960.
71. Aoki, M.: "Dynamic Programming and Numerical Experimentation as Applied to Adaptive Control," doctoral thesis, University of California at Los Angeles, November, 1959.
72. Kelley, H. J.: Gradient Theory of Optimal Flight Paths, *ARS J.*, vol. 30, no. 10, October, 1960.
73. Bryson, A. E., F. J. Carrol, K. Mikami, and W. Denham: Determination of the Lift and Drag Program That Minimizes Re-entry Heating with Acceleration or Range Constraints Using a Steepest Descent Computation Procedure, presented at IAS 29th Annual Meeting, New York, January, 1961.
74. Desoer, C. A.: The Bang-Bang Servo Problem Treated by Variational Techniques, *Inform. Control*, no. 4, pp. 333–348, 1959.
75. Pontryagin, L. S., V. G. Boltyanskii, R. V. Gamkrelidze, and E. F. Mishchenko: "The Mathematical Theory of Optimal Processes," Interscience Publishers, Inc., New York, 1962.
76. Berkovitz, L. D.: On Control Problems with Bounded State Variables, *Rand Corp. Mem.* RM-3207-PR, July, 1962.
77. Ho, Y. C.: A Successive Approximation Technique for Optimal Control Systems Subject to Input Saturation, *J. Basic Eng.*, March, 1962, pp. 33–40.
78. Ho, Y. C.: A Computational Procedure for Optimum Control Problems with State Variable Inequality Constraints, *Rand Corp. Rept.* P-2402, September, 1961.

appendix C

Statistical Signals and Linear Systems

A number of control-system design problems necessarily involve the treatment of signals with random components. Such signals occur with the presence of measurement noise, random load disturbances, and also uncertainties in the desired response of the system. This appendix is written to serve as a ready reference for well-known manipulations of statistical signals and their properties in linear systems. Also, this appendix introduces the notation used in this book for statistical signals. Finally, a brief discussion of gaussian signals is included.

C-1 Ensemble and time averages

A conventional method for computing average properties of signals is based on an ensemble of signals. Figure C-1 depicts an emsemble of signals for $v(\sigma)$, where the mth member of the ensemble is denoted as ${}_mv(\sigma)$. If the amplitude of the mth member at $\sigma = t$ is denoted as ${}_mv_t$, then the average amplitude of the ensemble of signals at $\sigma = t$ is defined notationally as

$$\langle v(t)\rangle = \frac{1}{M}\sum_{m=1}^{M} {}_mv_t \tag{C-1}$$

It is based on an equal weighting of each member of the ensemble.

Generally, the averages used in the mathematical analysis of statistical signals are based on ensemble averages where the number of ensemble members is taken to be infinite. For this case, the ensemble average is defined as

$$\overline{v(t)} = \lim_{M\to\infty}\left(\frac{1}{M}\sum_{m=1}^{M} {}_m v_t\right) \tag{C-2}$$

The average $\overline{v(t)}$ is referred to in the main body of the book simply as the ensemble average of $v(t)$.

Ensemble averages of algebraic combinations of signals often appear in the mathematical analysis of statistical signals in systems. For instance, the ensemble

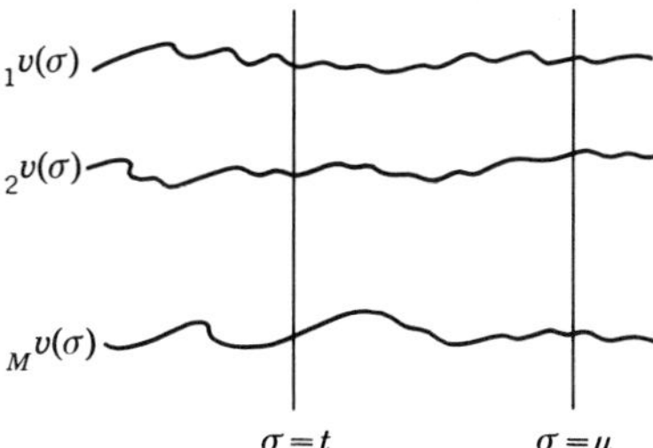

Fig. C-1 Ensemble of M signals for $v(\sigma)$.

average $\overline{v(t)v(\mu)}$ occurs and is defined as

$$\overline{v(t)v(\mu)} = \lim_{M\to\infty}\left[\frac{1}{M}\sum_{m=1}^{M} {}_m v_t\, {}_m v_\mu\right] \tag{C-3}$$

More generally, the average of $f[v(t), i(\mu)]$ over the ensembles of both $v(\sigma)$ and $i(\sigma)$ is written as

$$\overline{f[v(t), i(\mu)]} = \lim_{M\to\infty}\left[\frac{1}{M}\sum_{m=1}^{M} f({}_m v_t, {}_m i_\mu)\right] \tag{C-4}$$

When more than one signal occurs in system analysis, frequently difficulty is experienced in indicating the ensembles from which the average is computed. If the ensemble average of $f[v(t), i(\mu)]$ is computed for the ensemble of $v(\sigma)$ only and $i(\sigma)$ is not a statistical signal, then the average is written as

$$\overline{f[v(t), i(\mu)]}^{v} = \lim_{M\to\infty}\left\{\frac{1}{M}\sum_{m=1}^{M} f[{}_m v_t, i(\mu)]\right\} \tag{C-5}$$

The variable $i(\mu)$ is treated as a fixed parameter in (C-5) and hence is a constant in the averaging process. When the simple bar notation appearing in (C-4) is used, then the assumption is made that the averaging takes place for all the statistical signals appearing in the function f.

A second conventional basis for computing average properties of statistical signals is based upon a time average of a particular member of the ensemble. The time average of $_m v(\sigma)$ is defined as

$$\widetilde{{}_m v(\sigma)} = \lim_{T\to\infty}\left[\frac{1}{2T}\int_{-T}^{T} {}_m v(\sigma)\, d\sigma\right] \tag{C-6}$$

where the signal used in the integration is the mth member of the ensemble of $v(\sigma)$. Also, time averages analogous to ensemble averages given in (C-3) through (C-5) can be defined.

The time average given in (C-6) is equal to the ensemble average given in (C-2) under some circumstances. First, if each member of the ensemble possesses identical statistical characteristics, then the time average is not dependent upon which member of the ensemble is chosen. Secondly, if the statistical characteristics of the ensemble members are independent of time t, then the ensemble average is independent of time t. Under these two conditions, the ensemble average and the time average are equal so that $\bar{v} = \tilde{v}$. When these two conditions exist, the signal v is said to be an ergodic signal.[25] For ergodic signals, time and ensemble averages, in general, are equal.

In this book, ensemble averages are used exclusively because the second condition required for an ergodic signal can be eliminated with no conceptual difficulties in manipulating these averages in system analysis. In particular, a number of design problems involve nonstationary signals where statistical characteristics are dependent upon time. However, if the engineer is faced with computing averages for signals generated from some physical processes, then time averaging may be more convenient than ensemble averaging.

C-2 Probability densities and averages

The fundamental representation of the characteristics of statistical signals is in terms of probability densities. The averages presented in Sec. C-1 are just parameters of the corresponding probability densities. The probability that the amplitude of the signal v at $\sigma = t$ is less than some number V is defined in function form as

$$P(-\infty < v_t \leqq V; t)$$

In this function definition, the variable v_t is the amplitude of v, and the variable t is used to denote the evaluation of the probability at $\sigma = t$. The probability-density function $p(V; t)$ is defined as the derivative of P so that

$$p(V; t) = \lim_{\Delta V \to 0} \frac{P(-\infty < v_t \leqq V + \Delta V; t) - P(-\infty < v_t \leqq V; t)}{\Delta V} \tag{C-7}$$

Conversely, the probability P is written in terms of the probability density p as the integral

$$P(-\infty < v_t \leqq V; t) = \int_{-\infty}^{V} p(v_t; t)\, dv_t \tag{C-8}$$

The amplitude of v averaged over the ensemble of v can be related to the probability density from the following elementary considerations. The probability that v_t is in the interval

$$V - \frac{\Delta V}{2} \leqq v_t \leqq V + \frac{\Delta V}{2}$$

is

$$P\left(V - \frac{\Delta V}{2} \leqq v_t \leqq V + \frac{\Delta V}{2}; t\right) = \int_{V - \Delta V/2}^{V + \Delta V/2} p(v_t; t)\, dv_t \tag{C-9}$$

This probability is approximated for $\Delta V \to 0$ as

$$P\left(V - \frac{\Delta V}{2} \leqq v_t \leqq V + \frac{\Delta V}{2}; t\right) \approx p(V; t)\, \Delta V \tag{C-10}$$

assuming that p is continuous over the ΔV increment. Furthermore, the contribution to the average amplitude of v in the range $V - \Delta V/2 \leqq v_t \leqq V + \Delta V/2$ is the probability of occurrence in this range times the amplitude assigned to this range. If this contribution $Vp(V;\ t)\,\Delta V$ is summed over V, then the average amplitude of v at $\sigma = t$ is found. This average is the ensemble average and is written as

$$\overline{v(t)} = \int_{-\infty}^{\infty} v_t p(v_t;\ t)\, dv_t \tag{C-11}$$

by letting ΔV go to zero so that the sum limits to an integral.

The other ensemble averages introduced in Sec. C-1 also can be written in terms of probability densities as a result of similar arguments. The probability density $p(v_t, i_\mu;\ t, \mu)$ is defined as the probability density of the joint occurrence of the amplitude v_t for the signal v at $\sigma = t$ and the amplitude i_μ for the signal i at $\sigma = \mu$. For this definition, the ensemble average given in (C-4) is equivalent to

$$\overline{f[v(t), i(\mu)]} = \int_{-\infty}^{\infty}\int_{-\infty}^{\infty} f(v_t, i_\mu)p(v_t, i_\mu;\ t, \mu)\, dv_t\, di_\mu \tag{C-12}$$

The joint averaging of v and i appearing in (C-12) can be replaced by an average of v followed by an average of i when the conditional-probability density $p(v_t \mid i_\mu;\ t, \mu)$ is introduced. This conditional-probability density is the probability density of the amplitude v_t of v at $\sigma = t$ given the amplitude i_μ of i at $\sigma = t$. The joint-probability density is expressed in terms of the conditional-probability density as

$$p(v_t, i_\mu;\ t, \mu) = p(v_t \mid i_\mu;\ t, \mu)p(i_\mu;\ \mu) \tag{C-13}$$

The average appearing in (C-12) then is rewritten as

$$\overline{f[v(t), i(\mu)]} = \int_{-\infty}^{\infty}\left[\int_{-\infty}^{\infty} f(v_t, i_\mu)p(v_t \mid i_\mu;\ t, \mu)\, dv_t\right] p(i_\mu;\ \mu)\, di_\mu \tag{C-14}$$

The inner integration occurring in (C-14) is the average value of v at $\sigma = t$ given the amplitude i_μ of i at $\sigma = \mu$. This particular average is called the *conditional mean* of v given i. The shorthand notation $\overline{f[v(t), i(\mu)]}^{v|i}$ is adopted to denote the averaging associated with this conditional mean. With this shorthand notation, the joint average $\overline{f[v(t), i(\mu)]}$ is written as

$$\overline{f[v(t), i(\mu)]} = \overline{\overline{f[v(t), i(\mu)]}^{v|i}}^{i} \tag{C-15}$$

For signals with stationary statistical characteristics, the probability densities discussed in this section simplify. Joint- and conditional-probability densities become functions of time shift only. Also, the amplitude-probability density becomes independent of time. The probability densities of stationary signals are written as

$$\begin{aligned} p(v_t;\ t) &= p(v_t) \\ p(v_t \mid i_\mu;\ t, \mu) &= p(v_t \mid i_\mu;\ \mu - t) \\ p(v_t, i_\mu;\ t, \mu) &= p(v_t, i_\mu;\ \mu - t) \end{aligned} \tag{C-16}$$

Another form of simplification of the joint- and conditional-probability densities generally occurs for large time shifts $\mu - t$. If the signals v and i contain no periodic or d-c components, statistical independence of the amplitude v_t and the amplitude i_μ is assumed as the time shift $\mu - t$ becomes very large. In the limit for this restriction, these probability densities become

$$\lim_{(\mu - t)\to\infty} p(v_t \mid i_\mu; t, \mu) = p(v_t; t)$$

and

$$\lim_{(\mu - t)\to\infty} p(v_t, i_\mu; t, \mu) = p(v_t; t)p(i_\mu; \mu) \tag{C-17}$$

C-3 Correlation and spectral-density functions

The ensemble averages $\overline{v(t)v(\mu)}$ and $\overline{v(t)i(\mu)}$ occur very frequently in system analysis and hence are assigned additional names and function notation. The autocorrelation and cross-correlation functions are defined respectively as

$$\phi_{vv}(t, \mu) = \overline{v(t)v(\mu)}$$

and

$$\phi_{vi}(t, \mu) = \overline{v(t)i(\mu)} \tag{C-18}$$

Properties of these correlation functions can be deduced directly from their associated probability densities. For stationary signals, these correlation functions become functions of time shift only so that

$$\phi_{vv}(t, \mu) = \phi_{vv}(\mu - t)$$

and

$$\phi_{vi}(t, \mu) = \phi_{vi}(\mu - t) \tag{C-19}$$

Furthermore, statistical independence as a result of large time shift results in

$$\lim_{(\mu - t)\to\infty} \phi_{vv}(t, \mu) = \overline{v(t)}\,\overline{v(\mu)}$$

and

$$\lim_{(\mu - t)\to\infty} \phi_{vi}(t, \mu) = \overline{v(t)}\,\overline{i(\mu)} \tag{C-20}$$

Certain structural properties of correlation functions are also useful in system analysis. Obvious properties resulting from the function definitions are

$$\phi_{vv}(\mu - t) = \phi_{vv}(t - \mu)$$

and

$$\phi_{vi}(\mu - t) = \phi_{iv}(t - \mu) \tag{C-21}$$

for stationary signals. Two additional properties are established from the inequalities

$$\overline{[v(t) \pm v(\mu)]^2} \geqq 0$$

and

$$\overline{\left[\frac{v(t)}{\phi_{vv}(0)} \pm \frac{i(\mu)}{\phi_{ii}(0)}\right]^2} \geqq 0 \tag{C-22}$$

for stationary signals. These properties are, respectively,

$$|\phi_{vv}(\mu - t)| \leqq \phi_{vv}(0)$$

and

$$|\phi_{vi}(\mu - t)| \leqq \sqrt{\phi_{vv}(0)\phi_{ii}(0)} \tag{C-23}$$

Because the autocorrelation and cross-correlation functions are functions defined over all positive and negative values of $\tau = \mu - t$ for stationary signals, the Fourier transform must be introduced for frequency-domain studies. Therefore, the transform of the correlation function $\phi(\tau)$ is defined as

$$\Phi(s) = \frac{1}{2\pi} \int_{-\infty}^{\infty} \phi(\tau) e^{-s\tau}\, d\tau \qquad s = j\omega \tag{C-24}$$

The amplitude function $\Phi(s)$ is called the *spectral density*. The inverse transform corresponding to the definition in (C-24) is

$$\phi(\tau) = \frac{1}{j}\int_{-j\infty}^{j\infty} \Phi(s)e^{s\tau}\, ds \tag{C-25}$$

This transform pair is used for both autocorrelation and cross-correlation functions.

The factor $1/2\pi$ used in the transform definition given in (C-24) is introduced in order to modify the dimensions of the spectral density. The units of bandwidth are radians for the spectral density defined in (C-24). If, instead, the $1/2\pi$ factor had appeared in (C-25), then the units of bandwidth would be cycles. The choice of radians as the units of bandwidth is arbitrary but is more compatible with control-engineering practice.

C-4 Statistical signals in linear systems

The analysis of statistical signals in linear systems is common in control-system design. The useful relationships for such analyses can be found by considering the response of the linear system shown in Fig. C-2. The response of this linear system

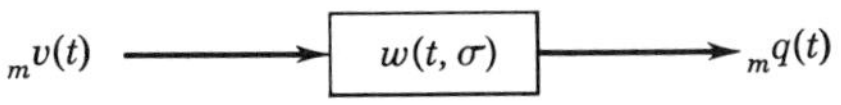

Fig. C-2 Linear system with impulse response $w(t, \sigma)$.

$_m v(t)$ → $w(t, \sigma)$ → $_m q(t)$

for an input signal that is the mth member of the ensemble of v is defined as $_m q(t)$. This response signal is the mth member of the ensemble of q. The ensemble of q is generated by exciting the linear system with each member of the input-signal ensemble. The response of the linear system shown in Fig. C-2 is written in terms of the superposition integral

$${}_m q(t) = \int_{-\infty}^{t} w(t, \sigma)\, {}_m v(\sigma)\, d\sigma \tag{C-26}$$

The ensemble averages defined in Sec. C-1 are computed now for the linear system. The ensemble average of the response signal is

$$\langle q(t)\rangle = \frac{1}{M}\sum_{m=1}^{M}\left[\int_{-\infty}^{t} w(t, \sigma)\, {}_m v(\sigma)\, d\sigma\right] \tag{C-27}$$

By interchanging the order of summation and integration, (C-27) reduces to

$$\langle q(t)\rangle = \int_{-\infty}^{t} w(t, \sigma)\langle v(\sigma)\rangle\, d\sigma \tag{C-28}$$

The limit $M \to \infty$ reduces (C-28) to the commonly used ensemble average

$$\overline{q(t)} = \int_{-\infty}^{t} w(t, \sigma)\overline{v(\sigma)}\, d\sigma \tag{C-29}$$

Similar steps are used to derive the joint averages $\overline{q(t)q(t+\tau)}$ and $\overline{q(t)v(t+\tau)}$ so that

$$\overline{q(t)q(t+\tau)} = \int_{-\infty}^{t+\tau}\int_{-\infty}^{t} w(t+\tau, \sigma)w(t, \mu)\overline{v(\sigma)v(\mu)}\, d\mu\, d\sigma$$

and $$\overline{q(t)v(t+\tau)} = \int_{-\infty}^{t} w(t, \sigma)\overline{v(\sigma)v(t+\tau)}\, d\sigma \tag{C-30}$$

Alternatively, these joint averages are written in terms of correlation functions as

$$\phi_{qq}(t, t+\tau) = \int_{-\infty}^{t+\tau}\int_{-\infty}^{t} w(t+\tau, \sigma)w(t, \mu)\phi_{vv}(\sigma, \mu)\, d\mu\, d\sigma$$

and
$$\phi_{qv}(t, t+\tau) = \int_{-\infty}^{t} w(t, \sigma)\phi_{vv}(\sigma, t+\tau)\, d\sigma \tag{C-31}$$

If the input signal v is stationary and if the linear system is time-invariant so that $w(t, \sigma) = w(t - \sigma)$, then the response signal is also stationary. This fact can be verified by introducing appropriate variable changes in the integrations appearing in (C-31). These manipulations yield

$$\phi_{qq}(\tau) = \int_{0}^{\infty}\int_{0}^{\infty} w(\alpha)w(\beta)\phi_{vv}(\tau + \alpha - \beta)\, d\alpha\, d\beta$$

and
$$\phi_{qv}(\tau) = \int_{0}^{\infty} w(\alpha)\phi_{vv}(\tau + \alpha)\, d\alpha \tag{C-32}$$

The spectral densities corresponding to the correlation functions appearing in (C-32) are easily shown to be

$$\Phi_{qq}(s) = W(s)W(-s)\Phi_{vv}(s)$$

and
$$\Phi_{qv}(s) = W(-s)\Phi_{vv}(s) \tag{C-33}$$

where the transfer function $W(s)$ is defined as

$$W(s) = \int_{0}^{\infty} w(\alpha)e^{-s\alpha}\, d\alpha \tag{C-34}$$

The time-domain and frequency-domain relationships used for stationary statistical signals in linear time-invariant systems are summarized in (C-32) and (C-33), respectively.

C-5 Gaussian signals

An important class of statistical signals called gaussian occur frequently in engineering design problems. Typical gaussian signals arise as shot noise in vacuum tubes or as radar glint noise. A signal is called *gaussian* when the statistical characteristics of the signal are described by probability densities that are normal in form. The normal form of the probability density of y is

$$p(y) = \frac{1}{\sqrt{2\pi}\sigma} \exp\left[-\frac{(y-m)^2}{2\sigma^2}\right] \tag{C-35}$$

The average value of y for this probability density is

$$m = \int_{-\infty}^{\infty} yp(y)\, dy \tag{C-36}$$

and the variance for this probability density is

$$\sigma^2 = \int_{-\infty}^{\infty} (y-m)^2 p(y)\, dy \tag{C-37}$$

Engineers frequently visualize a gaussian signal as a signal that is a sum of pulses which occur randomly in time. This signal $y(t)$ is written as

$$y(t) = \sum_{n} f(t - t_n) \tag{C-38}$$

and is depicted in Fig. C-3. The signal $y(t)$ has normal probability densities under certain assumptions. Grossly stated, the pulses must occur independently so that occurrence of a pulse does not determine the occurrence of the next pulse. Also, any number of pulses can occur in an increment of time Δt, and this number is described by a probability density. A typical probability density for the number of pulses occurring independently in a time increment Δt is the Poisson density.[32] Under these conditions, a fundamental theorem in statistical theory, known as the central limit theorem,[32] states that the signal possesses normal probability densities.

By using the model for gaussian signals depicted in Fig. C-3, two important properties of gaussian signals can be deduced. First, a linear algebraic combination of gaussian signals is also a gaussian signal. The signal resulting from the linear combination is the weighted sum of the independently occurring pulses of all signals

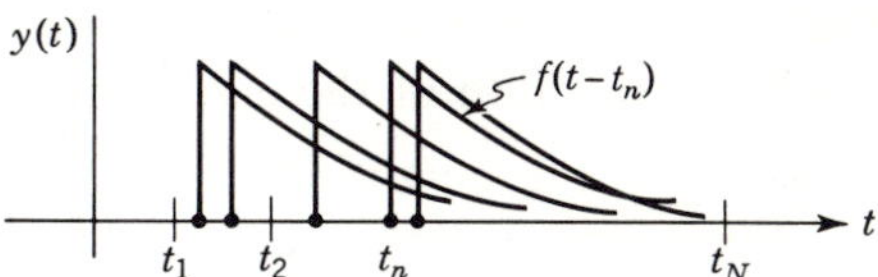

Fig. C-3 Component pulses comprising a gaussian signal.

and hence satisfies this same model even though the signal may contain pulses of different shape. Second, a linear weighting of the amplitude of a gaussian signal over a period of time, such as the superposition integral, is also a gaussian signal. This superposition in time simply modifies the shape of the pulses comprising the original gaussian signal. In fact, gaussian signals with specific statistical parameters can be generated as the output of a corresponding linear filter with a white gaussian input signal. The white gaussian signal is a signal such as $y(t)$, depicted in Fig. C-3, but the component pulses are pulses of varying height and vanishingly small width,§ $f(t - t_n) = a_n p_0(t - t_n)$. The parameter a_n is described by a normal probability density.

As indicated in Chap. 3, gaussian signals play an important role in impulse-response optimization. This role results from the function form of the conditional mean of the desired response. For a gaussian signal, the amplitude of $y(\sigma)$ at $\sigma = t_1$ is defined as y_1, and the amplitude of $y(\sigma)$ at $\sigma = t_2$ is defined as y_2. The joint-probability density for these amplitudes is given by the normal density

$$p(y_2, y_1; t_2, t_1) = \frac{\exp\left[-\dfrac{\sigma_1^2(y_2 - m_2)^2 - 2\sigma_1\sigma_2\rho_{12}(y_2 - m_2)(y_1 - m_1) + \sigma_2^2(y_1 - m_1)^2}{2\sigma_1^2\sigma_2^2(1 - \rho_{12}^2)}\right]}{2\pi\sigma_1\sigma_2(1 - \rho_{12}^2)^{\frac{1}{2}}} \tag{C-39}$$

The mean of y_n is

$$m_n = \int_{-\infty}^{\infty}\int_{-\infty}^{\infty} y_n p(y_2, y_1; t_2, t_1)\, dy_1\, dy_2 \qquad n = 1, 2 \tag{C-40}$$

and the variance of y_n is

$$\sigma_n^2 = \int_{-\infty}^{\infty}\int_{-\infty}^{\infty} (y_n - m_n)^2 p(y_2, y_1; t_2, t_1)\, dy_1\, dy_2 \qquad n = 1, 2 \tag{C-41}$$

§ The mathematical limiting process associated with white noise requires considerable rigor. White noise is analogous mathematically to impulse functions associated with deterministic signals.

Also, the correlation coefficient ρ_{12} of y_1 and y_2 is

$$\rho_{12} = \frac{1}{\sigma_1 \sigma_2} \int_{-\infty}^{\infty} \int_{-\infty}^{\infty} (y_2 - m_2)(y_1 - m_1) p(y_2, y_1; t_2, t_1) \, dy_1 \, dy_2 \tag{C-42}$$

he conditional-probability density of y_2 given y_1 is found from (C-13) to be

$$p(y_2 \mid y_1; t_2, t_1) = \frac{p(y_2, y_1; t_2, t_1)}{p(y_1; t_1)} = \frac{1}{\sqrt{2\pi}\sigma_{12}} \exp\left[-\frac{(y_2 - m_{12})^2}{2\sigma_{12}^2}\right] \tag{C-43}$$

where

$$m_{12} = \left(m_2 - \frac{\sigma_2}{\sigma_1}\rho_{12} m_1\right) + \left(\frac{\sigma_2}{\sigma_1}\rho_{12}\right) y_1 \tag{C-44}$$

and

$$\sigma_{12}^2 = \sigma_2^2(1 - \rho_{12}^2) \tag{C-45}$$

The mean m_{12} of the conditional-probability density, called the conditional mean, is a linear function of the given amplitude y_1. In terms of the shorthand notation for averages, this conditional mean is denoted as $\overline{y(t)}^{y_2|y_1}$

Furthermore, this result can be generalized[34] to an arbitrary number of given amplitudes $y_1, y_2, \ldots, y_{N-1}$. The conditional mean of y_N given these amplitudes is of the form

$$\overline{y(t)}^{y_N|y_1,\ldots,y_{N-1}} = w_N + \sum_{n=1}^{N-1} w_{N,n} y_n \tag{C-46}$$

The linear weighting of given samples in the conditional mean for gaussian signals is the property that makes the linear Wiener filter the optimum filter for this class of signals. In fact, the conditional mean, for gaussian signals, of $Q(t)$ given $v(\sigma)$ on the interval $-\infty < \sigma \leqq t$ is

$$\overline{Q(t)}^{Q|v} = \int_{-\infty}^{t} w^*(t, \sigma) v(\sigma) \, d\sigma \tag{C-47}$$

where $w^*(t, \sigma)$ is the impulse response of the Wiener filter.

Wiener[35] has shown that within certain mathematical restrictions, the conditional mean for nongaussian signals can be represented by linear energy-storage filters plus nonlinear no-energy-storage gains. However, this representation requires an infinite number of filters and nonlinear gains. Bose[36] has developed a method of approximating the conditional mean, but no theory of error exists for these approximations, which is a distinct disadvantage from a conceptual standpoint. Also, the representation used by Bose is difficult to instrument even when the approximations are rather crude in nature.

appendix d

State-determined Dynamic Processes

The treatment of minimization problems by dynamic programming requires the notion of a state-determined dynamic process. This appendix presents the pertinent concepts of state-determined dynamic processes without excessive mathematical detail.

D-1 Definitions

An energy-storage system generally is referred to as a system with responses that are dependent not only on the present values of the system excitations but also on the past values of these excitations. Figure D-1 depicts an energy-storage system where the notation is chosen in order to correspond to dynamic processes used in optimization problems. Also, the dummy time variable σ is chosen in order to correspond to the variable of integration occurring in the error index. Mathematically, an energy-storage system is characterized by a *functional*,[46] which is written as

$$\mathbf{q}(\sigma) = \mathscr{W}[\mathbf{m}(\rho), -\infty < \rho \leqq \sigma] \tag{D-1}$$

for the system shown in Fig. D-1. The dependence of $\mathscr{W}$ on $\mathbf{m}(\rho)$ over the interval $-\infty < \rho \leqq \sigma$ refers to the contributions to the response of the dynamic process due to past and present excitations.

The exclusion of the point $\rho = -\infty$ in (D-1) implies that the dynamic process is stable. In the case of linear dynamic processes, the functional $\mathscr{W}$ is the commonly used superposition integral. In the case of nonlinear dynamic processes, the functional $\mathscr{W}$ is a valid characterization even though the functional in general cannot be specified explicitly.

A dynamic process is defined to be *state-determined* if there is a finite set of variables $\mathbf{x}(t)$ such that the response $\mathbf{q}(\sigma)$ is determined uniquely by

$$\mathbf{q}(\sigma) = \mathscr{G}[\mathbf{x}(t), \sigma, t; \mathbf{m}(\rho), t < \rho \leqq \sigma] \tag{D-2}$$

The exclusion of the point $\rho = t$ in the dependence of $\mathscr{G}$ on $\mathbf{m}(\rho)$ implies that all paths from input to output involve energy storage. The explicit dependence of $\mathscr{G}$ on both σ and t implies that the dynamic process is time-varying and hence

Fig. D-1 Energy-storage system characterized in terms of a functional.

dependent on both the time of observing $\mathbf{q}$ and the time of observing $\mathbf{x}$. If the dynamic process is time-invariant, then the functional $\mathscr{G}$ simplifies to

$$\mathbf{q}(\sigma) = \mathscr{G}[\mathbf{x}(t), \sigma - t; \mathbf{m}(\rho), t < \rho \leqq \sigma] \tag{D-3}$$

Any finite set of variables $\mathbf{x}(t)$ possessing the property expressed in (D-2) is defined as a *state vector*. Fortunately, the choice of a state vector is not unique. This fortunate occurrence gives some freedom in selecting a set of state variables that are measured with the most available sensors. The lack of a unique state vector can be demonstrated easily. For instance, the vector $\mathbf{z}(t)$ is also a state vector, defined as

$$\mathbf{z}(t) = A\mathbf{x}(t) \tag{D-4}$$

if the inverse of the matrix A exists uniquely.

D-2 Equations defining the dynamic process

The response equation occurring in the generally used representation of the dynamic process is deduced readily from (D-2). If the time when $\mathbf{q}$ is observed is taken to be $\sigma = t + \epsilon$, then (D-2) becomes

$$\mathbf{q}(t + \epsilon) = \mathscr{G}[\mathbf{x}(t), t + \epsilon, t; \mathbf{m}(\sigma), t < \rho \leqq t + \epsilon] \tag{D-5}$$

This equation can be expanded in a Taylor series for ϵ arbitrarily small, thereby obtaining

$$\mathbf{q}(t + \epsilon) = \mathbf{g}[\mathbf{x}(t), t] + \epsilon \cdots \tag{D-6}$$

The zeroth-degree term in (D-6) does not involve $\mathbf{m}(t)$ because of the exclusion of the point $\rho = t$ in (D-5). For $\epsilon = 0$, the *response equation*

$$\mathbf{q}(t) = \mathbf{g}[\mathbf{x}(t), t] \tag{D-7}$$

is obtained from (D-6).

Also, the state equation of the dynamic process is deduced readily from (D-2). For a state-determined dynamic process, the dependence of $\mathbf{q}(\sigma)$ on the state vector must hold for any time where this vector is observed. Therefore, (D-2) is written as

$$\mathbf{q}(\sigma) = \mathscr{G}[\mathbf{x}(\mu), \sigma, \mu; \mathbf{m}(\rho), \mu < \rho \leqq \sigma] \tag{D-8}$$

for an arbitrary time μ on the interval $t < \mu \leqq \sigma$. A comparison of (D-2) and (D-8) shows that the contribution to (D-8) of $\mathbf{x}(t)$ and $\mathbf{m}(\rho)$ on the interval $t < \rho \leqq \mu$ must be specified uniquely by $\mathbf{x}(\mu)$. Hence, the state vector itself must be the response of a state-determined dynamic process. The state vector must be defined uniquely by

$$\mathbf{x}(\mu) = \mathscr{F}[\mathbf{x}(t), \mu, t; \mathbf{m}(\rho), t < \rho \leqq \mu] \tag{D-9}$$

The Taylor series expansion of (D-9) for $\mu = t + \epsilon$ is

$$\mathbf{x}(t + \epsilon) = \mathbf{x}(t) + \epsilon\mathbf{f}[\mathbf{x}(t), \mathbf{m}(t), t] + \epsilon^2 \cdots \tag{D-10}$$

If the state vector is assumed to be a set of continuous functions, then the definition of a derivative

$$\mathbf{x}'(t) = \lim_{\epsilon \to 0} \frac{\mathbf{x}(t + \epsilon) - \mathbf{x}(t)}{\epsilon} \tag{D-11}$$

yields the *state equation*

$$\mathbf{x}'(t) = \mathbf{f}[\mathbf{x}(t), \mathbf{m}(t), t] \tag{D-12}$$

D-3 Example

The state variables many times are referred to as the outputs of the integrators appearing in an analog-computer simulation of the dynamic process. This is an overly limited view of the state concept, as is indicated in the following example of a linear dynamic process.

The superposition integral is written as

$$q(\sigma) = \int_{-\infty}^{\sigma} w(\sigma, \rho)m(\rho)\, d\rho \tag{D-13}$$

for a dynamic process with one input and one output. This integral is equivalent to the functional $\mathscr{W}$ in (D-1). If the impulse response $w(\sigma, \rho)$ is taken as

$$w(\sigma, \rho) = \sum_{n=1}^{N} F_n(\sigma)G_n(\rho) \tag{D-14}$$

then (D-13) can be written as

$$q(\sigma) = \sum_{n=1}^{N} \left[\frac{F_n(\sigma)}{F_n(\mu)} \int_{-\infty}^{\mu} F_n(\mu)G_n(\rho)m(\rho)\, d\rho\right] + \int_{\mu}^{\sigma} w(\sigma, \rho)m(\rho)\, d\rho \tag{D-15}$$

Furthermore, a set of state variables is defined as

$$x_n(\mu) = \int_{-\infty}^{\mu} F_n(\mu)G_n(\rho)m(\rho)\, d\rho \qquad n = 1, 2, \ldots, N \tag{D-16}$$

Therefore, (D-15) becomes

$$q(\sigma) = \sum_{n=1}^{N} \frac{F_n(\sigma)}{F_n(\mu)} x_n(\mu) + \int_{\mu}^{\sigma} w(\sigma, \rho)m(\rho)\, d\rho \tag{D-17}$$

which is equivalent to the functional $\mathscr{G}$ appearing in (D-8). For $\sigma = \mu$ and $\mu = t$, (D-17) reduces to

$$q(t) = \sum_{n=1}^{N} x_n(t) \tag{D-18}$$

which is the response equation equivalent to (D-7).

The state equation is developed from (D-16), which can be written as

$$x_n(\mu) = \frac{F_n(\mu)}{F_n(t)} x_n(t) + \int_t^\mu F_n(\mu)G_n(\rho)m(\rho)\,d\rho \tag{D-19}$$

This equation is equivalent to the functional $\mathscr{F}$ appearing in (D-9), and the state equation given in (D-12) becomes

$$x_n'(t) = \frac{F_n'(t)}{F_n(t)} x_n(t) + F_n(t)G_n(t)m(t) \tag{D-20}$$

for this example. In general, the state variables defined here would correspond neither to the integrator outputs in an analog simulation nor to physical variables.

appendix e

Case Study for Linear Optimum Controls—The Design of an Aircraft Landing System§

The purpose of this appendix is to demonstrate the application of optimization theory to the design of a nontrivial control system. An attempt is made to discuss the various considerations and decisions made at each step in the design procedure.

Many control problems are characterized by the necessity of satisfying multiple performance requirements and constraints. In such cases, the design of the control by conventional trial-and-error methods may prove exceedingly difficult. If, in addition, the performance requirements are such that the need for a time-varying control is indicated, the control engineer is faced with the difficult task of selecting not only the configuration of the control system but also the values of the time-varying parameters or gains of the system. For such problems, design methods based on optimization theory offer distinct advantages over trial-and-error methods. In particular, the selection of the time-varying parameters or gains is replaced essentially by the selection of constant parameters which are the magnitudes of the weighting factors. These weighting

§ This material was prepared jointly with Mr. F. J. Ellert of the GE General Engineering Laboratory.

factors appear in an error measure formulated from the performance requirements. In addition, the absolute stability of the system is not a consideration in the selection of the weighting factors, whereas in the selection of time-varying gains by conventional methods, consideration of absolute stability assumes a primary importance.

E-1 Design-problem definition

The landing of an aircraft consists of several phases.[57] First, the aircraft is guided toward the airport with approximately the correct heading by radio direction-finding equipment. Within a few miles of the airport, radio contact is made with the radio beam of the instrument landing system (ILS). In following this beam, the pilot guides the aircraft along a glide-path angle of approximately $-3°$ toward the runway. Finally, at an altitude of approximately 100 ft, the flare-out phase of the landing begins. During this final phase of the landing, the ILS radio beam is no longer effective for guiding the aircraft, because of electromagnetic disturbances. Also, the $-3°$ glide-path angle is not particularly desirable from the viewpoint of safety and comfort. Hence, the pilot must guide the aircraft along the desired flare path by making visual contact with the ground.

The landing problem described here is concerned only with the final phase of the landing, that is, with the last 100 ft of the aircraft's descent. The aircraft is guided to the proper location by air traffic control, and altitude and rate of ascent at the beginning of this flare-out phase range from 80 to 120 ft and from -16 to -24 ft/sec, respectively. For values outside this range, the aircraft is waved off. Finally, only the longitudinal motion (motion in a vertical plane) need be considered in this final phase of the landing. Lateral motion of the aircraft is required primarily to point the aircraft down the runway. For the most part, this lateral motion is accomplished prior to the final flare-out phase of the landing.

During the flare-out, the aircraft may be subjected to both steady winds and wind gusts. Wind gusts are of primary importance, because they tend to be random. On the other hand, steady winds parallel to the ground can be counteracted merely by a steady-state change in the heading of the aircraft. In this problem, the assumption is made that the wind gusts cannot be measured and have a zero mean value.

E-2 Aircraft equations for longitudinal motion

A first step in the design of a control system is the development of a suitable mathematical description of the process or plant to be controlled. This description may take the form of a single differential equation of some order, a set of first-order differential equations, or a transfer function. In this landing problem, a description of the aircraft relating the response variables, measurable state variables, and control variables is necessary.

In the past, a great deal of attention has been directed toward the development of the equations of motion of an aircraft.[58] This development proceeds from a consideration of the aerodynamic forces and moments and from the application of the fundamental laws of mechanics. The resulting equations then are linearized based on the assumption that the deviation from the equilibrium flight condition is small. In addition, the assumption often is made that the glide-path angle γ is sufficiently small that the small-angle approximations $\sin \gamma = \gamma$ and $\cos \gamma = 1$ can be made. This assumption is valid here because of the landing geometry. Finally, the assumption is made that the velocity V of the aircraft is maintained essentially constant during the landing by utilizing throttle control. Thus, the longitudinal motion of the aircraft is governed entirely by the elevator deflection $\delta_e(t)$, which is the only control variable. The use of these assumptions leads to the so-called "short-period" equations of motion of the aircraft. These equations generally are rewritten in terms of the following transfer function relating elevator deflection and pitch angle:

$$\theta(s) = \frac{K_s(T_s s + 1)}{s(s^2/\omega_s^2 + 2\zeta s/\omega_s + 1)} \delta_e(s) \tag{E-1}$$

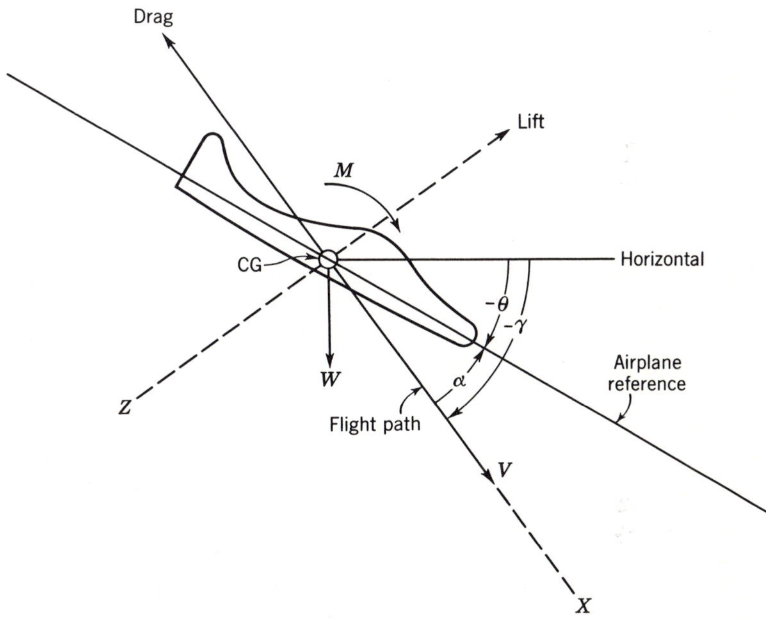

Fig. E-1 Definition of aircraft coordinates and angles.

The pitch angle θ and the stability axes x and z of the aircraft are defined in Fig. E-1.

The aircraft parameters K_s, T_s, ω_s, and ζ often are referred to as follows:

K_s = short-period gain
T_s = path time constant
ω_s = short-period resonant frequency
ζ = short-period damping ratio

These parameters are assumed to be time-invariant, and the numerical values used in this study are

$$K_s = -0.95 \text{ sec}^{-1}$$
$$T_s = 2.5 \text{ sec}$$
$$\omega_s = 1 \text{ rad/sec}$$
$$\zeta = 0.5$$

In order to complete the description of the aircraft, an equation relating pitch angle θ and altitude is required. Such a relationship exists and is given by

$$h(s) = \frac{V}{s(T_s s + 1)} \theta(s) \tag{E-2}$$

For this problem, the velocity V of the aircraft is assumed to be constant during the flare-out at a value of 256 ft/sec. Therefore, the overall aircraft transfer function relating altitude and elevator deflection is given by

$$h(s) = \frac{K_s V}{s^2(s^2/\omega_s^2 + 2\zeta s/\omega_s + 1)} \delta_e(s) \tag{E-3}$$

This relationship also can be written in terms of the fourth-order linear differential equation

$$\frac{d^4h(t)}{dt^4} + 2\zeta\omega_s \frac{d^3h(t)}{dt^3} + \omega_s^2 \frac{d^2h(t)}{dt^2} = K_s V \omega_s^2 \delta_e(t) \tag{E-4}$$

Equation (E-4) describes the behavior of the aircraft in terms of the four state signals $h(t), h'(t), h''(t), h'''(t)$. This fact could be demonstrated by drawing a block diagram for this equation containing four integrators, where $\delta_e(t)$ is the input to the diagram and $h(t)$ is the output. The outputs of the four integrators are the state signals.

Because the state signals ultimately appear as measured variables in the feedback portion of the optimum control system, the state signals selected must be measurable with available instrumentation in order to construct the system. For instance, as indicated above, a complete set of state signals for the aircraft is $h(t)$, $h'(t)$, $h''(t)$, and $h'''(t)$. The altitude $h(t)$ can be measured with a radar altimeter, and the rate of ascent $h'(t)$ can be measured with a barometric rate meter. However, the acceleration $h''(t)$ and its derivative $h'''(t)$ are not readily measurable. Another complete set of state signals is $h(t)$, $h'(t)$, $\theta(t)$, and $\theta'(t)$. Pitch angle $\theta(t)$ and pitch rate $\theta'(t)$ are readily measurable with gyros. Hence, this second set of state signals is preferable. In order to use this set, the relationships giving $\theta'(t)$ and $\theta(t)$ in terms of $h'''(t)$ and $h''(t)$ must be utilized in order to obtain from (E-4) an equation in terms of the new state signals. The equations relating $h''(t)$ and $h'''(t)$ to $\theta'(t)$ and $\theta(t)$ are found from (E-2) to be

$$h''(t) = \frac{V}{T_s} \theta(t) - \frac{1}{T_s} h'(t) \tag{E-5}$$

$$h'''(t) = \frac{V}{T_s} \theta'(t) - \frac{1}{T_s} h''(t) \tag{E-6}$$

$$h''''(t) = \frac{V}{T_s} \theta''(t) - \frac{1}{T_s} h'''(t) \tag{E-7}$$

The substitution of (E-5) to (E-7) into (E-4) gives

$$\frac{V}{T_s}\theta''(t) - \left(\frac{V}{T_s^2} - 2\zeta\omega_s\frac{V}{T_s}\right)\theta'(t) - \left(2\zeta\omega_s\frac{V}{T_s^2} - \omega_s^2\frac{V}{T_s} - \frac{V}{T_s^3}\right)\theta(t)$$
$$- \left(\frac{1}{T_s^3} - 2\zeta\omega_s\frac{1}{T_s^2} + \frac{\omega_s^2}{T_s}\right)h'(t) = \omega_s^2 K_s V\delta_e(t) \qquad \text{(E-8)}$$

Equation (E-8) also is a valid description of the behavior of the aircraft and is used here.

A block diagram for the aircraft may be drawn from (E-8) and is shown in Fig. E-2. Because the aircraft description is of fourth order, this diagram contains four integrators, and the output of each integrator is a state signal.

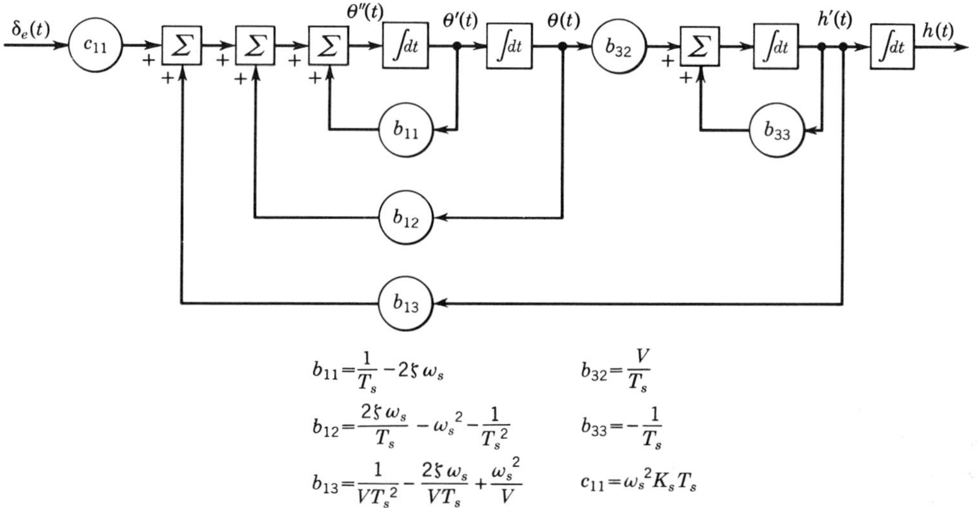

Fig. E-2 Aircraft block diagram in terms of measurable state signals.

In carrying out the design of the system by using optimization theory, the behavior of the aircraft is described conveniently in terms of four first-order differential equations rather than one fourth-order differential equation. This is accomplished by using (E-8) and assigning new symbols to the four state signals and the control signal. Thus, if

$$x_1(t) = \theta'(t) \qquad x_2(t) = \theta(t) \qquad x_3(t) = h'(t) \qquad x_4(t) = h(t) \qquad \text{(E-9)}$$

and $\quad m_1(t) = \delta_e(t) \qquad$ (E-10)

(E-8) can be rewritten as

$$\begin{aligned} x_1'(t) &= b_{11}x_1(t) + b_{12}x_2(t) + b_{13}x_3(t) + c_{11}m_1(t) \\ x_2'(t) &= x_1(t) \\ x_3'(t) &= b_{32}x_2(t) + b_{33}x_3(t) \\ x_4'(t) &= x_3(t) \end{aligned} \qquad \text{(E-11)}$$

Equations (E-11) can be written as a single equation using matrix notation by letting

$$\mathbf{x}(t) = \begin{bmatrix} x_1(t) \\ x_2(t) \\ x_3(t) \\ x_4(t) \end{bmatrix} \equiv \text{state-signal vector} \tag{E-12}$$

$$\mathbf{m}(t) = [m_1(t)] \equiv \text{control-signal vector} \tag{E-13}$$

$$B = \begin{bmatrix} b_{11} & b_{12} & b_{13} & 0 \\ 1 & 0 & 0 & 0 \\ 0 & b_{32} & b_{33} & 0 \\ 0 & 0 & 1 & 0 \end{bmatrix} \equiv B \text{ matrix} \tag{E-14}$$

$$C = \begin{bmatrix} c_{11} \\ 0 \\ 0 \\ 0 \end{bmatrix} \equiv C \text{ matrix} \tag{E-15}$$

Thus, (E-11) becomes

$$\mathbf{x}'(t) = B\mathbf{x}(t) + C\mathbf{m}(t) \tag{E-16}$$

which is the vector-matrix form of the state equation for the aircraft landing problem. In the more general case where wind gusts must be considered, an additional term appears in this equation to include the effects of load disturbances on the response of the process.

E-3 Performance requirements and constraints for aircraft landing

An aircraft landing system is satisfactory only if certain prescribed performance requirements and constraints are satisfied. Often these are described in terms of desired-response and desired-control signals and in terms of limits on these signals. The following requirements and constraints are considered to be of primary importance and are treated in this problem.

1. The desired altitude $h_d(t)$ of the aircraft at each instant of time during the landing is described by the flare path shown in Fig. E-3*a*. This path consists of an exponential function followed by a linear function and is described by

$$h_d(t)\ (\text{ft}) = \begin{cases} 100e^{-t/5} & 0 \leqq t < 15 \\ 20 - t & 15 \leqq t \leqq 20 \end{cases} \tag{E-17}$$

An exponential-linear path of this form ensures a safe and comfortable landing. The desired duration of the flare-out is 20 sec, including 5 sec over the runway. This value is appropriate for an aircraft flying at 175 mph and beginning to flare out at an altitude of 100 ft.

2. The desired rate of ascent $h_d'(t)$ of the aircraft is given by the time derivative of (E-17) and is shown in Fig. E-3*b*. That is,

$$h_d'(t)\ (\text{ft/sec}) = \begin{cases} -20e^{-t/5} & 0 \leqq t < 15 \\ -1 & 15 \leqq t \leqq 20 \end{cases} \tag{E-18}$$

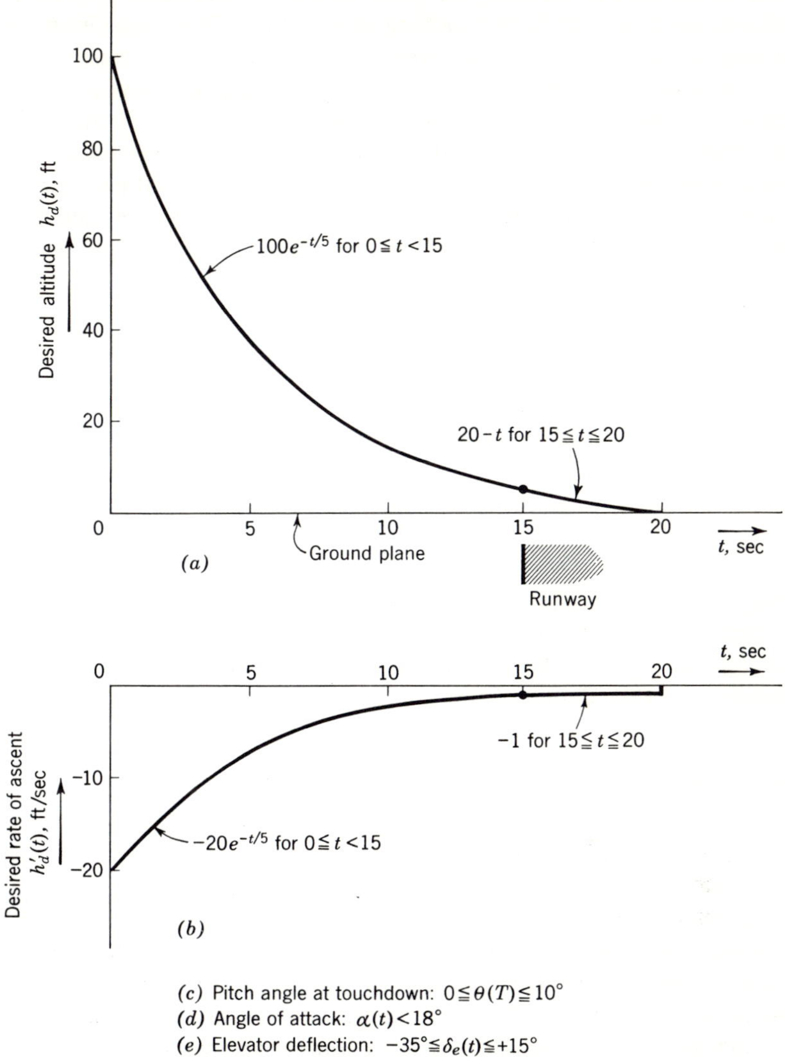

Fig. E-3 Performance requirements and constraints.

The magnitude of the rate of ascent of the aircraft is important primarily at touchdown. A value other than zero is desirable to prevent the aircraft from "floating" down the runway and hence perhaps overshooting the runway. A very large negative value for the rate of ascent may overstress the landing gear and thus is equally undesirable. The value of -1 ft/sec at touchdown given in (E-18) is equal to -60 ft/min, which is well below the maximum permissible value for a modern aircraft.

3. The pitch angle $\theta(t)$ of the aircraft at the desired touchdown time $t = T$ must lie between 0 and $+10°$. That is,

$$0° \leqq \theta(T) \leqq +10° \tag{E-19}$$

The lower limit is necessary in order to prevent the nosewheel of an aircraft with a tricycle landing gear from touching down first. The upper limit on the pitch angle is required to prevent the tail section of the aircraft from touching the runway.

4. During the entire landing, the angle of attack $\alpha(t)$ of the aircraft must remain below the stall value. This value varies from aircraft to aircraft. For this problem, the stall value is assumed to be $+18°$. The aircraft enters the flare-out phase in equilibrium with an angle of attack of approximately 80 per cent of the stall value. Hence, the permissible positive increment in the angle of attack during the flare-out is equal to 20 per cent of the stall value. Thus, the restrictions on the angle of attack are

$$\alpha(t) < 18° \tag{E-20}$$

and

$$\Delta\alpha(t) < 3.6° \tag{E-21}$$

5. The elevator, which controls the longitudinal response of the aircraft, is restricted to motion between mechanical stops. For this problem, these stops are taken to be at -35 and $+15°$ (elevator trailing edge down). Hence, during the flare-out, the elevator deflection $\delta_e(t)$ is constrained so that

$$-35° \leqq \delta_e(t) \leqq +15° \tag{E-22}$$

For a linear controller, the elevator is not permitted to attain these limiting values except instantaneously. That is, for linear operation, the elevator is not permitted to operate against the mechanical stops for nonzero periods of time during the landing. In the design of this aircraft landing system, saturation effects are avoided.

E-4 Selection of error measure and index

An important step in the design of a control system using optimization theory is the formulation of a suitable mathematical error index. The error index is important because it, to a large degree, determines the nature of the resulting optimum control. That is, the resulting control system may be linear, nonlinear, time-invariant, or time-varying, depending on the form of the error index. Thus, the control engineer is able to influence the nature of the resulting system by the manner in which this index is formulated in terms of the requirements of the problem at hand. In general, these requirements may include not only the performance requirements but also any constraints on the structure of the optimum control system.

A first step in the formulation of an error index is the establishment of the error measure, which takes into account all the important performance requirements and constraints. In general, the error measure consists of the sum of the weighted response-signal and control-signal errors. In this landing problem, the important response-signal errors are the deviation of the aircraft altitude from the desired altitude given in (E-17), the deviation of the aircraft rate of ascent from the desired rate of ascent given in (E-18), the deviation of the pitch angle at touchdown from some desired value within the prescribed range given in (E-21), and the deviation of the angle of attack from its initial value at $t = 0$. In addition, the pitch rate of the aircraft may be important. The aircraft equations of motion indicate that the pitch rate of the aircraft governs the magnitude of the angle of attack because the pitch-rate term enters the angle-of-attack equation as the driving function. Hence, the magnitudes of both the angle of attack and the pitch rate are governed by introducing a weighted pitch-rate error term into the error measure. Of course, both pitch-rate and angle-of-attack error terms could be included in this measure. However, this then requires the selection of two weighting factors instead of one. In this problem, the use of only the weighted pitch-rate error term is found to be completely satisfactory for controlling the magnitudes of both the angle of attack and the pitch rate. The final term in the error measure is the control-signal error, or, in this case, the elevator-deflection error. In order to maintain the elevator deflection within the prescribed limits, the value of the desired-control signal is selected conveniently to be zero. Thus, for the landing problem, an appropriate error measure is

$$e_m(t) = \phi_h(t)[h_d(t) - h(t)]^2 + \phi_{h'}(t)[h'_d(t) - h'(t)]^2 + \phi_\theta(t)[\theta_d(t) - \theta(t)]^2 + \phi_{\theta'}(t)[\theta'_d(t) - \theta'(t)]^2 + [\delta_e(t)]^2 \quad \text{(E-23)}$$

where $\phi_h(t)$, $\phi_{h'}(t)$, $\phi_\theta(t)$, and $\phi_{\theta'}(t)$ are the time-varying weighting factors which indicate the relative importance of the various terms in the error measure. In carrying out the design of the control system, these weighting factors must be selected so that the performance requirements and constraints are satisfied.

As already mentioned, the error measure may assume a variety of forms, depending on the nature of the problem to be solved and the requirements to be met. For this landing problem, a weighted quadratic error measure is selected because this form is felt to be sufficient to ensure satisfactory system behavior. In addition, the use of a weighted quadratic error measure in conjunction with the linear aircraft description results in a linear control system with feedback and feedforward gains which are not a function of the state of the aircraft but merely functions of the time-to-go before touchdown. Such a system is relatively simple to construct and is a practical airborne landing system.

The control engineer is interested not so much in an instantaneous value of the error measure as in the cumulative effect of this error measure throughout an interval of time. Hence, the error index $e(t)$ is expressed as the time integral of the error measure over a suitable future interval of time $t \leqq \sigma \leqq T$, throughout which the system performance is of interest. That is,

$$e(t) = \int_t^T e_m(\sigma)\, d\sigma \quad \text{(E-24)}$$

where σ is a dummy time variable. The lower limit of integration $\sigma = t$ is present or real time. For the landing problem, the future interval of time of importance is the time remaining before touchdown occurs. Hence, the upper limit of integration $\sigma = T$ is set equal to 20 sec. The error index then can be written as

$$e(t) = \int_t^{20} \{\phi_h(\sigma)[h_d(\sigma) - h(\sigma)]^2 + \phi_{h'}(\sigma)[h'_d(\sigma) - h'(\sigma)]^2 + \phi_\theta(\sigma)[\theta_d(\sigma) - \theta(\sigma)]^2 + \phi_{\theta'}(\sigma)[\theta'_d(\sigma) - \theta'(\sigma)]^2 + [\delta_e(\sigma)]^2\}\, d\sigma \quad \text{(E-25)}$$

The performance requirements indicate that the altitude and rate-of-ascent errors should be small at the desired touchdown point $\sigma = 20$ to ensure actual touchdown very close to this point. Even small altitude and rate-of-ascent errors at $\sigma = 20$ may result in touchdown prior to the start of the runway or touchdown so far down the runway that the aircraft cannot be brought to a stop. Hence, impulse weighting of these errors at $\sigma = 20$, in addition to weighting of the errors during the flare-out, appears to be desirable. Thus, $\phi_h(\sigma)$ and $\phi_{h'}(\sigma)$ are written as

$$\phi_h(\sigma) = \phi_4(\sigma) + \phi_{4,T}u_0(20 - \sigma) \tag{E-26}$$

and

$$\phi_{h'}(\sigma) = \phi_3(\sigma) + \phi_{3,T}u_0(20 - \sigma) \tag{E-27}$$

Here, $u_0(20 - \sigma)$ is an impulse of unit area occurring at $\sigma = 20$, and $\phi_{4,T}$ and $\phi_{3,T}$ are the areas of the impulse functions. However, fixed-point terminal-boundary conditions are not felt to be necessary in this problem, thereby avoiding considerable difficulty in constructing and computing the parameters of the optimum control equation. The nature of the time functions $\phi_4(\sigma)$ and $\phi_3(\sigma)$ is treated in a following paragraph.

The performance requirements indicate that the pitch-angle error is important only at the touchdown point. Hence, the mathematical form of the weighting factor $\phi_\theta(\sigma)$ is selected so that only the pitch-angle error at touchdown contributes to the value of the error index, thereby requiring an impulse function. Thus, $\phi_\theta(\sigma)$ is taken to be

$$\phi_\theta(\sigma) = \phi_{2,T}u_0(20 - \sigma) \tag{E-28}$$

Because the angle of attack and the pitch rate are important throughout the entire landing interval, $\phi_{\theta'}(\sigma)$ is taken to be

$$\phi_{\theta'}(\sigma) = \phi_1(\sigma) \tag{E-29}$$

Finally, the performance requirements suggest the mathematical forms of $\phi_4(\sigma)$, $\phi_3(\sigma)$, and $\phi_1(\sigma)$. Because altitude and pitch-rate errors are important throughout the flare-out, these weighting factors are taken to be constants. On the other hand, the rate-of-ascent errors are not of primary interest prior to the runway. Thus, $\phi_3(\sigma)$ is taken to be zero prior to the start of the runway and a constant over the runway. The resulting expressions for these weighting factors are

$$\phi_4(\sigma) = \phi_4 \tag{E-30}$$

$$\phi_3(\sigma) = \begin{cases} 0 & 0 \leqq \sigma < 15 \\ \phi_3 & 15 \leqq \sigma \leqq 20 \end{cases} \tag{E-31}$$

$$\phi_1(\sigma) = \phi_1 \tag{E-32}$$

The selection of the numerical values of ϕ_4, $\phi_{4,T}$, ϕ_3, $\phi_{3,T}$, $\phi_{2,T}$, and ϕ_1, and the effects of these values on the resulting aircraft performance, are discussed in Sec. E-6.

E-5 Optimum control equation

Once the dynamic-process equations and the error index are formulated, optimization theory can be applied to obtain directly both the configuration and the parameters of the control system. In this design problem, the dynamic process is linear and the error measure is quadratic. Therefore, the material presented in Chap. 7 applies here and the optimum control equation is linear.

The optimum control equation and the differential equations defining the k parameters are found by specializing (7-163), (7-160), and (7-161) to this particular design problem. This specialization is accomplished by using the B and C matrices given in (E-14) and (E-15), respectively. Also needed is the A matrix, which is given by

$$A = \begin{bmatrix} 1 & 0 & 0 & 0 \\ 0 & 1 & 0 & 0 \\ 0 & 0 & 1 & 0 \\ 0 & 0 & 0 & 1 \end{bmatrix} \tag{E-33}$$

for this design problem. For the aircraft landing problem, the optimum control equation becomes

$$\delta_e(t) = \omega_s{}^2 K_s T_s[k_1(t) - k_{11}(t)\theta'(t) - k_{12}(t)\theta(t) - k_{13}(t)h'(t) - k_{14}(t)h(t)] \tag{E-34}$$

Furthermore, the k parameters appearing in (E-34) are defined by

$$-k_1'(\mu) = \phi_1\theta_d'(\mu) + b_{11}k_1(\mu) + k_2(\mu) - c_{11}{}^2k_1(\mu)k_{11}(\mu) \tag{E-35}$$

$$-k_2'(\mu) = b_{12}k_1(\mu) + b_{32}k_3(\mu) - c_{11}{}^2k_1(\mu)k_{12}(\mu) \tag{E-36}$$

$$-k_3'(\mu) = \phi_3(\mu)h_d'(\mu) + b_{13}k_1(\mu) + b_{33}k_3(\mu) + k_4(\mu) - c_{11}{}^2k_1(\mu)k_{13}(\mu) \tag{E-37}$$

$$-k_4'(\mu) = \phi_4h_d(\mu) - c_{11}{}^2k_1(\mu)k_{14}(\mu) \tag{E-38}$$

$$-k_{11}'(\mu) = \phi_1 + 2b_{11}k_{11}(\mu) + 2k_{12}(\mu) - c_{11}{}^2k_{11}{}^2(\mu) \tag{E-39}$$

$$-k_{12}'(\mu) = b_{11}k_{12}(\mu) + b_{12}k_{11}(\mu) + k_{22}(\mu) + b_{32}k_{13}(\mu) - c_{11}{}^2k_{11}(\mu)k_{12}(\mu) \tag{E-40}$$

$$-k_{13}'(\mu) = b_{11}k_{13}(\mu) + b_{13}k_{11}(\mu) + k_{23}(\mu) + b_{33}k_{13}(\mu) + k_{14}(\mu) - c_{11}{}^2k_{11}(\mu)k_{13}(\mu) \tag{E-41}$$

$$-k_{14}'(\mu) = b_{11}k_{14}(\mu) + k_{24}(\mu) - c_{11}{}^2k_{11}(\mu)k_{14}(\mu) \tag{E-42}$$

$$-k_{22}'(\mu) = 2b_{12}k_{12}(\mu) + 2b_{32}k_{23}(\mu) - c_{11}{}^2k_{12}{}^2(\mu) \tag{E-43}$$

$$-k_{23}'(\mu) = b_{33}k_{23}(\mu) + b_{12}k_{13}(\mu) + b_{13}k_{12}(\mu) + b_{32}k_{33}(\mu) + k_{24}(\mu) - c_{11}{}^2k_{12}(\mu)k_{13}(\mu) \tag{E-44}$$

$$-k_{24}'(\mu) = b_{12}k_{14}(\mu) + b_{32}k_{34}(\mu) - c_{11}{}^2k_{12}(\mu)k_{14}(\mu) \tag{E-45}$$

$$-k_{33}'(\mu) = \phi_3(\mu) + 2b_{33}k_{33}(\mu) + 2b_{13}k_{13}(\mu) + 2k_{34}(\mu) - c_{11}{}^2k_{13}{}^2(\mu) \tag{E-46}$$

$$-k_{34}'(\mu) = b_{33}k_{34}(\mu) + b_{13}k_{14}(\mu) + k_{44}(\mu) - c_{11}{}^2k_{13}(\mu)k_{14}(\mu) \tag{E-47}$$

$$-k_{44}'(\mu) = \phi_4 - c_{11}{}^2k_{14}{}^2(\mu) \tag{E-48}$$

The boundary conditions for the solutions of these differential equations are

$$\begin{aligned} &k_1(20) = k_4(20) = 0 \qquad k_2(20) = \phi_{2,T}\theta_d(20) \qquad k_3(20) = -\phi_{3,T} \\ &k_{11}(20) = k_{12}(20) = k_{13}(20) = k_{14}(20) = k_{23}(20) = k_{24}(20) = k_{34}(20) = 0 \\ &k_{22}(20) = \phi_{2,T} \qquad k_{33}(20) = \phi_{3,T} \qquad k_{44}(20) = \phi_{4,T} \end{aligned} \tag{E-49}$$

where the nonzero conditions are due to the impulse weighting factors at the terminal point.

The block diagram of the system is drawn with the aid of (E-34) and is shown in Fig. E-4. The optimum system contains four feedback loops with a measured state

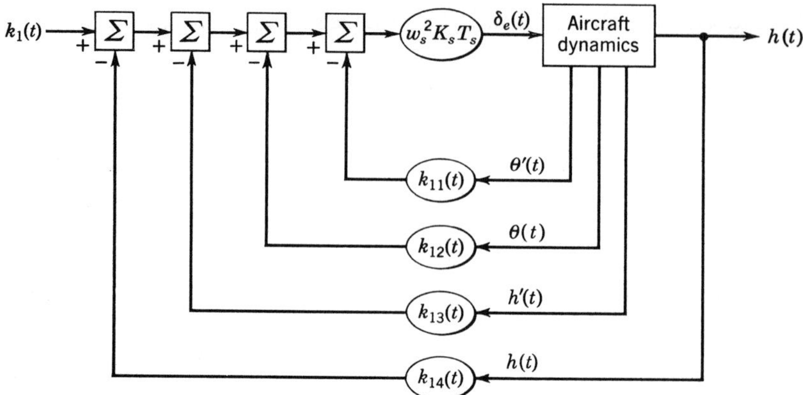

Fig. E-4 Block diagram of optimum aircraft landing system.

signal fed back in each loop. The gain in each feedback loop is time-varying and the input to the system also is time-varying.

E-6 Computation of the k parameters and landing-system trajectories

As a result of the use of optimization theory, the design procedure is reduced to the solution of the k equations. In general, the solution of these equations requires the use of an analog or digital computer. The parameters of the systems described in this appendix were found numerically by using an IBM 704 digital computer. The computer program consists of two major parts. In the first part, the k equations are solved. In the second part, the aircraft equations of motion are solved to obtain simulated landing trajectories. This two-part program is illustrated in Fig. E-5. The integration algorithm used in this study is the Gill modification of the Runge-Kutta fourth-order extrapolation algorithm. However, a number of algorithms

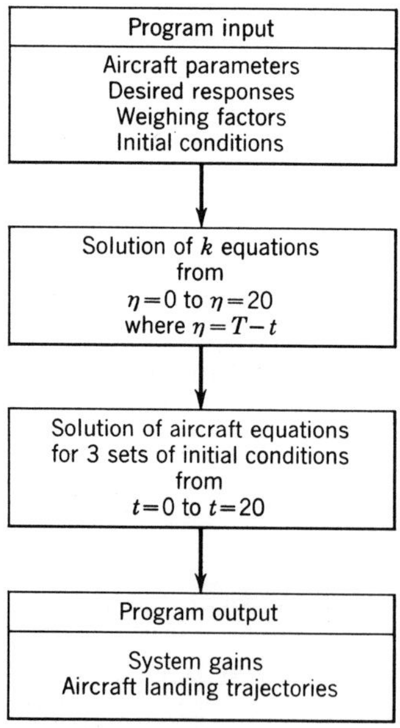

Fig. E-5 Simplified flow diagram of digital-computer program.

possessing similar properties also could be utilized. The basis for this computer program is outlined in Appendix F.

As indicated in Fig. E-5, the landing trajectories are computed for three sets of initial conditions. These conditions are tabulated in Fig. E-6. The assumption is made that the initial vertical acceleration $h''(0)$ is zero and the initial altitude, vertical velocity, and pitch rate are as given in the table. For these values, the initial pitch angle is computed from the aircraft equations of motion.

Sets 1 and 3 of the initial conditions correspond to the worst possible combinations of the largest permissible altitude and rate-of-ascent errors at the beginning of the flare-out. That is, in set 1 the aircraft is 20 per cent above the desired landing trajectory and is descending at a rate 20 per cent slower than desired. Hence, the

Set	Altitude $h(0)$, ft	Vertical velocity $h'(0)$, ft/sec	Pitch angle $\theta(0)$, rad	Pitch rate $\theta'(0)$, rad/sec
1	120	−16	−0.0625	0
2	100	−20	−0.0781	0
3	80	−24	−0.0938	0

Fig. E-6 Aircraft initial conditions.

aircraft is diverging from the desired trajectory at the largest permissible rate. A similar situation occurs in set 3, where the aircraft is below the desired trajectory and is descending at the maximum permissible rate. In set 2 the aircraft is on the desired trajectory and is descending at the rate prescribed by the performance requirements. This wide range in initial conditions serves as a severe test for the aircraft landing system.

The solution of the k equations cannot be carried out until the desired responses, aircraft parameters, and weighting factors are specified. In the preceding sections, the desired responses and the aircraft parameters are specified. Hence, the design procedure reduces to just the selection of a set of weighting factors which result in satisfactory system performance for the three sets of initial conditions.

As discussed in Chap. 8, there are two important aspects associated with each weighting factor $\phi(\sigma)$: the mathematical form $f(\sigma)$ and the magnitude Φ. That is, $\phi(\sigma)$ may be written as

$$\phi(\sigma) = \Phi f(\sigma) \tag{E-50}$$

The mathematical form $f(\sigma)$ usually is suggested by the performance requirements for the system; hence, only the magnitude Φ remains to be selected. For example, the pitch-angle requirement at touchdown leads to the selection of an impulse function as the mathematical form of $\phi_\theta(\sigma)$. However, the area $\phi_{2,T}$ is yet to be specified. Thus, the design procedure ultimately is reduced to just the selection of the magnitudes of a set of weighting factors.

Even before a detailed analysis of the design problem is performed, a rough characterization of the landing problem can be obtained by using the procedures discussed in Chap. 8. For $N = 4$, the normalizing factor $\mathcal{N}$ is found to be

$$\mathcal{N} = 3.95\phi_4^{\frac{1}{8}} \tag{E-51}$$

from (8-54) and the numerical values of the dynamic-process coefficients. Furthermore, the maximum altitude error is approximately 20 ft and the maximum elevator deflection is 0.262 rad. Because there is only one control variable $M = 1$ and four response signals $Q = 4$, (8-110) is used to obtain $\phi_4 = 0.428 \times 10^{-4}$. From (E-51), $\mathcal{N} = 1.13$ is obtained. Also, (8-57) and (8-58) are used to obtain $\mathcal{N}_1 = 0.0268$, $\mathcal{N}_{11} = 0.200$, $\mathcal{N}_{12} = 0.226$, $\mathcal{N}_{13} = 0.0025$, and $\mathcal{N}_{14} = 0.00282$. The value of $\mathcal{N}_1$ is found by using $Q_4 = 100$. Finally, if t_i is defined so that $t < t_i$ corresponds to the infinite-interval problem, then (8-85) is used to show that $t_i \approx 16$ sec.

These rough calculations yield considerable information. First, the approximate value of ϕ_4 is found, which, at first glance, may seem to be a surprisingly small number. Second, during the interval $0 \leqq t \leqq 16$ the aircraft-landing-control problem is essentially an infinite-interval problem, which implies that the terminal conditions are of little importance prior to $t = 16$. Finally, the normalizing factors $\mathcal{N}_{11}$, $\mathcal{N}_{12}$, $\mathcal{N}_{13}$, and $\mathcal{N}_{14}$ give the order of magnitude of the feedback gains on the interval $0 \leqq t \leqq 16$. Furthermore, one may expect that these feedback gains are nearly constant during this interval because the dynamic process is time-invariant and the weighting factors are also constants over much of this interval.

The approximate values of ϕ_1 and ϕ_3 also can be found by using (8-105). The maximum error in rate of ascent is roughly 4 ft/sec and hence $\phi_3 = 1.07 \times 10^{-3}$. Also, the equation for angle of attack is

$$\alpha'(t) = b_{33}\alpha(t) + \theta'(t) \tag{E-52}$$

so that $\theta'(t) = -b_{33}\alpha(t)$ under constant aerodynamic lift conditions. This relationship gives a maximum pitch rate of about $\theta'(t) = 0.0251$ rad/sec, which corresponds to (E-21), and hence $\phi_1 = 27.2$. On the other hand, the weighting-factor selection procedure used thus far is not adequate to obtain approximate values of $\phi_{2,T}$, $\phi_{3,T}$, and $\phi_{4,T}$, which are the areas of the impulse weighting factors. According to the reasoning used to arrive at (8-105), the relationship between the weighting factors is taken to be

$$\phi_{2,T}\epsilon_2{}^2(T) = \phi_{3,T}\epsilon_3{}^2(T) = \phi_{4,T}\epsilon_4{}^2(T) \tag{E-53}$$

The magnitudes of these weighting factors are found by placing a maximum value on $m'(t)$ at the touchdown point $t = T$. If the optimum control equation is differentiated and the terminal boundary conditions are imposed, then this derivative is found to be

$$\delta_e'(T) = -\omega_s{}^2 T_s K_s\{\phi_1[Q_1(T) - q_1(T)] + \phi_{2,T}[Q_2(T) - q_2(T)]\} \tag{E-54}$$

Furthermore, the "worst case" value of this derivative is defined as

$$|\delta_1'(T)| = |\omega_s{}^2 T_s K_s|\ [\phi_1\ |\epsilon_1(T)| + \phi_{2,T}\ |\epsilon_2(T)|] \tag{E-55}$$

If $\delta_1'(T) = 1.0$ rad/sec, $\epsilon_1(T) = 0$, $\epsilon_2(T) = 0.035$ rad, $\epsilon_3(T) = 0.5$ ft/sec, and $\epsilon_4(T) = 1$ ft are used, then (E-53) and (E-55) give $\phi_{2,T} = 12.0$, $\phi_{3,T} = 0.0588$, and $\phi_{4,T} = 0.0147$. The parameter $\epsilon_1(T) = 0$ is chosen because nearly equilibrium flight conditions should occur near touchdown owing to the straight-line portion of $h_d(t)$.

E-7 Evaluation of aircraft landing systems

In this section, three types of aircraft landing systems are evaluated. The first two types are an altitude control system and a terminal altitude control system. These systems are included merely for comparison purposes, and they demonstrate the consequences of neglecting important performance requirements.

Case I. Constant weighting of altitude error and elevator deflection

In this case, altitude error and elevator deflection are weighted by a constant over the entire 20-sec landing interval, and all other weighting factors are zero. That is,

$$\phi_h(\sigma) = \phi_4 = \text{const} \tag{E-56}$$

and

$$\phi_1 = \phi_{2,T} = \phi_3 = \phi_{3,T} = \phi_{4,T} = 0 \tag{E-57}$$

Then the error index becomes

$$e(t) = \int_t^{20} \{\phi_4[h_d(\sigma) - h(\sigma)]^2 + \delta_e^2(\sigma)\}\, d\sigma \tag{E-58}$$

The performance of the aircraft for $\phi_4 = 0.0001$ is shown in Fig. E-7. This figure contains four graphs. From top to bottom, these are:

1. Altitude error vs. time
2. Rate-of-ascent error vs. time
3. Angle of attack vs. time
4. Elevator deflection vs. time

Each graph, in turn, contains three curves, one for each of the three sets of initial conditions. In addition, the values of the pitch angle $\theta(T)$ at $t = 20$ sec are tabulated at the bottom of the figure for the three sets of initial conditions. The $\theta(T)$

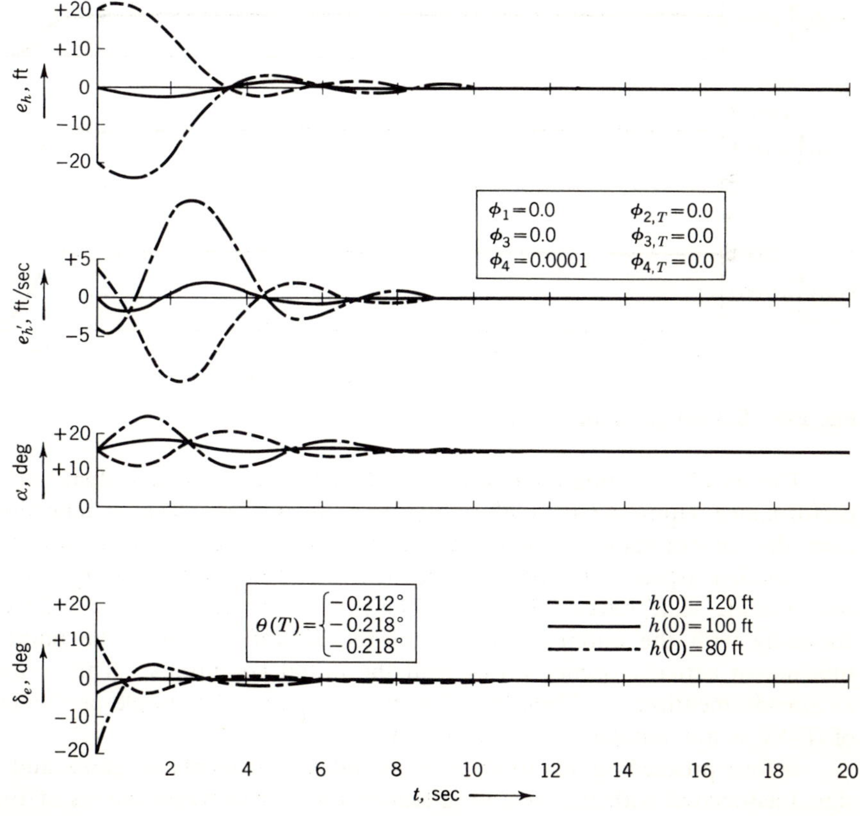

Fig. E-7 System performance for case I.

values correspond, from top to bottom, to the three sets of initial conditions given in Fig. E-6. The values of the weighting factors also are listed in Fig. E-7.

For this figure, e_h and $e_{h'}$ are given by

$$e_h(t) = h(t) - h_d(t) \qquad \text{ft} \tag{E-59}$$

and

$$e_{h'}(t) = h'(t) - h'_d(t) \qquad \text{ft/sec} \tag{E-60}$$

When the aircraft is at an altitude of 120 ft at the beginning of the flare-out, $e_h =$ 20 ft. The rate of ascent of the aircraft, for this same condition, is -16 ft/sec, giving $e_{h'} = 4$ ft/sec. Also at $t = 0$, the angle of attack α is at its equilibrium value of $+14.4°$ (80 per cent of the stall value) for all three sets of initial conditions.

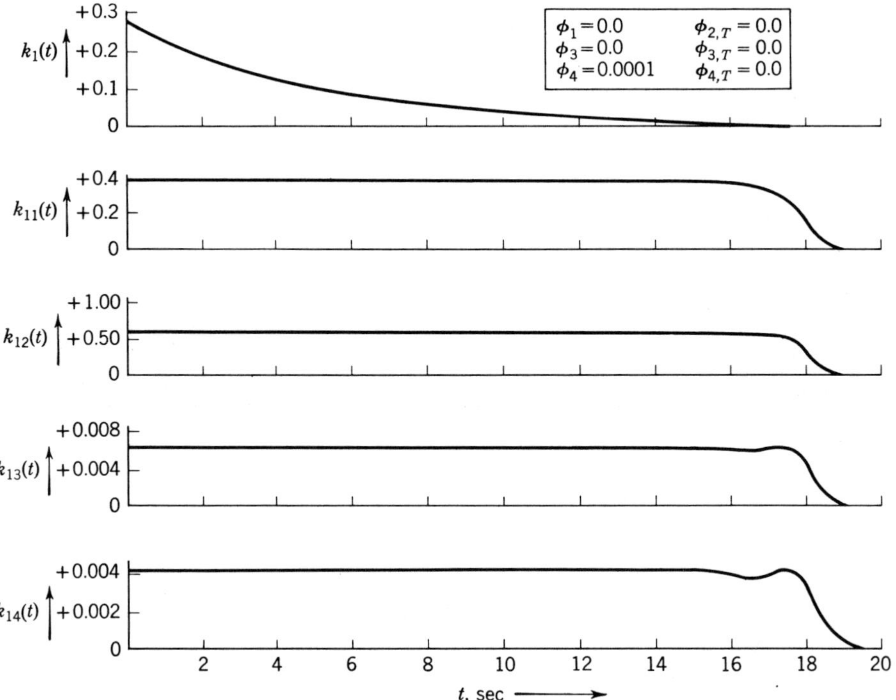

Fig. E-8 System gains for case I.

The set of weighting factors listed in Fig. E-7 results in a system that meets the performance requirements on altitude, rate of ascent, and elevator deflection. However, the constraints on angle of attack and pitch angle at touchdown are violated.

Another aspect of the performance is not evident from these figures, because of the ordinate scale selected for $e_{h'}$. The rate-of-ascent error at $t = 20$ varies substantially with the initial conditions of the aircraft. Hence, the point at which touchdown actually occurs varies with the initial conditions. This is undesirable, as already mentioned. Therefore, for many reasons, the simple performance index of (E-58) is not suitable for this problem.

Before proceeding to another error index, the feedback gains and the input signal associated with the weighting factors and performance curves of Fig. E-7 are worthwhile to consider. These gains are shown in Fig. E-8. The input signal

$k_1(T)$ appears to be similar in shape to the desired altitude $h_d(t)$ although substantially reduced in amplitude. The feedback gains $k_{11}(t)$, $k_{12}(t)$, $k_{13}(t)$, and $k_{14}(t)$ are essentially constant until the last 3 sec of the landing, when they gradually decrease to zero. Hence, just prior to touchdown, the landing system is operating essentially open-loop. This should cause no concern, because the trajectory of the aircraft cannot be altered significantly during this short interval with the finite control effort available.

Case II. Impulse weighting of altitude error at touchdown

A great deal has been written in the literature about so-called terminal control systems.[18,19,59] The purpose of such systems is to reduce error to zero at a future

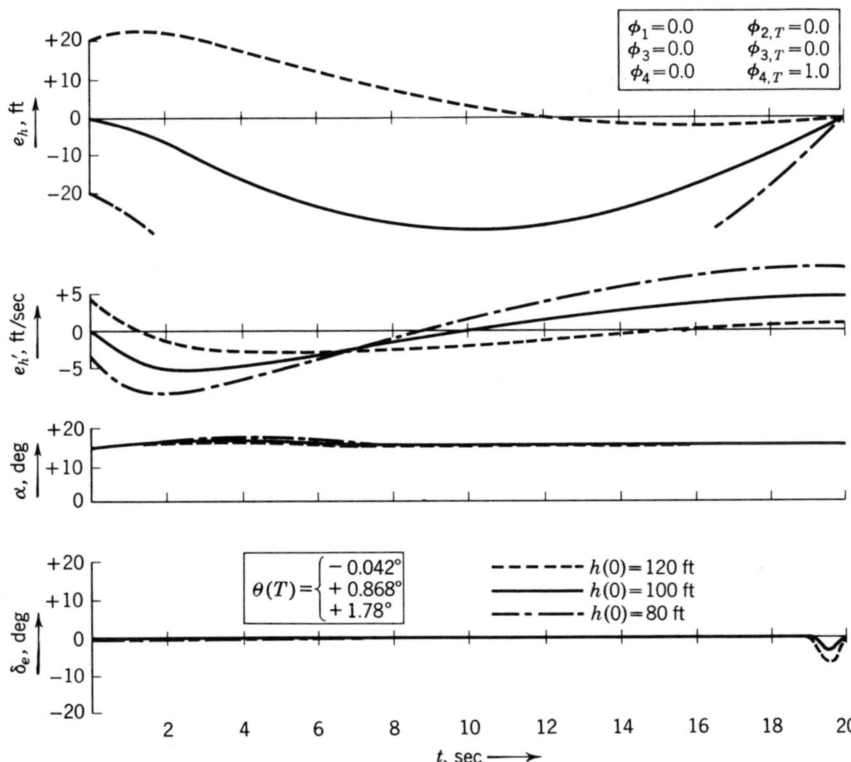

Fig. E-9 System performance for case II.

fixed point in time called the terminal point. A system with this property also can be designed using optimization theory by selecting an impulse weighting factor associated with this future error. Hence, if the performance requirement for this landing system is to reduce the altitude error at the desired touchdown point to the smallest possible value without exceeding the elevator deflection limits, then the error index becomes

$$e(t) = \phi_{4,T}[h_d(20) - h(20)]^2 + \int_t^{20} \delta_e^{\,2}(\sigma)\, d\sigma \tag{E-61}$$

The resulting performance for $\phi_{4,T} = 1.0$ is shown in Fig. E-9.

The altitude error at $t = 20$ is quite small, but unfortunately the aircraft crashes for two sets of initial conditions because of the large altitude errors occurring during the earlier part of the landing interval. In addition, the rate-of-ascent error is undesirably large at touchdown.

The system input signal and the feedback gains for this case are shown in Fig. E-10. The input signal $k_1(t)$ is zero during the entire landing interval, as might be expected. The feedback gains are negligibly small until the aircraft begins to approach the desired touchdown point. These gains then become quite large in

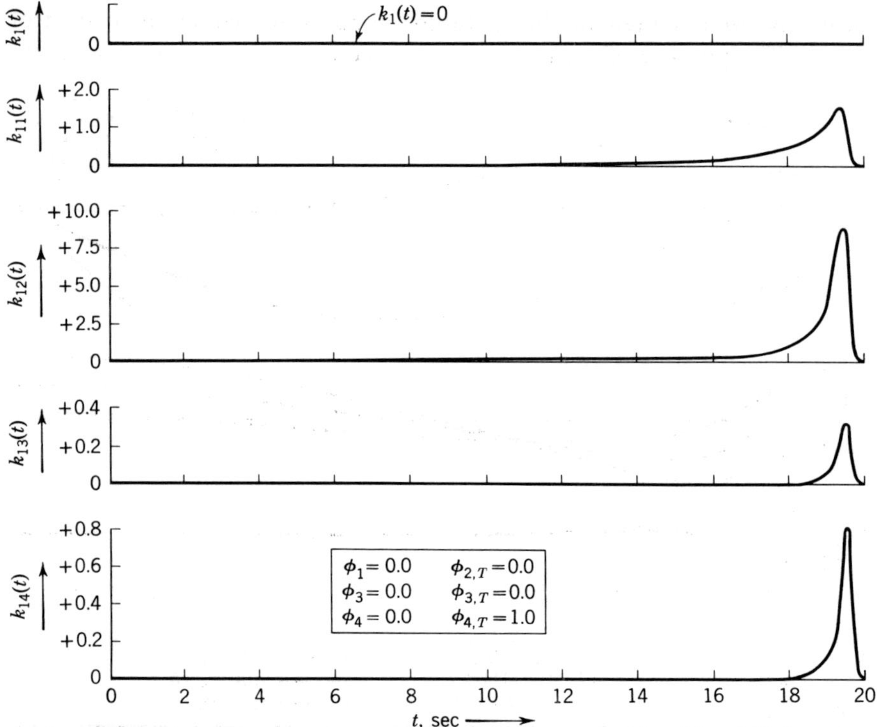

Fig. E-10 System gains for case II.

order to reduce the altitude error to a small value at $t = 20$. Because $k_{14}(t)$ essentially is zero for a large part of the landing interval, the altitude feedback loop in essence is open during this time and large altitude errors arise, as already demonstrated.

Case III. Constant weighting of altitude error, rate-of-ascent error over the runway, pitch-rate error, and elevator deflection; impulse weighting at touchdown of pitch-angle error, altitude error, and rate-of-ascent error

The error index for this case is

$$e(t) = \phi_{4,T}[h_d(20) - h(20)]^2 + \phi_{3,T}[h_d'(20) - h'(20)]^2 + \phi_{2,T}[\theta_d(20) - \theta(20)]^2 + \int_t^{20} \{\phi_4[h_d(\sigma) - h(\sigma)]^2 + \phi_3(\sigma)[h_d'(\sigma) - h'(\sigma)]^2 + \phi_1[\theta_d'(\sigma) - \theta'(\sigma)]^2 + \delta_e^2(\sigma)\}\, d\sigma \qquad \text{(E-62)}$$

Here the design parameters $\theta_d'(t) = 0$ and $\theta_d(20) = 2°$ are used in addition to the weighting factors specified in Fig. E-11. Also, the resulting system performance, which essentially meets all the performance requirements, is shown in Fig. E-11. The extremely small negative pitch angle at touchdown existing for one set of initial conditions could have been avoided by a slight increase in the value of $\phi_{2,T}$.

Figure E-11 indicates that very little margin exists between the performance obtained and the performance requirements for the system. In other words, the performance requirements are very difficult to satisfy with the given aircraft dynamics

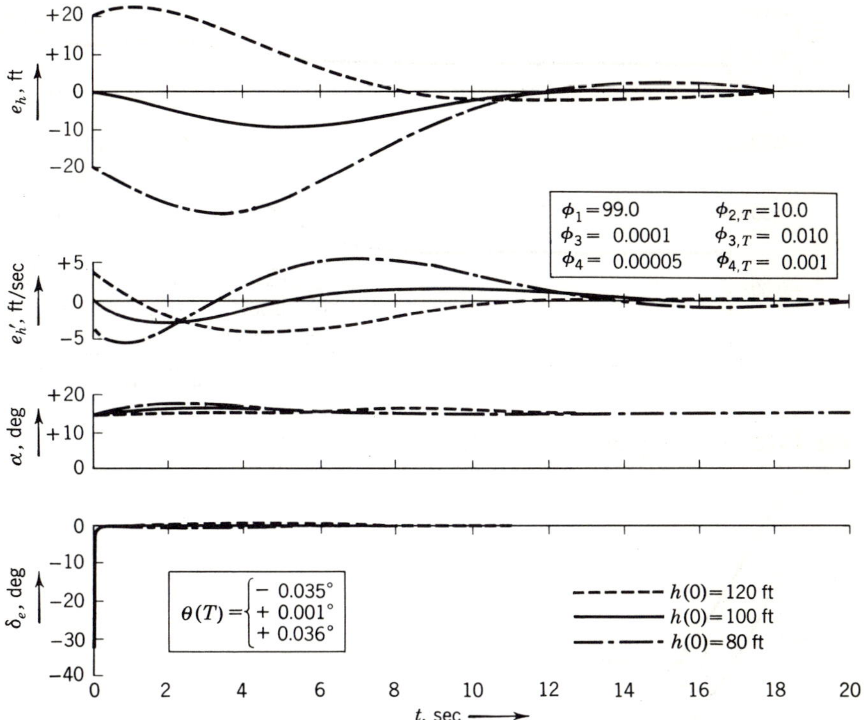

Fig. E-11 System performance for case III.

and the assumed initial conditions. This difficulty leads to some trial and error in the selection of the weighting factors. However, a portion of the trial and error required in this design problem results because angle of attack is not weighted directly in the error measure. A great many control problems are characterized by the need to satisfy a set of very restrictive performance requirements. Optimization theory is of particular value in the solution of such problems.

The input signal and the feedback gains associated with the performance of Fig. E-11 are shown in Fig. E-12. The increases in the gains $k_{12}(t)$, $k_{13}(t)$, and $k_{14}(t)$ near the touchdown point are due to the use of the impulse weighting factors $\phi_{2,T}$, $\phi_{3,T}$, and $\phi_{4,T}$. An important point to note in this case is that the input signal $k_1(t)$ is negative during the final phase of the landing and is positive again just prior to $t = 20$. The negative and then positive excursions evidently are required to satisfy the pitch-angle constraint at touchdown. In problems where multiple

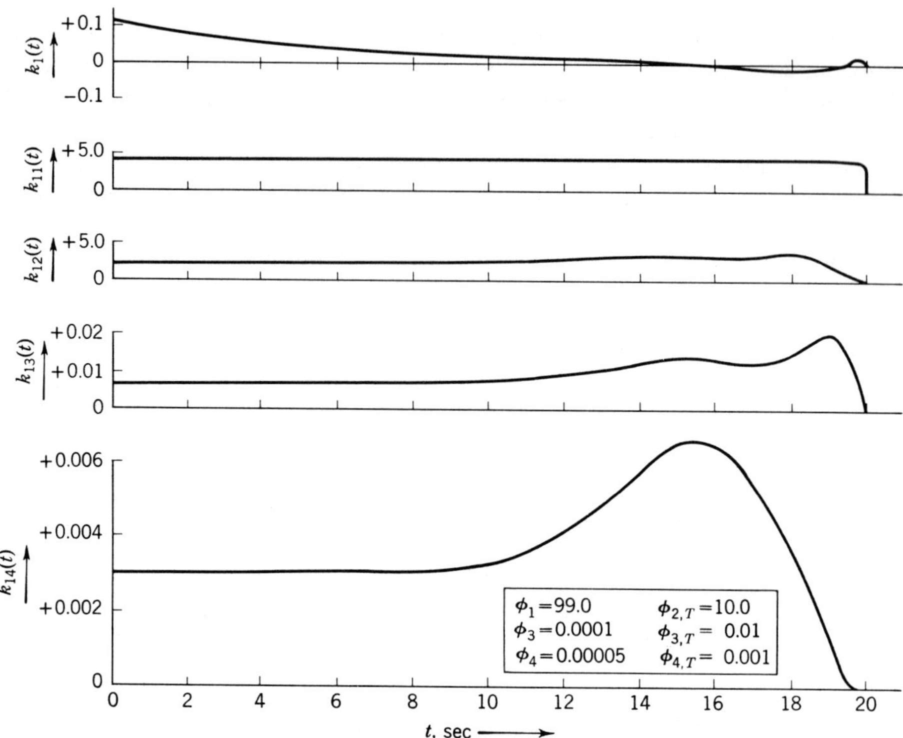

Fig. E-12 System gains for case III.

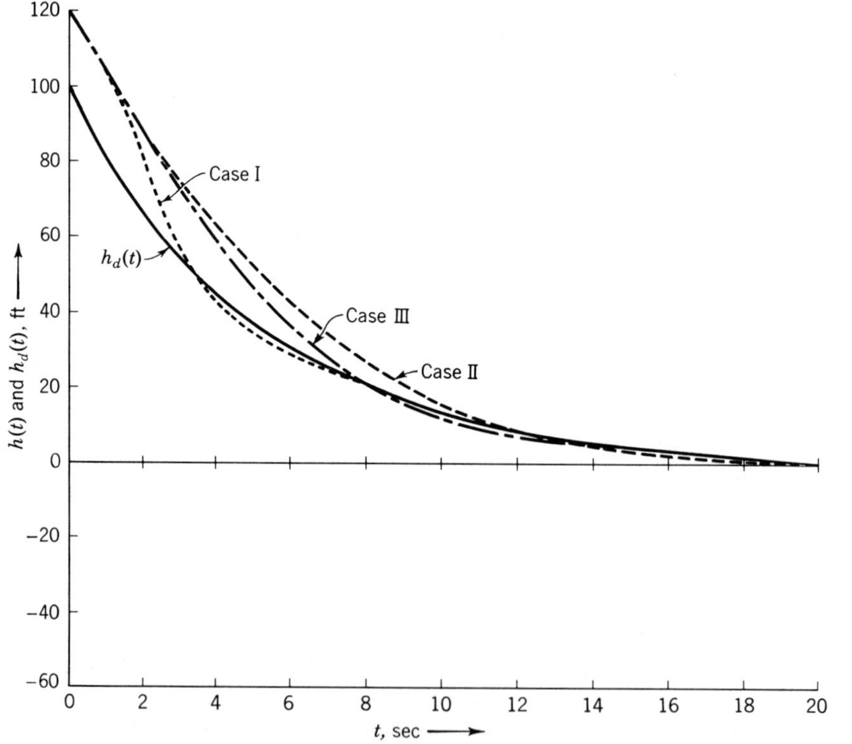

Fig. E-13 Aircraft landing trajectories for $h(0) = 120$ ft.

performance requirements exist, the system input signal, in essence, is a composite function derived from all the performance requirements.

The behavior of the systems of cases I to III also may be compared by displaying the aircraft landing trajectories on three separate graphs, one for each set of initial conditions. These trajectories are shown in Figs. E-13 to E-15 along with the desired altitude $h_d(t)$. The system of case I follows the desired flare path more closely than do the other systems. This is due primarily to the greater weighting of altitude

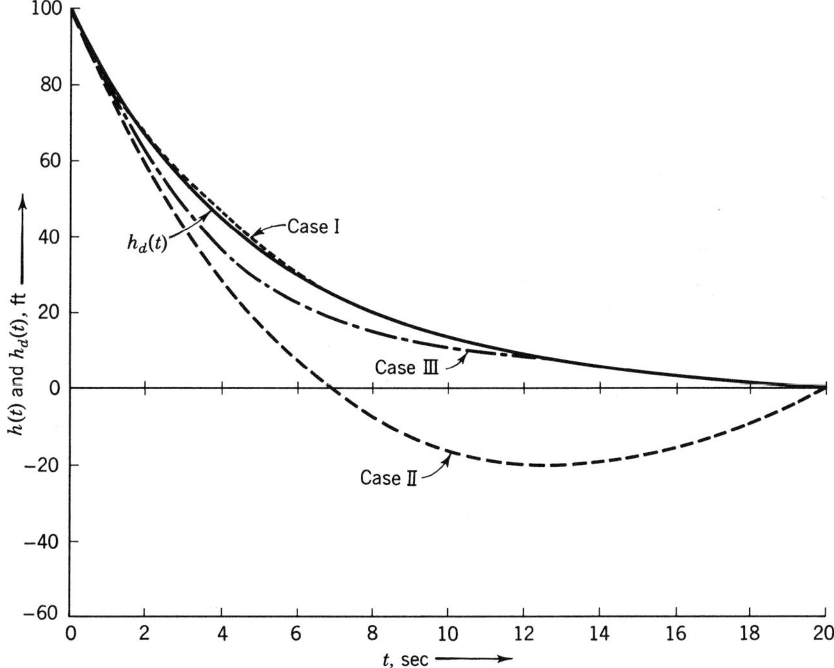

Fig. E-14 Aircraft landing trajectories for $h(0) = 100$ ft.

errors throughout the landing interval in the design of the system of case I. Of course, these curves do not indicate that other performance requirements are violated by the system of case I.

As already mentioned, the system of case II causes the aircraft to crash for two sets of initial conditions and therefore is completely unsatisfactory.

The system of case III meets all the performance requirements, but a substantially longer time than for the system of case I is required to correct the initial-condition errors. The persistence of the initial-altitude errors for a longer period of time is the price paid for satisfying the other performance requirements.

In order to mechanize the aircraft landing system, the feedback gains and the input signal must be programmed as functions of time-to-go $(20 - t)$. In other words, time-to-go is a quantity that must be available throughout the landing phase in order to construct the control equation specified by the theory. Undoubtedly, the measurement of time-to-go is the most difficult measurement associated with the physical construction of the proposed landing system. First of all, the aircraft velocity is not absolutely constant as assumed. However, appropriate thrust control generally keeps perturbations in the aircraft velocity small enough to be considered a second-order effect. On the other hand, the measurement of the total aircraft velocity is difficult. This aircraft velocity is considered to be the velocity of the aircraft with respect to an inertial frame, or rather with respect to the ground. In other words, the aircraft velocity in question is groundspeed. Because landing situations with nonzero winds occur, groundspeed cannot be measured from the aircraft itself. Therefore, the success of this landing system requires the successful

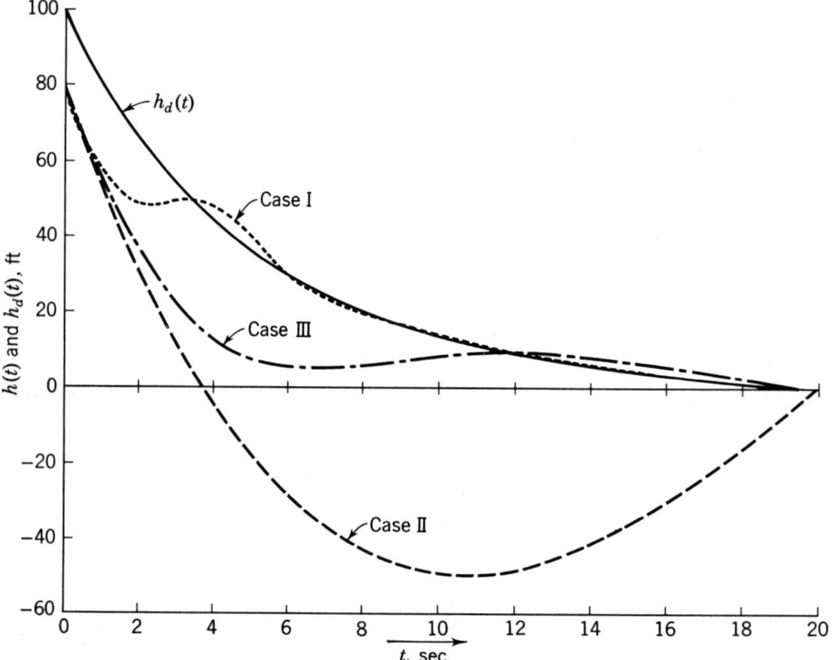

Fig. E-15 Aircraft landing trajectories for $h(0) = 80$ ft.

measurement of groundspeed from radar. A number of possible radar systems exist. For instance, a system of ground-based radar beacons could be used in order to estimate the time required to pass over two fixed points on the earth's surface. Another possibility is the use of ground-based doppler radar. In this case, the velocity information would be relayed to the aircraft. Because of these practical considerations, the measurement of time-to-go could be obtained from

$$\tau \approx \frac{d(t)}{V(t)} \tag{E-63}$$

where the time-to-go τ is defined by

$$\tau = 20 - t \tag{E-64}$$

In the estimation of time-to-go, the quantity $d(t)$ is defined as the distance from the aircraft to the desired point of touchdown on the runway, and $V(t)$ is the groundspeed of the aircraft. Both distance $d(t)$ and velocity $V(t)$ could be measured from a system of ground-based radar.

The engineering version of the control system can be constructed in terms of analog components. Therefore, some brief comments concerning the analog construction of the control system are presented here. The block diagram shown in Fig. E-16 indicates one possible realization of the time-varying gains appearing in the control system. If time-to-go τ is available as an angular shaft position, then the time-varying multiplications are obtained from a set of loaded potentiometers. Such potentiometers have multiple taps, and the appropriate k-parameter functions of time-to-go can be constructed by connecting appropriate voltage sources and/or resistors to the taps on the potentiometer. The signal on the wiper of such a potentiometer then is the appropriate function of the shaft angle τ times the excitation

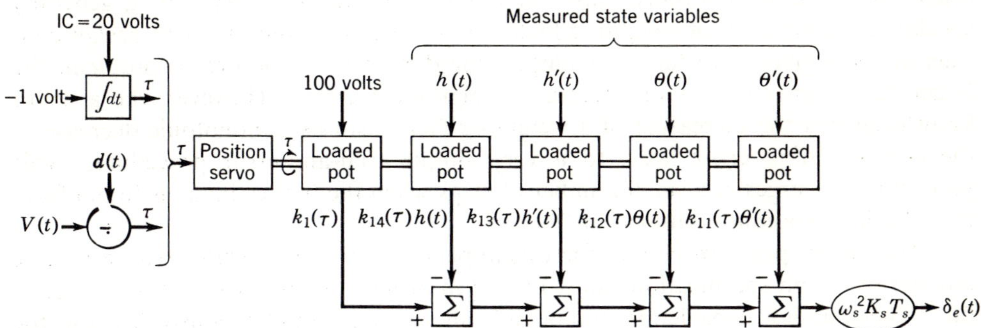

Fig. E-16 Mechanization of aircraft landing system.

voltage applied to the loaded pot. The summing and gain functions indicated in Fig. E-16 are achieved from standard analog-computer components. The shaft angle τ is achieved by placing an input signal τ to a position servo. The input signal τ could be generated from the expression for time-to-go given in (E-63) by the appropriate division of distance to the touchdown point by indicated groundspeed of the aircraft.

Another possibility for the generation of the signal τ arises under ideal situations where the groundspeed is a known constant and the initial distance to the touchdown point is known. Under these ideal conditions, a simple clock constructed by an integrator with a constant input can be used for the generation of time-to-go. Both these possibilities for the generation of the signal τ are indicated in Fig. E-16. The system indicated in Fig. E-16 is highly suited for airborne applications because this system is light, cheap, and reliable. Therefore, if the measurements of the state variables and time-to-go are readily available and reasonably accurate, then the theoretical results of this example form the basis for a practical landing system that exhibits high performance.

If the time-to-go measurement is in error during the initial portion of the flareout interval but is corrected before the desired touchdown point is reached, then the control system described in this appendix is still an optimum system after the correction. The initial error in time-to-go gives rise to a different dynamic state of the aircraft at the time of the correction than would be obtained with a precise measurement. However, the control system is optimum for all existing states of the aircraft within the limits required for linearity. Therefore, only precise measurements for small time-to-go are needed.

E-8 Conclusions

A number of important advantages of using optimization theory as the basis for the design of feedback control systems are pointed out by the example of the aircraft landing system. In particular, multiple design considerations associated with difficult dynamic situations can be readily included in the system design. The mathematical forms of the weighting factors appearing in the error index are chosen directly from considerations of the time-domain performance requirements. Except for certain mathematical restrictions, the mathematical forms of these weighting factors can be chosen without consideration of the system stability per se. On the other hand, one of the disadvantages of using optimization theory is that, generally speaking, a unique set of weighting factors does not exist when a set of design considerations is given. In fact the simpler the dynamics of the design problem, the larger the number of sets of acceptable weighting factors. However, the a priori knowledge that the increasing of a weighting factor causes a monotonic decrease in the corresponding error leads to a very simple trial-and-error procedure. This procedure converges in a small number of steps when the first estimate is found from the selection procedure outlined in Chap. 8.

The design problem chosen for this appendix is made difficult because a large number of design specifications must be met in order to consider the results successful. These design considerations are difficult to meet simultaneously because the trajectory requirements for flare-out are somewhat incompatible with the dynamic characteristics of the aircraft. In particular, the trajectory corrections required for passing from the glide phase to the flare-out phase are large in comparison with the amount of control effort allowable during the flare-out portion of the trajectory. Roughly speaking, the amount of control effort available is considered to be the magnitude of the control signal times the amount of time during which this control signal can be applied to the aircraft. Optimization theory assumes a most useful purpose under the conditions where the potential response errors are large relative to the amount of control effort that can be applied to the system.

The results of this particular example indicate that the design specifications for the aircraft landing system can be met. The digital-computer simulation of the aircraft trajectories shows that successful automatic aircraft landings can be achieved under ideal conditions. However, because a number of engineering questions are neglected in the study presented here, additional work is required in order to build a practical landing system from the theoretical results presented thus far. In particular, the effects of wind gusts and measurement noise should be included in trajectory studies in order to ascertain their effect on the performance of the system. These statistical disturbances could be included in a digital-computer simulation. However, these additional studies can be performed more efficiently on an analog computer.

Another area that should be investigated before the proposed landing system is considered checked out in an engineering sense is the area of sensitivity analysis. In particular, the aircraft parameters assumed in the previous results are only approximations, and therefore the sensitivity of the system trajectories to perturbations around these assumed parameters should be investigated. Also, the effects of approximations to the k parameters should be investigated because the engineering system may be able to generate these time-varying gains in only an approximate sense. Some of this work has been performed, and the results indicate that a successful landing system can be built.[64]

E-9 Nomenclature for Appendix E

$h(t)$	Aircraft altitude
$h'(t), h''(t), h'''(t)$	Time derivatives of aircraft altitude
$h_d(t)$	Desired aircraft altitude
$h'_d(t)$	Desired aircraft rate of ascent
K_s	Aircraft short-period gain
T_s	Aircraft path time constant
V	Aircraft total velocity
W	Weight of the aircraft
$\alpha(t)$	Aircraft angle of attack
$\Delta\alpha(t)$	Change in angle of attack from the equilibrium value
$\gamma(t)$	Aircraft glide-path angle
$\delta_e(t)$	Aircraft elevator deflection
ζ	Aircraft short-period damping ratio
$\theta(t)$	Aircraft pitch angle
$\theta(T)$	Aircraft pitch angle at $\sigma = T$
$\theta'(t)$	Aircraft pitch rate
$\theta_d(t)$	Desired aircraft pitch angle
$\theta'_d(t)$	Desired aircraft pitch rate
τ	Time-to-go before touchdown
ϕ_1	Weighting factor for pitch-rate errors
ϕ_3	Weighting factor for rate-of-ascent errors
ϕ_4	Weighting factor for altitude errors
$\phi_{2,T}$	Area of the impulse weighting factor for pitch-angle errors
$\phi_{3,T}$	Area of the impulse weighting factor for rate-of-ascent errors
$\phi_{4,T}$	Area of the impulse weighting factor for altitude errors
$\phi_h(t)$	Weighting factor for altitude errors
$\phi_{h'}(t)$	Weighting factor for rate-of-ascent errors
$\phi_\theta(t)$	Weighting factor for pitch-angle errors
$\phi_{\theta'}(t)$	Weighting factor for pitch-rate errors
ω_s	Aircraft short-period resonant frequency

appendix f

Numerical Integration of Ordinary Differential Equations with Discontinuous Time Functions

The use of optimization theory for control-system design generally requires the numerical solution of ordinary differential equations. Voluminous material is available on numerical integration,[60] and of course, many digital-computer programs are available. This appendix is included only to make available a computer program for those who do not have access to programs which are already written. Also, step functions and impulse functions many times occur in the k equations, and most existing programs are not written for these discontinuous functions.

F-1 Gill method

The Gill method is chosen here as the basic method of numerical integration.[61] This method is a modification of the Runge-Kutta procedure, which is especially suited for digital computers. The Gill algorithm efficiently uses computer memory, and the initialization of dependent variables is convenient.

The Gill method is summarized here for the single first-order equation

$$y' = f(y, x) \tag{F-1}$$

Suppose that this equation is integrated over an increment e of x and the initial conditions are $x = x_0$ and $y = y_0$. Then the integration over the increment e is performed in four steps, which are given in (F-2).

Step 1:
$$g_1 = f(y_0, x_0)$$
$$y_1 = y_0 + \frac{1}{2} e g_1$$
$$q_1 = g_1$$

Step 2:
$$g_2 = f\left(y_1, x_0 + \frac{e}{2}\right)$$
$$y_2 = y_1 + \left(1 - \sqrt{\frac{1}{2}}\right) e(g_2 - q_1)$$
$$q_2 = (2 - \sqrt{2})g_2 - \left(2 - 3\sqrt{\frac{1}{2}}\right) q_1 \tag{F-2}$$

Step 3:
$$g_3 = f\left(y_2, x_0 + \frac{e}{2}\right)$$
$$y_3 = y_2 + \left(1 + \sqrt{\frac{1}{2}}\right) e(g_3 - q_2)$$
$$q_3 = (2 + \sqrt{2})g_3 - \left(2 + 3\sqrt{\frac{1}{2}}\right) q_2$$

Step 4:
$$g_4 = f(y_3, x_0 + e)$$
$$y_4 = y_3 + \tfrac{1}{6} e g_4 - \tfrac{1}{3} e q_3$$

The final quantity is the approximate value of y such that $y_4 \approx y$ at $x = x_0 + e$. The variables q_1, q_2, and q_3 are introduced for the use of the method on a digital computer. Specifically, only three storage registers are needed for the g's, y's, and q's because the variables in any step are dependent only upon the corresponding variables appearing in the previous step.

The Gill method defined in (F-2) readily extends to the set of N first-order differential equations

$$y_i' = f_i(\mathbf{y}, x) \qquad i = 1, 2, \ldots, N \tag{F-3}$$

For N dependent variables, each equation is replaced by N equations. For instance, the equation $g_1 = f(y_0, x_0)$ is replaced by

$$g_{1i} = f_i(\mathbf{y}_0, x_0) \qquad i = 1, 2, \ldots, N \tag{F-4}$$

The flow diagram of a computer program for the Gill method is shown in Fig. F-1. The notation used in this flow diagram is chosen to correspond closely to the notation used in FORTRAN[62] because FORTRAN programming can be assimilated readily by engineers without previous programming experience. The notation is summarized by:

1. FORTRAN equality (=), summation (+ and −), multiplication (*), division (/), hierarchy of parentheses, subscripted variables [Y(I) corresponds to y_i, etc.], and integer variables (symbols with the first letter I, J, K, L, M, or N) are used.

2. COMPUTE indicates computation from a formula that is supplied by the program user.

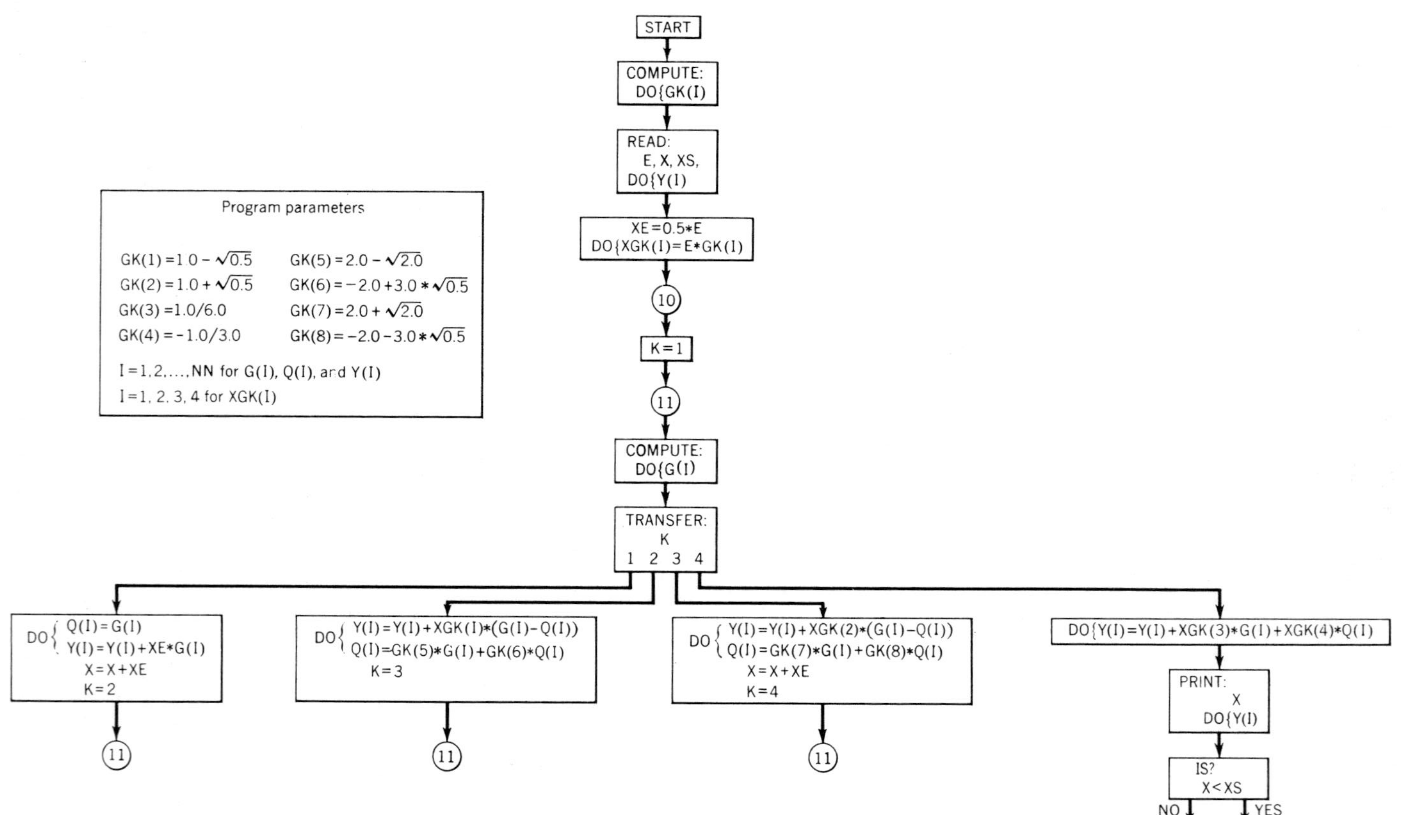

Fig. F-1 Flow diagram of the numerical-integration algorithm.

3. READ indicates program input. Program parameters not appearing in READ statements are entered when programming.

4. PRINT indicates program output.

5. IS indicates a test that is equivalent to the FORTRAN IF statement.

6. TRANSFER indicates a program transfer that is equivalent to the FORTRAN computed GO TO statement.

7. Numbered nodes indicate common points in the program.

8. The symbol DO{ is equivalent to the FORTRAN DO statement.

9. Order of computations in a block (top to bottom of page) must be observed generally.

The program shown in Fig. F-1 stops for $x \geqq x_s$.

F-2 Selection of increment size

The selection of increment size is important in the solution of ordinary differential equations on a digital computer. On the one hand, the increment size should be large enough that solutions do not consume excessive computer time. On the other hand, the increment size should be small enough that accurate solutions are obtained. Generally, speaking, a fixed increment size during the solution is found to be inadequate for achieving these goals simultaneously. Therefore, a method of automatically changing the increment size during the solution is needed.

The "rule of thumb" suggested by Collatz[63] as an approximate indicator of truncation errors possesses a number of suitable features for automatically changing increment size. This indicator is written in terms of the variables appearing in (F-2) as $|g_3 - g_2|/|g_2 - g_1|$. Some of the properties of this ratio can be examined readily by eliminating the q's in (F-2) so that

$$\begin{aligned}
g_1 &= f(y_0, x_0) \\
g_2 &= f\left(y_0 + \frac{1}{2} e g_1, x_0 + \frac{e}{2}\right) \\
g_3 &= f\left[y_0 + \frac{1}{2} e(2g_2 - g_1) - \sqrt{\frac{1}{2}}\, e(g_2 - g_1), x_0 + \frac{e}{2}\right] \\
g_4 &= f\left[y_0 + e g_3 + \sqrt{\frac{1}{2}}\, e(g_3 - g_2), x_0 + e\right] \\
y_4 - y_0 &= \frac{e}{6}(g_1 + 2g_2 + 2g_3 + g_4) + \frac{e}{6}\sqrt{2}(g_3 - g_2)
\end{aligned} \tag{F-5}$$

First, the special case where f is independent of y results in $g_3 = g_2$ so that

$$\frac{|g_3 - g_2|}{|g_2 - g_1|} = 0$$

and

$$y_4 - y_0 = \frac{e}{6}(g_1 + 4g_2 + g_4) \tag{F-6}$$

Second, a particular point on the solution may give rise to $g_2 = g_1$, in which case the ratio limits to

$$\frac{|g_3 - g_2|}{|g_2 - g_1|} = \left(1 - \sqrt{\frac{1}{2}}\right) e \left|\frac{\partial g_2}{\partial y_0}\right| \tag{F-7}$$

From these cases, certain conclusions can be drawn. This ratio does not indicate the truncation errors associated with the dependence of y' upon the independent variable x. This is demonstrated for a special case where the ratio is always zero, and the increment $y_4 - y_0$ is given by Simpson's rule. The ratio does, however, indicate the truncation errors associated with the dependence of y' upon y. The case resulting in (F-7) supports the assumption that the ratio increases monotonically with increasing increment size. Also, (F-7) demonstrates that the ratio possesses a limit on portions of the solutions where y' is a constant. These conclusions suggest the criterion

$$e_{mi} < \frac{|g_3 - g_2|}{|g_2 - g_1|} < e_{ma} \tag{F-8}$$

for adjusting the increment e in order to maintain truncation errors within acceptable bounds. The parameter e_{mi} is chosen so that the time required for solution is no longer than necessary. The parameter e_{ma} is chosen so that the required solution accuracy is obtained.

The use of this criterion possesses two advantages. First, if the increment must be decreased in order to satisfy (F-8), only approximately one-half of the steps in the Gill method must be repeated before the Gill method proceeds normally. Second, the parameters e_{mi} and e_{ma} are essentially constants for different sets of equations within a class. For the matrix-Riccati class formed by the k equations, acceptable values for e_{mi} and e_{ma} have been found,

$$e_{mi} \approx 0.05 \qquad \text{and} \qquad e_{ma} \approx 0.2 \tag{F-9}$$

which give rise to about four-place accuracy in most circumstances. The relative invariance of these parameters is due to the use of a ratio of derivatives as opposed to just derivatives. The ratio is independent of the amplitudes of the dependent variables and also independent of the time scale of the transients associated with the dependent variables. Therefore, the ratio is mainly dependent upon the shape of these transients.

The criterion given in (F-8) must be used with care because not all truncation errors determining solution accuracy are taken into account. The criterion is used to limit the rate at which truncation error accumulates and thereby to maintain the stability of the discrete form of the differential equation used in the Gill method.§ This criterion is used under the assumption that f is a slowly varying function of x in comparison with the rate at which f varies with y and hence the rate at which y varies with x. Generally, this assumption is true in the solution of the k equations provided that the weighting factors, desired signals, and parameters of the dynamic process are continuous functions of time. For impulse weighting factors and other discontinuities, f is continuous with respect to y but not with respect to x, and additional considerations must be introduced.

Basically, discontinuous functions are included by not allowing the increment e to span a point of discontinuity. In other words, an additional increment alteration is included so that the Gill method may start or stop at a discontinuity but never spans the discontinuity.

§ The discrete form of the differential equation is stable in general only when the differential equation itself is stable.

F-3 Program for numerical integration with discontinuous functions

The program shown in Fig. F-2 is written for a set of first-order differential equations which have coefficients that may be constants, continuous functions of time, step functions of time, impulse functions, and tabulated functions of time. Notationally, these functions are defined as

CC(J) = value of the constant jth coefficient
CI(J, I) = area of the ith impulse in the jth coefficient
CS(J, I) = value of the ith step in the jth coefficient
CT(J, I) = value of the ith tabulated point in the jth coefficient

The points in time where the discontinuities occur are defined as

TI(I) = time of ith impulse
TS(I) = time at the change between the ith and $(i + 1)$st steps
TT(I) = time of the ith tabulated point

The basic concept in the program is the construction of a composite list T(I), which includes all the points of discontinuities included in the lists TI(I), TS(I), and TT(I). The program is written in such a way that all coefficients are computed as continuous coefficients except at T(I). The list T(I) represents a small number of points compared with the usual number of iterations of the Gill method required in a solution. Therefore, the occurrence of discontinuous functions does not significantly increase computer time. The program shown in Fig. F-2 imposes the following restrictions:

1. All coefficients containing impulses must have an equal number of impulses. The ith impulse of all coefficients containing impulses occurs at the same point in time. The impulses are ordered so that TI(I + 1) > TI(I). Similar restrictions are placed on tabulated coefficients and step-function coefficients.
2. Points of discontinuities satisfy TT(1) ≦ TI(1), TT(1) < TS(1), TT(LTN) ≧ TI(LIN), and TT(LTN) > TS(LSN), where LTN, LIN, and LSN are the number of tabulated points, the number of impulses, and the number of step-function changes, respectively.
3. The number of step-function amplitudes is LSN + 1.
4. The integration is performed on the interval TT(1) ≦ X ≦ TT(LTN).
5. The interpolation of tabulated data is straight-line.
6. The symbol DO{ indicates indexing of I and J only.
7. XK(I) are algebraic combinations of the Y(J)'s.
8. The increment is decreased at the start of an iteration of the Gill method if X < T(L) ≦ X + E so that the iteration terminates at X = T(L). The increment is halved during an iteration until |G3(I) − G2(I)| < EMA* |G2(I) − G1(I)| for all I.
9. The increment is doubled once at the end of an iteration if

$$X + 2.0 * E \leqq T(L)$$

and if |G3(I) − G2(I)| < EMI* |G2(I) − G1(I)| for all I during the previous iteration. If the second condition for doubling is true but not the first, then the increment is increased to a value that results in X = T(L) at the end of the iteration.
10. When the increment is decreased because of a point of discontinuity, the

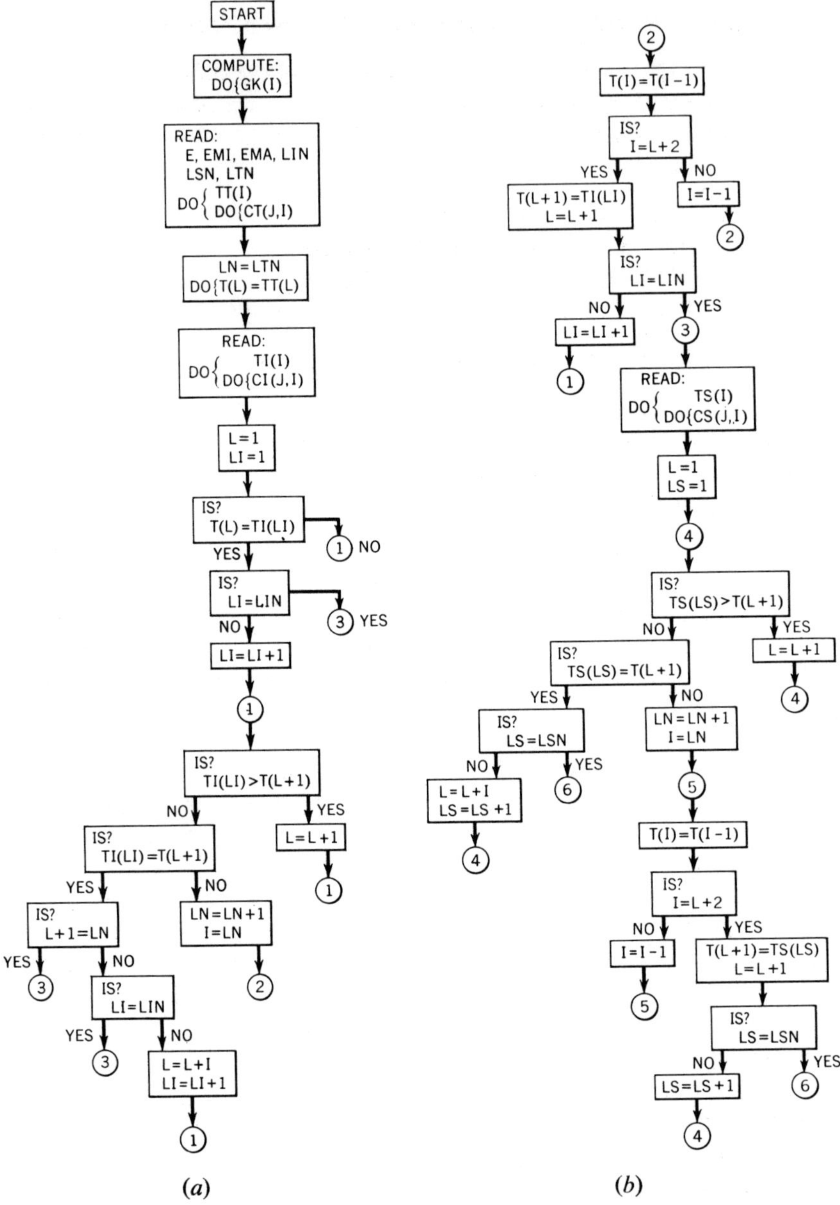

Fig. F-2 Flow diagram for numerical integration of differential equations with discontinuous time functions.

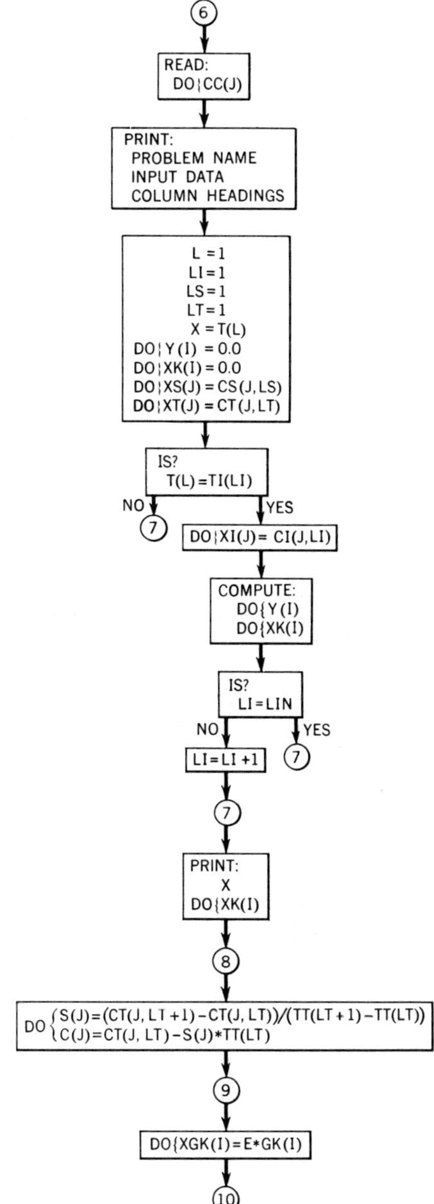

Fig. F-2c Flow diagram for numerical integration of differential equations with discontinuous time functions.

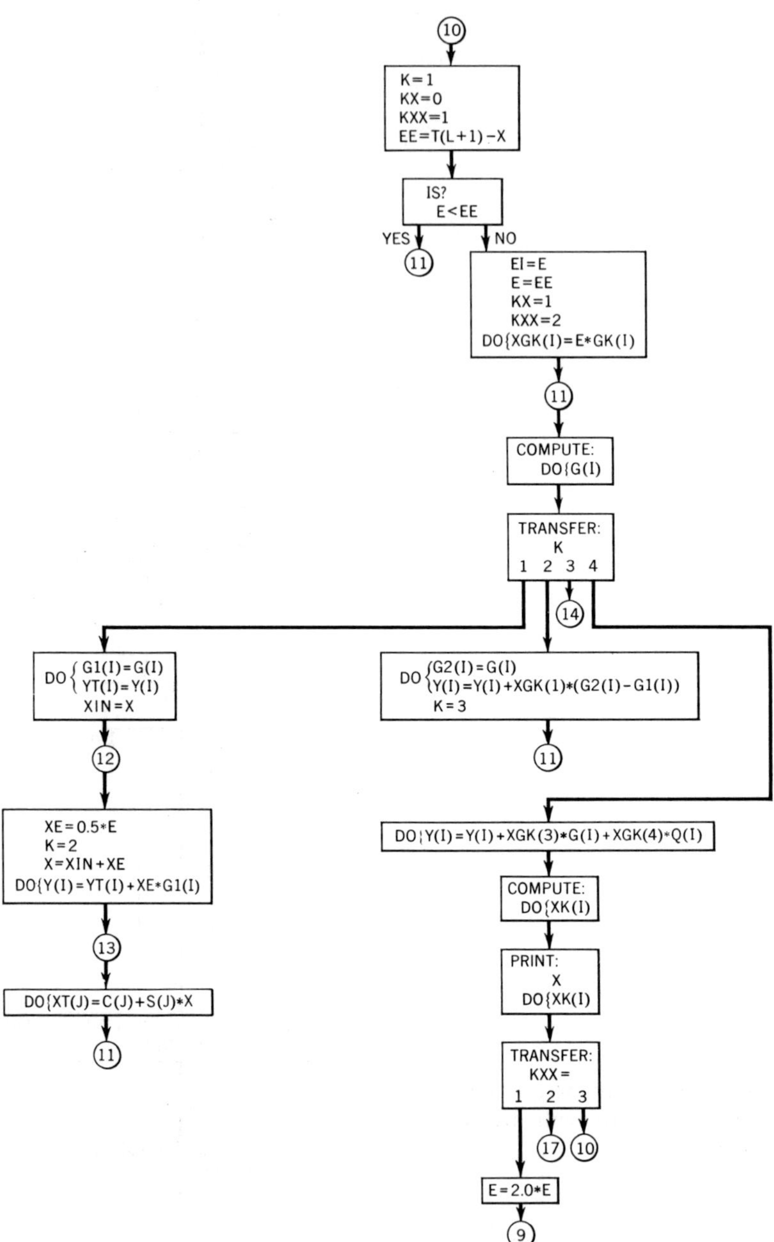

Fig. F-2*d* Flow diagram for numerical integration of differential equations with discontinuous time functions.

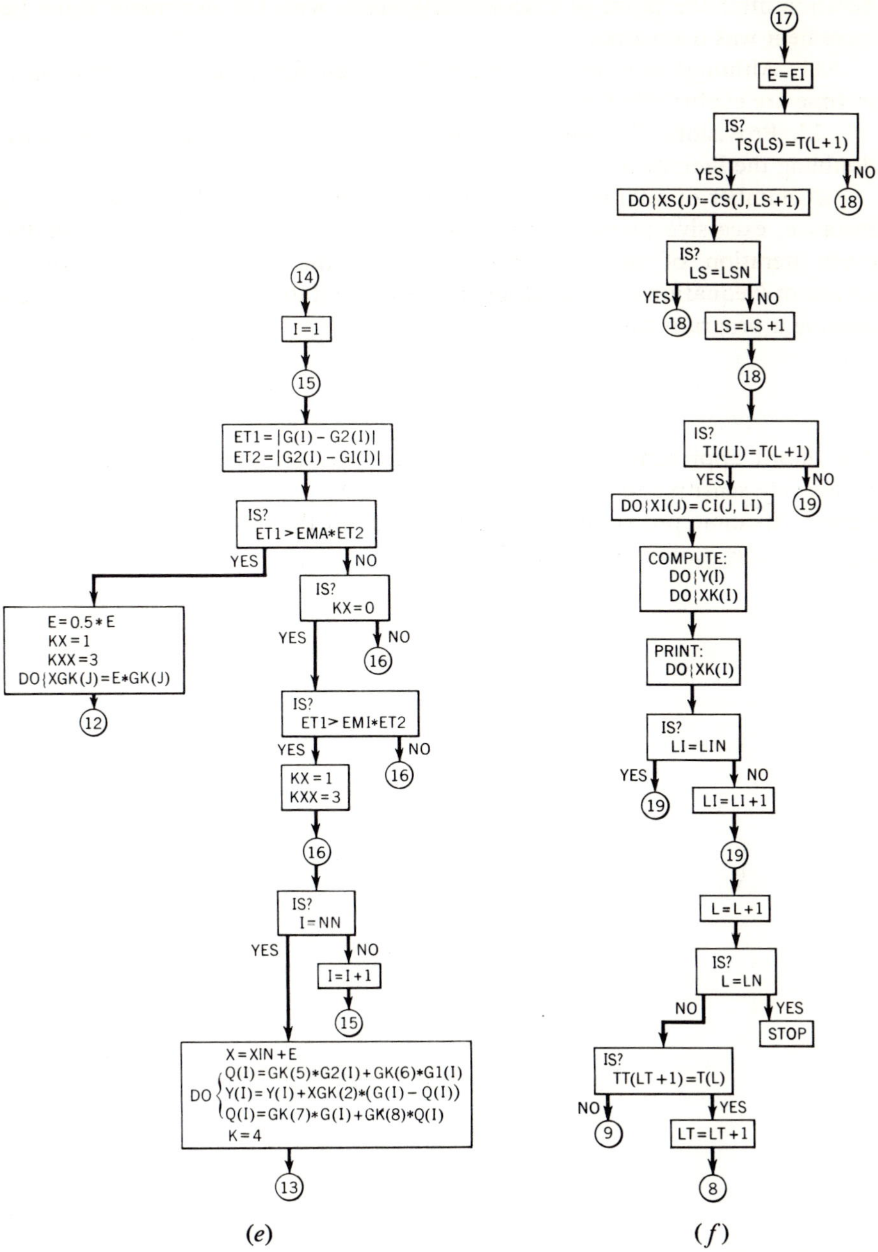

14

I=1

15

ET1=|G(I) − G2(I)|
ET2=|G2(I) − G1(I)|

IS?
ET1>EMA*ET2

YES

E=0.5*E
KX=1
KXX=3
DO{XGK(J)=E*GK(J)

12

NO

IS?
KX=0

YES

NO

16

IS?
ET1>EMI*ET2

YES

KX=1
KXX=3

NO

16

16

IS?
I=NN

YES

NO

I=I+1

15

X=XIN+E
DO {Q(I)=GK(5)*G2(I)+GK(6)*G1(I)
Y(I)=Y(I)+XGK(2)*(G(I) − Q(I))
Q(I)=GK(7)*G(I)+GK(8)*Q(I)}
K=4

13

(*e*)

17

E=EI

IS?
TS(LS)=T(L+1)

YES

DO{XS(J)=CS(J, LS+1)

NO

18

IS?
LS=LSN

YES

18

NO

LS=LS+1

18

IS?
TI(LI)=T(L+1)

YES

DO{XI(J)=CI(J, LI)

NO

19

COMPUTE:
DO{Y(I)
DO{XK(I)

PRINT:
DO{XK(I)

IS?
LI=LIN

YES

19

NO

LI=LI+1

19

L=L+1

IS?
L=LN

NO

YES

STOP

IS?
TT(LT+1)=T(L)

NO

9

YES

LT=LT+1

8

(*f*)

Fig. F-2 Flow diagram for numerical integration of differential equations with discontinuous time functions.

iteration after the point of discontinuity starts with the increment value before the increment was decreased.

11. Although not indicated in Fig. F-2, coefficients that are continuous functions of time are evaluated after node 12.

12. Provisions for underflow should be made in the test for halving and doubling the increment size.

A number of additions to the program shown in Fig. F-2 may be desired. For instance, excessive print-out often occurs when output is called for at the end of every iteration of the Gill method. Also, the steady-state solutions of time-invariant k equations may be desired. An acceptable test for steady-state conditions causing a program stop is

$$\frac{|y_i(x+e) - y_i(x)|}{e} < e_s \qquad \text{for all } i \tag{F-10}$$

The chord approximation to the derivative, as opposed to the derivative, is not excessively sensitive to round-off errors, thereby avoiding false indications of transients. In some problems, limiting the range of the integration increment, such that $e_{\min} \leqq e \leqq e_{\max}$, may be desirable.

appendix g

Expansion of the Hamiltonian Function

In this appendix, the problem of expressing $\mathscr{H}^*$ in terms of $\mathscr{H}_P^\dagger$, which is the truncated hamiltonian function, is treated. In order to simplify the notation, the first-degree expansion with a first-order dynamic process is considered initially. Also, for the sake of simplicity, both the subscript and time-function notations are dropped whenever possible.

G-1 First-degree expansion for a first-order dynamic process

For these conditions, (9-21) becomes

$$\mathscr{H}^*(x, \mu) = H(x, m^*, \mu) + f(x, m^*, \mu)\frac{\partial E(x, \mu)}{\partial x} \tag{G-1}$$

which is rewritten as

$$\begin{aligned}\mathscr{H}^*(x, \mu) = H(x, m^*, \mu) + f(x, m^*, \mu)\frac{\partial E_1(x, \mu)}{\partial x} \\ + 2\hat{P}(\mu)f(x, m^*, \mu)(x - \hat{x}) + \frac{\partial \hat{P}(\mu)}{\partial x}f(x, m^*, \mu)(x - \hat{x})^2 \end{aligned} \tag{G-2}$$

by substituting (9-37) into (G-1). Also, the truncated hamiltonian function given in (9-38) becomes

$$\mathscr{H}_1^\dagger(x, \mu) = H(x, m^\dagger, \mu) + f(x, m^\dagger, \mu)\frac{\partial E_1(x, \mu)}{\partial x} \tag{G-3}$$

The equations normally associated with the minimization of the hamiltonian and truncated hamiltonian functions are written conveniently in a somewhat different form from those used previously. Specifically, the functions $\mathscr{M}^*$ and $\mathscr{M}^\dagger$ are defined as

$$\frac{\partial H(x, m^*, \mu)}{\partial m^*} + \frac{\partial f(x, m^*, \mu)}{\partial m^*}\frac{\partial E_1(x, \mu)}{\partial x} + 2\hat{P}(\mu)\frac{\partial f(x, m^*, \mu)}{\partial m^*}(x - \hat{x}) + \frac{\partial \hat{P}(\mu)}{\partial x}\frac{\partial f(x, m^*, \mu)}{\partial m^*}(x - \hat{x})^2 = \mathscr{M}^*(x, \mu) \tag{G-4}$$

and
$$\frac{\partial H(x, m^\dagger, \mu)}{\partial m^\dagger} + \frac{\partial f(x, m^\dagger, \mu)}{\partial m^\dagger}\frac{\partial E_1(x, \mu)}{\partial x} = \mathscr{M}^\dagger(x, \mu) \tag{G-5}$$

These functions possess the properties that

$$\mathscr{M}^*(x, \mu) = \begin{cases} = 0 & m^* \text{ not on, entering, and leaving the boundary of } \mathscr{M} \\ \neq 0 & m^* \text{ on the boundary of } \mathscr{M} \end{cases} \tag{G-6}$$

and
$$\mathscr{M}^\dagger(x, \mu) = \begin{cases} = 0 & m^\dagger \text{ not on, entering, and leaving the boundary of } \mathscr{M} \\ \neq 0 & m^\dagger \text{ on the boundary of } \mathscr{M} \end{cases} \tag{G-7}$$

The transition points on and off the boundary of $\mathscr{M}$ correspond to the condition that the boundary value is also the minimizing value of the control signal without a saturation constraint. Because the third and fourth terms in (G-4) vanish when $x = \hat{x}$, the two additional properties

$$\hat{m}^*(\mu) = \hat{m}^\dagger(\mu) = \hat{m}(\mu) \tag{G-8}$$

and
$$\mathscr{M}^*(\hat{x}, \mu) = \mathscr{M}^\dagger(\hat{x}, \mu) \tag{G-9}$$

are obtained.

Before the expansion process for $\mathscr{H}^*$ is developed, three additional definitions are introduced. First, a closed region $\hat{\mathscr{R}}^*(\mu)$ is defined as that region of state space where $\mathscr{M}^*(x, \mu) = 0$ given a value of $\hat{x}(\mu)$. Similarly, a closed region $\hat{\mathscr{R}}^\dagger(\mu)$ is defined as that region of state space where $\mathscr{M}^\dagger(x, \mu) = 0$ given a value of $\hat{x}(\mu)$. Finally, a closed region $\hat{\mathscr{R}}(\mu)$ is defined as the largest region of state space where all the required series converge uniformly given a value of $\hat{x}(\mu)$. The size and shape of these regions are dependent upon the value of $\hat{x}(\mu)$, and these regions serve to define four different cases that must be considered in the expansion of $\mathscr{H}^*$. The material presented here must be regarded as the outline of a proof where mathematical rigor is not pursued.§

§ The regions $\hat{\mathscr{R}}^*(\mu)$ and $\hat{\mathscr{R}}^\dagger(\mu)$ are taken to be simply connected and are constructed from a closed set of points. The region $\hat{\mathscr{R}}(\mu)$ is taken to be simply connected and is constructed as an open set of points. Also, the point $\hat{x}(\mu)$ must be contained in $\hat{\mathscr{R}}(\mu)$.

The basic procedure is to expand $\mathscr{H}^*$, given in (G-2), in a Taylor series about $m^\dagger$. If only the first two terms in (G-2) are expanded in a first-degree series, then the result is

$$\mathscr{H}^*(x, \mu) = \mathscr{H}_1^\dagger(x, \mu) + \mathscr{M}^\dagger(x, \mu)(m^* - m^\dagger) + \hat{\mathscr{R}}_1(\mu)(m^* - m^\dagger)^2 + 2\hat{P}(\mu)f(x, m^*, \mu)(x - \hat{x}) + \frac{\partial \hat{P}(\mu)}{\partial x} f(x, m^*, \mu)(x - \hat{x})^2 \quad \text{(G-10)}$$

when the definitions given in (G-3) and (G-5) are utilized. The function $\hat{R}_1(\mu)$ is the Taylor-series remainder term. Now the function f appearing in the fourth term of (G-10) is expanded in a zeroth-degree series about the value $\hat{x}(\mu)$. When this series is introduced, (G-10) becomes

$$\mathscr{H}^*(x, \mu) = \mathscr{H}_1^\dagger(x, \mu) + 2\hat{P}(\mu)\hat{x}'(\mu)(x - \hat{x}) + \mathscr{M}^\dagger(x, \mu)(m^* - m^\dagger) + \hat{R}_1(\mu)(m^* - m^\dagger)^2 + \hat{R}_2(\mu)(x - \hat{x})^2 \quad \text{(G-11)}$$

if the definition given in (9-28) is utilized. The coefficient $\hat{R}_2(\mu)$ is due to the sum of the Taylor-series remainder term and the last term in (G-10). The conclusion of the expansion process requires a demonstration that the third and fourth terms appearing in (G-11) are of at least second degree in $(x - \hat{x})$.

Case I. $\hat{m}$ not on the boundary of $\mathscr{M}$

In this situation, the point $\hat{x}$ must lie within both the region $\hat{\mathscr{R}}^*(\mu)$ and the region $\hat{\mathscr{R}}^\dagger(\mu)$. Therefore, an open set of points must exist for x, which is contained by the intersection of $\hat{\mathscr{R}}^*(\mu)$, $\hat{\mathscr{R}}^\dagger(\mu)$, and $\hat{\mathscr{R}}(\mu)$. These points, excluding the point $\hat{x}$,

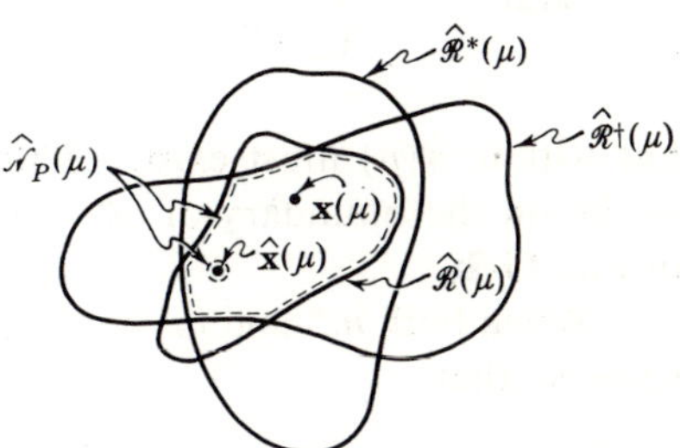

Fig. G-1 Construction of $\hat{\mathscr{N}}_P(\mu)$ for case I.

form a neighborhood $\hat{\mathscr{N}}_P(\mu)$ where neither m^* nor $m^\dagger$ lies on the boundary of $\mathscr{M}$. In the notation $\hat{\mathscr{N}}_P(\mu)$, $P = 1$ is used for the first-degree expansion of the minimum-error function. The construction of this neighborhood is depicted in Fig. G-1 for the case with N dimensions. The point $\hat{x}$ is excluded from this neighborhood because the desire is to show that (9-56) and (9-57) are necessary conditions for satisfying (9-55).

For $x \in \hat{\mathscr{N}}_1(\mu)$, (G-7) yields $\mathscr{M}^\dagger(x, \mu) = 0$ so that (G-11) reduces to

$$\mathscr{H}^*(x, \mu) = \mathscr{H}_1^\dagger(x, \mu) + 2\hat{P}(\mu)\hat{x}'(\mu)(x - \hat{x}) + \hat{R}_1(\mu)(m^* - m^\dagger)^2 + \hat{R}_2(\mu)(x - \hat{x})^2 \quad \text{(G-12)}$$

Furthermore, (G-4) reduces to

$$\frac{\partial H(x, m^*, \mu)}{\partial m^*} + \frac{\partial f(x, m^*, \mu)}{\partial m^*}\frac{\partial E_1(x, \mu)}{\partial x} + 2\hat{P}(\mu)\frac{\partial f(x, m^*, \mu)}{\partial m^*}(x - \hat{x}) + \frac{\partial \hat{P}(\mu)}{\partial x}\frac{\partial f(x, m^*, \mu)}{\partial m^*}(x - \hat{x})^2 = 0 \quad \text{(G-13)}$$

If (G-13) is expanded in a zeroth-degree Taylor series about $m^\dagger$, then the result is

$$\hat{R}_3(\mu)(m^* - m^\dagger) + \hat{R}_4(\mu)(x - \hat{x}) = 0 \tag{G-14}$$

when (G-5), under the condition $\mathscr{M}^\dagger(x, \mu) = 0$, is utilized and where $\hat{R}_3(\mu)$ and $\hat{R}_4(\mu)$ are the Taylor-series remainder terms. When (G-14) is substituted into (G-12), the result is

$$\mathscr{H}^*(x, \mu) = \mathscr{H}_1^\dagger(x, \mu) + 2\hat{P}(\mu)\hat{x}'(\mu)(x - \hat{x}) + \hat{Q}(\mu)(x - \hat{x})^2 \tag{G-15}$$

where§

$$\hat{Q}(\mu) = \hat{R}_1(\mu)\left[\frac{\hat{R}_4(\mu)}{\hat{R}_3(\mu)}\right]^2 + \hat{R}_2(\mu) \tag{G-16}$$

Equation (G-15) is the result appearing in (9-41) except that a first-order case is treated here.

Case II. $\hat{m}$ on the boundary of $\mathscr{M}$

In this situation, the point $\hat{x}$ must lie outside both the region $\hat{R}^*(\mu)$ and the region $\hat{R}^\dagger(\mu)$. Therefore, an open set of points for x that lie outside $\hat{\mathscr{R}}^*(\mu)$ and $\hat{\mathscr{R}}^\dagger(\mu)$

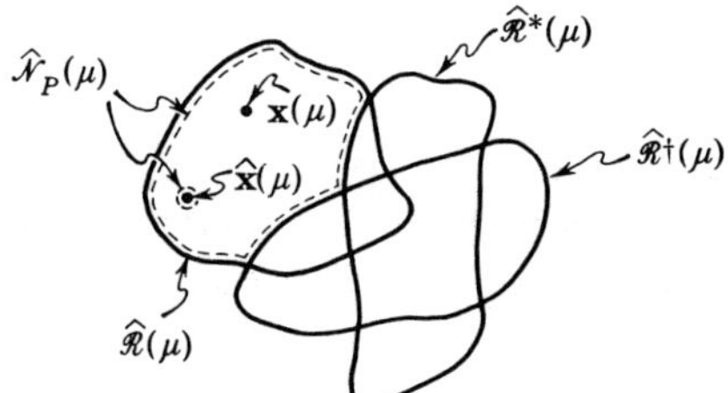

Fig. G-2 Construction of $\hat{\mathscr{N}}_P(\mu)$ for case II.

but within $\hat{\mathscr{R}}(\mu)$ must exist. This set of points forms $\hat{\mathscr{N}}_1(\mu)$, where both m^* and $m^\dagger$ lie on the boundary of $\mathscr{M}$. The construction of $\hat{\mathscr{N}}_1(\mu)$ in this case is depicted in Fig. G-2.

When both m^* and $m^\dagger$ lie on the boundary of $\mathscr{M}$, they are equal to the boundary value so that

$$m^*(\mu) = m^\dagger(\mu) \qquad x \in \hat{\mathscr{N}}_1(\mu) \tag{G-17}$$

Therefore, (G-11) reduces directly to (G-15), where

$$\hat{Q}(\mu) = \hat{R}_2(\mu) \tag{G-18}$$

Case III. $\hat{m}$ on the boundary of $\mathscr{M}$ but $\mathscr{M}^*(\hat{x}, \mu) = 0$, $x \in \hat{\mathscr{R}}^*(\mu)$, and $x \notin \hat{\mathscr{R}}^\dagger(\mu)$

This case corresponds to the transition point at the boundary of $\mathscr{M}$. The point $\hat{x}$ lies on the boundary of both $\hat{\mathscr{R}}^*(\mu)$ and $\hat{\mathscr{R}}^\dagger(\mu)$ due to (G-9) and hence lies at a point of intersection of the boundaries of $\hat{\mathscr{R}}^*(\mu)$ and $\hat{\mathscr{R}}^\dagger(\mu)$. The construction of $\hat{\mathscr{N}}_1(\mu)$ is depicted in Fig. G-3 for this case, where x is taken to be in a region where m^* is not on the boundary of $\mathscr{M}$.

For $x \in \hat{\mathscr{N}}_1(\mu)$, $\mathscr{M}^*(x, \mu) = 0$ but $\mathscr{M}^\dagger(x, \mu) \neq 0$ according to the definitions given in (G-6) and (G-7). First, the left-hand side of (G-5) is expanded in a zeroth-degree expansion about $\hat{x}(\mu)$. This expansion reduces (G-5) to

$$\mathscr{M}^\dagger(x, \mu) = \hat{R}_5(\mu)(x - \hat{x}) \tag{G-19}$$

§ Note that $\hat{\mathscr{R}}_3(\mu) > 0$ is due to the Legendre condition for a local minimum.

when $\mathscr{M}^\dagger(\hat{x}, \mu) = 0$ is utilized and $\hat{R}_5(\mu)$ is the Taylor-series remainder term. Now, if the expansion of (G-13) is repeated for this case, then the result is

$$\hat{R}_3(\mu)(m^* - m^\dagger) + \mathscr{M}^\dagger(x, \mu) + \hat{R}_4(\mu)(x - \hat{x}) = 0 \tag{G-20}$$

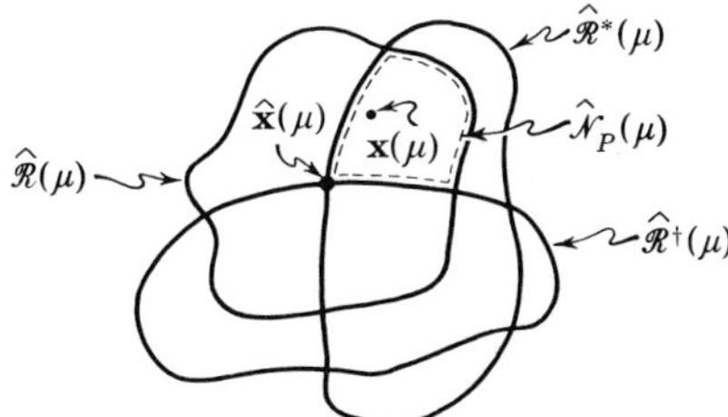

Fig. G-3 Construction of $\hat{\mathscr{N}}_P(\mu)$ for case III.

Finally, the substitution of (G-19) and (G-20) into (G-11) results in (G-15), where

$$\hat{Q}(\mu) = -\frac{\hat{R}_5(\mu)}{\hat{R}_3(\mu)}[\hat{R}_5(\mu) + \hat{R}_4(\mu)] + \frac{\hat{R}_1(\mu)}{\hat{R}_3^{\,2}(\mu)}[\hat{R}_5(\mu) + \hat{R}_4(\mu)]^2 + \hat{R}_2(\mu) \tag{G-21}$$

Case IV. $\hat{m}$ *on the boundary of* $\mathscr{M}$ *but* $\mathscr{M}^*(\hat{x}, \mu) = 0$, $x \notin \hat{\mathscr{R}}^*(\mu)$, *and* $x \in \hat{\mathscr{R}}^\dagger(\mu)$

This case is similar to case III, but x is chosen so that $\mathscr{M}^*(x, \mu) \neq 0$ and $\mathscr{M}^\dagger(x, \mu) = 0$, as indicated in Fig. G-4.

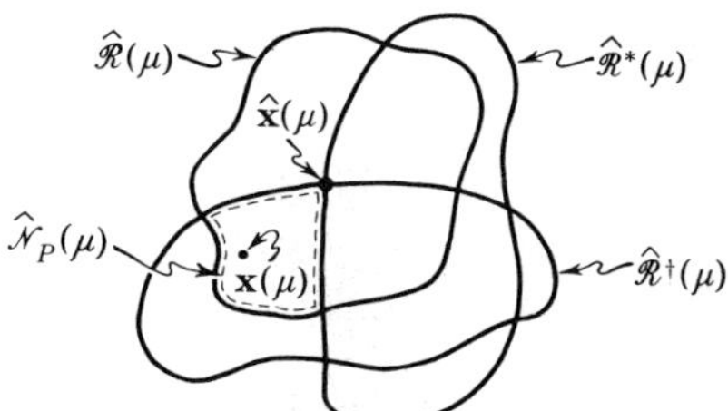

Fig. G-4 Construction of $\hat{\mathscr{N}}_P(\mu)$ for case IV.

For $x \in \mathscr{N}_1(\mu)$, the function $\mathscr{H}^*$ reduces to (G-12), and $\mathscr{M}^*(x, \mu)$ is given by

$$\mathscr{M}^*(x, \mu) = \hat{R}_6(\mu)(x - \hat{x}) \tag{G-22}$$

when steps similar to those leading to (G-19) are used. Now, if the expansion of (G-4) is repeated for this case, then the result is

$$\hat{R}_3(\mu)(m^* - m^\dagger) + \hat{R}_4(\mu)(x - \hat{x}) = \mathscr{M}^*(x, \mu) \tag{G-23}$$

Finally, the substitution of (G-22) and (G-23) into (G-12) yields (G-15), where

$$\hat{Q}(\mu) = \frac{\hat{R}_1(\mu)}{\hat{R}_3^{\,2}(\mu)}[\hat{R}_6(\mu) - \hat{R}_4(\mu)]^2 + \hat{R}_2(\mu) \tag{G-24}$$

The treatment of these four cases completes the first-degree expansion of the hamiltonian for a first-order dynamic process. However, the reader should note that there are two additional regions, not covered in cases III and IV, when $\hat{x}(\mu)$ lies on the boundary of $\hat{\mathscr{R}}^*(\mu)$ and $\hat{\mathscr{R}}^\dagger(\mu)$. These regions are defined by $x \in \hat{\mathscr{R}}^*(\mu)$, $x \in \hat{\mathscr{R}}^\dagger(\mu)$ and $x \notin \hat{\mathscr{R}}^*(\mu)$, $x \notin \hat{\mathscr{R}}^\dagger(\mu)$. However, these two regions correspond to special cases of case I and case II, respectively.

G-2 First-degree expansion for an Nth-order dynamic process

For these conditions, (9-21) becomes§

$$\mathscr{H}^*(\mathbf{x}, \mu) = H(\mathbf{x}, \mathbf{m}^*, \mu) + \sum_{n=1}^{N} f_n(\mathbf{x}, \mathbf{m}^*, \mu) \frac{\partial E_1(\mathbf{x}, \mu)}{\partial x_n} + 2 \sum_{n=1}^{N} \sum_{m=1}^{N} \hat{P}_{nm}(\mu) f_n(\mathbf{x}, \mathbf{m}^*, \mu)(x_m - \hat{x}_m) \quad \text{(G-25)}$$

and (9-38) becomes

$$\mathscr{H}^\dagger(\mathbf{x}, \mu) = H(\mathbf{x}, \mathbf{m}^\dagger, \mu) + \sum_{n=1}^{N} f_n(\mathbf{x}, \mathbf{m}^\dagger, \mu) \frac{\partial E_1(\mathbf{x}, \mu)}{\partial x_n} \quad \text{(G-26)}$$

Also, functions equivalent to $\mathscr{M}^*(x, \mu)$ and $\mathscr{M}^\dagger(x, \mu)$ are defined as

$$\frac{\partial H(\mathbf{x}, \mathbf{m}^*, \mu)}{\partial m_j^*} + \sum_{n=1}^{N} \frac{\partial f_n(\mathbf{x}, \mathbf{m}^*, \mu)}{\partial m_j^*} \frac{\partial E_1(\mathbf{x}, \mu)}{\partial x_n} + 2 \sum_{n=1}^{N} \sum_{m=1}^{N} \hat{P}_{nm}(\mu) \frac{\partial f_n(\mathbf{x}, \mathbf{m}^*, \mu)}{\partial m_j^*} (x_m - \hat{x}_m) = \mathscr{M}_j^*(\mathbf{x}, \mu) \quad \text{(G-27)}$$

and

$$\frac{\partial H(\mathbf{x}, \mathbf{m}^\dagger, \mu)}{\partial m_j^\dagger} + \sum_{n=1}^{N} \frac{\partial f_n(\mathbf{x}, \mathbf{m}^\dagger, \mu)}{\partial m_j^\dagger} \frac{\partial E_1(\mathbf{x}, \mu)}{\partial x_n} = \mathscr{M}_j^\dagger(\mathbf{x}, \mu) \quad \text{(G-28)}$$

These functions possess the properties

$$\mathscr{M}_j^*(\mathbf{x}, \mu) = \begin{cases} = 0 & m_j^* \text{ not on, entering, and leaving the boundary of } \mathscr{M} \\ \neq 0 & m_j^* \text{ on the boundary of } \mathscr{M} \end{cases} \quad \text{(G-29)}$$

and

$$\mathscr{M}_j^\dagger(\mathbf{x}, \mu) = \begin{cases} = 0 & m_j^\dagger \text{ not on, entering, and leaving the boundary of } \mathscr{M} \\ \neq 0 & m_j^\dagger \text{ on the boundary of } \mathscr{M} \end{cases} \quad \text{(G-30)}$$

The expansions which correspond to (G-10) and (G-11) are also found to be

$$\mathscr{H}^*(\mathbf{x}, \mu) = \mathscr{H}_1^\dagger(\mathbf{x}, \mu) + \sum_{j=1}^{M} \left[\mathscr{M}_j^\dagger(\mathbf{x}, \mu) + \sum_{i=1}^{M} \hat{R}_{1ij}(\mu)(m_i^* - m_i^\dagger) \right] (m_j^* - m_j^\dagger) + 2 \sum_{n=1}^{N} \sum_{m=1}^{N} \hat{P}_{nm}(\mu) f_n(\mathbf{x}, \mathbf{m}^*, \mu)(x_m - \hat{x}_m) \quad \text{(G-31)}$$

and

$$\mathscr{H}^*(\mathbf{x}, \mu) = \mathscr{H}_1^\dagger(\mathbf{x}, \mu) + 2 \sum_{n=1}^{N} \sum_{m=1}^{N} \hat{P}_{nm}(\mu) \hat{x}_n'(\mu)(x_m - \hat{x}_m) + \sum_{j=1}^{M} \left[\mathscr{M}_j^\dagger(\mathbf{x}, \mu) + \sum_{i=1}^{M} \hat{R}_{1ij}(\mu)(m_i^* - m_i^\dagger) \right] (m_j^* - m_j^\dagger) + \sum_{n=1}^{N} \sum_{m=1}^{N} \sum_{i=1}^{N} \hat{P}_{im}(\mu) \hat{R}_{2ni}(\mu)(x_n - \hat{x}_n)(x_m - \hat{x}_m) \quad \text{(G-32)}$$

As in the case of the first-order system, the terms $\mathscr{M}_j^\dagger(\mathbf{x}, \mu)$ and $(m_j^* - m_j^\dagger)$ must be shown to be of at least first degree in $(x_n - \hat{x}_n)$. Case I, defined in Sec. G-1, requires no changes in the expansion except for the additional dimensionality.

§ Note that a sum of terms which correspond to the last term in (G-2) is neglected here. As seen in Sec. G-1, this term merely changes the value of $\hat{Q}(\mu)$. Therefore, the equalities in (G-25) and subsequent equations in this section are valid to the zeroth and first degree in $(x_n - \hat{x}_n)$.

However, cases II to IV, which involve the boundaries of $\mathscr{M}$, require a change in the definitions of $\hat{\mathscr{R}}^*(\mu)$ and $\hat{\mathscr{R}}^\dagger(\mu)$ when the number of control signals is greater than one. Specifically, the control signals must be divided into subsets. One subset includes all signals on the boundary of $\mathscr{M}$, another subset includes all signals not on the boundary of $\mathscr{M}$, and the final subset includes the remaining signals that are at transition points. With the appropriate definitions of $\hat{\mathscr{R}}^*(\mu)$ and $\hat{\mathscr{R}}^\dagger(\mu)$ in this situation, the expansions are analogous to those for cases II to IV. The results of the method outlined here indeed demonstrate that (G-32) reduces to

$$\mathscr{H}^*(\mathbf{x}, \mu) = \mathscr{H}_1^\dagger(\mathbf{x}, \mu) + 2\sum_{n=1}^{N}\sum_{m=1}^{N} \hat{P}_{nm}(\mu)\hat{x}_n'(\mu)(x_m - \hat{x}_m) + \sum_{n=1}^{N}\sum_{m=1}^{N} \hat{Q}_{nm}(\mu)(x_n - \hat{x}_n)(x_m - \hat{x}_m) \quad \text{(G-33)}$$

for $\mathbf{x} \in \hat{\mathscr{N}}_1(\mu)$. This is the result appearing in (9-41).

G-3 *P*th-degree expansion for a first-order dynamic process

In Sec. 9-4, the Pth-degree expansion of the minimum-error function is introduced. For this case, the expansion of the hamiltonian is outlined for a first-order dynamic process. The first-order dynamic process is treated here in order to avoid the lengthy notation and expressions that would arise with higher-order dynamic process. Extensions of this treatment follow along the lines outlined in Sec. G-2.

The Pth-degree Taylor series of the minimum-error function is written as

$$E_P(x, \mu) = \sum_{p=0}^{P} \hat{a}_p(\mu)(x - \hat{x})^p \quad \text{(G-34)}$$

where

$$\hat{a}_p(\mu) = \frac{1}{p!}\frac{\partial^p E(x, \mu)}{\partial x^p}\bigg|_{x=\hat{x}} \quad \text{(G-35)}$$

In (G-34), the notational changes from the $\hat{p}$ parameters used previously to $\hat{a}$ parameters is obvious. Therefore, the minimum-error function is

$$E(x, \mu) = E_P(x, \mu) + \hat{P}(\mu)(x - \hat{x})^{P+1} \quad \text{(G-36)}$$

where the remainder term is given by

$$\hat{P}(\mu) = \frac{\partial^{P+1}E(\xi, \mu)}{\partial \xi^{P+1}} \quad \text{(G-37)}$$

and

$$\hat{x} < \xi < x \quad \text{(G-38)}$$

Also, the required partial derivatives of the minimum-error function are

$$\frac{\partial E(x, \mu)}{\partial \mu} = \frac{\partial E_P(x, \mu)}{\partial \mu} - (P + 1)\hat{P}(\mu)\hat{x}'(\mu)(x - \hat{x})^P + \hat{P}'(\mu)(x - \hat{x})^{P+1} \quad \text{(G-39)}$$

and §

$$\frac{\partial E(x, \mu)}{\partial x} = \frac{\partial E_P(x, \mu)}{\partial x} + (P + 1)\hat{P}(\mu)(x - \hat{x})^P \quad \text{(G-40)}$$

Finally, the truncated hamiltonian is defined as

$$\mathscr{H}_P^\dagger(x, \mu) = H(x, m^\dagger, \mu) + f(x, m^\dagger, \mu)\frac{\partial E_P(x, \mu)}{\partial x} \quad \text{(G-41)}$$

§ Again, the term corresponding to the final term in (G-2) is neglected so that the equality in (G-40) is valid only through the Pth degree.

and the function $\mathscr{M}^{\dagger}(x, \mu)$ is defined for this case by (G-5) when $\partial E_1(x, \mu)/\partial x$ is replaced by $\partial E_P(x, \mu)/\partial x$.

The expansions, which are equivalent to (G-10) and (G-11), for the Pth-degree expansion of the minimum-error function are

$$\mathscr{H}^*(x, \mu) = \mathscr{H}^{\dagger}_P(x, \mu) + \mathscr{M}^{\dagger}(x, \mu)(m^* - m^{\dagger}) + \hat{R}_1(\mu)(m^* - m^{\dagger})^2 + (P + 1)\hat{P}(\mu)f(x, m^*, \mu)(x - \hat{x})^P \quad \text{(G-42)}$$

and

$$\mathscr{H}^*(x, \mu) = \mathscr{H}^{\dagger}_P(x, \mu) + (P + 1)\hat{P}(\mu)\hat{x}'(\mu)(x - \hat{x})^P + \mathscr{M}^{\dagger}(x, \mu)(m^* - m^{\dagger}) + \hat{R}_1(\mu)(m^* - m^{\dagger})^2 + \hat{R}_2(\mu)(x - \hat{x})^{P+1} \quad \text{(G-43)}$$

Furthermore, if the expansions used in Sec. G-1 are applied here, then the function $\mathscr{M}^{\dagger}(x, \mu)$ is found to be of at least first degree in $(x - \hat{x})$, and $(m^* - m^{\dagger})$ is found to be of at least Pth degree in $(x - \hat{x})$. The final expansion of the hamiltonian, therefore, is

$$\mathscr{H}^*(x, \mu) = \mathscr{H}^{\dagger}_P(x, \mu) + (P + 1)\hat{P}(\mu)\hat{x}'(\mu)(x - \hat{x})^P + \hat{Q}(\mu)(x - \hat{x})^{P+1} \quad \text{(G-44)}$$

The expansions resulting in (G-44) suffice for demonstrating the validity of the material presented in Secs. 9-3 and 9-4. If (G-39) and (G-44) are substituted into the Hamilton-Jacobi equation given in (9-22), then the resulting Pth-degree expansion of the Hamilton-Jacobi equation becomes

$$\frac{\partial E_P(x, \mu)}{\partial \mu} + \mathscr{H}^{\dagger}_P(x, \mu) + [\hat{P}'(\mu) + \hat{Q}(\mu)](x - \hat{x})^{P+1} = 0 \quad \text{(G-45)}$$

This is the result used in (9-102) but specialized here for $N = 1$.

appendix h

Problems

Chapter 1

1-1. (*a*) Derive the equation defining the extrapolated response $q_e(T, t)$ in Fig. 1-2 for the missile equation

$$q''(t) + aq'(t) = bm'(t) + m(t)$$

The answer should be expressed in terms of $q(t)$, $q'(t)$, and $m(t)$.

(*b*) Repeat part *a* but use the expression for the extrapolated response given in (1-4).

(*c*) Compare parts *a* and *b* with engineering considerations, such as wind gusts, in mind.

(*d*) For the coefficient value $b = 0$, what are the structural restrictions on the gain F, appearing in Fig. 1-2, that give stability in the sense mentioned for terminal control?

(*e*) For $a = b = 0$, $q(0) = q'(0) = 0$, and $Q_p(T, t) = Q$, compute and sketch $e_c(t)/Q$ for $F[e_c(t)] = (M/Q)e_c(t)$ and $T^2 = 8Q/M$.

(*f*) Repeat part *e* for

$$F[e_c(t)] = \begin{cases} M & e_c(t) > 0 \\ 0 & e_c(t) = 0 \\ -M & e_c(t) < 0 \end{cases}$$

(*g*) Compare the performance of the systems in parts *e* and *f*.

1-2. Consider the process shown in Fig. P 1-2, which consists of a d-c generator, a d-c motor with fixed field current, and an inertia load. The input to the process is the generator field voltage e_F, and the output is the motor shaft position θ.

Assume that the process is linear and that the following relations apply:

$$e_G(t) = K_F i_F(t)$$

$$e_B(t) = K_B \frac{d\theta(t)}{dt}$$

$$\tau_M(t) = K_M i_M(t)$$

where $\tau_M(t)$ is the torque developed by the motor.

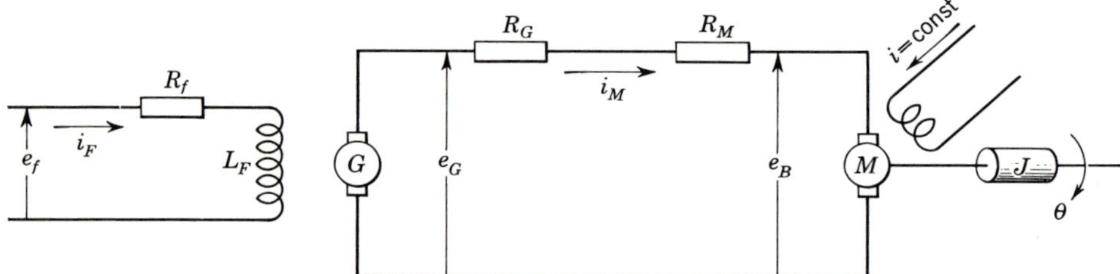

Fig. P 1-2

It is required that this process be controlled in such a manner that the displacement of the output shaft $\theta(t)$ follows, as closely as possible, a desired angular displacement $\theta_d(t)$. At the same time, it is desirable to maintain the generator field voltage e_F and the motor speed within certain physical limits. That is, it is desirable to keep these two quantities small.

Assume that the synthesis of a suitable control system is to be carried out by using optimization theory. The following steps to be performed illustrate the usual procedure.

(*a*) Describe the process in terms of a single differential equation relating the process input to time derivatives of the output.

(*b*) Specify a suitable set of state signals.

(*c*) Describe the process in terms of a set of first-order differential equations using the state signals selected.

(*d*) Write the process equations resulting from task *c* in matrix form. Specify the components of the vectors $\mathbf{x}'$, $\mathbf{x}$, and $\mathbf{m}$ and of the matrices B and C. Draw a block diagram of the process from the matrix form of the equation of state.

(*e*) Formulate a suitable error measure and express it in terms of the general notation defined in Chap. 1.

(*f*) Formulate a suitable error index.

(*g*) Write the response equations of the process.

(*h*) Write the response equations in matrix form and specify the components of the vector $\mathbf{q}$ and the matrix A. Draw a block diagram from the matrix form of the response equation.

1-3. For a dynamic process defined by

$$x^{(N)}(t) = m(t)$$

$$\sum_{n=0}^{N-1} x^{(n)}(t) = q(t)$$

suppose that the state signals $x(t), x'(t), \ldots, x^{(N-1)}(t)$ cannot be measured directly. Instead, the state-signal measurement is accomplished by measurement of q and m and their derivatives.

(*a*) Express $x^{(n)}(t)$ in terms of measurements on q and m.

(*b*) What is the total number of measurements required to define the state of the dynamic process?

(*c*) Consider the difficulties in this method of state measurement.

1-4. An Nth-order linear dynamic process is to be used in a terminal-control system, but all state-signal measurements are delayed by τ sec because of sensor limitations. The response is given by

$$q(t) = \sum_{n=1}^{N} q_n(t)$$

$$q_n(t) = \int_{-\infty}^{t} w_n(t, \sigma)m(\sigma)\, d\sigma$$

and

$$w_n(t, \sigma) = F_n(t)G_n(\sigma)$$

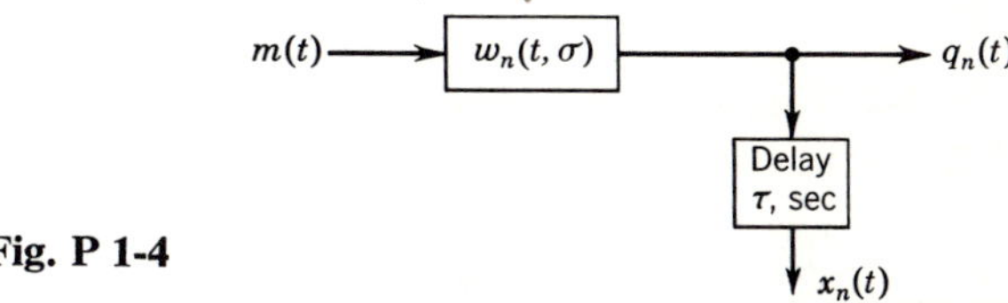

Fig. P 1-4

The sensor placement for the N feedback loops is defined by the diagram in Fig. P 1-4. The signal $x_n(t)$ is the output of the sensor in the nth feedback loop.

(*a*) Express $q_e(T, t)$ in terms of F_n, G_n, $x_n(t)$, and a past segment of m.

(*b*) Compare the construction problems of the response extrapolator with those for the case $\tau = 0$.

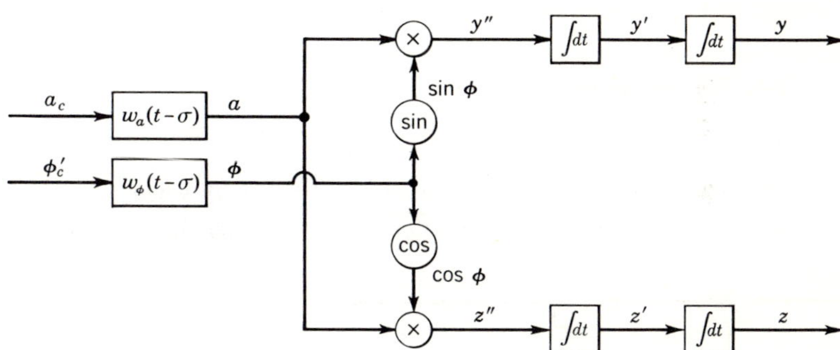

Fig. P 1-5

1-5. A "roll-to-turn" missile is to be used in a terminal-control application. The missile dynamics are represented by the block diagram in Fig. P 1-5, where the definitions

a_c = acceleration command
a = acceleration
ϕ'_c = roll-rate command
ϕ = roll angle
y, z = roll-stabilized coordinates of missile measured from an inertial reference

are used. Also, the transfer functions corresponding to w_a and w_ϕ are

$$W_a(s) = \frac{a_1}{s^2 + a_2 s + a_3} \qquad W_\phi(s) = \frac{b_1}{s(s + b_2)}$$

(*a*) Discuss the problem of constructing the missile extrapolator for the coordinates y and z with analog components.

(*b*) Draw the block diagram of the missile extrapolator depicting the measured state variables when $\phi'/b_2 \ll 1$.

Chapter 2

2-1. Find the mean-square value of the output signal of the system shown in Fig. P 2-1.

Fig. P 2-1

The signal $v(t)$ is a stationary gaussian process with the spectral density

$$\Phi_{vv}(s) = \frac{1}{2\pi} V^2$$

2-2. Find the autocorrelation function of the output signal of the system shown in Fig. P 2-2. The dynamic process is described by

Fig. P 2-2

$$q'(t) + a(t)q(t) = m(t)$$

where
$$a(t) = \frac{1}{t + 1/a_0}$$

Assume that $v(t)$ is a stationary gaussian process with

$$\phi_{vv}(\tau) = u_0(\tau)$$

2-3. For the system shown and

$$\phi_{vv}(\tau) = V^2 e^{-|\tau|/a} \qquad \phi_{nn}(\tau) = N^2 e^{-|\tau|/a} \qquad \phi_{vn}(\tau) = 0$$

(*a*) Find the value of k that minimizes $\overline{[q(t) - v(t)]^2}$ when $N^2/V^2 = a = \frac{1}{5}$.

(*b*) Find the minimum value of $\overline{[q(t) - v(t)]^2}$ under the conditions of part *a*.

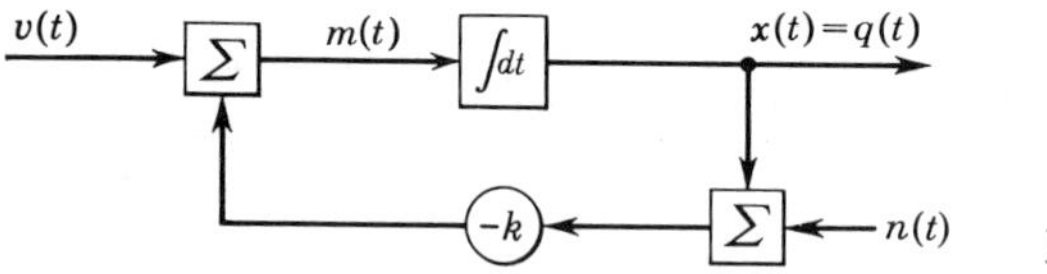

Fig. P 2-3

Chapter 3

3-1. Suppose an error index is chosen to be

$$e = \overline{[Q(t) - q(t)]^2}^{Q|v}$$

where the averaging is the conditional mean of $Q(t)$ given $v(\sigma)$ on $-\infty < \sigma \leq t$.

(*a*) What is the minimizing estimate $q^*(t)$?

(*b*) How does this index differ from

$$e = \overline{[Q(t) - q(t)]^2}$$

for optimum linear filters? Consider both the case of gaussian signals and the case of non-gaussian signals.

3-2. Suppose an error index is chosen to be

$$e = \overline{F[Q(t) - q(t)]}$$

The function F is arbitrary except that

$$F(x) = F(-x)$$
$$F(0) = 0$$
$$F'(x) = \begin{cases} > 0 & x > 0 \\ = 0 & x = 0 \\ < 0 & x < 0 \end{cases}$$

What is the minimizing estimate $q^*(t)$ in terms of $v(\sigma)$ on the interval $-\infty < \sigma \leqq t$? Both v and Q are gaussian, and Q is statistically dependent upon v.

3-3. Assume that the optimum control of a chemical plant requires the minimum mean-square estimate of a signal $q_n(\mu)$, where

$$q_n(t) = [i(t)]^n$$

so that $\mu \geqq t$. Assume that the past segment $-\infty < \sigma \leqq t$ of a signal v is measurable, where

$$v(t) = i(t) + n(t)$$

and $n(t)$ is a noise signal. The signals i and n are gaussian signals.

(*a*) Determine the minimum mean-square predictor for $q_n(\mu)$ in terms of the required average statistical properties of the signals.

(*b*) Discuss the construction of the predictor (linear, nonlinear, energy-storage, no-energy-storage, etc.).

3-4. Find the minimum-mean-square-error filter for the system shown in Fig. P 3-4, where

$$v(t) = V + n(t)$$
$$e = \overline{[V - q(t)]^2}$$

and

$$\phi_{nn}(t, t + \tau) = \frac{a}{a + t} u_0(\tau)$$

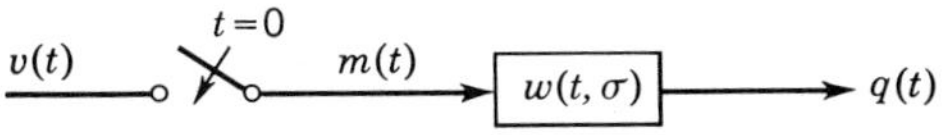

Fig. P 3-4

3-5. Find the transfer function of the compensation filter when the system shown in Fig. P 3-5 is designed to be a minimum mean-square predictor. The fixed member has the transfer function

$$W_f(s) = \frac{K}{s(T_m s + 1)}$$

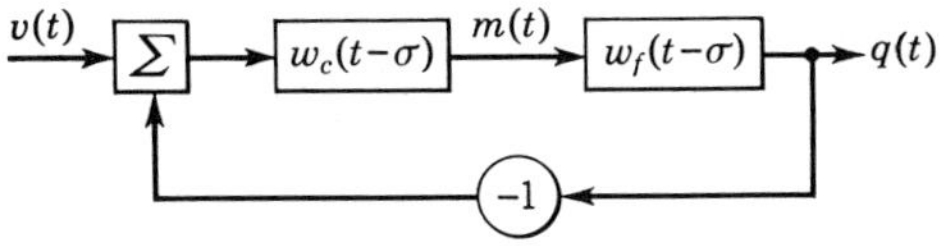

Fig. P 3-5

The input signal is given by

$$v(t) = Q(t) + n(t)$$

and the spectral densities

$$\Phi_{QQ}(s) = \frac{Q^2}{a^2 - s^2} \qquad \Phi_{nn}(s) = N^2 \qquad \Phi_{Qn}(s) = 0$$

3-6. Under what conditions is the so-called distortionless filter, discussed in Chap. 3, the optimum filter without a configuration constraint?

3-7. Determine the optimum filtering for two gaussian input signals

$$v_1(t) = Q(t) + n_1(t) \qquad \text{and} \qquad v_2(t) = Q(t) + n_2(t)$$

where $\quad \phi_{QQ}(\beta) = Q^2 e^{-a|\beta|} \qquad \phi_{n_1n_1}(\beta) = N_1^2 u_0(\beta) \qquad \phi_{n_2n_2}(\beta) = N_2^2 u_0(\beta)$

$$\phi_{n_1n_2}(\beta) = \phi_{n_1Q}(\beta) = \phi_{n_2Q}(\beta) = 0$$

Is there improvement in this case over the filtering of a single input signal?

3-8. Find the transfer function of the compensation that minimizes

$$\overline{[Q(t) - q(t)]^2}$$

subject to the constraint $\overline{m^2(t)} \leq M$ for the system shown in Fig. P 3-8. Assume that $W_f(s) = 1/s$, $\Phi_{QQ}(s) = (1/2\pi)[2aQ^2/(s + a)(-s + a)]$, and $a^2Q^2/M^2 = 8$.

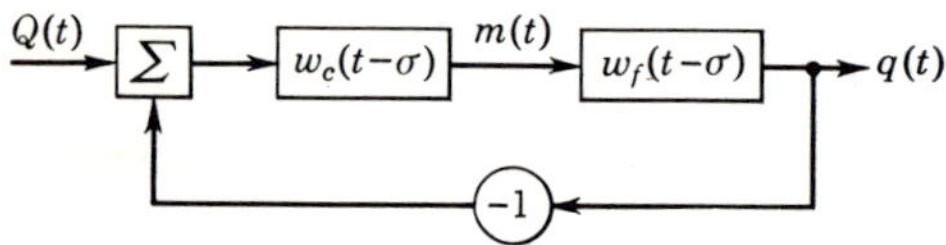

Fig. P 3-8

Chapter 4

4-1. Assume that the error index is given by

$$e(t) = \int_t^T \{[Q - q(\sigma)]^2 + \lambda[m(\sigma)]^2\}\, d\sigma$$

and the dynamic process by

$$q''(\sigma) = m(\sigma)$$

(*a*) Determine the optimum control equation for $T = \infty$ and sketch the block diagram.

(*b*) Demonstrate that the results of part *a* are equivalent to the Wiener results for certain initial conditions. What are these initial conditions?

(*c*) Is it possible to find a closed-form expression in terms of $T - t$ for the optimum control equation when $T - t$ is finite?

4-2. Suppose that the error index is

$$e(t) = \int_t^\infty \{[Q - q(\sigma)]^2 + [\cosh m(\sigma) - 1]\}\, d\sigma$$

and the dynamic process is

$$m(\sigma) = [q'(\sigma)]^3$$

(*a*) Find an expression that defines the optimum control signal $m_*(t)$ in terms of $q(t)$ and Q.

(*b*) Sketch $m_*(t)$ versus $[Q - q(t)]$ over the region where $m_*(t)/[Q - q(t)] \geq \frac{1}{10}$.

(*c*) What are the maximum and minimum values of the incremental gain, defined as $\partial m_*(t)/\partial[Q - q(t)]$, in this region?

4-3. Assume that the dynamic process is given by $q(t) = [x(t)]^2$, $x'(t) = m(t)$. If the error measure is given by

$$h[q(\sigma), m(\sigma), \sigma] = [Q - q(\sigma)]^2 + \rho m^2(\sigma)$$

obtain the Euler-Lagrange condition for minimum error. Show that the solution of this differential equation is an elliptic function. Discuss the difficulties of obtaining a solution if the initial state is $x(t)$ and the final state is $x_*(T) = Q$. Under what set of initial- and/or final-state values can a solution readily be obtained?

4-4. Assume that the error index is given by

$$e(t) = \int_t^T \{\lambda(\sigma)[Q(\sigma) - q(\sigma)]^2 + [m(\sigma)]^2\}\, d\sigma$$

and the dynamic process by

$$\sum_{n=0}^{N} d_n q^{(n)}(\sigma) = \sum_{n=0}^{N-1} c_n m^{(n)}(\sigma) \qquad d_N = 1$$

(*a*) Derive the Euler-Lagrange condition.

(*b*) What are the free-point terminal-boundary conditions at $\sigma = T$?

(*c*) Draw a block diagram that shows the configuration of the optimum controller which requires a minimum number of sensors.

(*d*) Specialize parts *a* to *c* for $\lambda(\sigma) = \lambda$, $Q(\sigma) = Q$, $T = \infty$, and

$$q''(\sigma) + d_1 q'(\sigma) = c_1 m'(\sigma) + c_0 m(\sigma)$$

4-5. Assume an error index

$$e(t) = \int_t^T H[\mathbf{x}(\sigma), \mathbf{m}(\sigma)]\, d\sigma + F[\mathbf{x}(T)]$$

and a state equation

$$\mathbf{x}'(t) = \mathbf{f}[\mathbf{x}(t), \mathbf{m}(t)]$$

Suppose that the error index is to be minimized with respect to both $\mathbf{m}(\sigma)$ and T subject to the free-point terminal-boundary condition. Derive the first necessary condition for a minimum.

Chapter 5

5-1. You are the quarterback of a team that has just secured a first down inside the opponent's 10-yard line. Suppose that the only courses of action open to you are those given in the following table. Suppose also that you have at most four downs to score.

Play	Yards gained if successful	Probability of success
Line buck	4	0.8
End run........	6	0.5
Pass	12	0.3

Determine the optimal choices which maximize your probability of scoring if play commences from any integral yard stripe inside the 10. Specify both the choices to be made and the resultant maximum probability of scoring.

Note that Bellman's principle of optimality provides a direct method for solving this problem.

5-2. Given the definition

$$E_N(a) = \max_{0 \leq x_i} \left(\prod_{i=1}^{N} x_i \right)$$

such that

$$\sum_{i=1}^{N} x_i = a$$

determine the dynamic-programming equation for E_N in terms of E_{N-1}. Note that a is treated as the state variable.

5-3. Given the definition

$$E_N(a) = \min_{0 \leq x_i} \left(\sum_{i=1}^{N} x_i^{p_i} \right)$$

such that $p_i > 0$ and $\sum_{i=1}^{N} x_i = a$, determine the dynamic-programming equation for E_N in terms of E_{N-1}.

5-4. Find the control law that minimizes the function

$$E_N(c) = \min_{y_i} \left[\sum_{i=0}^{N} (x_i^2 + y_i^2) \right]$$

where
$$x_{i+1} = x_i + \alpha y_i$$

and
$$x_0 = c \qquad y_i \text{ free}$$

The control law is to be expressed as a function $y_i(x_i)$.

5-5. Let $x = c$ be the initial capital and x be the capital at any given time t. Let the growth rate for capital be a, so that at the end of an interval $t_0 \leq t \leq t_0 + dt$, the capital will have grown from $x(t_0)$ to $x(t_0) + ax(t_0)\,dt$. Let y = profit rate taken at any instant subject to the constant

$$0 \leq y \leq ax$$

Find the policy that maximizes the integrated profits for an interval of T units of time. Thus you are asked to find the policy $y(x, t)$ that maximizes the integrated profit

$$P(c) = \max_{0 \leq y} \int_0^T y\,dt$$

subject to
$$\frac{dx}{dt} = ax - y \qquad x_0(0) = c$$

Your policy $y(t)$ is going to be one of the "inertia-less" or "bang-bang" type. SUGGESTION: Treat the problem in time segments.

5-6. Assume an error index

$$e(t) = \int_t^T h[q(\sigma), m(\sigma), \sigma]\,d\sigma$$

and a dynamic process defined by the impulse response

$$w(t, \sigma) = \sum_{n=1}^{N} F_n(t)G_n(\sigma)$$

so that
$$q(\sigma) = \int_{-\infty}^{t} w(t, \sigma)m(\sigma)\,d\sigma$$

(*a*) Derive the dynamic-programming condition for the minimum-error function. This condition should be written explicitly in terms of F_n and G_n only.

(*b*) Write the complete set of equations and appropriate boundary conditions defining the optimum control equation for $N = 1$ and

$$h[q(\sigma), m(\sigma), \sigma] = \lambda(\sigma)[Q(\sigma) - q(\sigma)]^2 + [M(\sigma) - m(\sigma)]^2.$$

(*c*) Sketch the block diagram for the control equation of part *b*.

(*d*) Consider the desirability of optimum controllers designed from the impulse-response model of the fixed member from the standpoint of sensors, etc.

5-7. Assume the error index

$$e(t) = \int_t^T h[q(\sigma), m(\sigma), \sigma]\,d\sigma$$

and the dynamic process

$$q(t) = F[x(t), x'(t), \ldots, x^{(N)}(t), t]$$
$$m(t) = G[x(t), x'(t), \ldots, x^{(N)}(t), t]$$

(*a*) Derive the dynamic-programming condition for minimum error by using the definition $h[q(\sigma), m(\sigma), \sigma] = H[x(\sigma), x'(\sigma), \ldots, x^{(N)}(\sigma), \sigma]$. The result should be written explicitly in terms of H.

(*b*) Specialize the results for

$$q(t) = \sum_{n=0}^{N} a_n(t)x^{(n)}(t)$$

$$m(t) = \sum_{n=0}^{N} b_n(t)x^{(n)}(t)$$

and $$h[q(\sigma), m(\sigma), \sigma] = \lambda(\sigma)[Q(\sigma) - q(\sigma)]^2 + m^2(\sigma)$$

Also write the optimum value of the control signal explicitly in terms of the minimum-error function.

5-8. Consider the formulation of a dynamic-programming condition for minimum error when the state of the dynamic process is not known deterministically.

5-9. Suppose that the error index is given by

$$e(t) = \int_t^T \{\lambda[Q - q(\sigma)]^2 + m^2(\sigma)\}\, d\sigma$$

and the dynamic process by

$$q''(t) + a_1 q'(t) = m(t)$$

(*a*) Write the complete set of differential equations that describe the minimum-error function in parametric form.

(*b*) Find the parameters of the optimum control equations when $T = \infty$ and $a_1 = \sqrt{2}\lambda^{\frac{1}{4}}$.

5-10. Repeat Prob. 5-9 for

$$q'''(t) = m(t)$$

and $$q''''(t) = m(t)$$

5-11. Discuss the dynamic-programming formulation of the free-terminal-time problem given in Prob. 4-5. In particular, identify the arguments of the minimum-error function, and derive the dynamic-programming condition for minimum error.

Chapter 6

6-1. Find the solution of

$$2y(z - 3)\frac{\partial z}{\partial x} + (2x - z)\frac{\partial z}{\partial y} = y(2x - 3)$$

subject to the boundary condition $z = 0$ at $y^2 + x^2 - 2x = 0$.

6-2. Solve

$$x(x + y)\frac{\partial z}{\partial x} - y(x + y)\frac{\partial z}{\partial y} = (y - x)(2x + 2y + z)$$

6-3. Find the complete integral of

$$y\left[\left(\frac{\partial z}{\partial x}\right)^2 + \left(\frac{\partial z}{\partial y}\right)^2\right] - z\frac{\partial z}{\partial y} = 0$$

Also, find the solution which passes through the curve

$$x = 0 \qquad y = t \qquad z = \sqrt{t}$$

6-4. Consider

$$z^2 = xy\frac{\partial z}{\partial x}\frac{\partial z}{\partial y}$$

Find the equation of and sketch the characteristic curve whose vertex is (1, 1, 1). Also, find the complete solution of this equation.

6-5. Consider a system defined by the state equations

$$\frac{dx_1(t)}{dt} = m(t) \qquad x_1(0) = a$$

$$\frac{dx_2(t)}{dt} = \frac{1}{2} m^2(t) + e^{x_1(t)} \qquad x_2(0) = 0$$

If $m(t)$ is a restricted control signal

$$m(t) \leqq 0 \qquad 0 \leqq t \leqq T$$

find the optimum $m(t)$ and the states $x_1(t)$ and $x_2(t)$ under the condition that the final value of x_2 [that is, $x_2(T)$] is to be minimized. Give curves of numerical results for the values of $x_1(t)$, $x_2(t)$, and $m(t)$ with $a = 0$ and $T = 1$.

6-6. A conical scan radar during the transition from search to track is exceedingly nonlinear. A block-diagram representation of the system during this phase is shown in Fig. P 6-6 with

θ_i = target angular position
θ_0 = antenna angular position
α = dish tilt
ϵ = angular tracking error

Also, $$\theta_i(t) = w_i t + \theta_{i_0}$$

and $$V_1(\epsilon, \alpha) = 2K_1^2 I_1(4m\epsilon\alpha)e^{-2m(\epsilon^2+\alpha^2)}$$

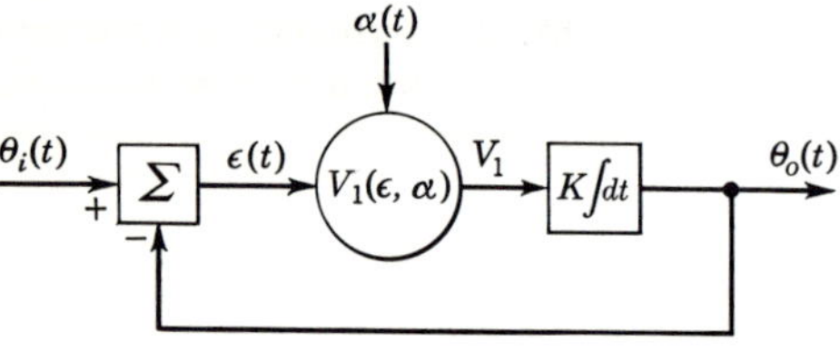

Fig. P 6-6

where I_1 is the modified Bessel function of the first kind and order unity. The constants are

$$2K_1^2 = 50 \text{ volts}$$
$$K = 1°/(\text{sec})(\text{volt})$$
$$m = 0.040 \text{ deg}^{-2}$$
$$w_i = 5°/\text{sec}$$

and the initial values of $\theta_i(0)$ and $\theta_0(0)$ are such that the initial error $\epsilon_0 = \epsilon(0)$ has values

$$-25° \leqq \epsilon_0 \leqq 25°$$

Find the optimal dish-tilt program α as a function of error ϵ where the maximum adjustable range of dish tilt is

$$3° \leqq \alpha \leqq 22°$$

when the system is required to minimize any arbitrarily weighted combination of time to settle and average squared error. In other words, the performance criterion is

$$\min_{\alpha} \int_0^T [1 + \lambda\epsilon^2(t)]\, dt$$

Chapter 7

7-1. Assume that a performance index is given by

$$e(t) = \int_t^\infty \overline{\{\phi[Q(\sigma) - q(\sigma)]^2 + m^2(\sigma)\}}^{Q|Q} d\sigma$$

and a dynamic process by

$$q(\sigma) = ax(\sigma)$$
$$x'(\sigma) = m(\sigma)$$

(*a*) Specify explicitly the optimum control equation in parametric form when the optimum-mean-square-error estimate is

$$Q_e^*(\sigma) = [e^{-(\sigma-t)} \cos(\sigma - t)]Q(t)$$

(*b*) Draw the block diagram of the optimum control system.

7-2. Suppose that the error index is

$$e(t) = \int_t^T \{\lambda(\sigma)[Q - q(\sigma)]^2 + m^2(\sigma)\} d\sigma$$

where

$$\lambda(\sigma) = \lambda_1 + \lambda_2 u_0(\sigma - T)$$

and the dynamic process is

$$q(t) = x(t)$$
$$x'(t) = m(t)$$

(*a*) Find the parameters of the optimum control equation for $\lambda_1 = 1$ and $\lambda_2 \geqq 1$.

(*b*) Specialize part *a* for $\lambda_2 = \infty$.

(*c*) Discuss the physical significance of part *b* from the standpoint of boundary conditions.

7-3. Assume that the error index is

$$e(t) = \int_t^\infty \{\phi_2[Q_2 - q_2(\sigma)]^2 + \phi_1 q_1^2(\sigma) + m_1^2(\sigma)\} d\sigma$$

and the linear dynamic process is defined by

$$A = \begin{bmatrix} 1 & 0 \\ 0 & 1 \end{bmatrix} \qquad B = \begin{bmatrix} 0 & 0 \\ 1 & 0 \end{bmatrix} \qquad C = \begin{bmatrix} 1 \\ 0 \end{bmatrix}$$

(*a*) Determine the parameters for the optimum control equation.

(*b*) Express the closed-loop-system damping ratio and resonant frequency in terms of ϕ_1 and ϕ_2.

(*c*) If no velocity feedback can be used in the controller because of the unavailability of a sensor, what is the transfer function of the optimum feedforward compensation element? Under what conditions is this system optimum?

7-4. For a quadratic error measure and a linear dynamic process, assume that all state-signal measurements involve a time delay of τ sec. In other words, the observable state signals $\mathbf{z}(t)$ are related to the state signals so that $\mathbf{z}(t) = \mathbf{x}(t - \tau)$.

(*a*) Formulate a suitable error index and corresponding definition for the minimum-error function.

(*b*) Derive the optimum control equation in terms of $\mathbf{z}(t)$. In this derivation, define suitable functions associated with the dynamic process, such as impulse responses. The result can be left in matrix form.

(*c*) Obtain explicit results for the special case for the state equation defined by

$$B = \begin{bmatrix} \dfrac{-1}{\tau_c} & 0 \\ 1 & 0 \end{bmatrix} \qquad C = \begin{bmatrix} \dfrac{k}{\tau_c} \\ 0 \end{bmatrix}$$

7-5. Under what conditions does the optimum system, with the pure-time-delay dynamic processes treated in Sec. 7-10, have a response that is independent of τ?

Chapter 8

8-1. Suppose that a dynamic-process configuration is defined by the logical arrays

$$[a_{nm}{}^0] = \begin{bmatrix} 1 & 0 & 0 & 0 \\ 0 & 1 & 0 & 0 \\ 0 & 0 & 1 & 0 \\ 0 & 0 & 0 & 1 \end{bmatrix} \qquad [b_{nm}{}^0] = \begin{bmatrix} 1 & 0 & 0 & 1 \\ 1 & 0 & 0 & 0 \\ 1 & 0 & 0 & 1 \\ 0 & 0 & 1 & 0 \end{bmatrix} \qquad [c_{nm}{}^0] = \begin{bmatrix} 1 \\ 0 \\ 1 \\ 0 \end{bmatrix} \qquad [u_n{}^0] = \begin{bmatrix} 0 \\ 0 \\ 0 \\ 0 \end{bmatrix}$$

and an error index is defined by the logical vectors

$$[\phi_n{}^0] = \begin{bmatrix} 1 \\ 0 \\ 1 \\ 1 \end{bmatrix} \qquad [Q_n{}^0] = \begin{bmatrix} 0 \\ 0 \\ 0 \\ 0 \end{bmatrix} \qquad [M_n{}^0] = \begin{bmatrix} 0 \\ 0 \\ 0 \\ 0 \end{bmatrix}$$

(*a*) What is the order of the dynamic process?
(*b*) What is the number of dimensions of the minimum-error function?
(*c*) Write the optimum control equations showing only the nonzero terms.

8-2. In a practical aircraft landing system, the desired altitude trajectory may vary in shape from airport to airport. Furthermore, timing problems due to air traffic control may require additional adjustments in this trajectory. Therefore, the instrumentation of an alternative landing system is required.

Specifically, assume that the desired altitude and the desired rate of ascent are telemetered to the aircraft during the landing and are available as continuous electric signals. Of course, these two signals are not known exactly at the time of design, but assume that their shapes roughly correspond to $h_d(t)$ and $h_d'(t)$ given in Appendix E.

(*a*) Treating the telemetered signals $h_d(t)$ and $h_d'(t)$ as separable signals, discuss the design of the aircraft landing system using all other design considerations given in Appendix E.

(*b*) Draw the block diagram of this system and specify the equations required to find the system gains.

(*c*) Discuss semiquantitatively the degradation in system performance resulting from the approximations due to the use of these separable signals.

Chapter 9

9-1. Assume that the error index is

$$e(t) = \sum_{n=1}^{N} \int_t^{T} \{[Q_n(\sigma) - q_n(\sigma)]^2 + \psi[m_n(\sigma)]^2\}\, d\sigma$$

and the dynamic process is

$$q_n(\sigma) = \sum_{m=1}^{N} a_{nm}(\sigma)x_m(\sigma)$$

$$x_n'(\sigma) = \sum_{m=1}^{N} [b_{nm}(\sigma)x_m(\sigma)] + c_{nn}(\sigma)m_n(\sigma)$$

$$-M_n \leqq m_n(\sigma) \leqq M_n$$

(*a*) Derive the quasi-optimum control equation and the differential equations defining the parameters appearing in this equation. Use the second-degree power-series expansion of the minimum-error function.

(*b*) By finding an appropriate normalizing factor, specialize the results of part *a* for the case $\psi = 0$.

9-2. Assume that the error index is

$$e(t) = \int_t^T [q^4(\sigma) + m^2(\sigma)]\, d\sigma$$

and the dynamic process is

$$q'(\sigma) = m(\sigma)$$

(*a*) Find the quasi-optimum control equation and the differential equations defining the parameters in this equation for the $P = 1, 2, 3,$ and 4 power-series expansion of the minimum-error function.

(*b*) For the characteristic curve defined by $q(t) = 0$, compare the approximating properties of these expansions in the neighborhood of this characteristic curve.

9-3. Assume that the error index is

$$e(t) = \sum_{n=1}^{N} \int_t^T \left[2 \cosh \frac{q_n(\sigma)}{Q_n} + 2 \cosh \frac{m_n(\sigma)}{M_n} - 4\right] d\sigma$$

and the dynamic process is

$$q_n(\sigma) = \sum_{m=1}^{N} a_{nm}(\sigma)x_m(\sigma)$$

$$x_n'(\sigma) = \sum_{m=1}^{N} [b_{nm}(\sigma)x_m(\sigma)] + m_n(\sigma)$$

(*a*) Derive the quasi-optimum control equation and the differential equations defining the parameters appearing in this equation. Use $P = 2$.

(*b*) Simplify the results of part *a* for

$$\left|\frac{\hat{m}_n(\sigma)}{M_n}\right| \ll 1 \qquad \text{and} \qquad \left|\frac{\hat{q}_n(\sigma)}{Q_n}\right| \ll 1$$

9-4. Assume that the error index is

$$e(t) = \sum_{n=1}^{N} \int_t^T \left\{\left[\frac{q_n(\sigma)}{Q_n}\right]^2 + \psi_1\left[\frac{m_n(\sigma)}{M_n}\right]^2 + \psi_r\left[\frac{m_n(\sigma)}{M_n}\right]^{2r}\right\} d\sigma$$

and the dynamic process is

$$q_n(\sigma) = \sum_{m=1}^{N} a_{nm}(\sigma)x_m(\sigma)$$

$$x_n'(\sigma) = \sum_{m=1}^{N} [b_{nm}(\sigma)x_m(\sigma)] + m_n(\sigma)$$

(*a*) Derive the quasi-optimum control equation and the differential equations defining the parameters appearing in this equation. Use the second-degree power-series expansion of the minimum-error function.

(*b*) Discuss possible difficulties in obtaining numerical solutions in general and for the case $\psi_1 = 0$ and $\hat{m}_n(\sigma) \approx 0$.

(*c*) Discuss possible difficulties in the construction of the quasi-optimum control equation for the case $\psi_1 \neq 0$ and $\psi_r \neq 0$.

(*d*) Derive the optimum linear control equation for the conditions of part *c*.

9-5. Suppose that there exists a suitably small region of state space so that

$$E[x(\mu), \mu] = \sum_{n=0}^{\infty} \hat{a}_n(\mu)[x(\mu) - \hat{x}(\mu)]^n$$

converges uniformly.

(*a*) Derive the differential equations that define the parameters $\hat{a}_n(\mu)$ in terms of the hamiltonian function for a first-order dynamic process.

(*b*) Discuss the relationships between these results and the truncated series used in the text material.

9-6. Whenever approximations such as the expansions of the minimum-error function are introduced, the problem of computing error bounds always arises.

(*a*) Discuss the computation of upper and lower bounds for the minimum-error function given a single characteristic curve in a specific region of state space. HINT: Consider the Taylor remainder term.

(*b*) Given a collection of characteristic curves and the first-degree expansion of the minimum-error function for each curve, discuss the computation of upper and lower bounds for a first-order dynamic process.

HINT: Use the properties of a strictly convex function. For this purpose, these properties are stated in two convenient forms:

$$f(x) - f(\hat{x}) - (x - \hat{x})\frac{\partial f(\hat{x})}{\partial \hat{x}} > 0 \qquad x \neq \hat{x}$$

$$f[(1-\rho)\hat{x}_1 + \rho\hat{x}_2] < (1-\rho)f(\hat{x}_1) + \rho f(\hat{x}_2) \qquad 0 < \rho < 1$$

(*c*) Do the results of part *b* extend to an Nth-order dynamic process?

(*d*) Can similar methods be used to compute error bounds for $\partial E/\partial x$ with a first-order dynamic process?

(*e*) Do the results of part *d* extend to Nth-order dynamic processes?

(*f*) What error bounds are of interest in the construction of the quasi-optimum control equation?

9-7. Rederive (9-209) and (9-214) when (9-197) is not valid for a first-order dynamic process.

9-8. Suppose that the minimum-error function is defined as

$$E(\mathbf{x}, t) = \min_{\mathbf{m}} \int_t^T H(\mathbf{x}, \mathbf{m}, \sigma)\, d\sigma$$

subject to the boundary condition

$$E(\mathbf{x}, T) = 0$$

Also, suppose that the hamiltonian function is defined as

$$\mathscr{H}^*(\mathbf{x}, t) = \min_{\mathbf{m}} \left[H(\mathbf{x}, \mathbf{m}, t) + \sum_{n=1}^{N} p_n f_n(\mathbf{x}, \mathbf{m}, t) \right]$$

where p_n is defined as

$$p_n = \frac{\partial E(\mathbf{x}, t)}{\partial x_n}$$

Then the minimum-error function satisfies

$$\frac{\partial E(\mathbf{x}, t)}{\partial t} + \mathscr{H}^*(\mathbf{x}, t) = 0$$

Suppose that the characteristic equation is

$$x_n' = f_n(\mathbf{x}, \mathbf{m}, t)$$

(*a*) Derive the other characteristic equation for p_n' by direct differentiation of the appropriate equations given here.

(*b*) Derive the equation for p_{kl}' by direct differentiation of the appropriate equations, where

$$p_{kl} = \frac{1}{2}\frac{\partial p_k}{\partial x_l} = \frac{1}{2}\frac{\partial p_l}{\partial x_k}$$

Chapter 10

10-1. Consider the dynamic process

$$x'(t) = m(t)$$

and the error index

$$e(t) = \int_t^T [x^2(\sigma) + m^2(\sigma)]\, d\sigma$$

Also, suppose that the required initial-boundary condition is $\hat{x}(0) = 1$, and the priming control signal $m^{(0)}(t) = 0$ for $0 \leqq t \leqq T$ is used.

(*a*) Demonstrate the one-step convergence property of the relaxation method based on second variations.

(*b*) Plot the value of the error index obtained after the first iteration as a function of step size used in this iteration.

10-2. Demonstrate to your satisfaction the statements in Sec. 10-6 concerning the method based on second variations and control-signal saturation.

10-3. Discuss the possibilities of using third variations and higher variations in an iterative procedure for solving two-point boundary-value problems.

10-4. Derive the differential equations corresponding to (10-122) through (10-131) but for case II discussed in Sec. 10-8.

10-5. The concept of first variations is often discussed in terms of hill climbing, where the hill is considered to be the function $\mathscr{H}$ defined in (10-39) for a first-order dynamic process. Explain in these terms the properties of the relaxation method based on second variations which give rise to rapid convergence. Remember that a functional, not a function, actually is being minimized.

10-6. For the auxiliary minimization problem which leads to the results given in Sec. 10-7,

(*a*) Derive the appropriate characteristic equations by introducing the incremental costate vector $\boldsymbol{\lambda}^{(i)}(\sigma)$.

(*b*) From the properties of linear ordinary differential equations, the solutions of these characteristic equations can be written as

$$\mathbf{X}^{(i)}(\sigma) = \Phi^{(i)}\mathbf{X}^{(i)}(T) + \boldsymbol{\phi}^{(i)}(\sigma)$$

and

$$\boldsymbol{\lambda}^{(i)}(\sigma) = \Psi^{(i)}\mathbf{X}^{(i)}(T) + \boldsymbol{\psi}^{(i)}(\sigma)$$

where $\Phi^{(i)}$ and $\Psi^{(i)}$ are transition matrices with elements that are functions of σ and T. Determine the matrix differential equations and boundary conditions which define $\Phi^{(i)}$, $\Psi^{(i)}$, $\boldsymbol{\phi}^{(i)}$, and $\boldsymbol{\psi}^{(i)}$.

(*c*) Show that

$$\boldsymbol{\lambda}^{(i)}(\sigma) = P^{(i)}\mathbf{X}^{(i)}(\sigma) + \mathbf{g}^{(i)}(\sigma)$$

where $P^{(i)}$ is a matrix with elements that are functions of σ. Give the properties of $P^{(i)}$ and $\mathbf{g}^{(i)}$ and their dependence upon the matrices and vectors introduced in part *b*. Determine the differential equations and boundary conditions that define the elements of $P^{(i)}$ and $\mathbf{g}^{(i)}$.

(*d*) Give the explicit mathematical condition that causes a conjugate point and hence the finite escape time of the matrix Riccati equations. Assume that T is finite and that the coefficients of the differential equations are continuous and bounded.

10-7. Give a physical interpretation of a conjugate point. The results obtained in part *c* of Prob. 7-2 are pertinent.

10-8. For the minimization problem treated in Secs. 10-4 and 10-7, assume that a relaxation method is to be constructed from an auxiliary minimization problem. This auxiliary minimization problem is constructed from the linearized state equation and the index.

$$v^{(i)} = \int_t^T \left[\sum_{n=1}^{M} \left(S_n{}^{(i)} M_n{}^{(i)} + \sum_{m=1}^{M} \tfrac{1}{2} \xi_n{}^{(i)} W_{nm}{}^{(i)} \xi_m{}^{(i)} \right) \right] d\sigma$$

where $$\xi_n{}^{(i)} = M_n{}^{(i)} + \sum_{k=1}^{N} G_{nk}{}^{(i)} X_k{}^{(i)}$$

The functions $W_{nm}{}^{(i)}$ and $G_{nk}{}^{(i)}$ are arbitrary.

(*a*) Derive the iteration algorithm which minimizes $v^{(i)}$.

(*b*) Give the conditions that the matrices $[W_{nm}{}^{(i)}]$ and $[G_{nk}{}^{(i)}]$ must satisfy for guaranteed convergence subject to a restricted step size.

(*c*) Under what conditions does this algorithm specialize to first variations only?

(*d*) Discuss the numerical advantages of introducing the asymptotic stability constraint when using this algorithm and state how this is accomplished.

(*e*) Discuss practical methods for circumventing the difficulties caused by the Legendre and Jacobi conditions when using second variations.

10-9. Derive a boundary-condition iteration algorithm based on second variations. Assume that the boundary conditions $p_n{}^{(i)}(t)$ are iterated and that all the equations which comprise the first necessary condition for a minimum are solved in the forward-time direction. Determine the appropriate backward-time equations such that the iteration algorithm is given by $p_n{}^{(i+1)}(t) = p_n{}^{(i)}(t) + \varepsilon g_n{}^{(i)}(t)$. Note that methods of the type given in Sec. 4-4 are not required in the forward-time direction. Under what conditions is one-step convergence obtained?

10-10. Derive a relaxation algorithm based on first variations for the free-terminal-time problem given in Prob. 4-5.

10-11. Derive a relaxation algorithm based on second variations for the free-terminal-time problem given in Prob. 4-5.

10-12. Discuss the selection of the appropriate penalty functions in conjunction with the free-terminal-time problem for obtaining an approximation to the minimum-time problem. Keep in mind the saturation constraints, the desire to obtain $\mathbf{x}(T) \approx \mathbf{0}$, the convexity requirements for the guaranteed convergence of the relaxation method, and suitable methods for initializing the iterations.

Chapter 11

11-1. Derive the equations that would be used in a sequential optimization procedure when disturbances are assumed to be appreciable. Specifically, the relaxation procedure should be based on first variations, and the control equation should be based on the second-degree expansion of the error function under the constraint of asymptotic stability. Draw a simplified flow diagram for the optimization computer.

11-2. Develop a suitable sequential optimization procedure for the floating-interval problem under the condition of negligible disturbances. Draw a simplified flow diagram for the optimization computer. Discuss the problem of selecting a suitable value for $\tau = T - t$. In this regard, consider the problems of convergence to minimum error and control-system stability.

11-3. Discuss the system advantages of using the algorithm derived in Prob. 10-8 for sequential optimization. Specifically, consider the case where the functions $G_{nk}^{(i)}$ are in fact the feedback gains of an autopilot which has been predesigned and is fixed.

Index